Energy and Climate Change

Energy and Climate Change

Creating a Sustainable Future

David A. Coley
Centre for Energy and the Environment,
University of Exeter

John Wiley & Sons, Ltd

Other Wiley Editorial Offices

John Wiley & Sons Inc., 111 River Street, Hoboken, NJ 07030, USA

Jossey-Bass, 989 Market Street, San Francisco, CA 94103-1741, USA

Wiley-VCH Verlag GmbH, Boschstr. 12, D-69469 Weinheim, Germany

John Wiley & Sons Australia Ltd, 42 McDougall Street, Milton, Queensland 4064, Australia

John Wiley & Sons (Asia) Pte Ltd, 2 Clementi Loop #02-01, Jin Xing Distripark, Singapore 129809

John Wiley & Sons Canada Ltd, 6045 Freemont Blvd, Mississauga, ONT, L5R 4J3, Canada

Wiley also publishes its books in a variety of electronic formats. Some content that appears in print may not be
available in electronic books.

Library of Congress Cataloging-in-Publication Data

Coley, David A.
    Energy and climate change: creating a sustainable future / David A. Coley.
      p.   cm.
    Includes bibliographical references and index.
    ISBN 978-0-470-85312-2 (cloth) – ISBN 978-0-470-85313-9 (pbk.)
1. Power resources–Textbooks.   2. Climatic changes–Textbooks.   I. Title.

    TJ163.2.C625 2008
    621.042–dc22   2007046622

British Library Cataloguing in Publication Data

A catalogue record for this book is available from the British Library

ISBN 978-0-470-85312-2 (H/B)   ISBN 978-0-470-85313-9 (P/B)

Typeset in 10.5/12.5pt Minion by Aptara Inc., New Delhi, India
Printed and bound in Great Britain by Antony Rowe Ltd, Chippenham, Wiltshire
This book is printed on acid-free paper.

To Helen, Scarlett and Theo

Contents

Preface

Please visit the book's website www.wileyeurope.com/college/coley
for additional teaching resources, web links and energy data

This book was written with a passionate belief that humanity needs to change the
way it treats the planet and treats many of the people who inhabit it. For millennia,
humankind has had an ever-increasing need for energy. Initially we relied on heat from
the sun and biomass as food and firewood. Then we learnt to use animals other than
ourselves as agricultural labour; by 100 BC we had harnessed the power of moving water,
then the winds. Up until this point our use of energy had been largely sustainable –
with the possible exception of excess forest cutting – and our impact on the planet
was only of a local nature. The industrial revolution brought a requirement for much
larger amounts of power in locations far from any natural resource, necessitating a
radical change. Fossil fuels (first coal, then oil) proved ideal for providing this power.
Unfortunately the emissions from their use have altered not only the local environment
but also the atmosphere itself, and the concentration of carbon dioxide has risen from
280 parts per million to 370 today – a level unknown for millions of years. Because
carbon dioxide acts as an insulator, this has slowly warmed the planet, in turn melting
ice and raising sea levels.

It has taken us a long time to realize the seriousness of the situation. The basic
phenomenon and its consequences were first described in 1859, and now terms such as
global warming and *climate change* appear regularly within the media, political debates
and dinner-table chitchat.

The common realization of the problem is proving to be the easy part. We want
energy and we want lots of it. The developed world uses the equivalent of 12 kilograms
of oil per person per day and this ensures a reasonably affluent existence for the vast
majority of its citizens, where starvation is non-existent, heating and lighting sufficient
and travel the norm. In some parts of the world energy use is equivalent to as little as 80
grams of oil per day. At this level it would appear impossible to meet the fundamental
needs of a society and individual opportunities are severely limited: child morbidity
high and life expectancy low.

For any degree of equality, humanity needs to be using more energy, not less. Yet
failure to reduce our emissions of greenhouse gases will lead to a level of climate change

that will affect the wealth and survival of many of the poorest people on the planet and harm the economies and landscapes of the wealthiest. The only sensible solution would appear to be that we use energy more efficiently in the short-term and that we give up our reliance on fossil fuels in the medium term.

This book discusses what energy is, why we need it, the harm we are doing to the planet and future generations, the current range of energy technologies and fuels (coal; oil; gas, including methyl hydrates, shale oil and tar sands; hydropower; and nuclear power), attempts by the international community to write treaties to reduce emissions, and future, *sustainable*, energy technologies (energy efficiency, solar, wind, wave, tidal, biomass, carbon sequestration and fusion). The text has been designed to be used as either a stand-alone course or as the major part of a course on traditional energy technologies, renewable energy, the history of energy use or climate change. It should appeal to, and be suitable for, those studying science, engineering, geography or politics (and hopefully other disciplines). Such a wide-ranging audience has meant some compromise has been necessary: the physicists may have liked more equations, the geographers fewer, and the political scientists more on international treaty arrangements. However, compromise has its rewards. The author strongly believes that scientists and engineers should study the history of their subject and its impact on the world, and that those in the humanities should not be short-changed when it comes to science. The book tries to take an international and inclusive approach. Real-world installations of the technologies and fuels studied are presented, and these are as likely to be sited in, say, Japan as the USA. The text is peppered with numerical problems (the end of each chapter contains essay-type alternatives), and again, these are as likely to involve data from India as well as the UK. Climate change is no respecter of national boundaries, and as we will see, only a global approach will provide the tools to solve the problem.

Many individuals and companies have helped with the production of this book, but in particular I would like to thank Helen Coley, Ronald Coley, Mark Brandon, Adrian Wyatt and Andy Forbes. I would also like to thank my colleagues at the Centre for Energy and the Environment, for putting up with the disruption writing any book inevitably causes.

David A. Coley
University of Exeter
January 2008

Corrections and additional material

It is hoped that you will enjoy studying (or teaching) the material presented, and appreciate solving some of the in-text problems. Like any work of this size that relies upon secondary sources and commercial data, it may contain a few errors and I hope that readers report any that they find via the book's website (www.wileyeurope.com/college/coley). Amendments can then be posted on the site for the benefit of all. If you have non copy-right material that might be of interest to others, please feel free to send it to me for inclusion in future editions of the book and the website – full acknowledgement will be given. The website also holds colour versions of most of the tables, graphs and photographs found in the book. These are for teaching purposes only. Please remember that this material is copyright-protected by those kind enough to provide it and that all the usual restrictions on its use apply.

1

Introduction

Energy is the single most important problem facing humanity today
                    –Richard Smalley (1996 Nobel laureate in Chemistry [SMA04])

Humankind currently uses 410×10^{18} joules of commercially traded energy per annum. This is equivalent to the energy content of over 90 000 billion litres of oil. We are addicted to energy, and as most of this comes from oil, gas and coal, it can be said that we are addicted to fossil fuels. There is a logical reason for the first addiction: without a large energy input, much of modern society would not be possible. We would have few lights, no cars, less warmth in winter and no division of labour. Our society would mirror any pre-industrial society, with the majority of us being subsistence farmers. Much of the world population no longer lives like this, and few would be willing to turn back the clock. Unfortunately our second addiction, that to fossil fuels, is proving to have severe consequences for both humankind and the planet's flora and fauna. The problem is climate change (often termed *global warming*) caused by a build-up of carbon dioxide and other gases in the atmosphere. The majority of these pollutants emanate from our use of carbon-based fossil fuels. This book is about breaking the second addiction, without compromising the first.

As has been reported in the world's media for over 10 years, there is clear evidence that we need to move away from fossil fuels. The last decade has been the warmest since records began: mean global temperature is up 0.6 °C since 1900; sea levels are rising by 1–2 mm per annum; summer arctic sea ice has thinned by 40 per cent since 1960 [ROT99, VIN99]; and the Thames barrier (which protects London from flooding) is now being raised on average six times a year, rather than the once every two years of the 1980s. In addition, carbon dioxide emissions are still rising and will rise faster as the developing world develops, suggesting climate change will accelerate as the century progresses. The added costs of flood defences and building damage caused by more extreme weather

Energy and Climate Change   David A. Coley
© 2008 John Wiley & Sons, Ltd

are likely to be extensive within the developed nations. The human costs of agricultural failures, water shortages and possibly political destabilization within the developing world could be much greater.

It has been estimated [ROY00] that the developed world needs to cut its emissions of carbon dioxide by 60 per cent if carbon dioxide levels in the atmosphere are to remain below 550 parts per million (ppm) – beyond which point irreversible damage will have been done. Changing technologies and changing the way we live to achieve this is likely to cost developed nations around one per cent of their gross domestic product (GDP) per annum by 2050 [AEA03]. Economic growth will mean that GDP will probably have tripled by then, suggesting that this sum is affordable. In addition, much of this cost will be offset by reductions in costs associated with increased flooding etc. However, the pre-industrial atmospheric concentration of carbon dioxide was approximately 280 ppm; it is now around 370 ppm (a level not witnessed for over a million years) and rising at more than half of one per cent per annum. As we are already starting to see the effects of climate change, the changes found at 550 ppm might well be considered unacceptable, implying greater cuts are necessary – possibly of the order of 90 per cent. This will require us to develop and deploy a whole new sustainable energy infrastructure.

Figure 1.1 shows the major energy transformations, fuels and groupings studied in this book. It is clear that there are many sources of energy from which we can choose. In Parts II and IV we will define some of these as *unsustainable* and others as *sustainable* and examine technologies for their exploitation, but clearly there are many alternatives to fossil fuels.

The central question is: why haven't we already made the switch to non-carbon fuels? There would seem to be two fundamental problems. Firstly, fossil fuels are cheap (crude has until recently traded at the same price as it did in 1880), and secondly, such fuels are highly energy dense. The first of these problems could in theory be solved by reducing income tax and other taxes, and taxing carbon instead. However, this might not find favour with voters, who are notoriously suspicious of new taxes, and it would also create difficulties for business unless the approach was adopted worldwide. Much more acceptable is probably the state subsidy of non-carbon alternatives and the pump priming of sustainable technologies to the point where they can compete with fossil fuels. In much of the developed world all three approaches are being applied to varying degrees and the cost of energy from alternative sources is falling rapidly. At the same time concern over climate change is growing. Together these signify a turning point both in the economic cost of renewables and the public's concern about the future of the planet. This makes it an extremely exciting time to be involved in, or studying, energy and its impact upon the environment. No longer is the use of alternative energy a theoretical opportunity; it is an imperative. It is also happening all around us in the form of wind turbines, hybrid motor vehicles and the introduction of energy efficient technologies.

The second problem, the question of energy density, will be harder to solve. Filling a car's petrol (gasoline) tank takes around one minute. This is a power transfer[1] of

---

[1] Technically, there is no flow of energy taking place here, just a relocation.

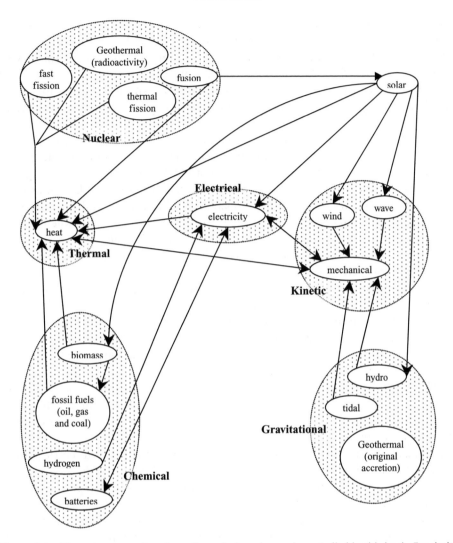

Figure 1.1 The major energy transformations, fuels and groupings studied in this book. For clarity many intermediate processes are not shown

35 million watts (MW). (For comparison a typical domestic light bulb draws 100 watts.) To charge an electric car with this much energy would take around eight days through a domestic socket. It also means that a large filling station capable of simultaneously recharging 30 electric vehicles in one minute would draw 30×35 MW $= 1050$ MW, approximately the output of a nuclear power station. Other estimates lead to equally depressing conclusions. It would take a land area of around 940 000 km² to grow energy crops capable of replacing all the UK's fossil fuel use, close to four times the country's total land area [CIA04a]. From this we can conclude that energy efficiency will have a major role in the energy policy of the future, but it is also likely that energy

production will require larger land areas and be much more visible than it has been for many generations. It is worth noting that in 1900 the USA used one quarter of its arable land for the growing of energy crops for the transportation and work systems of the day – horses [SMI94, p91]. Today sustainable generation is likely to meet stiff resistance unless such technologies are conceived to fit within the landscape or are placed out of sight: for example roof-mounted solar panels or sub-surface tidal stream generators.

Possibly we will simply have to accept change. In the early nineteenth century there were around 10 000 windmills operating in England [DEZ78]. An equivalent number of modern wind turbines each with a rated output of 2 MW would have a capacity equal to nearly 30 per cent of the UK's power stations. We could therefore take this number as a historical minimum that the landscape is capable of holding without undue degradation. This would also return us to the idea of large-scale embedded generation where there is less geographic separation between generation and use. This could be described as the *re-democratization* of supply since we would all experience both the benefits and the consequences of energy supply.

It would appear that it will be far more cost effective to make reductions in carbon emissions early [AEA03, p44], even though the costs of sustainable technologies are probably at their greatest due to their infancy and low production volumes. The main reason for this is that if we delay our reductions, then to have achieved the same total carbon emissions over any fixed period we will have to make much larger cuts. This point is worth emphasising. It is the total cumulative amount of carbon (and other greenhouse gases) emitted that is key. Therefore the later we leave it the greater the required reductions will need to be, and beyond a certain date we will face the reality of either making cuts of over 100 per cent, which is clearly impossible[2], or face a great degree of warming.

Unfortunately, environmental concerns often clash not just with our desire to minimize cost in the short term, but also with each other. The use of diesel as a fuel is common in buses and other public transport systems and it is used in great quantities within cities, creating a potential problem with local air quality. Its greater efficiency as a fuel though does mean its carbon dioxide emissions per km driven are typically lower. Its widespread adoption in private vehicles would therefore be beneficial in the fight against climate change, but would degrade local air quality. This is an example of an environmental dichotomy. We have a pair of options, neither of which is environmentally benign and, because they have differing environmental effects, it is very difficult to compare the relative magnitudes of the impacts. In this case the dichotomy is deepened because the impacts affect different groups. Changes in local air quality are largely 'democratic', in that those affected either own cars themselves, or derive wealth from the use of transport within the economy in which they live. However, the majority of greenhouse gas emissions are from more developed economies, whereas the majority

---

[2] Unless carbon storage technologies, such as increasing land or ocean biomass, were taken to extremes and could absorb more carbon than we emit each year.

of the future victims of climate change are likely to live in less developed economies. Climate change can therefore be seen as highly 'undemocratic'. This reduces the pressure for change.

In the industrialized countries, around two thirds of carbon dioxide emissions are from the energy used for transport (mostly the private car), and heating and lighting homes. This implies that climate change and the need to switch energy systems is not a problem created by industry, but is the responsibility of all of us. No faceless commercial giant is accountable for these emissions, but you and I.

There is also the problem of resource depletion. We are accustomed to plentiful supplies of our most valued hydrocarbons, oil and gas. Yet given the rate of use, reserves of such fuels may only last a further 40 years, and as much of it lies in geopolitically unstable locations, there are concerns over the security of supply[3]. There are fortunately much greater reserves of coal, shale oil and tar sands, all capable of being converted to oil and gas, but at a price. We will not run out of carbon for a long time. It could indeed be argued that it is our overwhelming wealth in carbon that is the problem, in that it is creating a stop to innovation. Even those most sceptical about whether manmade climate change is upon us would probably agree that we simply can not afford the risk of releasing all this carbon to the atmosphere. Clearly there are challenges ahead. Is it not surprising that in the twenty first century we still make the majority of our electricity – our most valued fuel – by essentially setting fire to a pile of coal, or other fossil product, using the heat from this to boil water, then using the steam to blow a giant fan around?

Rather than other important local pollution issues, the energy-related problems we will be concentrating upon in this book are global in nature. In particular we are interested in climate change. If climate change is to be tackled seriously then the whole international community will need to be involved. To emphasize the global nature of the problem and the spirit of international co-operation required in finding a solution, this book has tried to present an international view of energy and its use. Where possible, resource estimates and historic trends are given for the whole world; detailed, national figures then being used to illustrate various specifics. We also present example applications of technologies from around the world, and ask the reader to solve problems centred on a variety of countries.

The text is organized into four parts. Part I asks the question: what is energy? We then discuss the size of natural energy flows such as winds and tides that might become sources of sustainable energy and introduce the central environmental concerns that arise from how we currently provide our energy. Part II describes each of the current major energy technologies in turn, from coal to nuclear power. Part III returns to climate change to see what the future may hold for us and describes the work of the international community in trying to find agreement over carbon emissions. Part IV introduces the

---

[3] For example, by 2010 the UK will be a net importer of oil. By 2020 three quarters of UK energy will be imported and there is expected to be only one remaining nuclear plant operating in the country by 2025.

new sustainable energy technologies that will hopefully form the basis of future energy production, including solar, wind and wave power.

Throughout the book you will come across many problems within the text. It is strongly recommended that you try and complete each of these before moving on. They have been designed to encourage you to engage with the material and to practice accessing tables, manipulating important data and concepts and analysing the results of simple calculations. With the confidence such practice brings, it is hoped that you will try to rationalize and debunk some of the articles that are often published in newspapers and other media. Frequently, this can have amusing results, quite at odds with what the journalists were trying to say. The use of only approximate values for any data is encouraged. The idea is to show that back-of-the-envelope style calculations often allow one to answer important questions. If you don't know the values for some of the data, just guess. The same goes for how to actually answer the questions; just start manipulating the data you do know and the answer may just drop out.

As an example of the approach required, imagine one of the questions asks you to estimate the number of individuals who die in China (population 1.3 billion [CIA04b]) each year. Note we are not asking whether you already know the answer to this question, but to work it out using approximate methods. So, if the population is 1.3 billion and we assume most people die when they are, say, 70 years old then the answer is $1\,300\,000\,000/70 \approx 19$ million per annum. If we had assumed that average age at death was 60 or 80, the answer would have been 22 or 16 million, respectively. Either way it probably doesn't matter in the context in which the question is likely to be asked: it's a very large number of people. Some students find such questions relatively easy; others find it difficult to work approximately. It is the author's opinion that such simple calculations have a pedagogical role, in that they reinforce the learning process by helping students remember concepts, data and results.

Answers are given in an appendix, but please try to attempt each one before looking the answers up, or moving on.

The student exercises at the end of each chapter are a mix of numerical and essay-based ones, and are designed to test whether students have understood qualitative and quantitative concepts. Those wishing for additional quantitative exercises should extract and adapt relevant in-text problems.

PART I

Energy:
concepts, history and problems

"Shame on us if 100 or 200 years from now our grandchildren and great-grandchildren are living on a planet that has been irreparably damaged by global warming, and they ask, 'How could those who came before us, who saw this coming, have let this happen?'"

–Joe Lieberman

In this first part we will ask the question, what is energy? Despite being a concept central to the teaching of science in schools and universities, the concept of energy will prove to be difficult to tie down with a single satisfactory definition. We will then move on to discuss some of the environmental problems caused by the way we currently meet our energy needs. Central to this will be an introduction to climate change (often termed global warming), but other topics such as air pollution, acid rain and sustainability will also be covered.

The possibility of meeting all our energy requirements from sustainable renewable sources has been a dream of many for a long time. By reviewing the natural energy inputs to which the planet is exposed, we will see that there is in theory no reason why this cannot be achieved.

Another question that needs to be asked is: how important is energy to the functioning of society? In the past energy use was much lower, and only by considering how the evolution of society is connected to the development of new sources of energy will we realize the importance of the link. By appreciating the connection between energy use, wealth and development, we can see why poorer nations will, and must be allowed to, substantially expand their demand for energy. This connection is also possibly why many of the world's politicians have shied away from the issue of climate change. Without a change to the way we derive our energy, this expansion will have significant implications for the concentration of greenhouse gases in the atmosphere.

Finally we will examine some of the physical limits to the efficiency with which we can supply energy (or work). This will lead to an analysis of how our most valued form of energy – electricity – is generated and distributed.

2

Energy

"Our planet ... consists largely of lumps of fall-out from a star-sized hydrogen bomb ... Within our bodies, no less than three million atoms rendered unstable in that event still erupt every minute, releasing a tiny fraction of the energy stored from that fierce fire of long ago."

–James Lovelock, Gaia

Before embarking on our journey through current and future energy systems and the problems that their use might cause, we need to achieve a firm understanding of what energy is in all its guises, its units and any fundamental constraints that exist in transforming it from one form to another.

2.1 What is energy?

So, what is energy? If you have a background in physics or engineering this might seem rather a strange question to ask. Using terms such as *potential energy* and *kinetic energy* might almost seem second nature to you now, but pause for a moment and ask yourself the question, what actually is energy?

We know from common experience that a rotating wheel, a hot cup of tea, the current flowing through a wire or a crashing wave are all objects or systems displaying 'energy', whatever this might mean. We also know that we can make good use of energy in our lives: to cook with, warm us, light our homes, transport us and power our bodies. Yet this familiarity does little to help us with a definition.

In this book we will be particularly interested in understanding transformations of energy from one type to another. In order to do this, we need to equate an amount of energy of one type with the same amount of a very different type. We need an instinctive way to relate the amount of 'energy' contained in an electromagnetic wave with that of a spinning top.

Energy and Climate Change   David A. Coley
© 2008 John Wiley & Sons, Ltd

The most common way to relate such disparate systems is in terms of the amount of *work* they have the capacity to do. And indeed, the phrase *the capacity to do work* is often used as the standard textbook definition of energy. Unfortunately this still leaves us with the problem of defining *work*[1]. However, work is certainly less of an ethereal concept than energy and is much closer to what this book is about. Energy is not something we want per se. It is what we can do with it that is important. People want to drive from *a* to *b*; to keep their drinks cool; make their houses warm. It is the result of such energy transformations and the resultant amount of work we can achieve that is of interest, not energy itself.

If all this seems nothing more than an attempt to dodge the issue and avoid formally defining energy, then in some ways maybe it is. Some of the finest minds in science have tried to answer the question. The result of this process can be summed up by the comment of the Nobel prize-winning physicist Richard Feynman:

> 'It is important to realise that in physics today, we have no knowledge of what energy is' [FEY63].

What we do know is that the concept of energy allows us to quantitatively connect a wide variety of phenomena easily. By only considering changes in energy, we escape any definitional problems, and energy becomes a very easy and intuitive concept to work with.

Another way of considering what energy might be is by making use of Einstein's famous equation:

$$E = mc^2 \tag{2.1}$$

where energy E (in Joules, J) is related to mass m (in kg) and the speed of light c (in metres per second). By re-writing this as:

$$m = \frac{E}{c^2} \tag{2.2}$$

we see that energy is mass – i.e. it has weight! This in turn implies that as we transform energy from one system to another, or from one form to another, a small amount of mass is also transferred. The amount is actually very small indeed and undetectable in any engineering-sized system. However, as we shall see later, such almost infinitesimal sized mass transfers are key to an understanding of nuclear power, where Equation (2.1) is used to explain the transformation of small amounts of matter into very large amounts of energy.

Before we hurriedly conclude that Equation (2.2) gives a true definition of energy, we must remember that Equations (2.1) and (2.2) form little more than a tautological loop: energy has mass and mass can be converted to energy; neither truly defines the other.

---

[1] Physicists often define work as the product of force and the distance through which the force acts. The force being provided by an object such as a person, an electric motor or a falling weight.

Problem 2.1   Estimate the theoretical increase in mass of 1000 kg of steel as its temperature is raised from 20 °C to 200 °C. Hint: the energy required to warm 1 kg of steel by 1 °C (the specific heat capacity of steel) is 0.5 kJ kg^{-1} K^{-1}. The result shows that the change in mass is essentially immeasurably small.

Although we have perhaps failed to pin the concept of energy down as firmly as we might like, we can now start to think about the many forms it might come in and how to relate these through a single common unit.

2.2 Units

Throughout this book the metric unit of energy, the joule (J), is used. However for many large systems joules prove to be very small and we are forced to talk about mega-joules (1 000 000 J) and giga-joules (1 000 000 000 J), etc. The main advantage of relying on a single unit, even with numerous prefixes, is that it allows us easily to quantify the efficiency of conversions between different energy forms and numerically to compare the relative importance of different types of energy in our lives. This is a lot harder if, for example, energy in food is measured in calories whilst that derived from gas is in British Thermal Units.

Historically, each energy industry has rarely needed to make such comparisons and has therefore selected units that are directly related to how such energy is used and transported. Some of these are *true* energy units, such as the foot-pound (the amount of energy required to raise one pound in weight (approximately 0.45 kg) by one foot (approximately 30 cm). These can easily be converted into joules by applying a simple conversion factor. Others are not really energy units but mass or volume units, such as barrels of oil or tonnes of coal. These can also be converted to energy units if you know the density and the calorific value (the amount of energy stored in 1 kg) of the fuel. For energy sources such as coal, density and calorific values vary between fuels from different regions, but for approximate work, standard values have been developed and it is straightforward to make conversions to and from joules. Tables 2.1 and 2.2 list common conversion factors, densities and calorific values; the tables are repeated in Appendix 4 for easy reference. More detailed values, which account for the exact fuel composition, are given under the appropriate sections in Part II.

Although the joule will be our standard unit, the importance of oil is such that the energy industry more often uses tonnes-of-oil-equivalent (t_{oe}). This is the energy content of one tonne of oil, i.e. 42 billion joules ($= 42 \times 10^9$ J $= 42$ GJ).

Problem 2.2   As we will see in Problem 2.8, the burning of a single match releases 2720 J. Using Table 2.1, estimate how high a pound (0.45 kg) of potatoes might be lifted by this energy. (Ignore the efficiency of the machine that might do the lifting.) You may you find the answer surprising.

Table 2.1 Common energy units

Unit	Pronunciation	Equivalence	Notes
kWh	kilo-watt-hour	1 kWh = 3.6 × 10^6 J	Energy used by a single-bar electric fire left on for one hour
ft.lb	foot-pound	1 ft.lb = 1.356 J	Energy required to raise one pound in weight (≈0.45 kg) by one foot (≈0.3 m)
cal	calorie	1 cal = 4.19 J	Energy required to heat 1 g of water 1 °C (at atmospheric pressure)
kcal or Cal	kilo-calorie	1 Cal = 4.19 kJ	Unit used by the food industry and easily confused with the cal
Therm	therm	1 therm = 100 000 BTU	
BTU	British thermal unit	1 BTU = 1055 J	
t$_{oe}$	tonne of oil equivalent	1 t$_{oe}$ = 42 GJ	

Problem 2.3 Complete the following table of derived units.

Fuel	Heat (J) obtained by burning
oil	1 barrel =
gas	1 m^3 =
coal	1 t =

Scientific notation

Whichever units are used, a large range of values will be encountered. Your daily consumption of chemical energy (via food) is approximately 8 500 000 J. Over the same period a large power station will produce 86 400 000 000 000 J. To avoid the problems, and possible errors, of writing down large numbers of trailing zeros but still be able to cover such a massive range of numbers, a form of scientific notation is commonly used when carrying out the mathematical manipulation of energy data.

Table 2.2 Additional conversion factors

Fuel	Density (approx.)	Calorific value	Common unit
oil	0.84 kg/l	43 MJ/kg	Barrel = 159 litres
natural gas	0.76 kg/m^3 (422 kg/m^3 liquid)	49 MJ/kg	m^3 = 0.76 kg
coal	1.1–1.5 kg/m^3	29 MJ/kg approx.	tonne (t) = 1000 kg
wood: pine	0.35–0.66 kg/m^3	16 MJ/kg approx.	kg
oak	0.59–0.93 kg/m^3		

Table 2.3 Some common, and not so common, prefixes

Prefix	Symbol	Multiply by	Prefix	Symbol	Multiply by
deka	(da)	10^1	deci	(d)	10^{-1}
hecto	(h)	10^2	centi	(c)	10^{-2}
kilo	(k)	10^3	milli	(m)	10^{-3}
mega	(M)	10^6	micro	(u)	10^{-6}
giga	(G)	10^9	nano	(n)	10^{-9}
tera	(T)	10^{12}	pico	(p)	10^{-12}
peta	(P)	10^{15}	femto	(f)	10^{-15}
exa	(E)	10^{18}	atto	(a)	10^{-18}
zetta	(Z)	10^{21}	zepto	(z)	10^{-21}
yotta	(Y)	10^{24}	yocto	(y)	10^{-24}

This notation allows 7 000 000 to be written as 7×10^6 and 492 500 as 492.5×10^3, where the exponents (6 and 3) simply count the number of places the decimal point has moved. Of course 492 500 could be written as 4.925×10^5, but you will probably find it easier if the exponents follow the same standard series as the terms Mega, Giga, Tera etc. This is because it is much easier to say 'fourteen Mega-joules' than '1.4×10^7 joules'. This series is laid out in Table 2.3.

2.3 Power

The terms *energy* and *power* are often confused with each other, and not just by students. Sales literature from electricity and gas companies can sometimes fall into the trap of implying that the terms are interchangeable. *Power* is the rate of energy use or delivery, its unit is the Watt (W) and 1 W = 1 joule per second. It differs from energy in much the same way as speed does from distance, i.e. as kilometres per hour does from just kilometres. Part of the confusion comes from the use of kWh as a common unit of energy. This obviously has watts, i.e. power, in it, but also includes time. As one hour equals 3600 seconds, we see that 1 kWh = 3600 kW seconds, or:

$$3600 \text{ k}\frac{\text{J}}{\text{s}}\text{s}.$$

The two time units (the seconds) cancel to leave 3600 kJ. So, 1 kWh is in fact 3600 kJ, i.e. a measure of energy, not power.

Hopefully the following example will help to reinforce this important distinction: The average citizen of a developed country might use around 1 GJ of electricity in their home per month. The average modern power station produces around 1 GW or 1 GJ/s. The numeric equivalence of these two figures does not imply that each person needs a personal power station to provide their electricity. You use the energy over a whole month; the power station has to produce this amount every second.

Problem 2.4   If completed, this dramatic class-based exercise will hopefully always remind you of the difference between energy and power, and should only be completed under supervision. You will need a large clear space, two wine glasses (plastic ones would be safer), a baseball bat or equivalent (it needs to be fairly long), some wine (or water), a wooden broom handle, safety goggles, clothes you don't mind getting wine on and two tables. The exercise contrasts the effect on the broom handle of applying a large amount of energy to it, with that of applying a large amount of power. First, hold the broom handle almost vertically with one end on the floor. Press hard on the middle of the handle with your other hand. The handle should flex but not break, and will not break however long you press on it; i.e. no matter how much energy you apply, it will not break. This is because you are applying the energy over a long time, i.e. applying very little power. Now half-fill the wine glasses, place one on the edge of each table and balance one end of the broom handle on each glass. The space under the handle should be clear, ask anyone near to stand well back and put your safety goggles on. Apply a relatively small amount of energy, in a very short period of time (i.e. a large amount of power) to the centre of the broom handle by striking it very hard with the baseball bat (you will need to raise the bat high over your head). The broom handle will snap. If all went well, the wine glasses will, surprisingly, still be intact – making the demonstration more memorable. By missing out the glasses and the wine a slightly safer demonstration can still be made, but remember, the shattered broom handle can still fly off in unexpected directions.

Unfortunately, the term *power* is also used as a pseudonym for electricity. Hence terms such as power-cut, hydro-power and wind-power being used in place of electricity-cut, hydro-electricity or electricity from wind. Because this usage has become so common, we will also occasionally adopt it.

Power stations can be rated in either terms of their electrical output or their much greater thermal output, i.e. the heat raised to drive the generators. To distinguish these two alternatives, we use the symbols W_e or W_{th} respectively.

2.4 Energy in various disguises

As we discussed above, energy comes in many forms. Each form can be converted into others, sometimes in surprising ways. For example, the thermal (heat) energy stored in a hot cup of tea can be easily and directly converted into electrical energy in the following way. Form a loop from two suitable dissimilar metal wires and place one of the junctions in the tea and the other in cold water. A small electric current (which can be measured with an ammeter) will flow around the loop. Although this conversion from thermal to electrical energy is highly inefficient (about one per cent), it doesn't involve any moving parts and is therefore extremely robust. For this reason, it is the way NASA chose to power the Voyager spacecraft. (Although NASA opted for a small radioactive source to heat the junction, rather than warm tea.) The reverse effect, where electrical energy is transformed into heat, is much more common; an electric toaster being an obvious example.

In the following sections, energy is classified into the seven forms most useful for our later analysis of energy systems, and some of the most useful conversions introduced. These forms, or categories, often overlap and are in part conventions based on convenience. For example, a vibrating molecule is described as having vibrational energy, which is itself, at least in part, a form of kinetic energy.

Kinetic energy

Kinetic energy is the form of energy associated with mass in motion. Together with potential energy, it is often referred to as mechanical energy. The kinetic energy, $E_{kinetic}$, of a body of mass, m, moving at speed, v, is given by:

$$E_{kinetic} = \frac{1}{2}mv^2. \tag{2.3}$$

If mass is measured in kilograms (kg) and speed in metres per second (m/s) then $E_{kinetic}$ will automatically be in joules.

Kinetic energy is turned into thermal energy by friction and to electrical energy by generators in power stations.

Problem 2.5 Compare the kinetic energy of a 1000 kg car travelling at 100 km/h and that of a 10 g pen on-board a Saturn V rocket as it reaches its maximum speed of 40 064 km/h (24 900 miles/h).

As hopefully you discovered, the answer to Problem 2.4 is possibly surprising and comes about because of the power of 2 in Equation (2.3). This shows that for kinetic energy the speed, not the mass, is often the greater determinant of the amount of energy; an observation which will be critical to our later study of wind turbines, where the average wind speed at a location will prove to be the dominant factor of whether a particular wind farm will be economic.

Potential energy

Badly named, since it is as real as any other type of energy, potential energy arises from the position of an object within a potential field, or the tension in an elastic system such as a wound spring (elastic energy). Most commonly, the force field is the gravitational field that surrounds the earth: the higher we are, the greater the potential energy we have. More precisely, for an object of mass, m kg, at height, h metres, the potential energy, E_{pot} (in J) is given by:

$$E_{pot} = mgh \tag{2.4}$$

where g, the acceleration due to gravity $= 9.8 \, (m/s^2)$. Often potential energy is converted to kinetic energy by letting a mass fall. As it accelerates its potential energy will reduce, but its kinetic energy will increase. Through the law of energy conservation, which states energy can not be destroyed but only transformed from one type to another, we know that the sum $E_{pot} + E_{kinetic}$ will remain constant.

In a pendulum there is an alternation between maximum E_{pot} at the top of each swing and the maximum $E_{kinetic}$ at the bottom of the swing. The rapid cycling of potential and kinetic energy is often referred to as vibrational energy, with sound waves being an example.

Problem 2.6　Estimate the kinetic energy and final velocity of 1000 kg of water which has fallen from a height of 100 m. (Ignore air resistance.) Hint: the final kinetic energy will equal the original potential energy.

Thermal energy

This is the energy associated with vibrations at the molecular or atomic scale and is often called internal energy. As a body is heated it gains thermal energy and its temperature rises. It is important though to realise that temperature and thermal energy are not equivalent. 100 kg of water at 10 °C contains more thermal energy than 1 kg of water at 100 °C. The amount of matter is important, as is the *specific heat capacity* of the substance.

The *specific heat capacity* of a substance is defined as the heat needed to raise the temperature of a unit mass of the substance one degree, and in SI units is given in joules per kg per kelvin. The *thermal* or *heat capacity* of a substance is then given by the product of the specific heat capacity and the mass of the substance and is also given in joules per kg per kelvin.

Both kelvin (K) and the more common celsius or centigrade (°C) are used as temperature scales in this book. They are equivalent except for a shift by a constant 273, i.e. 0 K $= -273$ °C and 20 °C $= 293$ K. In equations where we are only interested in changes in temperature, either kelvin or celsius can usually be used. In some equations, for example the estimation of Carnot efficiencies (Chapter 7), it is essential that kelvin is used. If in doubt about which to use, select kelvin, as this will always give the correct answer.

The thermal energy, E_{th} (J) of mass m (kg) of a substance with specific heat capacity C_p (J/kgK) and at a temperature of T K is given by:

$$E_{th} = mC_p T. \tag{2.5a}$$

If the temperature of a mass, m, changes by δT, then the change, δE_{th}, in thermal energy is given by:

$$\delta E_{th} = mC_p \delta T. \tag{2.5b}$$

A gain in thermal energy is not always associated with an increase in temperature. If the heated substance goes through a phase change (i.e. from solid to liquid or liquid to gaseous), then during this change no increase in temperature will be seen, as energy is required to break the solid or liquid bonds. However the substance is still *storing* this energy despite the lack of a temperature rise. If the substance is later allowed to cool and return to its former state this *latent heat* can be recovered.

Most energy will finally transform itself into thermal energy. For example, the potential energy of a raised hammer will be converted to kinetic energy as the hammer falls, then to thermal energy in both the hammer and whatever it strikes. In many ways thermal energy can be seen as the lowest form of energy, or as energy with the lowest *quality* (and the lower the temperature, the lower the quality). Although other forms of energy can be converted with 100 per cent efficiency into thermal energy, transforming thermal energy to other forms is always less than 100 per cent efficient: often less than 50 per cent. The generation of electricity from the burning of fossil fuels such as coal, gas and oil to provide the thermal energy used to run a generating set can be as little as 30 per cent efficient. This implies that electricity is usually a very inefficient way to heat buildings. It would often be much more efficient and use fewer resources if the fossil fuels were burnt within the building to provide heat directly.

Because of the poor thermal efficiency of electricity generation there is often the need to distinguish between the thermal output of the power station and the smaller electrical output. As stated earlier, the subscripts *th* and *e* are commonly used to make this distinction, e.g. GW_{th} and GW_e for thermal and electrical power respectively.

For an ideal gas (i.e. one at a relatively low pressure) the translational kinetic energy of the molecules is only a function of the temperature of the gas:

$$E_{kinetic} = {}^3/_2 nRT \qquad (2.6a)$$

where n is the number of moles of gas and R is the universal gas constant (8.31 J/mol.K), or for a single molecule:

$$E_{kinetic} = {}^3/_2 (1/N_A)RT \qquad (2.6b)$$

where N_A is Avogadro's constant (the number of molecules in a mol, 6.02×10^{23}). If the gas has no other energy, for example in the form of molecular rotations or velocity due to bulk motion, then $E_{kinetic} = U$ (the *internal* or *thermal energy* of the gas), and Equation (2.6a) can be used in much the same way as Equation (2.5a).

Problem 2.7a   Estimate the increase in thermal energy of 10 kg of steel (specific heat capacity 0.5 kJ/kgK) and 10 kg of wood (specific heat capacity 1.7 kJ/kgK) if they are both raised from 20 to 50 °C.

Problem 2.7b   Use Equation (2.6b) to estimate the average kinetic energy and speed of a hydrogen molecule (mass = 3.35×10^{-27} kg) in a bottle of hydrogen at 50 °C.

Chemical energy

Chemical energy is exchanged whenever chemical bonds are broken or formed. For some reactions this process will be exothermic (thermal energy is released), for others it will be endothermic (a net in-flow of energy is required to allow the reaction to proceed).

Within the context of this book, the most important chemical reactions are those involving the burning of fossil fuels (oil, gas and coal) to produce mainly carbon dioxide (CO_2) and water. These reactions are highly exothermic, with approximately 35 MJ/kg of energy (the calorific value) being given off for every kg of fuel burnt[2]. Thus, the burning of fossil and other organic fuels, such as wood, releases carbon dioxide into the atmosphere. The growth of plants represents the reversal of this reaction: plants absorb carbon dioxide from the atmosphere and, through photosynthesis, use sunlight to break the carbon dioxide into carbon and oxygen atoms, which are then recombined with nitrogen and other elements to form carbohydrates, sugars and other complex organic molecules essential to the plant. When plants decay, or burn in forest fires, the carbon in these compounds reacts once more with atmospheric oxygen to form carbon dioxide, thereby forming a *carbon cycle*.

If C_v (J/kg) is the calorific value of a fuel, then the burning of m kg of fuel will produce:

$$E_{th} = C_v m \qquad (2.7)$$

joules of energy.

Problem 2.8   If the calorific value of wood is 16 MJ/kg and a match weighs approximately 0.17 g, how much energy in joules is released by burning the wooden part of a single match?

Electric energy

When a loop of wire is moved through a magnetic field some of the electrons in the wire are forced to travel along the wire. Thus the kinetic energy of the moving wire is transferred to the electrons, and a current flows. As energy must be conserved, the loop of wire will have lost energy and been slowed by the creation of the current. Such currents can also be formed by chemical reactions, most notably in batteries.

Electrons are not free to move without resistance. The resistance produces heat, which is harnessed to useful effect in electric fires and the filaments of light bulbs. Equally usefully, the force that slows the rotating wire in a generator will force a current-carrying wire to move if placed in a magnetic field, forming the basis of an electric motor. In many industrial applications, kinetic energy in the generator is converted to electric

---

[2] Each fossil fuel has a particular calorific value that depends on its exact composition. Thus the calorific value of oil is different from that of natural gas, and each type of oil will in turn have a different value. However, for many purposes the value of 35 MJ/kg for any fossil fuel is useful for very approximate calculations.

energy, which is converted back to kinetic energy within a motor at some distant factory. Electric energy is therefore really being employed as an intermediary, or carrier, that allows easy access to kinetic energy or heat at a distance. As we will see in Chapter 4, the ease with which electricity allows energy to be distributed has had a profound effect on industry and national wealth. As might be expected, this process is not 100 per cent efficient. Energy is lost as heat in the generator, the connecting wires and the motor.

The heating effect of an electric current is easy to calculate. If I ampere (A) of current is flowing along a wire of resistance r ohms (Ω) then

$$I^2 r \tag{2.8}$$

joules of heat energy will be released every second.

Problem 2.9   Calculate the mass of water per second that needs to fall from a height of 10 m in order to turn a generator that will produce 1 kJ of electrical energy per second (i.e. 1 kW). (Ignore all losses and engineering constraints.)

Nuclear, or mass, energy

As we have already seen, some chemical reactions are exothermic. In a similar way, some nuclear reactions also produce a net energy gain. Many elements will undergo reactions in which the total energy needed to hold together the final nuclei is less than that required to hold together the original nuclei. The 'missing' energy leaves the reaction either as electromagnetic radiation, or in the kinetic energy of the newly formed nuclei and sub-atomic particles. Through Equation (2.1) ($E = mc^2$) we can see that the new nuclei once slowed will also weigh less than the original ones, and such reactions can be viewed as a way of converting mass directly into energy.

There are three ways in which nuclei can undergo such a transformation. In *fission*, a nucleus can be prompted to split into smaller nuclei by forcing it to absorb an additional neutron. In *fusion*, a pair of nuclei can be forced together to create a new, larger, nucleus. In *radioactive decay*, nuclei spontaneously decay into other nuclei and sub-atomic particles. In all three cases, the total mass of the nuclei at the end of the process will be less than that at the start, implying energy has been released.

All current nuclear power stations use fission to create lighter, fast moving nuclei. These undergo collisions with other nuclei and the fabric of the reactor vessel thereby transferring their additional kinetic energy into thermal energy. This heat is then used to raise steam and operate an electrical generator in the same way as a coal, gas or oil fired plant would.

Problem 2.10   Only 0.7 per cent of natural uranium is of the correct form (^{235}U) to undergo fission. Given that the fission of a single uranium nucleus releases 3.2×10^{-11} J, what mass of natural uranium is required to produce 1 kWh of electrical energy? Hints: 1 g of uranium contains approximately 2.5×10^{21} nuclei; assume 100 per cent efficiency in the conversion to electrical energy; see Table 2.1 for the conversion factor for kWh to joules.

Table 2.4 An energy hierarchy

System	1 kg of 'fuel' produces	Fuel 'efficiency' ratio
Nuclear energy (fission of pure ^{235}U)	8.2×10^{13} J	Nuclear/chemical $= 2.3 \times 10^6$
↑	↑	↑
Chemical energy (oil being burnt)	35×10^6 J	Chemical/potential $= 36{,}000$
↑	↑	↑
Potential energy (mass falling 100 m)	981 J	N/A

Even after allowing for the operating inefficiencies of a real power station, which would increase the mass implied by Problem 2.9 by a factor of three, this is a surprisingly small amount of fuel. Comparing the result with the 0.36 kg of coal needed to do the same job, we see that a nuclear station will use approximately $1.35 \times 10^{-6}/0.36 = 1/270\,000^{th}$ of the fuel of a conventional plant. One can take this analysis further by estimating the energy released by a potential, rather than chemical or nuclear source: 1 kg of water falling 100 m will only provide ($m.g.h = 1 \times 9.8 \times 100 =$) 981 J of kinetic energy. Table 2.4 collects these relative efficiencies into a natural hierarchy of energy-producing reactions. The order of the hierarchy is also reflected in the historic order in which humankind has made use of these processes.

Electromagnetic radiation

Energetic electrons and other sub-atomic particles and nuclei can lose energy through the emission of electromagnetic radiation. Such radiation is mass-less (except for the mass given to it through its energy, i.e. as a consequence of $E = mc^2$) and consists of a stream of *photons* travelling at the speed of light (3×10^5 km/s). Light itself is electromagnetic radiation, as are x-rays, gamma-rays, radio-waves, micro-waves and infrared radiation from hot objects. As the stars in the night sky show, such radiation can travel across the universe, but is easily absorbed by matter: black card will stop light; glass will stop infrared; bone will attenuate x-rays. However, as the transmission of light through glass shows, this interaction with matter depends on the type of radiation, or more specifically its frequency.

The frequency, f (Hz or cycles per second), of radiation is inversely proportional to its wavelength, λ (m):

$$f = \frac{c}{\lambda} \tag{2.9}$$

where c is the speed of light (m).

The energy, E, of each photon in a beam of such radiation is given by:

$$E_{rad} = hf$$

where h is Planck's constant (6.625×10^{-34} J.s).

The energy emitted per second, i.e. the power, P, emitted from a hot object, of area A, by radiation is given by:

$$P = A\varepsilon\sigma T^4 \qquad (2.10)$$

where ε is the *emissivity* of the body and σ is *Stefan-Boltzmann's* constant (5.67×10^{-8} W/m^{2}K^4). For 'black bodies' ε takes its maximum value of 1. Many objects can be treated as 'grey bodies' for which ε is independent of the wavelength. At high temperatures, all objects can be considered as black bodies.

Problem 2.11   Which is more impressive as an energy source, the sun or you? Hint: calculate the energy output per second of the sun and of you, per kg of mass. The sun releases 4×10^{26} J/s; an adult human has a dietary intake of about 2000 kcal per day.

Problem 2.12   Try and complete the following table with common examples of energy transformations: one example is given. Hint: not all transformations are possible.

	To						
From	Chemical	Thermal	Nuclear	Kinetic	Potential	Electric	Electromagnetic
Chemical							
Thermal							
Nuclear							
Kinetic		car crash					
Potential							
Electric							
Electro-Magnetic							

2.5 Energy quality and exergy

The first law of thermodynamics states that:

The heat, Q, added to a system equals the change in the internal energy, U, of the system plus the work, W, done by the system [TIP99, p573]:

$$Q = \Delta U + W \qquad (2.11)$$

Q, U and W all have units of energy, so this implies that *energy is always conserved.* (The symbol Δ is used to imply a change.)

The second law (which can be phrased in several ways) states that:

It is impossible to remove thermal energy from a system at a single temperature and convert it to mechanical work without changing the system or surroundings in some other way [TIP99, p600].

Often when analyses of the efficiency of a process are undertaken consideration is only given to the first law. For example, if we use electricity to heat water, we know that the total amount of energy has not changed and the conversion of electricity to heat will be 100 per cent efficient. However, and as we will see in later chapters, we can not convert the energy in the hot water back into electricity with 100 per cent efficiency. The same is true of other exchanges. A weight held above the ground has a certain potential energy. If the weight falls to the ground it will lose this potential energy, gaining kinetic energy as it falls, which will be converted to internal energy (heat) as the weight strikes the ground. No energy has been lost, but again it would be impossible to use this heat energy to raise the weight with 100 per cent efficiency, i.e. to the same height as at the start. Heat itself demonstrates this effect: water at 100 °C is more useful than that at 20 °C. The former can be used to raise steam to drive generators, can be used within various industrial processes, and can easily raise with 100 per cent efficiency the temperature of another material from 70 to 80 °C. Water at 20 °C can not do this. Clearly there is a second energy hierarchy with low temperature heat energy (to which most energy is eventually converted) at the bottom of the pile. We therefore need a concept that expresses the *quality* of an energy resource, i.e. the likely usefulness of a fixed amount of it. The term *exergy* has been coined for this purpose. Exergy is defined as:

The useful work that can be extracted from a system which executes a loss-free process between its initial state and a dead state [UMN04].

The dead state is the equilibrium state with the surroundings, e.g. when the hot water cools to the same temperature as its surroundings, or when the weight is finally sitting on the ground. The unit of exergy is the joule (or any other unit of energy). We can see that, unlike energy, the quantity of exergy an object has depends on the state of the environment around it. This may seem reasonable now, and as will be formally explained later, the amount and variety of work we can obtain from 1 kg of material at 90 °C is likely to be greater if the temperature of the accessible surroundings is 10 °C, than if it is 85 °C.

Analyses based only on the first law fail to recognize the distinction between quantity and quality. Because of the second law we know that when a degradation in quality occurs the process is irreversible: we cannot get quality back in equal amounts. So a system, such as an electric water heater, may seem very efficient, but warm water is not a very useful substance and under an exergy, or second law analysis, it can be shown that the process is only one to five per cent efficient! [AMI75 and ECN04, p604]. Such

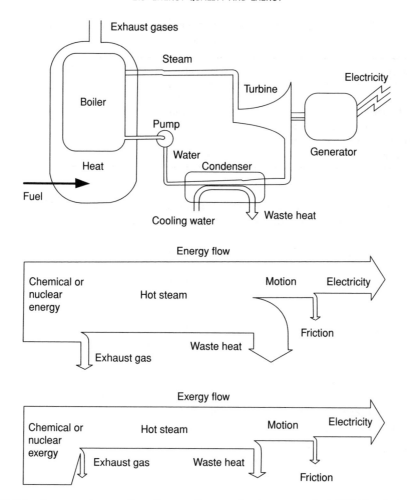

Figure 2.1 Energy and exergy flow diagrams for a condensing power plant [ECN04, vol. 2 p603]

inefficiencies can be greatly reduced if low-grade energy is converted to more low-grade energy, rather than using high-grade, high quality energy; an example being the use of waste water from an industrial process, at say 30 °C, to heat adjacent offices. The situation is illustrated in Figure 2.1 for electricity generation, where we can see that the major loss in exergy is when the fuel is burnt, but the major energy loss occurs when the steam is converted into the kinetic energy of the rotating generator. Table 2.5 compares exergy and energy efficiencies of other processes.

Exergy, X, can be expressed mathematically [UMN04] (see ECN04, pp593-640 for alternative equations; those without a science or engineering background may wish to skip the rest of this section):

$$X = (U + E_{kinetic} + E_{pot} - U_o) + p_o(V - V_o) - T_o(S - S_o). \qquad (2.12)$$

Table 2.5 Energy and exergy efficiencies of some typical systems [data from ECN04, vol. 2 p604]

Technology	Energy efficiency (%)	Exergy efficiency (%)
Oil furnace	85	4
Electric heater	100	5
Electric heat pump (see Chapter 7)	'300'	15
Combined heat and power plant (see Chapter 19)	85	40

Or, if we are considering a change between two states (state 1 and state 2), neither of which is the dead state, then the change in exergy is given by:

$$X_2 - X_1 = (U_2 - U_1) + (E_{kinetic,2} - E_{kinetic,1})$$
$$+ (E_{pot,2} - E_{pot,1}) + p_o(V_2 - V_1) - T_o(S_2 - S_1),$$

where:
   U is the internal energy of the system at its initial state,
   $E_{kinetic}$ is the kinetic energy of the system in its initial state,
   E_{pot} is the potential energy of the system in its initial state,
   V is the volume of the system in its initial state,
   S is the entropy[3] of the system in its initial state (units of J/K per kg),
   U_o is the internal energy of the system in the dead state,
   p_o is the pressure of the system in the dead state,
   V_o is the volume of the system in the dead state,
   T_o is the temperature of the system in the dead state, and
   S_o is the entropy of the system in the dead state.

The following example [adapted from UMN04] applies Equation (2.11) to the case of 1 kg of gaseous combustion products existing at $p = 7$ atmospheres and $T = 867\,°C$, when the dead state is assumed to be $p_o = 1$ atmosphere and $T_o = 27\,°C$. To simplify the calculation, the properties of air will be used instead of the properties of the combustion gases:

1. In this case, $E_{kinetic} = 0$ and $E_{pot} = 0$

2. Internal energy (from Equation (2.6))

---

[3] Mathematically, the change in entropy during a reversible process is given by the amount of heat supplied divided by the absolute temperature. Entropy measures the amount of disorder in a system and how far it is from equilibrium. It can been seen as providing the 'force' which causes heat to migrate from a hot object to a cold one when the two are placed in contact.

U at 867 °C (=1140 K) = 880.35 kJ
U at 27 °C (=300 K) = 214.07 kJ

3. Entropy: tabulated values
   $S - S_o = 0.8621$ kJ/K per kg

4. $T_o(S - S_o) = (300)(0.8621) = 258.63$ kJ/kg

5. $p_o(V - V_o) = p_o(nRT/p) - p_o(nRT_o/p_o)$
   $$= nR[(p_o T/p) - T_o], \text{ where } n \text{ is the number of moles of gas}$$

6. $nR = 1000\, R/$molecular weight of air, where the 1000 converts to kg.

   $$= (8.314)/(28.97) = 0.287 \text{ kJ/kg.K}$$

   $$p_o(V - V_o) = (0.287)((1.01325)(1140)/7 - 300)$$
   $$= -38.74 \text{ kJ/kg}$$

and therefore

$$X = (880.35 - 214.07) - 38.74 - (258.63) = 368.91 \text{ kJ/kg.}$$

Thus, the maximum amount of useful work that could be extracted from 1 kg of gas is 368.91 kJ.

Due to the complexity of carrying out calculations using exergy, we will concentrate on energy in the remaining text. However, readers are encouraged to keep the concept of exergy in their minds and to think about not only the quantity, but also the quality of energy supplied by various resources and technologies. One can then consider whether they are best matched quality-wise with what they are being used for; a classic mismatch being the use of high quality resources to provide lowly heat.

2.6 Student exercises

"We've arranged a civilization in which most crucial elements profoundly depend on science and technology. We have also arranged things so that almost no one understands science and technology. This is a prescription for disaster. We might get away with it for a while, but sooner or later this combustible mixture of ignorance and power is going to blow up in our faces."

–Carl Sagan

1. How much mass and energy did the sun need to lose in order to boil an electric kettle back on earth? Estimate the efficiency of this whole chain.

2. How can the equation $E = mc^2$ be used as a definition of energy? Why is this not as useful a thing to do as it might seem?

3. What is the difference between energy and power? Give an example of both.

4. List the seven forms of energy described above and give at least one equation involving each.

5. What is exergy and why is it important?

6. A ball weighing 15 kg sits on a ledge then falls 20 m to the floor. Ignoring air resistance:
(i) What are the initial and final values of the ball's potential and kinetic energy?
(ii) Estimate the kinetic energy, potential energy and velocity of the ball half way to the floor.

7. 10 litres of water is heated from 10 °C to 20 °C. What is the gain in thermal energy of the water?

8. A power station has a thermal output of 1500 MW. What is the thermal output in (i) kW, (ii) J/s, (iii) horse-power, (iv) foot-pounds per second? If it runs continuously for a year, how much heat energy will it produce in (v) joules, (vi) kWh? If its efficiency is 38%, what will its electrical output be in (vii) watts, in (viii) kWh per annum?

9. A wind turbine is rated at 2 MW. What is its rating in horsepower?

3

The planet's energy balance

"... by the late eighteenth century scientists knew very precisely the shape and dimensions of the earth and its distance from the Sun and planets; and now Cavendish, without even leaving home, had given them its weight. So you might think that determining the age of the Earth would be relatively straightforward. After all, the necessary materials were literally at their feet. But no. Human beings would split the atom and invent television, nylon, and instant coffee before they could figure out the age of their own planet."

–Bill Bryson, *A Short History of Nearly Everything

A central theme of this book is that we can at least in theory meet much of our energy demand from sustainable resources, such as wind and wave power. By sustainable we mean that the resource is being replenished at roughly the same rate as we are using it. This it is not so for fossil fuels, such as oil and gas. In order to be able to see if we can balance our usage with what the planet is capable of supplying, we need to know two things – the amount of energy we currently use and the size of the renewable resource. The first of these is discussed in the next chapter, the second is considered here.

In this chapter we will study a series of renewable energy flows that might feasibly meet some of our future energy needs, and estimate the size of each resource from a theoretical standpoint. In Part IV we expand on the form of each resource and discuss possible technologies aimed at unlocking what is a vast, constantly renewed, flow of sustainable energy.

3.1 The sun

Almost all the energy fluxes on Earth have their origin in the near constant warmth provided by the sun. Even the winds powering wind turbines and the coal burnt in power stations have their roots in the radiation provided by the ongoing nuclear reactions in the

Energy and Climate Change   David A. Coley
© 2008 John Wiley & Sons, Ltd

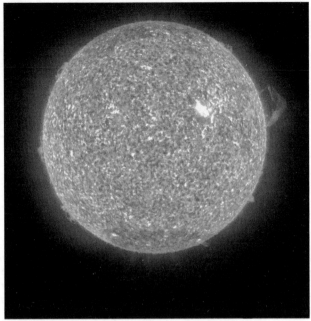

Figure 3.1   The sun at ultra-violet wavelengths [SOH05]. Note the massive ejection of material on the right hand side

core of the sun. Only two sources originate outside of this: tidal energy and geothermal power[1].

The sun (Figure 3.1) is approximately 4.5 billion years old and is an average looking, medium sized star. It has been identified in religious worship from the earliest times and, not unsurprisingly, much scientific effort has been applied to trying to work out how it and other stars function. An early suggestion was that it was formed of burning coal, but it wasn't until the 1930s that Hans Bethe, Charles Critchfield and Carl Friedrich von Weizsäcker suggested nuclear reactions as the source of such long lasting energy.

The sun is made of gas, and as such has no well-defined surface. Structurally there are three zones making up the body of the sun: the core, the radiative zone and the convective zone (see the upper half of Figure 3.2). Above the convective zone of the sun is the sun's atmosphere, which also consists of three zones – the photosphere, the chromosphere and the corona – as shown in the lower half of Figure 3.2.

The core extends to 25 per cent of the sun's radius [HOW05]. In this zone, gravity pulls all of the mass inward and creates intense pressure. The pressure is high enough to force atoms of hydrogen to come together via the fusion equations given below. These

---

[1] Being largely radioactive in origin, geothermal power can be classified as a form of nuclear power. It should not be confused with the solar heating that warms the Earth to a modest depth, which can be accessed as a renewable resource by ground-source heat pumps (Part IV).

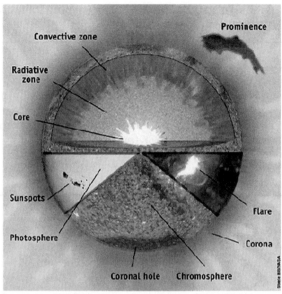

Figure 3.2 Basic overview of the sun's structure [SOH05]. The top half of the illustration is a cut-away showing the various inner layers; the bottom half, surface details visible at differing wavelengths

reactions account for 85 per cent of the sun's energy. The remaining 15 per cent comes from additional reactions that form beryllium and lithium.

The *radiative* zone extends to 55 per cent of the sun's radius. In this zone, energy, in the form of photons from the core, is carried outward. A photon typically travels about one millionth of a metre (one micron) before being absorbed by a gas molecule. Upon absorption, the gas molecule is heated and re-emits another photon of the same wavelength. The re-emitted photon travels another micron before being absorbed by another gas molecule and the cycle repeats itself. Approximately 10^{25} absorptions and re-emissions take place in this zone before a photon reaches the surface, so there is a significant time delay between a photon being made in the core and one reaching the surface [HOW05].

The final 30 per cent of the sun's radius is formed by the convective zone. Here convection currents of hot gas carry the energy outward to the surface. The convection currents carry photons outward to the surface faster than the radiative transfer that occurs in the core and radiative zone. Because of the numerous interactions occurring between photons and gas molecules in the radiative and convection zones, it takes a photon approximately 100 000 to 200 000 years to reach the surface [HOW05].

The photosphere is the lowest region of the sun's atmosphere and is the region that can be seen from Earth. It is 300–400 km deep with an average temperature of 5800 K. Above the photosphere lies the chromosphere reaching out to about 2000 km. The temperature rises across the chromosphere from 4500 K to around 10 000 K. The corona is the final

Table 3.1 Basic solar statistics [HOW05]. Data from HowStuffWorks, http://www.howstuffworks.com

Statistic	Value
Average distance from Earth	149 million km (93 million miles)
Radius	696 000 km (418 000 miles)
Mass	1.99×10^{30} kg (330 000 Earth masses)
Makeup (by mass)	74% hydrogen, 25% helium, 1% other elements
Average temperature	5800 K (surface), 15.5 million K (core)
Average density	1.41 grams per cm^3
Rotational period	25 days (centre) to 35 days (poles)
Solar output	3.846×10^{26} W

layer and extends several million kilometres outward from the photosphere and averages 2 million K [HOW05]. Table 3.1 lists some of the sun's statistics.

The core converts hydrogen nuclei (protons, H) into helium (He) nuclei in a three-stage reaction: First,

$$^{1}\mathrm{H} + {}^{1}\mathrm{H} \rightarrow {}^{2}\mathrm{H} + e^{+} + \text{radiation} \tag{3.1}$$

where one of the protons has been converted to a neutron and a positively charged electron, or positron (e^{+}); this is followed by

$$^{2}\mathrm{H} + {}^{1}\mathrm{H} \rightarrow {}^{3}\mathrm{He} + \text{radiation} \tag{3.2}$$

and finally

$$^{3}\mathrm{He} + {}^{3}\mathrm{He} \rightarrow {}^{4}\mathrm{He} + {}^{1}\mathrm{H} + {}^{1}\mathrm{H}. \tag{3.3}$$

This occurs at temperatures in excess of 13 million celsius.

Problem 3.1 The above process is essentially an illustration of the production of radiation from mass energy. Given that the sun has a luminosity of 3.846×10^{26} W [NAS05], estimate the loss of mass from the sun that is required every second to provide this energy.

The radiation flux from the sun varies on several time scales. If large numbers of sunspots are present the flux can increase by 0.2 per cent, although this change is not even across the whole solar spectrum and is more pronounced at ultraviolet wavelengths. Such wavelengths heat the Earth's upper atmosphere and can thereby have an impact on our climate. A sunspot cycle has been identified with a period of eleven years; on a longer time scale, it has been predicted that the solar flux was only 70 per cent of its current value 4.5 million years ago. This variation should have had a much larger effect on the Earth's climate than the record shows and thus provides strong evidence that the planet's climate is moderated in a benign way by something else, the most obvious candidate being our atmosphere.

Problem 3.2   The sun has a radius of 696 000 km. Given the above power production, estimate the power passing through each square metre of the sun's surface. Compare this (in terms of W/m²) with the power passing through the surface of a domestic light bulb.

3.2 The earth

Because of the elliptical shape of the Earth's orbit and the inclination of its axis, solar radiation does not strike the upper atmosphere equally. Furthermore, some is reflected by clouds and the surface of the planet, leaving only about 70 per cent of the incoming radiation to heat the planet. This radiation is not spread evenly. Latitude and local cloud cover affect the level received substantially. The average value of 240 W/m², or 122 PW in total (over 9000 times the world's primary energy[2] use), may vary by over 60 per cent over distances of only 10 km (Figure 3.3). This variability can have a large impact on the economics of likely locations for using solar radiation for either water heating or the production of electricity.

Sunlight

At the top of the atmosphere, the incoming radiation received every second onto every square metre, (or *flux*, F_S), must exactly equal the outgoing flux, or the planet would become infinitely hot. The radiation from the sun is mainly in the visible part of the spectrum, that from the Earth is a mix of directly reflected radiation (F_a), again mostly in the visible range, and longer wavelength, or infrared, radiation which is being re-radiated from the warmed planet (F_E). Mathematically:

$$F_S = F_E + F_a \tag{3.4}$$

The *fraction* reflected is termed the albedo, a (≈ 0.31), of the planet:

$$F_a = F_S \times a \tag{3.5}$$

These fluxes, when averaged over the planet and over a whole year have the values:

$$F_S = 342 \text{ W/m}^2,$$
$$F_a = 107 \text{ W/m}^2 \text{ and}$$
$$F_E = 235 \text{ W/m}^2.$$

---

[2] Primary energy is the term given to energy inputs, rather than outputs. So, for example, it includes the calorific value of the fuel burnt in a power station rather than the much lower amount of electricity generated. It usually (but not always) only includes commercially traded fuels, e.g. fossil fuels, hydroelectricity and nuclear power, not locally resourced biomass, food or animal power.

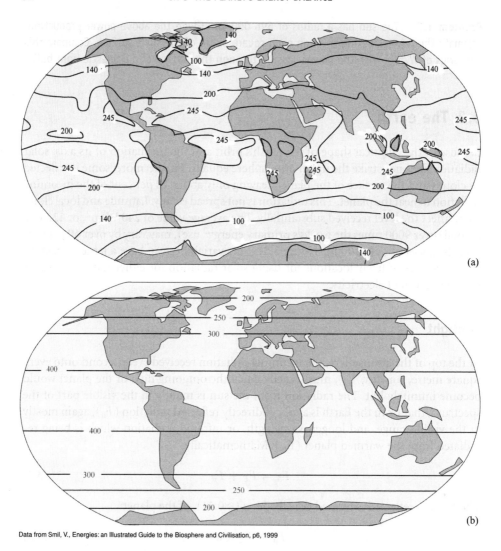

Figure 3.3 Annual average incoming radiation (a) at the surface of the planet (at smaller scales the variation is even greater; if radiation absorbed by the atmosphere is included, the average rises to 240 W/m²), and (b) at the top of the atmosphere [data from SMI99].

Problem 3.3   Calculate the annual energy received from the sun and radiated by the planet, both directly and indirectly. Hint: the radius of the Earth is 6370 km.

Rain

Although only 4.2 kJ are needed to raise the temperature of one kilogram of water by one degree centigrade, 2453 kJ are needed to evaporate the same mass [USA05] – an

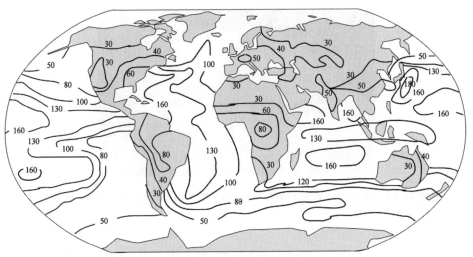

Data from Open University: http://openlearn.open.ac.uk/mod/resource/view.php?id=94988

Figure 3.4 Global patterns of latent heat flux (W/m²) [SMI99, p15]

amount equal to the latent heat of evaporation. This means that if water and sunlight are abundant, vast quantities of energy will be transported aloft by evaporation. One millimetre of evaporation per day is easily possible at all but the most Southerly or Northerly latitudes (greater than $70°$ N/S) and will provide 30 W/m² of heat to the atmosphere. In the Caribbean and similar locations average figures are 170 W/m², with the global mean latent heat flux being 80 W/m². A comparison of Figures 3.3 and 3.4 shows that the quantities of heat involved are of the same order as the original radiation balance. In many locations, for example the Sahara, the limiting factor will be the availability of water rather than the amount of sunlight. In total, an average of one metre of water is evaporated from the surface per annum.

Problem 3.4   Use the above figures to estimate the total latent heat flux (in watts) for the Earth.

The latent heat is re-released during precipitation, and in storms the amount of energy released by this route can be over 100 times that provided by any accompanying winds. When rain falls over land, much will return to the oceans, either directly or after falling as snow then melting (Figure 3.5). In doing so, it will lose height and therefore potential energy. Approximately 50 000 km³ of precipitation is returned this way. Assuming an average land elevation of 850 m, 400 EJ per annum are lost. Not only does this energy provide the principle erosive force for the planet, it sets an upper theoretical limit to the amount of hydroelectricity that could be generated. Figure 3.6 illustrates the general cycle and shows that most evaporation and rain occurs over the oceans and that the return flow to the sea through rivers is surprisingly small.

Figure 3.5   Heavy snows and avalanche in Northern Pakistan (photo by the author)

Problem 3.5   Estimate the theoretical world-wide limit to the amount of hydropower that can be generated (in watts).

As this water returns to the oceans it causes denudation. Locally, the loss can be substantial in river systems and mudslides; regionally the loss is much smaller when measured over human time scales. Denudation is usually measured in Bubnoff units (B). One Bubnoff is equivalent to 1×10^{-3} mm of loss per year, with rates between 1 B and 100 B being common, although the Nanga Parbat section of the Himalayas undergoes denudation at nearly 10 000 B, or 10 mm per annum. Figure 3.7 shows two examples.

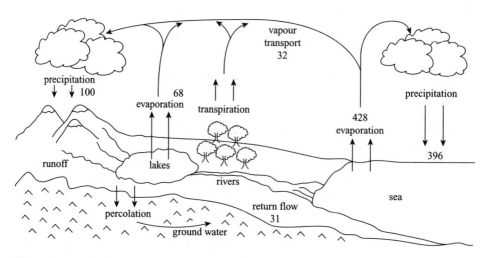

Figure 3.6   Major flows in the global water cycle for 100 units of precipitation on land [SMI99, p15]

(a)

(b)

Figure 3.7 Highly eroded mountainous landscapes: (a) Iceland, (b) Pakistan (photos by the author)

Problem 3.6 Estimate the total potential energy loss from denudation for India (land area = 2 973 000 km^2 [CIA05]). Assume an erosion rate of 50 B and an average density of material of 2.5 g/cm^3. Compare the answer with India's average annual oil use (see Appendix 1).

Averaging over the planet, the quantity of soil and rock moved, and the potential energy lost, is impressive, but the total energy exchange is tiny compared with other processes. Given an average rate of 50 B, an average density of material of 2.5 g/cm^3 and

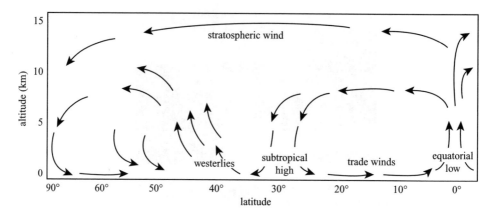

Figure 3.8   Section through the atmosphere showing large-scale winds [SMI99, p13]

a mean elevation of 850 m, the potential energy lost is approximately 133 TJ or 4.2 MW. As typical coal power stations are rated at 1 GW$_e$, this is truly small for a planet-wide phenomenon, especially when compared with the 400 EJ of potential energy lost by the water as it makes the same journey.

Wind

Although at times spectacular, winds provide a much smaller energy resource than evaporation and only about one per cent of incident solar radiation is used to maintain the planet's winds. On a planetary scale, the form of the winds is dominated by the interplay of the Coriolis force provided by the rotation of the Earth, and the direct heating of the surface by the sun. Figure 3.8 shows a section through the atmosphere illustrating the large-scale movements. At various points there are regular substantial surface winds, suggesting the possibility of using wind turbines to generate electricity. At other points, for example at the horse latitudes (approximately 30–35° N/S) – so named for the once common practice of disposing of live-stock over the side when sailing ships were becalmed there – air is slowly subsiding, leading to weak, variable winds, poor for turbines. At others, for instance between 0–5° N/S (termed the doldrums), air is gently rising. At a local scale, localized heating can produce winds such as sea and land breezes associated with the different heating and cooling rates of adjacent water and landmasses.

Unlike surface water movement, wind occurs in three-dimensional space. This means that it is very difficult to estimate the size of the global resource because the altitude of the wind is likely to be as important as its strength (we are unlikely to build 10 kilometre high wind turbines). However, 300 EJ is estimated to be flowing in the world's lower level winds, which is close to the world's primary energy consumption.

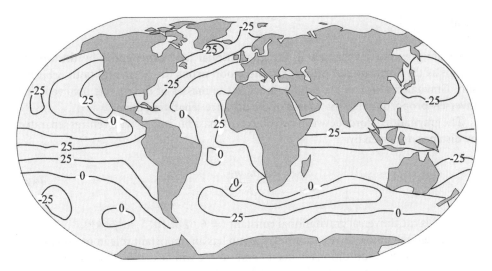

Figure 3.9 Heat flux (W/m²) from surface to deeper waters [Data from SI99]

The oceans

The oceans have a much lower albedo than the continents, and as they form over 70 per cent of the surface of the planet, they absorb far more radiation. The oceans are also very deep (3.8 km on average) and mobile, which means they have the ability to store, transport and release vast quantities of heat and thereby partially regulate the planet's climate.

Problem 3.7   Estimate the total thermal energy stored in the world's oceans and compare it to our annual use of oil (153 EJ). Assume a specific heat capacity of 4.2 kJ/kgK[3]. Hint: the answer is equal to the heat required to warm an equivalent mass of water from zero kelvin to the ocean average.

Water is a relatively poor conductor of heat, so warmer water near the surface tends not to transfer heat, or mix, with deeper water. Instead, energy is transported around the oceans by the mass movement of water in convective currents. At some points these will consist of sinking water, in others, rising water. The quantities of water involved can be quite impressive: between Iceland and Greenland five million cubic metres of water per second plunges to a final depth of over three kilometres. Figure 3.9 maps this exchange.

---

[3] A slightly more accurate answer can be obtained using a more realistic specific heat capacity for seawater: 3.8 kJ kg−1 K−1

Tides

Tides are powered by changes in the gravitational field experienced by the world's oceans as the Earth rotates on its axis, the moon orbits the Earth and both orbit the sun. Gravitational forces are proportional to the masses of the objects involved, and inversely proportional to the square of the distance separating them.

The gravitational force between two objects of mass m_1 and m_2 with centres separated by distance d is given by:

$$\frac{G m_1 m_2}{d^2}, \tag{3.6}$$

where G is the universal gravitational constant ($= 6.67 \times 10^{-11}$ m^3/s^{2}kg).

Although the moon drives our tides, the sun has an important role in modifying the tidal range. Such modifications in height and timing will be explored in Chapter 24.

Tides represent a vertical rising and falling of water – a lot of water. This implies a change in potential energy: energy we can use for other purposes. How large this energy source might be will depend on the height gained. We can get some idea of the minimum change in height by considering an object of mass m suspended at a distance r from the centre of the Earth (normally the surface) at an angle θ from a line joining the Earth and moon. Then, applying Equation (3.6) and some basic geometry, we find the change in height, Δh, is given by [SOR79, p149]:

$$(MGr^2/gR^3)\cos^2\theta \tag{3.7}$$

where R is the distance from the centre of the Earth to the centre of the moon, M the mass of the moon and g the local gravitational constant (9.81 m/s^2). Inserting values for the constants into Equation (3.7) we see that:

$$\Delta h = 0.36\cos^2\theta. \tag{3.8}$$

When an object is on the line joining the centres of the Earth and the moon, $\theta = 0$ and $\cos^2\theta = 1$. This implies that Δh will equal 0.36 m. This is a non-negligible height change. The associated change in potential energy will be given by:

$$mg\Delta h. \tag{3.9}$$

As the mass of the oceans is truly vast, the change in potential energy is considerable.

As the Earth rotates, θ will change and the height and potential energy of any mass of water will rise and fall in a sinusoidal manner as a tidal 'wave' sweeps across the oceans. This simple picture is complicated by the shape and position of the continents. However, Equations (3.8) and (3.9) indicate that the height will change in a cyclical manner through the day and that the corresponding change in energy will be considerable. This suggests that tidal forces could supply us with a large and highly predictable source of endlessly renewable energy.

The tidal bulge, or wave, produced will lose energy due to friction on the seabed and coastal strip. This transfer of potential energy into internal heat energy is the principle reason for the slowing of the Earth's rotation by 0.0016s per century [KIN71] and thereby a change in the Earth's kinetic energy of 95 EJ per annum.

Energy in biological systems

As Figure 3.13 indicates, there is a flow of 40TW from the sun into organic matter. This equates to approximately 0.02 per cent of the energy entering the atmosphere and has led to a standing biomass store of 1.5×10^{22} J [ODU72]. Still larger deposits of dead fossilized organic matter exist. These total in excess of 6×10^{23} J, which equates to roughly 1.4×10^{16} kg[4] of carbon [SOR79, p298]. Unfortunately (or not, in light of growing concerns over climate change) much of this is unlikely to be economically recoverable in the near-term.

As we will see in Chapter 4, humans initially relied on continuously renewed biomass for the majority of their energy. During the industrial revolution we started to switch to accessing the non-renewable fraction and thereby altered the global carbon cycle. We now face the challenge of either returning to using only the renewable fraction, or inventing alterative, non-biological, energy sources.

Internal energy

The Earth's core is at a temperature of approximately 4000 K [USE05] and slowly cooling via heat loss to the surface. There are also large amounts of radioactive elements (^{235}U, ^{238}U, ^{232}Th, and ^{40}K) which contribute at least as much heat. Figure 3.10 illustrates the global distribution at the surface and shows that compared with the radiation we receive from the sun the quantities are small, at only around 0.08 W/m^2, on average. Where the sea floor is expanding at its fastest, areas of a few square metres are found which are ejecting water at 350 °C and a flow of 300 MW, (cf. 64 MW/m^2 for the surface of the sun). About 60 per cent of the total heat flux is from the formation of new ocean floor.

Surprisingly, earthquakes account for only a small energy exchange, about 450 PJ, or 0.03 per cent of the total internal energy loss. The Richter magnitude, R, as reported by the news media for each major quake, can be converted into joules, $E_{kinetic}$, using the formula:

$$\log_{10} E_{kinetic} = a + bR \tag{3.10}$$

---

[4] Given recent discoveries of gas-hydrates and other non-standard fossil fuels, this figure may be far larger.
[5] Reprinted from *Renewable Energy*, Bert Sorensen, p296, 1979, with permission from Elsevier.

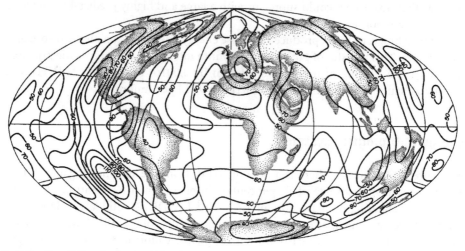

Figure 3.10   Geothermal heat flows (mW/m²) [SOR79][5]

where a is in the range 6.1 to 13.5 and b is in the range 1.2 to 2. The logarithmic form of Equation (3.10) implies that even a small difference in Richter magnitude implies a large energy difference between two earthquakes. Even though the total energy released by such events is a small fraction of the planetary internal energy loss, their brief duration, together with the relatively small area affected, gives rise to very high power and power densities (i.e. power per unit area), plenty large enough to topple buildings, destroy roads (Figure 3.11) and create tidal waves.

Problem 3.8   Estimate the ratio of the energy released between events with Richter magnitudes of six and eight.

As a demonstration of the energy that lies in wait under our feet, volcanoes are possibly even more spectacular than earthquakes. Although the explosive nature of such events can carry large amounts of kinetic energy, most of the exchange is in the form of heat. The Mount St. Helen's eruption (1980) lost 1.7 EJ over nine hours, thereby averaging 52 TW. Although much larger eruptions have occurred in recent history: Tambora (1815) released over 80 EJ. Again, however spectacular and deadly such events are, the total annual energy released is relatively small at about two per cent of the internal energy exchange. Other smaller outflowings of lava frequently occur in some parts of the word leaving a landscape of hot rocks and rising steam from ground water (Figure 3.12).

Problem 3.9   Compare the energy released in the Tambora eruption to the world's annual primary energy use.

Figure 3.11 Road blocked after an earth tremor, in Northern Pakistan (photo by author)

3.3 Comparisons

We can use the data above to complete a power audit of the planet (Figure 3.13). This shows that solar and evaporative flows dominate and that processes such as tides and winds show much smaller, but still significant, fluxes. Energy technologies have been developed which tap these vast resources: solar (photovoltaics, heat-pumps, passive and active solar); tidal power; rain (hydroelectricity); wind (wind turbines); internal energy (geothermal power); biological energy stores (wood burning for heat and power). It is also apparent that the following sustainable energy flows exceed our current level of

(a)   (b)

Figure 3.12 Iceland: (a) newly formed lava (b) rising steam from ground water (photos by the author)

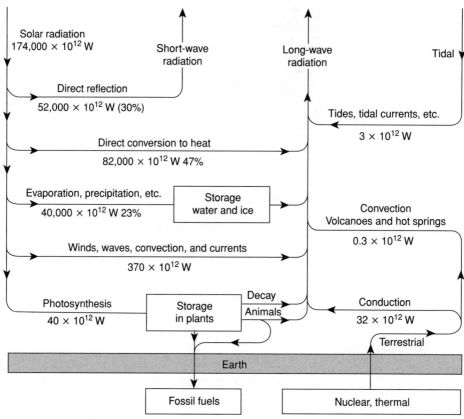

Figure 3.13   Power audit of the planet in terms of watts [RIS99, p126]. For comparison, world primary energy use is 13×10^{12} W

energy use: solar, photosynthesis and geothermal[6]. This implies that if we could capture a relatively small fraction of this total flow in a useful manner we might dispense with fossil fuels entirely. Our historic resistance to this path is not based on the lack of an alternative resource, but from a combination of the form of the resource, technological issues, sociological obstinacy and cost.

Problem 3.10   Figure 3.13 is a power audit; adapt this figure, or complete an equivalent table, to produce an annual energy audit of the planet in joules. For each energy flow, compare in percentage terms the answer with world primary energy use.

---

[6] All such flows cannot simply be added together to give the total renewable resource, as this would be double counting such flows as evaporation/hydro, which have already been accounted for under solar.

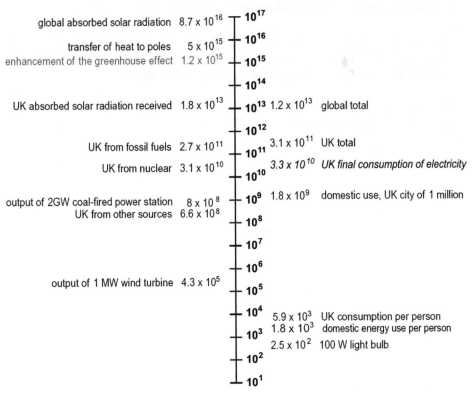

rate of energy supply   **rate of primary energy consumption**

global absorbed solar radiation 8.7×10^{16} — 10^{17}

transfer of heat to poles 5×10^{15} — 10^{16}
enhancement of the greenhouse effect 1.2×10^{15} — 10^{15}

— 10^{14}

UK absorbed solar radiation received 1.8×10^{13} — 10^{13} 1.2×10^{13} global total

— 10^{12}

UK from fossil fuels 2.7×10^{11} — 10^{11} 3.1×10^{11} UK total

UK from nuclear 3.1×10^{10} — 3.3×10^{10} UK final consumption of electricity
10^{10}

output of 2GW coal-fired power station 8×10^{8} — 10^{9} 1.8×10^{9} domestic use, UK city of 1 million
UK from other sources 6.6×10^{8} — 10^{8}

— 10^{7}

— 10^{6}

output of 1 MW wind turbine 4.3×10^{5} — 10^{5}

— 10^{4} 5.9×10^{3} UK consumption per person
10^{3} 1.8×10^{3} domestic energy use per person
2.5×10^{2} 100 W light bulb
— 10^{2}

— 10^{1}

Figure 3.14 Annual average rate of energy use/supply in watts [DTI00]

It is useful to compare these numbers with the energy or power typically provided by man-made energy technologies. Figure 3.14 does this on a logarithmic scale and provides a useful aid to grasping the relative magnitudes of the various flows/demands.

3.4 Student exercises

"Irrationally held truths may be more harmful than reasoned errors." – Thomas Henry Hux-ley (1825–1895), Friday evening discourse at the Royal Institution 1880, "On the Coming of Age of the Origin of Species"

–(Collected Essays, vol. 2, p. 227)

1. Produce a table categorizing energy flows on Earth as either solar in origin, or not.

2. Describe the major structures and processes within the sun.

3. Describe the key features of the following energy resources: sunlight, rain, wind, the oceans, tides, organic matter and the Earth's internal energy.

4. For each of the resources mentioned in Question 3, list any environmental, physical or technological problems that might restrict their use. (This has yet to be covered in the text, so use a mix of imagination and rational thought.)

5. The sun outputs 3×10^{32} J per day. How much solar energy is received by the Earth each day? (Hint: use geometry.)

4

A history of humankind's use of energy

"For a successful technology, reality must take precedence over public relations, for Nature cannot be fooled."

–Richard P. Feynman

In this chapter we will explore what we use energy for and how this use has changed over time. In doing so we will make extensive use of sources and material presented in Vaclav Smil's excellent *Energy in World History* [SMI94] – a recommended starting point for those interested in the history of energy use. In Part IV this analysis will be continued to answer the question of what level of energy resource will be needed in future years as technologies and economies change and world population grows.

The progression of human history and the development of society are inextricably linked to our ability to harness an increasing number of forms of energy on an ever-growing scale. This has allowed the production of materials such as metals, the transportation and supply of goods and services that makes, at least in the developed world, our existence and standard of living quite remarkable and wedded to energy use:

Since energy is an essential ingredient in all terrestrial activities, organic and inorganic, it follows that the history of the evolution of human culture must also be a history of man's increasing ability to control and manipulate energy
     – M. King Hubbert, Energy Resources (1962)

Energy and Climate Change   David A. Coley
© 2008 John Wiley & Sons, Ltd

4.1 Energy and society

All organisms capture chemical energy in the form of food and transform it into kinetic energy and stores of chemical energy. As history has unfolded humans have been able to continuously increase the size of these transformations and their diversity. Some have argued [WHI43, FOX88] that energy use defines the degree of cultural evolution possible and that history does not so much drive energy use, but energy use and innovation drive history.

In this chapter we will investigate how energy use has changed through human history from prehistoric times to the present day and how this use has to some degree played an important role in defining the size, complexity and wealth of society. We will finish by reviewing our current level of energy use before discussing in Part II how this energy is provided.

We will be looking at agriculture in depth because it was for a long time the source of most of our energy – first as the food for human power, then the same for agricultural animals. Today, food is only a minor energy input into our lives. The rest of the book will cover the fuels and energy flows that account for the majority of our energy use today and we leave in-depth study until then.

Prehistory

For almost all of our history humans have lived as simple foragers [WHI92], requiring solely that the average energy expenditure required to forage was less than the energy expended in foraging, living and breeding. From an energetic standpoint, the two most important adaptations for humankind were the adoption of bipedality 7.5 million years ago [LEA92], and the resultant growth in brain size to the point where it now accounts for one sixth of our basal (resting) metabolic rate [FOL92]. Two and a half million years before the present, humankind was known to be using simple stone tools such as knives made of sharp flakes [TOT87], and the hand axe was in use at least 1.5 million years ago. Such tools allowed the application of relatively small amounts of energy provided by muscles over very small areas and very short periods of time, i.e. they represented high *rates* of energy use, or power.

It is believed that humankind could control the use of fire by around 250 000 years ago [GOU92], allowing the species an alternative access route to the large stores of chemical energy found in plants.

Problem 4.1   Compare the daily resting energy expenditure of a human to the energy provided by burning 0.5 kg of firewood. (Assume the resting metabolic rate of a human to be 100 W and the calorific value of firewood to be 16 MJ/kg.)

Foraging societies were typically small and of low density – 0.01 to 3 individuals per square kilometre [JOC76], a reflection of the limited availability of resources (food).

Even the hunting of mammals could not provide enough excess energy to allow large fractions of the population to engage in non-food gathering occupations and the creation of complex societies in many locations. Most mammals in tropical forests are small and sparsely spread, implying small energy returns for substantial efforts. In other landscapes the existence of greater numbers of much larger mammals presented a different opportunity to early humans. As pointed out by Smil [SMI94], humankind can sweat more profusely than any other mammal, which in combination with bipedalism, allowed early hominids to chase herbivores to exhaustion in hot climates. Breathing in quadrupeds is limited to one breath per locomotor cycle because of the connection between the thorax and the front limbs. In humans there is no such limit and the breath frequency can be varied with stride frequency, allowing us to run at a wide variety of speeds. In addition, humans can sweat at rates in excess of 500 g/m^2 of surface area per hour, whereas horses, for example, can only manage 100 g/m^2 [HAN83]. Humans are also small compared to many large herbivores and therefore have a high surface area to mass ratio, meaning that our heat loss per kilogram of body weight is even more impressive and enough ($\approx$600 W) to regulate core body temperature even during extended periods of heavy work. This allowed, for example, deer, antelope, zebra and kangaroos to be driven to exhaustion. These large mammals and marsupials represented very good stores of energy. Whereas a large monkey might only have an energy content of 5–30 MJ per animal, a gazelle has 5–6 MJ/kg and a total of 25–180 MJ. 'Fatty' mammals represent an even more valuable opportunity, with seals providing 15–18 MJ/kg (500–1800 MJ per animal), bison 10–12 MJ/kg (2500–24 000 MJ per animal) and whales 25–30 MJ/kg (80 000–800 000 MJ per animal). Therefore one whale might be energetically equivalent to 800 000/30 = 26 000 monkeys [J. F. Eisenberg and others in SMI94]. Typically energy returns (the ratio of food energy acquired to energy expended in obtaining it) for gathering may have been 10 to 20-fold; that from killing baleen whales in excess of 20 000 [SMI94, p21–22]. Access to such large stores of chemical energy allowed the development of more complex societies [KIL54].

Early agriculture would have offered little energetic advantage over large mammal hunting, with energy returns of around 20 to 40-fold, and so its origins must have been for other reasons – possibly to combat food shortages, or to grow luxury items such as grains for beer making.

Traditional sustainable (?) societies

Agriculture

Even though agriculture was probably introduced for other reasons than energetic ones, it was still limited by the same energetic constraints. Foraging societies could only support small numbers of non-food producing individuals; traditional agriculture may have been able to support more but still a large percentage of the population was needed to work the land. The only way to change this state of affairs was to replace human beings with some other form of *prime mover* (the object that does the work).

Table 4.1 Typical weights, drafts, speeds and power of domestic animals [SM194, p86, from HOP69 and others]. Data from Smil, V., *Energy in World History*, Westview Press, 1994

Animals	Weights (kg)		Typical Draft (kg)	Usual Speed (m/s)	Power (W)[a]
	Common Range	large Sizes			
Horses	350–700	800–1000	50–80	0.9–1.1	500–850
Mules	350–500	500–600	50–60	0.9–1.0	500–600
Oxen	350–700	800–950	40–70	0.6–0.8	250–550
Cows	200–400	500–600	20–40	0.6–0.7	100–300
Buffaloes	300–600	600–700	30–60	0.8–0.9	250–550
Donkeys	200–300	300–350	15–30	0.6–0.7	100–200

[a]Power values are rounded to the nearest 50 W.

Domesticated animals filled this role for millennia across the world but were eventually limited by the same constraints when a large fraction of animals and of agricultural land was being used to produce food for other agricultural prime movers. The introduction of water and wind power and then of fossil fuels broke this constraint and today almost all the energy used to provide food in the developed world is produced by a very small percentage of the population – those working in the fossil fuel industry.

Traditional agriculturists relied heavily on grain. There were several reasons for this. The tubers sourced by foraging societies had too high a water content to be stored for long periods and grains, including rice and corn, yield more food energy per unit area than tubers or legumes (beans). The human energy required to grow and process such crops has been estimated to be around 800 kJ/hour [SMI94, p86]. Not all the grain grown could be consumed, there were losses during milling and storage and a fraction had to be set aside to act as seed for the next harvest. The labour requirement was very high, reaching 94 per cent or more of the available labour supply [BUC37].

Greater intensification could only be achieved by replacing some of the human work force with another prime mover, the adoption of irrigation and fertilization and the use of rotations and other techniques to stop the quality of the soil being reduced over time. The application of all these techniques allowed for increasing numbers of people to be removed from food production and permitted large complex societies to develop.

Although the tractive force provided by a draft animal is roughly proportional to its weight, other factors determine how useful it will be on the farm. As Table 4.1 shows, donkeys could provide little more power than humans, and horses would appear to be both the truck and the Ferrari of traditional agricultural practice. As a large horse could do the work of around eight people, the workforce could be released for other enterprises. However, they cannot usually feed themselves off wild food sources found in marginal areas around the farm, nor survive in wet tropical climates. Many such draft animals are still in use today (Figure 4.1).

A second use for such animals was found in raising water from wells or lower levels, or as prime movers in the construction of dams and leats used for irrigation. A tonne of wheat requires around 1500 tonnes of water during the growing season [SM194, p49],

Figure 4.1 Prime movers in Northern China (photograph by author)

or 15–30 cm of water for a growing density of 1–2 tonnes per hectare (10 000m^2), but in the Middle East and other areas rainfall could only provide much smaller amounts. Some areas could use gravity fed systems; others needed large quantities of water to be lifted from lower lying rivers. Even gravity-based systems required an energy input to build and maintain the extensive system of leats. Hand and foot-operated paddle wheels were common in India, Korea, Vietnam, and Japan. In China a water ladder was used (Figure 4.2). In the Middle East the counter-weighted lift was in operation from at least 2000 BC (Figure 4.2).

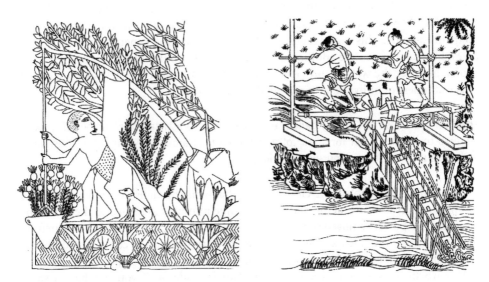

Figure 4.2 Counter-poised water lift (left) and Chinese water ladder (right). Adapted in [SMI94] from [MON66] and [TIA37]

Figure 4.3 The effect of irrigation on a landscape (photograph by author). Here an infertile mountain landscape in Pakistan has been transformed by extensive terracing and irrigation

Such systems could lift 5–22 m^3 of water per hour through 0.6 to 9 metres with efficiencies of up to 29 per cent [SMI94, p87] – the larger devices requiring animal power. A canal or leat-based approach removed the need for this energy input, but whereas lifts could be implemented by small village groups, large canals could only be constructed by well-organized societies able to wait for many years for the energy payback of higher yields.

Estimates have been made of the energy benefit/cost ratio of irrigation, and the following example is taken from [SMI94, p87].

A good late Qing dynasty wheat harvest was 1.5 t/ha and was reduced by 150 kg on a typical 0.2 ha field for a 10 cm (200 t) reduction in rainfall. If this shortage of rain is replaced by irrigation, 400 t of water might be needed because of losses. Using a traditional Chinese water ladder, two workers could lift 400 t of water around one metre in 80 hours. This would require 65 MJ of additional food energy, yet yield 2 GJ of digestible energy from the additional wheat: an energy multiplier of 30. As Figure 4.3 shows, however the water is provided, the results can be spectacular and allow much larger numbers of individuals to exist within a landscape.

Problem 4.2   Use the energy benefit example above to estimate the number of workers required to lift the water for 1 km^2 of agricultural land around a village if all this water had to be provided over a four-week period.

The solution to Problem 4.2 clearly shows that using humans as prime movers can be an extremely limiting factor, and requires large numbers to be tied to the land. The use of animal power removed some of these constraints but also had another benefit in that the animals provided a source of fertilizer: working animals could be fed grasses and their droppings placed on cropping land as a source of nitrogen.

It is worth pausing to consider what a plant is made of. If you purchase a small plant in a smallish pot and water it regularly it will, given the right species, grow into a substantial plant reaching the ceiling. Examining the pot will show that the soil mass will have appeared not to have changed, so where did the plant come from? The only inputs would seem to be water and light. Photons have no rest mass and it is hard to see how water, being just oxygen and hydrogen could form a woody stem. This conundrum led the ancient Greeks to believe in spontaneous creation of organic matter. Their mistake was to ignore the role of atmospheric carbon dioxide. Plants grow because they remove large quantities of carbon dioxide from the atmosphere and use the carbon (via photosynthesis) to build their structures. As we saw in Chapter 3, this carbon is returned to the atmosphere when the plant decays or is burnt. Plants therefore represent a solidified state of carbon dioxide as it passes around the global carbon cycle. But a plant is more than just carbon and water; it needs trace elements, particularly nitrogen, but also iron, copper, sulphur, silica, calcium, phosphorus and potassium. Most of these are replenished, at least in part, by rainwater and the deposit of uncultivated soils transported by winds. However, the much larger quantity of nitrogen needed by plants is not easily replenished at the required rate by these processes, and therefore is likely to be in short supply in continuously cropped land, leading to reduced yields.

Apart from manure, another way of replenishing the soil nitrogen was to grow species that could make use of atmospheric nitrogen, particularly edible legumes on a rotating basis with cereals. This practice greatly reduced the energy input required to gather and plough back in human and animal excreta or stalks and straw to provide this nitrogen. This left some of the dung and straw to be used as a fuel for heating and cooking.

Together, these innovations led to higher yields for less human energy expenditure and thereby lower numbers of agricultural workers per unit yielded. This allowed, for example, Egypt to produce 1.5 times as much food as it needed for its 5 million inhabitants during the Roman occupation, allowing Rome to maintain and expand its empire. During later periods the size and productivity of draft animals increased until, in the nineteenth century, horses weighing more than one tonne were being bred in Europe and crop yields were more than four times medieval levels. In 1800 most of a North American farmer's time was spent in hard manual labour. By 1900 every unit of food energy needed for farm labour produced 25 times more edible energy than in 1800, with the work being done by horses dragging sophisticated machinery. In 1900 the USA had 25 million horses; these could only be fed because of the abundance of farmland, the low population density (1.5 hectares per capita, against 0.15 hectares per capita in China) and with around 30 million of the USA's 120 million hectares of cropped land being used to grow food simply to feed these new prime movers [SMI94, p91].

Clearly some kind of limit was about to be reached as the land area required to feed prime movers grew: it already had been reached in countries such as China. Roman agriculture yielded an energy gain of approximately 45; by 1800 Europe had gains of 200, by 1900 gains of 500, with some farms in the USA having a gain of 2500! This is a massive leap from early subsistence farming, and as a result mega-cities became common. The Romans supplemented every unit of human energy expenditure with eight units of animal labour; nineteenth-century Europeans used 15 and some American farms 200, a point at which human labour input became negligible. The inhabitants of South East Asia and other areas could not make these energetic progressions and relied on cropping intensification with large human labour inputs for planting and irrigation. This allowed

Figure 4.4 Agriculture in the mountains of Northern Pakistan (photographs by author); (a) shows the type of marginal landscape where animate power has had little effect. (b) Although some areas could make use of animate power all available land is required to feed the local population, leaving none for animate prime movers

much greater human populations to be sustained – but only if the majority still worked in agriculture.

In several parts of the world the landscape conspires against the use of even these agricultural methods and human effort is still the dominant force – Figure 4.4.

Animate power was also used for a growing number of non-agricultural purposes. Figures 4.5 to 4.7 show examples from the mining and manufacturing industries. Size often restricted the power of such systems as only a limited number of workers or animals could fit around any wheel or machine. A treadmill with a single worker could provide no more than 150–200 W, and then not continuously. The largest eight-person wheel would produce possibly 1500 W. As a comparison today's typical domestic vacuum cleaner uses 1000 W of electrical power, which suggests that industrial development would be severely limited unless new prime movers could be found.

Water and wind power

Human innovation also allowed access to other sources of energy. Wind and water mills were developed initially for agricultural use and then for use in industry and mining. These systems could provide much higher work rates without the need to divert a large fraction of agricultural output of fuel to them.

Figure 4.5 Mid eighteenth-century vertical capstan powered by eight people. The capstan is being used to draw a gold wire through a die [D. Diderot and J. L. D'Alembert in Smil, V., *Energy in World History*, Westview Press, 1994]

Figure 4.6 Human-powered metalworking lathe [D. Diderot and J. L. D'Alembert in SMI94]

(a)   (b)

Figure 4.7 Large internal treadwheels being used to (a) remove water from a German mine, and (b), lift-building materials at a construction site [G. Agricola in Smil, V., *Energy in World History*, Westview Press, 1994]

It is known from first century BC writings, that watermills had been employed in milling grain for some time. We also know that vertical axis windmills were common by the first century AD in what is now Iran [FOR65]. The design used plaited reed sails placed behind narrow openings in mud walls and lasted within the region well into the twentieth century. The Doomsday book of 1086 lists 5624 water mills in Southern and Eastern England, one for every 350 people: a very high deployment of a renewable energy technology.

Problem 4.3   Assume Japan (population 127 million [CIA04]) had one 2 MW wind turbine for every 350 people. What fraction of its electricity demand (231 TWh) could be met?

Efficiency was initially low with only a few percent of the water's kinetic energy being converted to mechanical energy. Efficiency had risen to 20 per cent by 1800 and to 45 per cent 50 years later, greater than the efficiency of a modern power station. Such wheels were used for ore crushing, woodturning, sawing, powering bellows, wire pulling, stamping and polishing. The increase in industrial activity they allowed was staggering. For example, a US patent was granted in 1785 for a mechanism that could cut and head 200 000 nails a day [ROS75]. This brought down nail prices by 90 per cent. In the UK, Shaw's waterworks near Glasgow, Scotland, had 30 water wheels fed from a single reservoir and provided 1.5 MW of power.

Problem 4.4   How many horses would have been needed to provide the same power as of Shaw's waterworks continuously?

Figure 4.8 Traditional charcoal production [Smil, V., *Energy in World History*, Westview Press, 1994, p118]

In locations where either water was in short supply or the terrain could provide little by the way of a head, wind power provided an alternative prime mover to that of domesticated animals. Wind power was used for a variety of agricultural and non-agricultural purposes including milling, crushing, sawing and metalwork [HIL84]. In addition it could be used to drain low-lying coastal land, a common application in the Netherlands from the sixteenth century onwards. Pumping also proved a popular application in North America. Here small machines were powered by a large number of sails or blades and were sited on farms or adjacent to railway lines.

The number of machines deployed is impressive. In the USA several million such machines were installed during the second half of the nineteenth century [SMI94, p112]. The Netherlands had 8000 wind machines by 1650, England 10 000 in the early nineteenth century and Germany 30 000 in 1895. In 1900, around 30 000 mills representing an installed capacity of about 0.3 GW were working in the countries around the North Sea [DEZ78], each providing on average 10 kW of useful power.

Charcoal and dung

Charcoal has 1.5 times the calorific value of dry wood and is virtually pure carbon. This purity gives it a distinct advantage in metal production, allowing for finer control over the final product, with fewer impurities entering the metal. Unfortunately its production, by partial combustion of wood in earth-covered kilns, is highly wasteful with 60 per cent of the energy content of the wood being lost (Figure 4.8).

Dung was important in some societies and is still used in the home in many parts of the world. Llama dung was the principal fuel of the Altiplano of the Andes, the core of the Inca Empire [WIN77]. In the nineteenth century American buffalo and cattle dung was used and the Tibetans still make use of yak dung today.

Light and heat

Early fires for heating and cooking were simple piles of biomass. This approach provided efficiencies of only a few percent and drew in large amounts of cold external air – possibly causing a net heat loss rather than gain to the home. Early mud and brick stoves were around 15 per cent efficient. Much greater efficiencies were realized in systems with more controlled air flow and a large heat exchanger surface: the most well known being the Roman 'hypocaust' used from around 80 BC. This passed the hot flue gases beneath raised floors and up a chimney. In northern China the 'Kang' is still in use. This enables the waste heat from a cooking stove to warm a high mass-sleeping platform. Iron fireplaces and stoves were not deployed until iron production had reached the point where costs were low enough.

Anyone who has spent time in societies that do not have access to electricity can testify to the poor performance of traditional lighting systems. Liquid fuels based on animal fats for lighting were invented surprisingly early, with examples appearing in Europe from the Upper Palaeolithic, 40 000 years ago [DEB93]. Candles came much later – after 800 BC in the Middle East. Such lights converted only around 0.01 per cent of the chemical energy they consumed to light. The next major advance was the production and supply of gas made from coal to households. Those homes not converted to this *town* gas continued to use biomass, much of it oil rendered from the blubber of sperm whales.

Transport and construction

The earliest forms of human transport were walking and running. Depending on the speed, running requires a power output of 700 to 1400 W, or 10–20 times the basal metabolic rate. The main failing of running or walking is not its speed – it was a long time before more rapid approaches were developed – but the very small amount of additional material that can be carried. Having said that, loads in excess of 35 kg (half bodyweight) are still carried by porters in the world's high mountains where roads systems have yet to be developed (Figure 4.9).

In order to apply more power, more people were needed, necessitating the use of sledges or other devices to connect the workforce to the object to be moved. Figure 4.10 shows an example with 160 workers pulling a 50 tonne statue mounted on a sledge running on a lubricated path and providing a peak power production of 30 kW. The Incas would appear to have achieved even greater feats, with 2400 workers pulling blocks up inclined planes and producing peak powers in excess of 600 kW [PRO86]. This is greater than the power produced from most of the water and wind machines during the industrial age.

However unlike their inanimate counterparts, they could not work at this rate all day every day. They also needed food and housing.

It is believed from ancient sources that the construction of the Egyptian pyramids took 100 000 workers three months a year for twenty years, or 100 to 200 million days

Figure 4.9 Mountain porter in Northern Pakistan (photograph by the author)

of labour. Assuming an energy input of 800 kJ per hour per person suggests an energy investment between 0.8 and 1.6 PJ and a peak power of between 12 and 15 MW [SMI94, p155].

Problem 4.5 Compare the energy invested in building the pyramids and Italy's daily oil use.

Horseback riding possibly appeared around 4000 BC. If we define a vehicle as having one or more wheels, then the earliest known vehicles date from Uruk circa 3200 BC

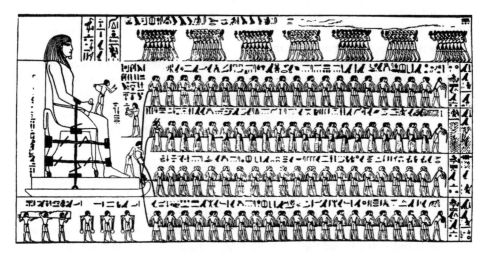

Figure 4.10 160 workers producing a peak power of 300 kW [G. Ruhlmann in SMI94]

Figure 4.11   Domesticated camels in Western China used for load carrying (photograph by author)

and used one metre diameter solid wooden wheels. This breakthrough later allowed ox-drawn carriages to pull loads of nearly 500 kg on Roman roads at 15 to 20 km per day. Even though this was a great advance over men and women carrying loads on their backs, it still implied water transport was much easier and cheaper. Hence cities tended to be built where possible near large waterways or the ocean. As an example, it cost more in AD 301 to move grain the 120 km by road from Rome's harbour (Ostia) to the city than to ship it to Ostia from Egypt [SMI94, p131]. As cities grew, other problems arose from relying on animate transportation systems. In 1900, London had a population of around 300 000 horses. This required a vast amount of feeding, stabling and waste removal. The arid regions of central and western Asia and northern Africa still make use of camels for load carrying (Figure 4.11).

Tonnage wise, shipping has always been the most impressive way to move goods or people. The ships that took the Greek troops to Troy used 50 oarsmen producing a peak power of 7 kW. 'Trieres' or 'triremes' with three tiers of 170 oarsmen could produce 20 kW and a speed of 20 km/h. The 'tessarakonter' built around 210 BC was 126 metres long with over 4000 oarsmen producing 5 MW and carrying 3000 troops – a phenomenal but unfortunately unmanoeuvrable machine.

Smaller scale ships and boats could still move great loads great distances. The 'da yunhe' or Grand Canal in China was 1800 km long in 1327, although it had taken 700 years to reach this length. The single heavy European horse could pull 30 to 50 tonnes on a canal, 10 times what it could pull on road. On open water sails could be deployed and speeds increased. Roman cargo ships had a top speed of 2 m/s. By 1890 the use of better hulls and sails allowed the 'Cutty Sark' to average 5.3 m/s over 6000 km.

Metal production

The production and processing of metals in large quantities was at the centre of the industrialization of the western world. This could only occur with access to equally large sources of energy for smelting. By definition, this had to be a source of thermal energy, implying that animate power, however highly amassed, would do little in pushing society forward in this direction.

The use of charcoal lay behind this achievement, but there was still the need to cut wood from which to make the charcoal and to mine and crush the ore and work the metal produced. This suggests that, before the adoption of waterpower, high animate energy inputs would still have been needed. The level of deforestation was substantial even in early history, refuting the view that pre-fossil fuelled societies were sustainable. For example, the Roman Rio Tinto furnaces produced around 60 000 t of copper, which at 90 kg of wood per kilogram of product indicates 40 000 hectares of deciduous forest were used – or every tree within a radius of 11 km [SMI94, p155].

The introduction of tin into the copper produced bronze, which not only had superior properties, but because the tin had a lower melting point (232 °C compared to 1083 °C for copper), made for lower energy costs. Iron smelting requires higher temperatures (> 1535 °C) and therefore required advances in furnace design. Charcoal burning can produce temperatures of 900 °C if burnt using natural convection, but nearly 2000 °C if a forced air supply is provided. The amount of air needed was so great that even eighteenth century water wheels could not always be relied upon to produce a large enough supply. Future developments awaited the introduction of the steam engine.

Deforestation was still a problem, even given the greater efficiency of later furnaces. In the early 1700s, English iron production consumed about 1100 km^2 of forest annually and competed with the local population for fuel used in domestic heating and cooking.

Fossil fuel based society

> *'I see two paramount themes in the history of the past century: the growth of human control over inanimate forms of energy; and an ever increasing readiness to tinker with social institutions and customs in the hope of attaining desired goals.'*
>
> —William H. McNeill, The rise of the West (1963)

The discovery and widespread adoption of fossil fuels transformed society. Much of the world now has access to abundant supplies of food and an ever-increasing number of people have access also to an extensive array of personal possessions. There are today a greater number of people living in some form of affluence than were living in total in 1930. For the first time in history this has led to a widespread polarization of our societies.

Pre-fossil fuel societies contained rich and poor within the same society, whether they lived in Europe, the Indian sub-continent or Africa. Now the central divide is between those living in certain countries and those not. In many of the industrialized countries the poorest citizens are wealthier than all but a small fraction of those living in the very poorest economies.

However, we know that this situation cannot last forever. Fossil fuels are a strictly limited resource and we are using, in a small number of generations, what has taken many millions of years to accumulate. The reality of climate change may not even support the luxury of this position, forcing us to curtail our use of such fuels before they run out. What we replace them with is as yet unknown, but to gain acceptance in the developed world they will need to be able to grant us the same degree of luxury we have become used to. For the developing world the adoption of sustainable energy systems might be less problematic and could offer several advantages: they would be accessing energy supplies produced within their national boundaries and using systems which create less pollution. For distributed technologies, such as small-scale solar power, there would also be less need for the development of nation-wide distribution systems. This might allow such countries to leapfrog the fossil fuel society, straight to a sustainable future.

Coal

Although some use had been made of fossil fuels before the sixteenth century the use had been marginal as the following quote illustrates:

> 'Until the 16th century coal was hardly ever burned, either in the family hearth or in the kitchen, at distances of more than a mile or two from the outcrops, and, even within the area thus circumscribed it was only by the poor who could not afford wood.'
> — J.U. Nef, The rise of the British coal industry, 1932

For the widespread adoption of coal an impetus was clearly needed. This occurred when the use of biomass by industry and in the home reached a level such that it was no longer sustainable, and wood or peat had to be collected from an ever-enlarging area – no easy task given the primitive state of transportation at the time. England was the first country to make the transition to coal due to the increasing cost of wood and charcoal during the sixteenth and seventeenth centuries [HAR74], and almost all of the country's coalfields were opened between 1540 and 1640. By 1650 output was two million tonnes per annum and 10 million tonnes by 1790. The easiest-to-reach coal was exploited first, but by 1765 depths of 200 metres or more were being worked. This increased the need to pump water out, ventilate mines and bring the coal to the surface. These tasks were completed using water wheels, horses and human labour.

Because coal contains large quantities of impurities it could not be used directly in iron production – a severe limit to its usefulness. However in much the same way as charcoal is produced, the heating of coal in the absence of air produces pure carbon, and in 1709 Abraham Darby produced pig iron with this new fuel. This removed the biomass restriction on iron production and truly opened the door to mass production.

Coal could also be used to produce gas, which in turn could be used in lighting installations. As early as 1812 a company had been formed to provide London with a centralized gas supply [SMI94, p160]. However, the most important breakthrough was the invention of the steam engine, which, with seemingly limitless quantities of fuel, freed the whole of industry from the cost and location difficulties of biomass and waterpower.

Steam engines

In around 1700 Thomas Savery produced a small steam-driven pump with an output of about 750 W. In 1712 Thomas Newcomen built a 3.75 kW machine to power mine pumps. James Watt increased the efficiency of these early machines by adding a separate condenser, an insulated steam jacket, a pump to retain the vacuum in the condenser and a governor to give constant speed at varying loads. By 1800 most of these machines had only just overtaken the power produced by windmills, with the largest (100 kW) hard pressed to match the greatest water mills. So given this relatively poor performance, why did they catch on, especially given the extra cost of providing the fuel? As explained previously, the answer lies in the flexibility of the new machines: they could be sited almost anywhere and were not dependent on the vagaries of wind or rainfall. They could also be used as mobile sources of power, and steam power vessels rapidly replaced their wind-powered alternatives.

After Watt's engine, George Stephenson's 'Rocket' is probably the most famous steam powered device and was employed on the world's first commercial railway from Liverpool to Manchester in 1830. By 1900, speeds of 100 km/h had been a reached and 200 km/h by 1930.

Oil

Steam engines suffered two major failings: they were massive and they were not powerful enough to drive the generators needed to produce electricity on a large scale. This first failing meant they were unlikely to form the basis of a personal transportation system; although some early low-powered steam engine cars were built. The introduction of oil, and the technologies it spawned, solved these problems.

Abraham Gesner distilled kerosene in 1853 and this cheaper fuel soon replaced whale oil in lamps. Nikolaus Otto constructed a four-stroke internal combustion engine that ran on coal gas in 1876. Gottlieb Daimler patented the first high-speed, gasoline-powered engine in 1885; Karl Benz built the first car in the same year; the Wright brothers the first aeroplane in 1903: a truly remarkable period of innovation.

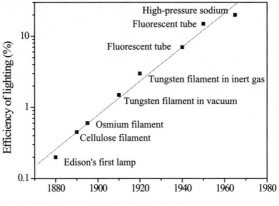

Data from Smil, V., Energy in World History, p 231, Westview Press,
The Perseus Books Group, 1994

Figure 4.12   Improvements in electric lighting technologies since Edison's first lamp – a 100-fold improvement is seen [Smil, V., *Energy in World History*, Westview Press, 1994, p231]

Electricity

Steam power did not end with the introduction of the internal combustion engine. (The room in which I write this is lit by lamps, the electricity for which is produced by steam power.) Edison produced the first durable, high vacuum bulb in 1879 and the first electricity generating plant (also built by Edison) began operating in 1882. However, the efficiency was not impressive (Figure 4.12). Although the first plants used steam engines, these were soon replaced by Charles Parson's more efficient steam turbine, patented in 1884. Parson installed a 75 kW machine in 1888, but only 12 years later had produced a 1 MW machine. To allow electricity to fulfil its promise two further innovations were required: the transformer (William Stanley, 1885) and the electric motor (Tesla, 1888).

The modern era

The adoptions of fossil fuels and electricity have led to far-reaching technological innovations that have transformed our societies. What is surprising to the author is that we seem to have got stuck with fossil fuels for so long and our love affair continues to grow. We still generate most of our electricity by setting fire to hydrocarbons, boiling water and blowing a fan around with the resultant steam. Later in this chapter we will examine the current worldwide demand for such fuels, but first we consider the scale of some of the changes and innovations of the modern era.

Figure 4.13 shows how the need to transport oil, our best loved fossil fuel, has led to a near exponential growth in the size of tankers plying the seas.

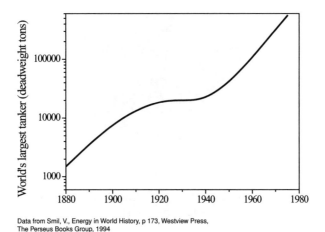

Data from Smil, V., Energy in World History, p 173, Westview Press,
The Perseus Books Group, 1994

Figure 4.13   Growth in tanker size: the curve is almost exponential and the largest vessels are now
400 m long [Smil, V., *Energy in World History*, Westview Press, 1994, p173]

Not all developments have centred on fossil fuels. Electricity is also produced using
hydropower and nuclear reactors; the latter undergoing a rapid expansion from the first
reactors in the 1950s, but then stalling due to high costs, concerns over safety and the
unresolved issue of waste disposal.

Gas turbines have been used in many applications including flying. Aeroplanes using
turbines were first built during the 1930s in England and Germany. Britain built the
first passenger jet, the 'Comet', in 1952, Boeing the 747 in 1969. And the supersonic
'Concorde' was completed in 1976.

Only rocket engines are capable of delivering more power per unit mass than gas
turbines. During World War II, Germany used ethanol-powered engines in its V2 rockets
to drop explosives on London. The Soviets launched the first satellite, 'Sputnik', in 1957.
America's Saturn V rocket (Figure 4.14), used on flights to the moon, fired its engines
for only 150 seconds producing a maximum of 119 GW (160 million horsepower) from
kerosene and hydrogen burning rockets. By comparison the capacity of all UK power
stations is only around 70 GW.

Metals

The use of coke and increases in the efficiency of furnaces led to an expansion of iron
production, from worldwide totals of 400 000 tonnes in 1800 to over half a billion
tonnes in 1990 (a 1300-fold increase). Access to large sources of electricity allowed
aluminium production to be commenced using the electrolysis of aluminium oxide –
a process first devised in 1886 but requiring six times as much energy as melting an
equivalent amount of iron.

Figure 4.14   Saturn V rocket [BOE05]

Agriculture

As we saw earlier, one of the major limits to agriculture was the availability of nutrients, the other the amount of land area needed to grow food for animate prime movers, particularly horses. Fossil fuels have solved both of these problems.

Chilean nitrates provided the only inorganic nitrogen fertilizers until the 1890s and it wasn't until 1913 that the Haber-Bosch process for ammonia synthesis was introduced, promising an inexpensive and energy-efficient route to nitrogen fertilization. Initially, due to the intervention of two world wars, uptake was slow and stayed below five million tonnes per annum until the late 1940s. By 1990 total use of synthetic nitrogen compounds had reached 85 million tonnes of nitrogen (Figure 4.15). Synthetic nitrogen now supplies around half the nutrients required by the world's crops, and because about three quarters of all nitrogen in food proteins come from arable land, one third of the protein in the global diet is from synthetic nitrogen. In countries such as China, where a higher reliance is placed on crops for proteins, the fraction is even greater [SMI92]. Natural gas is the leading feedstock for the process, the inference being that modern farming is to some extent a means of converting fossil fuels into food. Petrochemicals are also extensively used as feedstocks in the pesticide industry, which has grown from a single pesticide (DDT) to over 50 000 compounds today.

The direct use of fossil fuels within prime movers in the agricultural sector allowed labour inputs to be further reduced, continuing the trend started by the mass use of horses. For example, the average labour inputs per tonne of wheat grain in American farms was 30 hours in 1800, but had fallen to two hours per tonne in 1970 [SMI94, p189]. This led to further rural depopulation and growth in urbanization; with the rural labour falling from 60 per cent of the workforce in 1850 to just two per cent in 1975. By 1963 America's tractor power was nearly 12 times the record draft and animal capacity of 1920 [SMI94, p189].

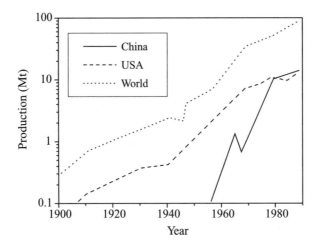

Figure 4.15 The exponential rise in nitrogenous fertilizer production. Note the log scale [data from Smil, V., *Energy in World History*, Westview Press, 1994, p183]

It is interesting to note that the global crop harvest now amounts to twice the nominal need of humanity, for a population more than three times that of 1900, yet we still have large populations who are underfed.

We will return to the question of nutrient shortages in Chapter 29 when we study proposals to increase biological activity in the oceans by adding nutrients, thereby sequestering atmospheric carbon dioxide and reducing global warming.

Transport

One of the most obvious changes has been in transportation. For most of human history the maximum travelling speed was defined by how fast one could run, then by how fast one's horse could gallop. Railways ended this period within a single generation in much of the world. The current train speed record now stands at 297 km/h. A combination of new fuels and steel hulls also increased the speed and size of ships dramatically. However the biggest personal change arrived with the car. Not surprisingly, the latest mode to have an impact has been air travel, with transatlantic flights starting in the 1930s. The invention of jet engines substantially reduced travel times and costs, to the point where air travel can be cheaper than rail and one may travel from one to almost any other of the world's major cities in less than a day.

Communication

The adoption of electricity has arguably had its greatest impact in the transfer of information between individuals and across the planet. The invention of the telegraph

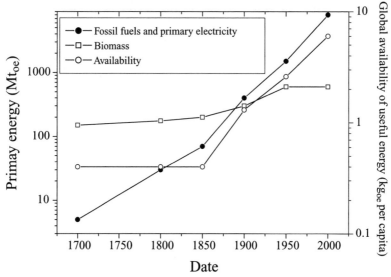

Data from Smil, V., Energy in World History, 1994, Westview Press, The Perseus Books Group.

Figure 4.16 Growth in world primary energy consumption and growth in useful energy per capita of the world population from 1700 [from data in SMI94, p187]

(a simple character sending and receiving device) and the subsequent laying of undersea links (the English Channel in 1851, the Atlantic in 1866) made the world a smaller place almost overnight. Alexander Graham Bell and Elisha Gray's invention of the telephone in 1876 had an even greater impact. Neither technology would have been possible without electricity, and the same is also true of today's communication technologies – the computer and the Internet.

Such technologies can have surprisingly high power densities. The integrated circuit within the central processor of a modern computer may have a power consumption of only a few hundred watts, but its very small size implies a power dissipation of 1 MW/m^2. This is greater than that of the outer surface of the space shuttle on re-entry [SMI99].

4.2 Wealth, urbanization and conflict

Historically, there had been growth in our use of animate and inanimate sources of power, but the rate of growth had always been limited by the question of availability; with the introduction of fossil fuels availability was no longer a problem and use grew exponentially. Since the 1800s fossil fuel extraction has grown from 10 million tonnes of oil equivalent (t_{oe}) to eight billion, whereas biomass use has only expanded from around 700 Mt_{oe} in 1700 to 1.8 Bt_{oe} today (Figure 4.16). At the same time, pressures and temperatures within heat engines have also increased, meaning that the growth in

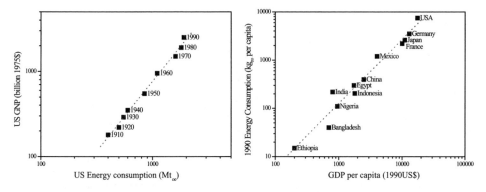

Data from Smil, V., Energy in World History, Westview Press, The Perseus Books Group, 1994

Figure 4.17   US energy consumption (left): wealth would appear to be a clear function of energy use across many decades. Gross domestic product and energy consumption of various nations (right): once more a clear trend is seen. Note, the plot is a log/log one and therefore there is a vast difference in both the wealth per capita of the energy rich/poor nations and the amounts of energy their respective citizens use [both SMI94]

'useful'[1] energy has been even greater (Figure 4.16.). This massive increase in energy use in 150 years is unprecedented. For millennia many societies hardly changed their energy use per capita, hence it is not surprising that the world we find ourselves in looks very different from that of one and a half centuries ago. This expansion has also been achieved with very little thought for the possible environmental and social costs.

Economically, non-fossil fuelled societies were either largely stationary or managed to grow by a few per cent per decade [SMI94, p204]. By contrast, the industrialising nations of the nineteenth century sustained growths of between two and six per cent per year. This allowed British economic output to grow 10-fold between 1800 and 1900, and the USA's to double between 1880 and 1900 alone. Growth has continued, with world gross product rising from US$1 trillion in 1900 to US$4 trillion in 1950 and US$14 trillion in 1973 (all constant 1990 dollars) – a four-fold increase within a single generation.

Not all nations have fared so well. There would appear to be a strong relationship between energy consumption and national wealth, in both the time domain and between nations at any one time (Figure 4.17).

This growth in national wealth and energy is of course mirrored in individual wealth and living patterns, and it is at the level of the individual that we really see the impact of fossil-fuelled economic expansion. It would seem that below 0.1 t_{oe} of consumption per person per annum it is impossible for a country to guarantee the basic needs of its people. (Large parts of Western Europe had this level of energy consumption prior to 1800.) At about 1 t_{oe} industrialization is well underway and at 2 t_{oe} what we would

---

[1] See Chapter 30 for a definition of useful energy.

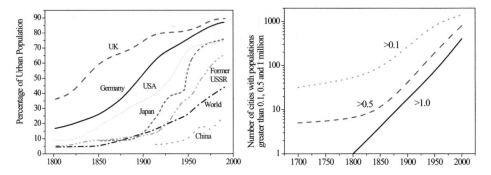

Figure 4.18 Urban population growth [Smil, V., *Energy in World History*, Westview Press, 1994, p209]

count as widespread affluence is possible. As we said before, the speed of this change can be dramatic, as can be the beneficial consequences. For example, in 1954 less than 60 per cent of French households had running water, by the mid-1970s 60 per cent had central heating; during this same period energy consumption doubled.

Not only do we live a wealthier lifestyle due to fossil fuels, we tend to live closer together (much closer together if we consider travel times). Traditional, sustainable societies could only support a minority of their population within cities. Such cities may have needed a land area 100 times their size in order to provide biomass for fuel consumption. Industrialized, fossil fuelled, cities need perhaps a land area of less than 0.1 per cent to provide their fuel – and this is with a much greater energy consumption per capita. Improvements in agricultural yields have also allowed the area required to feed such cities to be greatly reduced. In 1800 there was only one urban area containing over a million inhabitants, now there are more than 300. In the same period we have moved from around five per cent of the world's population living an urban existence to around 40 per cent, with the figure being closer to 90 per cent in many developed nations (Figure 4.18).

It has been commented that:

> '*The strategic importance of oil for military operations has become so great that is impossible to envisage wars, conducted with today's weapons, without oil*'
> – Awai S. Al-Ani, The challenge of energy (1981)

However, not only is oil needed by today's armies to power tanks, aircraft and other machines, but the geographical distribution of oil can of itself create political tensions that in the extreme lead to bloodshed. Such tensions also have the ability to affect the price of oil on the world markets, and for the reasons outlined above, thereby have an impact upon economic structures within the industrialized nations. One of the major reasons for this is that many of the most industrialized nations import large quantities of oil from less politically stable regions. This leaves a large amount of power with hands of a few states or individuals. During the 1970s, Saudi Arabia and Iran control more than one third of the world's oil reserve and their decisions to restrict the supply of oil quintupled the cost of a barrel of oil. This led to reduced economic growth and high unemployment around the world. Currently, the concentration of reserves is equally extreme with three quarters of the world's oil residing in just two per cent of the world's oil fields (i.e. in only five of the 300 carbon basins): Venezuela, the Volga-Ural region, the Gulf of Mexico, western Siberia and the Persian Gulf; the latter containing two thirds of the known reserves.

It is therefore not surprising that there has been outside intervention in the Middle East since the end of World War II. These interventions have included: the Soviet attempt to control northern Iran (1945–46); America landing troops in Lebanon (1958 and 1982); Western arms sales to Iran (prior to 1979) and continuously to Saudi Arabia; Soviet arms sales to Egypt, Syria and Iraq; then Western arms sales to Iraq (1980–88); the Western invasion of Iraq in 1990 and again in 2003.

Fossil fuels have led to ever more sophisticated forms of weaponry, although the most energetic have relied upon nuclear energy. The atomic bomb that was dropped on Hiroshima in 1945 exploded 580 metres above the ground creating a temperature of several million celsius, a blast velocity of 440 m/s, a pressure of 3.5 kg/cm^2 and the release of 52.5 TJ – equivalent to 12.5 kt of TNT. Since this date nuclear weapons have increased in capacity considerably.

Problem 4.6   The largest nuclear weapons developed are equivalent to 100 Mt of TNT, and the total US and Soviet nuclear arsenal in 1990 was in excess of 10 000 Mt. By converting these Figures to joules, compare the energy produced by the use of either to world annual primary energy consumption.

Energetically, we have come a long way from the traditional solar-powered societies discussed at the beginning of this chapter, and some of this change is summarized in Figures 4.19 to 4.22. The benefits of this transition have been substantial, but at the cost of sustainability. Our current use of fossil fuels is unsustainable from both a resource and an impact perspective. Only time will tell the price future generations and the planet will pay for this.

Problem 4.7   Estimate the quantity of heat (in watts) produced by a hypothetical computer chip of area 1 km^2. Compare this to world electricity consumption.

4.3 Our current level of energy use

If we ignore non-commercially or locally traded biomass fuels, such as wood and dung, and sources of 'free' energy such as light passing through windows, the world

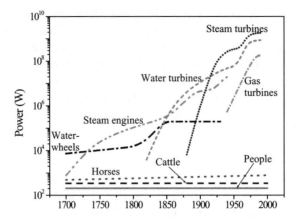

Figure 4.19 Maximum capacities of prime movers 1700 to 1990. Modern steam turbines are seen to provide the power of more than a million draught animals [Smil, V., *Energy in World History*, Westview Press, 1994, p228]

uses 409,000 million GJ (9741 Mt$_{oe}$) of primary energy per annum. (The UK uses 9374 million GJ (223 Mt$_{oe}$).) By primary energy we mean the energy content of the fuels that we use. So, for example, it includes the energy content of the fossil fuels used to generate electricity, rather than the lower figure of the amount of electricity generated. For nuclear power this includes a factor to represent the loss of energy in going from the heat in the reactor to steam and then electricity. It also ignores other conversion inefficiencies; i.e. it measures the energy content of automotive fuels, rather then the kinetic energy of all automotive movements. Figure 4.23 shows how this vast amount of energy

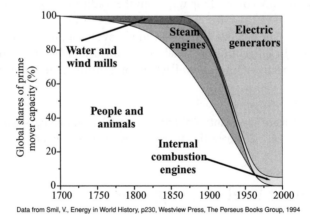

Data from Smil, V., Energy in World History, p230, Westview Press, The Perseus Books Group, 1994

Figure 4.20 Global share of common prime movers 1700 to 1990. In 1700 the state of affairs was almost identical to that 2000 years before. By 1950 the world had been transformed and almost all power was provided by electricity and internal combustion engines [SMI94, p230]

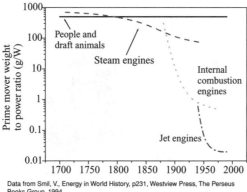

Data from Smil, V., Energy in World History, p231, Westview Press, The Perseus Books Group, 1994

Figure 4.21 Weight/power ratios of modern prime movers. Technological innovation has not only increased the power of prime movers but also greatly decreased their size. The inability of steam engines to keep pace with internal combustion engines in this respect led to their obsolescence [SMI94, p231]

is split across the world. Obvious inequalities can be observed in how the world's primary energy is shared out, with North America (mainly the USA) using almost ten times more primary energy than the whole of Africa. This point is re-emphasized if we divide national primary energy use by national population, to give the energy use per capita – Figure 4.24. Much of the world is seen to use less than 63 GJ (1.5 Toe) per capita per annum, whereas others use many times this.

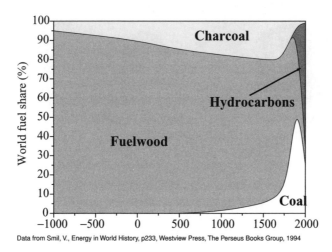

Data from Smil, V., Energy in World History, p233, Westview Press, The Perseus Books Group, 1994

Figure 4.22 Fuel shares 1000 BC to AD 2000. Fuelwood has provided the majority of our fuel for most of human history and the non-coal hydrocarbon era can be viewed as a mere blip on the time axis [SMI94, p233]

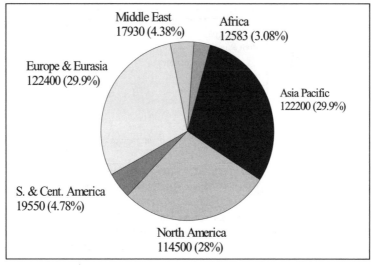

Data adapted from, Energy in Focus, BP Statistical Review of World Energy, BP Plc, 2004, bpdistributionservices@bp.com

Figure 4.23   World annual primary energy use (million GJ) [data from ENE04]

If we separate this energy use into its constituent parts (Figure 4.25) we see that: oil is the most popular energy source; coal and gas share a similar, but smaller popularity; and hydroelectricity and nuclear power each supply around one sixth of that of oil. As was the case with total energy use, there are regional discrepancies (Figure 4.26), with

Reprinted from Energy in Focus, BP Statistical Review of World Energy, BP Plc , 2004, bpdistributionservices@bp.com

Figure 4.24   Annual primary energy consumption per capita (t$_{oe}$) [ENE04]

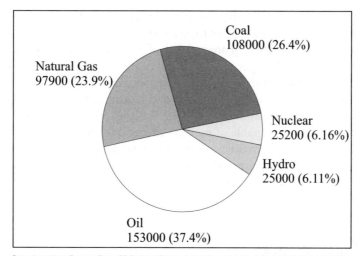

Figure 4.25   World fuel split (million GJ) [data from ENE04]

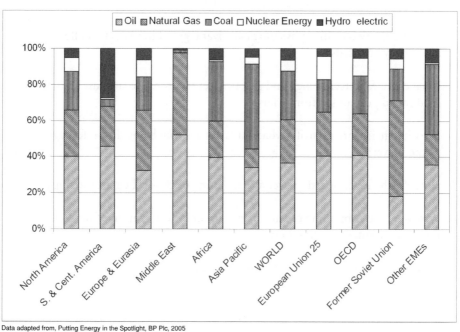

Figure 4.26   Regional consumption pattern, per cent [data from BP05]

the greatest variability being in the fraction of nuclear, hydro and coal used, whereas the fraction of oil and gas used would seem to be more uniform. In part this is a reflection of the greater ease of transport and adaptability of these two fuels. Most importantly, we see that the majority ($\approx$87 per cent) of primary energy is provided by fossil fuels, i.e. by fuels that contribute to anthropogenic climate change.

For the non-renewable sources[2] – oil, coal and natural gas – we can estimate the ratio of reserves to annual production (the R/P ratio). This shows that world oil and gas resources are only capable of meeting demand for the next 50 or so years[3], coal however is capable of providing power for a far longer period. This suggests that we are unlikely to face an immediate energy 'crisis' resulting from resource, but one arising from usage, namely climate change, a point discussed in greater detail in Part II.

4.4 Student exercises

"It has become appallingly obvious that our technology has exceeded our humanity."

–Albert Einstein

1. In what ways do energy use and innovation drive history?

2. What are the two central physiological advantages of humans that allow us to run at a variety of speeds?

3. What changes in technology and practice were required to increase the amount of food raised per human from a farm?

4. Ignoring their work capacity, what limits horses from being the perfect agricultural prime mover?

5. Describe the form, speed, power and manning of classical Greek warships.

6. Why were the combining of various metals and the invention of the steam engine important to metal production?

7. Why was coal not initially used in iron production?

---

[2] As we discover in later chapters, hydro and current nuclear technologies should probably not be classified as truly renewable sources of energy. However, it would also be difficult to estimate the R/P ratio for such fuels.

[3] As extraction technologies evolve, the world's fossil fuel reserves are likely to grow; however our rate of consumption is also rising, suggesting that the conception of an R/P ratio is still valid.

8. What key advantages did steam engines offer over water and wind machines?

9. How did changes in agricultural practice allow cities to increase in size?

10. How much primary energy do we use today? How is this use shared amongst fuels and regions? Is the current situation equitable?

5

Sustainability, climate change and the global environment

'*the impacts of global warming are such that I have no hesitation in describing it as a "weapon of mass destruction"*'

–John Houghton, Former head of the UK Meteorological Office,
and co-chair of scientific assessment for the
UN intergovernmental panel on climate change [HOU03].

Many texts leave the discussion of the environmental consequences of energy use until after the introduction of the common energy systems in use today. However, acknowledging that for many it is these consequences that are the reason for their interest in energy science, we will introduce the greatest of these concerns – climate change – together with the idea of sustainability, at this point. Other energy related environmental problems, such as local air quality, depend more on the details of specific energy systems, and are dealt with later, as are the detailed predicted consequences of climate change and the political discussions it engenders amongst the international community (Chapter 18).

5.1 Sustainability

Sustainability is usually defined as:

> *Meeting the needs of the present generation without compromising the ability of future generations to meet their own needs* [BRU87].

Clearly, exhausting a natural resource, leaving large costs for future generations or doing irreversible harm to the planet conflicts with the idea of sustainability.

Energy and Climate Change   David A. Coley
© 2008 John Wiley & Sons, Ltd

The above definition of sustainability allows us to place energy technologies into one of two categories: *sustainable* or *unsustainable*. For the purposes of this text, an energy technology is considered sustainable if:

1. it contributes little to manmade climate change

2. it is capable of providing power for many generations without significant reduction in the size of the resource, and

3. it does not leave a burden to future generations

The term 'sustainable' is often defined more broadly than this and it would seem sensible at least to consider whether a technology entails the emission of large quantities of pollutants, or is the cause of health problems or social injustice. However, we will mainly concern ourselves with the physical global environment.

Such sustainable technologies are often termed *alternative* or *renewable* technologies and are largely undeveloped compared to their unsustainable brethren. As almost anything we do has some impact on the environment, it is difficult to say that a technology is truly sustainable. For example, wind power is considered an un-environmentally friendly option by some, in that it has a visual impact upon the landscape; tidal power can alter tidal habitats and the production of photovoltaics requires the use of large amounts of hazardous chemicals. Yet by concentrating our focus on the global environment and in particular on climate change, we can make use of the above categorization to separate the various energy technologies as follows:

Unsustainable
 Fossil fuels, i.e. coal, oil and gas
 Large-scale hydropower
 Thermal nuclear reactors
Sustainable
 Solar
 Wind
 Wave
 Tidal
 Small-scale hydropower
 Biomass
 Geothermal
 Fast nuclear reactors
 Nuclear fusion

5.2 Climate change

'Technology', said the physicist C.P. Snow, 'brings you great gifts with one hand, and stabs you in the back with the other'. For us the technology in question is our use of fossil fuels, the gift a rapid rise in our standard of living, and the stab global climate change, which threatens us, and our economies.

Few environmental topics have had the degree of sustained media coverage that global warming, or more accurately, climate change has. Often this coverage has been highly sensationalist, suggesting everything from a Mediterranean future for London, to New York being inundated by rising sea levels. This author for one has learned to live with being contacted by the press each time we have a long period without rain, or conversely, a few weeks of heavy downpours, and asked if this is final proof of climate change. Given the stochastic nature of the weather this is hardly surprising and records of the form hottest/ driest/ wettest/ coldest/ windiest month/ week/ year since x, y or z are almost bound to be broken on a regular basis. Indeed, much of the confusion about the topic arises from a misunderstanding of the terms *weather* and *climate*. Weather refers to short-term variability at a specific location; climate refers to longer-term statistics of a region. So both Sydney, Australia and Reykjavik, Iceland can have exactly the same weather, i.e. temperature, wind speed and precipitation on a particular day. However, they have very different climates, as testified by the flora found in the two localities. The science of climate change can tell us much about the future climate of the planet, or of a region, but very little about the future weather in a specific city on a specific date.

If one writes a simple computer program or spreadsheet that randomly assigns a number of rainy days to each month in the last 300 years, then calculates how many years one has to look back to find the same month (January, February, ...) that has at least as many rainy days, you will see that one would expect climate records to be broken often – without the need to invoke climate change (Figure 5.1). Conversely, this means that if such records were not being regularly broken, then it would be a strong indication of something amiss.

So, if short-term changes in weather are unlikely to provide evidence of climate change, what evidence do we have that humankind might be affecting the planet's climate? The evidence we have takes five forms: firstly we know of a *process* that strongly suggests that increasing the concentration of carbon dioxide and other gases in the atmosphere will affect the climate; secondly, we have *historic evidence* of a link between climate and carbon dioxide concentration; thirdly, we know that the *concentration* of carbon dioxide in the atmosphere is rapidly increasing; fourthly, precision *measurements* of the global mean temperature and sea level indicate that temperatures and levels are rising. In addition there is growing evidence of changes in flora and fauna. Finally, *computer models* of the current climate would appear only to give the correct answers if man-made emissions of certain gases are included. In the following sections some of the mechanisms behind the climate will be discussed and the evidence for anthroprogentic (man-made) climate change introduced. (Part III considers what a warmer world might

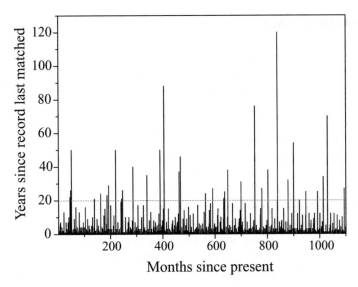

Figure 5.1 Frequency of the wettest month on record. The 20-year record (dotted line) is seen to be broken every few years. The weather model used is very simplistic, but given there are many other statistics: max and min temperatures, snowfall etc., which in turn can be calculated over various periods: years, months, seasons, we see that one should expect weather records to be broken frequently

mean for all of us and how the international community is reacting to these challenges.) Finally some other components of sustainability will be discussed in relation to energy systems.

Evidence of a process: the natural greenhouse effect

All objects at a temperature above absolute zero (−273 °C) emit radiation. The hotter the object, the shorter the mean wavelength of this radiation. The wavelength, λ_{max} (m), of the peak of this radiation is given by Wien's displacement law:

$$\lambda_{max} = \frac{3 \times 10^{-3}}{T} \tag{5.1}$$

where, T (in kelvin, K) is the temperature of the object. For the sun, with a surface temperature of over 6000 °C, the peak is at 500×10^{-9}m, which is in the visible region. For the earth, with its much lower temperature, the radiation is of longer wavelength and within the infrared part of the electromagnetic spectrum.

A hot object will radiate with an intensity of I (W/m^2) given by:

$$I = \sigma T^4 \tag{5.2}$$

where σ is the Stefan-Boltzmann constant ($5.67 \times 10^{-8}\,\mathrm{Wm^{-2}\,K^{-4}}$), so the sun radiates ($\sigma T^4 =$) $7.35 \times 10^7\,\mathrm{W/m^2}$. The solar radius is 6.96×10^8 m, so the sun in total must emit 3.91×10^{26} W. To estimate how much of this reaches the earth, we can imagine this radiation smeared across the inside of a sphere centred on the sun and just enclosing the earth. The earth will occupy $1/4.6 \times 10^{-10}$ of the area of this sphere[1] and therefore receive only this fraction of the sun's radiation, i.e. 1.8×10^{17} W. Some of this is absorbed by the planet and warms it, but about a third is simply reflected back directly into space. Thus the planet receives, very approximately, 1.2×10^{17} W or 232 $\mathrm{W/m^2}$ of heat from the sun. All radiation received by the planet must be re-radiated back into space, or the planet's temperature would continuously escalate: so there must be a radiation balance.

Applying Equation 5.2 to this value (232 $\mathrm{W/m^2}$) suggests that the temperature of the earth should be 254 K, or $-19\,°\mathrm{C}$. Clearly, the planet is warmer than this, but why?

The answer, which all scientists agree on without controversy, lies in the composition of the atmosphere, which through the natural greenhouse effect ensures that the planet is about $33\,°\mathrm{C}$ warmer than it otherwise would be, and that life as we know it can exist. The basis of the source of this effect can be gleaned from a consideration of the relative wavelengths emitted by the sun and the earth and how they interact with the individual gases in the atmosphere (Figure 5.2).

In general, these gases are far more transparent (i.e. less absorbing) to visible light than to infrared radiation. So, visible light travels relatively easily through the atmosphere to warm the surface. But it is much harder for the long wavelength heat to re-radiate back up through the atmosphere and into space and it therefore warms the planet to a greater degree than expected.

The atmospheric absorption curve shown in Figure 5.2b is complex and arises from the absorption curves of all the gases in the atmosphere, the most important of which are shown in Figure 5.2c. (Note Nitrogen is not a greenhouse gas.) The greenhouse gases are able to absorb the outgoing infrared radiation and convert this into vibrational energy stored in molecular bonds and the motion of their constituent atoms. This energy is slowly transferred to the other, non-greenhouse gases, principally nitrogen and oxygen, leading to a bulk warming of the atmosphere. The emission spectra of the sun and earth are shown in Figure 5.2a.

So, although the picture of a simple radiation balance is true for the upper atmosphere, for the surface and the rest of the atmosphere we need a more complex model, which takes account of this absorption, or cycling of heat within the atmosphere (Figure 5.3). The gases in Figure 5.2c must themselves re-radiate all the energy they absorb (or they would continuously increase in temperature); but unlike radiation from the surface, this radiation will be in all directions and some will return towards the earth and re-heat the surface. It is this re-heating, or *natural greenhouse effect* that gives rise to an average surface temperature of $14\,°\mathrm{C}$: warm enough for water to be liquid and life to exist. As we have said, without this natural greenhouse effect the average surface temperature would

---

[1] This can be worked out from the radius of the earth (6.4×10^6 m) and the distance to the sun (1.49×10^{11} m).

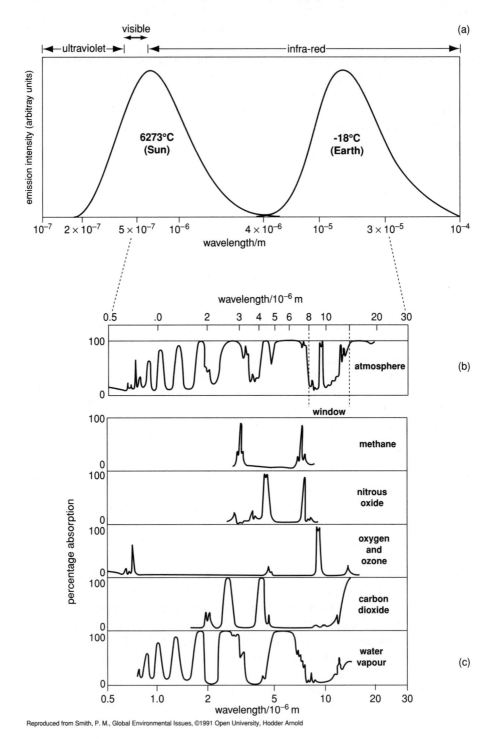

Figure 5.2 Absorption spectra for the atmosphere and some of the important greenhouse gases. Note how the absorption is relatively weak in the visible part of the spectrum ($0.4 - 0.7 \times 10^{-6}$ m), allowing light into the atmosphere, but highly absorbing in the infrared ($> 0.7 \times 10^{-6}$ m); there is only a small window where infrared energy can leave the planet. Without this window the earth's temperature would be even higher [OU91]

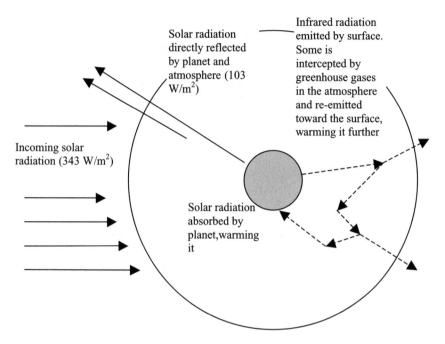

Incoming solar radiation (343 W/m²)

Solar radiation directly reflected by planet and atmosphere (103 W/m²)

Infrared radiation emitted by surface. Some is intercepted by greenhouse gases in the atmosphere and re-emitted toward the surface, warming it further

Solar radiation absorbed by planet, warming it

Net incoming radiation (240 W/m²) = net outgoing radiation

Figure 5.3 Simple radiation balance of the planet allowing for the cycling of heat by the greenhouse gases. The atmosphere is represented by the larger circle and is not to scale. Note the incoming figure of 343 W/m² assumes the radiation is spread equally over the whole planet, i.e. $\pi r^2/4\pi r^2 = 1/4$ of the value of the solar constant (1370 W/m²) which is the amount intercepted by a disk of radius equal to the earth's at the same distance from the sun as the earth

be $-19\,^\circ$C and water would not exist in a liquid form. Table 5.1 shows the amount of warming provided by each of the major greenhouse gases and demonstrates that most of the warming is due to water vapour. It is worth noting that the term 'greenhouse effect' is slightly misleading. The main reason a real greenhouse warms up is that the air inside is trapped; therefore heat cannot be lost by convection.

If we re-plot the solar emission spectrum given in Figure 5.2a, after allowing for the absorption provided by the various gases which make up the atmosphere (Figure 5.2b), we will have the spectrum of light reaching the surface of the planet: Figure 5.4.

Figure 5.5 shows the radiation budget for the earth and its atmosphere in more detail, with the role of evaporation and clouds included.

Problem 5.1 Mars has very little in the way of an atmosphere. This means that a simple radiation balance alone defines its temperature. Repeat the analysis above to estimate the average surface temperature of Mars. Hint: the radius of Mars is 3385 km, it directly reflects 29 per cent [THE05] of the light that strikes it, and it lies at a distance of 228×10^6 km from the sun [NAS05a].

Table 5.1 The warming provided by the most important greenhouse gases [data from C. Schonwiese and B. Diekmann, in BOE99]. Data from chonwiese, C. and Diekmann, B., Der Treibhauseffekt, Deutsche Verlags-Anstalt, Stuttgart, 1987

Gas	Present concentration (ppm_v^2)	Present warming effect (°C)
Water vapour	5000	20.6
CO_2	358	7.2
Tropospheric O_3	0.03	2.4
N_2O	0.3	0.8
CH_4	1.7	0.8
Others (HFCs etc.)	Approx. 20 ppt_v^3	0.6
Total		33.0

A radiative model of the atmosphere and the greenhouse effect

Although a true model of the atmosphere would need to include mathematical descriptions of a multitude of processes such as convection and cloud formation, together with

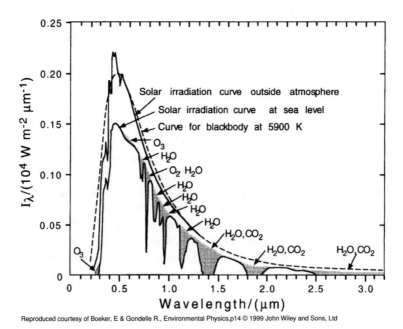

Reproduced courtesy of Boeker, E & Gondelle R., Environmental Physics,p14 © 1999 John Wiley and Sons, Ltd

Figure 5.4 The Solar spectrum at the top and bottom of the atmosphere with the compounds responsible for various dips in the received intensity shown [BOE99, p14]

---

[2] Parts per million by volume.
[3] Parts per trillion by volume.

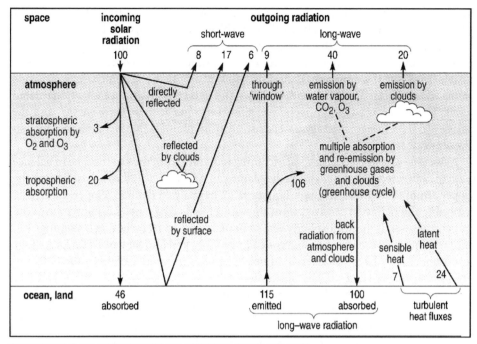

Figure 5.5 Radiation balance for 100 units of incoming radiation [OU91, p76]

a description of the chemistry of the various layers of the atmosphere, we can use the above arguments as the basis for a surprisingly useful, and reasonably accurate model of the atmosphere, the greenhouse effect and climate change, be it natural or anthropogenic. Those with the necessary skills are greatly encouraged to implement this model on a computer (See Problems 5.2 and 5.4, and Exercises 5.15 and 5.16). This will result in a tool that will allow you to explore how changes in solar output, surface reflectivity and greenhouse gas concentrations have the potential to lead to drastic climate change. More detailed models are considered in Chapter 17; the structure of the atmosphere and its history are covered at the end of this chapter in Section 5.4.

To all extent and purpose space is a vacuum; therefore the planet can only lose heat via radiation, not conduction or convection. However, as Figure 5.3 shows, at many wavelengths almost all of the long-wave radiation emanating from the surface is absorbed by greenhouse gases and clouds before it reaches the upper atmosphere. In essence, the atmosphere is all but opaque to infrared photons at many wavelengths, so a disproportionate number of the photons released into space must emanate in the *upper* atmosphere rather than travel directly through it from below—in fact from the upper troposphere, where the atmosphere starts to become transparent to such radiation, at a height termed *the emission level* (around 5.5 km [HAD07]).

Figure 5.27 at the end of this chapter reinforces this point. In (a) we see the Earth viewed at visible wavelengths (i.e. light): the planet can be seen in great detail, therefore visible photons emanating at the surface must have been able to travel the full height of the atmosphere with little absorption. In (b) we see an image of the planet formed from infrared photons with a variety of wavelengths covering much of the infrared spectrum: much less detail is observed, but the outlines of the continents are still just visible. However, if we observe (c) the Earth only at around 6 μm, which lies within one of the water vapour absorption peaks shown in Figure 5.3, we can see no surface detail at all. This is because the atmosphere is almost opaque at these wavelengths, and the image was therefore formed almost exclusively from photons that were released high in the atmosphere.

As we have already seen, the total incoming radiation from the sun must be in balance with the outgoing radiation, and as the planet can only lose heat through radiation, the upper atmosphere must, through Equation 5.2 have a well defined temperature. And, assuming the radiation received from the sun remains constant, this temperature can not change even if the atmosphere were to undergo a change, including becoming even more opaque because of increasing concentrations of greenhouse gases. However, if the atmosphere does become generally more opaque to infrared radiation, the emission level must rise. We can see this by considering a bank of fog and a torch beam. The point at which the beam will be visible from outside the bank will depend on both how close the torch is to the edge of the bank and how dense (or opaque) the fog is. If the fog becomes denser the closer the torch will need to be to the edge for an appreciable number of the photons in the torch beam to escape, i.e. the "emission level" of the beam will have moved outwards.

As anyone who has gone skiing or hill walking knows, air temperature decreases with height. The reason the temperature falls as one ascends is because the higher one goes the less weight of atmosphere remains above, which in turns means the atmospheric pressure decreases with height. Hence as a parcel of air travels up through the atmosphere the pressure around it gradually reduces, causing it to expand. The ideal gas law[4] shows us that such an expanding gas will cool. This reduction is surprisingly constant on average, at around 6.5°C per kilometre, until one reaches the stratosphere (10–15 km) and is termed the *environmental lapse rate*[5]. The corollary is that as one descends from the upper troposphere, with its fixed temperature given by the need for radiative balance, the temperature must rise by 6.5°C per kilometre. If because of an increase in the concentration of greenhouse gases the emission level has risen, then because the environmental lapse rate is a constant 6.5°C/km, the final temperature upon reaching the ground must be higher as the distance descended would have been greater. This is the position we find ourselves in today: increases in greenhouse gas concentrations have made the atmosphere even more opaque to infrared photons, this has raised the height of the emission level, which, through the environmental lapse rate, means the surface temperature has risen.

---

[4] $PV = nRT$.

[5] The exact rate of cooling will also depend on the moisture content of the air in question, as it can also lose heat via condensation. The instantaneous (rather than the mean) environment lapse rate will therefore vary with location, altitude and weather.

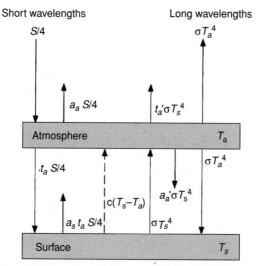

Short wavelengths Long wavelengths

Reproduced from Boeker, E &Gondelle R., Environmental Physics, p31, © 1999 John Wiley and Sons, Ltd

Figure 5.6 A zero-dimensional atmospheric model. Incoming solar radiation is shown on the left as 'short wavelength' with magnitude $S/4$ and the emitted long wavelength infrared radiation is shown on the right. Parameters a and t represent the fractions reflected or transmitted, respectively. A prime indicates the values at short wavelengths [B. Sorensen in BOE99, p31]

We will now use the above qualitative model to construct a simple, but highly useful, mathematical model of a planet with an atmosphere from a consideration of the amount of sunlight and infrared radiation transmitted by the atmosphere, and the albedos of the planet and the atmosphere (Figure 5.6). In the following the subscript a refers to the atmosphere, s to the surface and primes are used to indicate infrared values. If σ is Stefan-Boltzmann's constant, c another constant[6], S the solar constant, t_a the transmittance of the atmosphere, a an albedo[7] and T_a the temperature in kelvin, then for the surface [BOE99, p31]:

$$(-t_a)(1 - a_s)\frac{S}{4} + c(T_s - T_a) + \sigma T_s^4(1 - a_a') - \sigma T_a^4 = 0. \qquad (5.3)$$

The first term describes the absorption, the second the non-radiative (i.e. convective etc.) interaction between the surface and the atmosphere, the third the emitted radiation minus the backscattered and the final term the incoming heat radiation from the atmosphere. For the atmosphere we have:

$$-(1 - a_a - t_a + a_s t_a)\frac{S}{4} + c(T_s - T_a) - \sigma T_s^4(1 - t_a' - a_a') + 2\sigma T_a^4 = 0 \qquad (5.4)$$

---

[6] We are creating a first-order model, where the interaction between the atmosphere and the earth is proportional to the temperature difference between them, and c is the constant of proportionality.
[7] i.e. the fraction of light or infrared radiation reflected directly back without absorption.

where the first term represents the absorption of sunlight, the second the non-radiative interaction, the third the absorption by the atmosphere of radiation from the earth and the last, the atmospheric emission.

Solving Equation (5.3) for T_a and inserting the result in Equation (5.4) gives an equation for the surface temperature, T_s, of the planet. If the following reasonable values are assumed: $a_s = 0.11$, $t_a = 0.53$, $a_a = 0.30$, $t'_a = 0.06$ and $a'_a = 0.31$, then c equals 2.7 Wm^{-2}K^{-1} and $T_s = 288$ K.

Problem 5.2   This problem is more difficult and more time consuming than others in the book and some readers may wish to skip it. Solve Equations (5.3) and (5.4) for T_s. Using the values of a_s etc., given above, check that the model gives a surface temperature of 288 K. This and the following calculations might best be completed using a spreadsheet. Assuming you obtained the correct value, re-run the calculation for the scenarios given below, estimating T_s each time.

Equations (5.3) and (5.4) represent a very simple model of the earth and caution should be used in drawing conclusions from applying it. However it is complex enough to give qualitative estimates of what might happen to the temperature of the planet if some of the parameters took different values.

The following scenarios [in part adapted from BOE99, p32–3] use this model to show the likely impact of such changes.

- The *white, or snowball, earth.* If the earth were covered in snow the surface albedo, a_s, might have a value of around 0.75. T_s would then equal 272 K, which is below the freezing point of water. This suggests a stable and long-lasting state.

- *Nuclear winter.* It has been postulated that a large-scale nuclear exchange would create fires on such a scale that the sky would transmit much less light. If this resulted in t_a falling to 0.43, T_s would equal 284 K, a cooling of 4 °C: assuming the change in t_a is short-term, this would not be a catastrophic temperature drop.

- *Solar world.* If we covered one-third of the planet with solar collectors a_s would fall to 0.10 [BOE99, p32], T_s would then increase by only 0.2 °C. This implies that we can extract large quantities of renewable energy without overly perturbing the planet.

- The *young earth.* Two billion years ago the sun radiated less light and S would have been only 85 per cent of its current value, implying T_s was around 275 K. A value that was unlikely to be conducive to the origin of life. It is believed that a greater concentration of greenhouse gases existed at the time, implying that t'_a was smaller, a'_a greater and T_s higher than this value. This indicates that not only is the greenhouse effect important to life today, but that it was critical for the formation of life.

Table 5.2 lists the albedo of various surfaces. Readers may like to run the simple greenhouse model from Problem 5.2 (or download it from the book's web site) for

Table 5.2 Albedo of various surfaces for visible light ($\lambda = 500 \times 10^{-9}$m) [data from I. Campbell, and J. Peixoto and A. Oort, in BOE99. Reproduced from Boeker, E & Gondelle R., Environmental Physics, ©1999 John Wiley and Sons, Ltd]

Surface	Albedo	Clouds	Albedo	Planets	Albedo
Water	0.05	Cumulus	0.7−0.9	Earth	0.34−0.42
Snow	0.85	Stratus	0.6−0.85	(Moon)	0.06−0.07
Desert	0.3	Altostratus	0.4−0.6	Venus	0.76
Green meadow	0.15	Cirrostratus	0.4−0.5	Jupiter	0.73
Deciduous forest	0.15				
Coniferous forest	0.1				
Crops	0.1				
Dark soil	0.1				
Dry earth	0.2				

various albedos to get a feel for the impact that other changes to the planet's surface (some caused by global warming itself) might have on surface temperatures. In Chapter 17 more complex, and accurate climate models will be discussed. Some of these have taken many years to produce and make use of the world's fastest computers.

The anthropogenic or enhanced greenhouse effect

We have seen that the natural greenhouse effect is critical for our existence and that its effect depends on the composition of the atmosphere. Given the very large volume of the atmosphere, initially it seems unlikely that man's activities could alter this composition significantly and lead to an enhanced, anthropogenic (man-made) greenhouse effect. Unfortunately this is not so. Figure 5.7 shows how the concentration of carbon dioxide and other gases in the atmosphere has been changing since the industrial revolution. This increase (of about 30 per cent for carbon dioxide and 230 per cent for methane) arises mostly from the burning of fossil fuels for the production of energy. That the excess carbon is anthropogenic in origin is known from carbon dating of the atmosphere.

Carbon in the atmosphere is a mix of 98.89% carbon-12, 1.11% carbon-13 and trace amounts of carbon-14. The first two are stable, but carbon-14 is radioactive with a half-life of 5730 years. This half-life is so short that any carbon-14 that existed when the planet was formed has long since vanished. So, fresh carbon-14 must be being constantly made. This occurs when energetic neutrons produced by cosmic rays strike nitrogen in the atmosphere:

$$^{1}n + {}^{14}N \rightarrow {}^{14}C + {}^{1}p,$$

with the greatest rate of production taking place at an altitude of between 9 and 15 km. The carbon-14 combines with oxygen to form carbon dioxide and thereby enters living organisms at a concentration of 1 part in a trillion (10^{12}) of carbon-14 to all carbon.

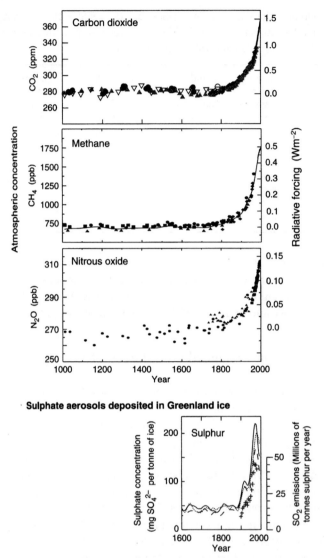

Figure 5.7   Change in atmospheric greenhouse gas concentrations over the last 1000 years. The term *radiative forcing* indicates the amount of additional heating (in watts per square metre of the planet's surface) provided by a change in concentration of a greenhouse gas. Sulphate particles are not a greenhouse gas, but are anthropogenic emissions that have an impact on the planet's climate and are discussed later [IPC01]

This concentration will be maintained in the organism by respiration and food intake until it dies.

Fossil fuels were formed so long ago that almost all the carbon-14 they contained has also decayed. When such fuels are burnt, the carbon dioxide produced therefore also contains very little carbon-14 and dilutes the carbon-14 populated carbon dioxide in the atmosphere. We can estimate the resultant reduction in the concentration of carbon-14 in the atmosphere over time from samples taken from tree rings, and we find it has indeed been falling exactly in-line with our increasing use of fossil fuels since the start of the industrial revolution. So the increasing levels of carbon dioxide shown in Figure 5.7 must be due to our use of fossil fuels rather than some biological process such as the more rapid decay of vegetation.

What effect on the earth's temperature might we expect this increase to have? It has been estimated that the long wave radiation directly leaving the surface would be reduced by 4 W/m^2 if the carbon dioxide concentration were to double from pre-industrial values (i.e. to 560 ppm), and that this in turn would lead to an increase in temperature for the planet of 1.2 °C[8]. The term *radiative forcing* is used to quantify the direct effect of a particular concentration of a greenhouse gas. Thus the radiative forcing of a doubling of carbon dioxide concentration is 4 W.

The natural greenhouse effect was first described in 1859 by the British scientist John Tyndall, when he discovered that the most common components of the atmosphere – nitrogen and oxygen – were transparent to both visible and infrared radiation. Whereas gases such as carbon dioxide, methane and water vapour were not transparent in the infrared. He concluded that such gases must have a great influence on our climate. In 1894 the Swedish chemist Svanta Arrhenius showed that anthropogenic emissions had the potential to alter the climate by further reducing the transparency in the infrared. He also concluded that it would take humankind 3000 years of coal burning to double the concentration of carbon dioxide in the atmosphere; in this last point he was off by around 28 centuries.

Historic evidence

Given that the current concentration of carbon dioxide in the atmosphere is around 30 per cent greater than pre-industrial levels, one might expect to see the corresponding radiative forcing reflected in elevated average global temperatures. Figure 5.8a shows the temperature record over this period, which indeed demonstrates an increase. 1990 was the warmest year since instrumented records were kept, until 1991 proved hotter. Most subsequent years have been even hotter. 1998 still holds the record, but closely

---

[8] More complex models, which include the effects that a warmer world would have on evaporation, the oceans and other processes, suggest the final temperature would be even higher. As there must still be a radiative balance for the planet as a whole, this 4 W/m^2 has to be re-radiated; it is just that this cannot now occur until the lower atmosphere is 1.2 °C warmer, because hotter objects radiate more intensely – Equation (5.2).

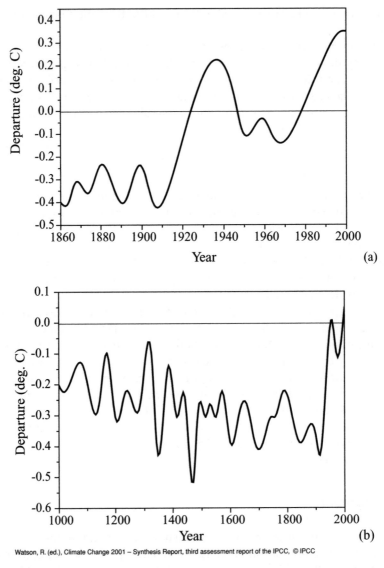

Figure 5.8 Mean global temperature, (a) since the pre-industrial era [IPC01], (b) over the last one thousand years. Plotted as a departure from the 1961 to 1990 average. Data from thermometers from 1860 and from tree rings, ice cores and corals prior to 1860

followed by 2001, 2002, 2003 and 2004. Figure 5.8b shows the temperature record over a longer period.

Clearly we are seeing much more than the odd single anomalous year: the world is now warmer than it has been at any point in the past two millennia, and if the trend continues will be hotter than at any point in the last two million years by the end of the century.

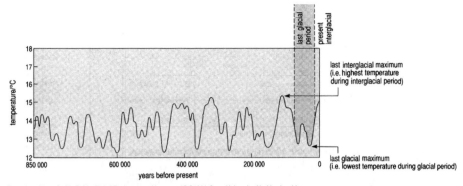

Figure 5.9   Long-term global mean temperature record [OU91, p85]

Looking at the temperature record over a longer period (Figure 5.9) we find that the evidence initially appears more mixed. It is clear from Figure 5.9 that the planet has gone through many cycles of warming and cooling. Most of the pattern in this record can be explained by variations in the earth's orbit and orientation. Over a period of 100 000 years, our orbit around the sun cycles slowly between an elliptical path and a near circular one and back again. This variation gives rise to one of the main rhythms in Figure 5.9. Another arises from the 41 000 year periodicity in the tilt of the earth on its axis of rotation. Finally the axis wobbles with a periodicity of 21 000 years. These periods are termed *Milankovic cycles* in honour of the astronomer who first proposed that such cycles could be used to explain variations in global ice cover.

Milankovic cycles can be used to explain changes in the severity of the seasons. For example, if the orbit is more elliptical, the planet will be either closer or further from the sun than usual, leading to heightened seasonal variations. However, the total average radiation received by the whole planet over a complete year will be the same whether the orbit is more, or less, elliptical. All that will change is its distribution over the planet's surface and its timing. The long-term variations in temperature must have some other contributing factor that uses the Milankovic cycles as a trigger, but then amplifies the response of the planet. Such factors are termed *feedback processes*. There are probably a large number of these feedbacks, some positive – in that they amplify any small temperature change – others negative, in that they suppress any small variations and stop them growing out of control.

Figure 5.10 shows data from Antarctic ice cores (which contain bubbles of trapped air) and gives a record of carbon dioxide concentration and temperature[9] over

---

[9] Ice in glaciers is found to have a higher than normal abundance of heavy oxygen if it was deposited during relatively warm periods. Glacier ice originates from the oceans as vapour, later falling as snow and becoming compacted in ice. When water evaporates, a higher proportion of the heavy water (H_2O^{18}) is left behind and the water vapour is slightly enriched in light water (H_2O^{16}). This is because it is harder for the heavier molecules to overcome the barriers to evaporation. Glaciers are therefore relatively enhanced in O^{16}, while

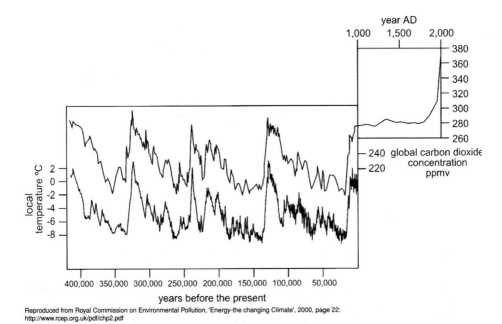

Figure 5.10   Correlation between temperature (bottom line) and carbon dioxide concentration (top line) – from ice-core data [DTI00]

a 400 000 year period. Clearly, there is a high degree of correlation. But did the rise in carbon dioxide cause the increase in temperature, or was it the rise in temperature that caused carbon dioxide levels to increase? Possibly it doesn't matter. What we do know is that the planet maintains a well-balanced carbon cycle on both geological and biological time scales and that this would appear to work as a thermostat regulating the temperature of earth. From which it seems sensible to conclude that interfering with this balance might be unwise unless there is overwhelming evidence to show otherwise.

Evidence from observed changes

During the last one hundred years there have been numerous changes to our world indicating that anthropogenic atmospheric and climatic change is a reality. We can separate these indictors into three categories:

1. changes in anthropogenic concentrations – which imply from a theoretical standpoint that climate change is likely

---

the oceans are relatively enriched in O^{18}. This imbalance is more marked for colder climates than for warmer climates. A decrease of one part per million of O^{18} in ice reflects a 1.5 °C drop in air temperature at the time the water originally evaporated from the oceans.

Figure 5.11   Glacier in Northern Pakistan retreating due to climate change (photo by author)

2. measured changes in the climate – which indicate climate change is real – and

3. observed changes in physical and biological systems – which show that the size of the climatic changes are great enough to lead to measurable consequences for the landscape, humankind and other species (Figure 5.11)

Table 5.3 lists such indicators and Figure 5.12 shows, as an example, how the cost of extreme weather events has been changing. What is of greatest concern is that we have as yet only made modest alterations to the atmospheric concentration of carbon dioxide, yet we can identify global climate change. Carbon dioxide concentrations would continue to rise even if we do not expand our use of fossil fuels, yet we are planning to expand the use of such fuels. In light of the positive feedback processes we will be

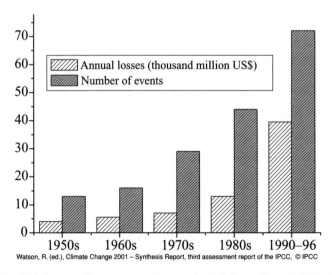

Watson, R. (ed.), Climate Change 2001 – Synthesis Report, third assessment report of the IPCC, © IPCC

Figure 5.12   Global costs of extreme weather events (inflation adjusted) [data from IPC01]

Table 5.3 Indicators of climate change (not exhaustive) [data from IPC01]. Data from Watson, R. (ed.), Climate Change 2001 – Synthesis Report, third assessment report of the IPCC, ©IPCC

Indicator	Observed changes
Concentration indicators	
Atmospheric concentration of CO_2	280 ppm for the period 1000–1750 to 368 ppm in year 2000 ($31 \pm 4\%$ increase)
Terrestrial biospheric CO_2 exchange	Cumulative source of about 30 Gt C between the years 1800 and 2000; but during the 1990s, a net sink of about 14 ± 7 Gt C
Atmospheric concentration of CH_4	700 ppb for the period 1000–1750 to 1750 ppb in year 2000 ($151 \pm 25\%$ increase)
Atmospheric concentration of N_2O	270 ppb for the period 1000–1750 to 316 ppb in year 2000 ($17 \pm 5\%$ increase)
Tropospheric concentration of O_3	Increased by $35 \pm 15\%$ from the years 1750 to 2000; varies with region
Stratospheric concentration of O_3	Decreased over the years 1970 to 2000; varies with altitude and latitude
Atmospheric concentrations of HFCs, PFCs, and SF_6	Increased globally over the last 50 years
Weather indicators	
Global mean surface temperature	Increased by 0.6 ± 0.2 °C over the twentieth century; land areas warmed more than the oceans
Northern hemisphere surface temperature	Increase over the twentieth century greater than during any other century in the last 1000 years; 1990s warmest decade of the millennium
Diurnal surface temperature range	Decreased over the years 1950 to 2000 over land; night time minimum temperatures increased at twice the rate of daytime maximum temperatures
Hot days/heat index	Increased
Cold/frost days	Decreased for nearly all land areas during the twentieth century
Continental precipitation	Increased by 5–10% over the twentieth century in the northern hemisphere, although decreased in some regions (e.g., north and west Africa and parts of the Mediterranean)
Heavy precipitation events	Increased at mid- and high northern latitudes
Frequency and severity of drought	Increased summer drying and associated incidence of drought in a few areas. In some regions, such as parts of Asia and Africa, the frequency and intensity of droughts have been observed to increase in recent decades
Biological indicators	
Global mean sea level	Increased at an average annual rate of 1 to 2 mm during the twentieth century
Duration of ice cover of rivers and lakes	Decreased by about 2 weeks over the twentieth century in mid- and high latitudes of the northern hemisphere
Arctic sea-ice extent and thickness	Thinned by 40% in recent decades in late summer to early autumn and decreased in extent by 10–15% since the 1950s in spring and summer
Non-polar glaciers	Widespread retreat during the twentieth century

Table 5.3 (*Continued*)

Indicator	Observed changes
Snow cover	Decreased in area by 10% since global observations became available from satellites in the 1960s
Permafrost	Thawed, warmed, and degraded in parts of the polar, sub-polar, and mountainous regions
El Nino events	Became more frequent, persistent, and intense during the last 20 to 30 years compared to the previous 100 years
Growing season	Lengthened by about 1 to 4 days per decade during the last 40 years in the northern hemisphere, especially at higher latitudes
Plant and animal ranges	Shifted poleward and up in elevation for plants, insects, birds, and fish
Breeding, flowering, and migration	Earlier plant flowering, earlier bird arrival, earlier dates of breeding season, and earlier emergence of insects in the northern hemisphere
Coral reef bleaching	Increased frequency, especially during El Nino events
Economic indicators	
Weather-related economic losses	Global inflation-adjusted losses increased by an order of magnitude over the last 40 years. (Part of the observed upward trend is linked to socio-economic factors and part is linked to climatic factors)

discussing later and the potential for non-linear rapid switching of atmospheric and oceanic systems that might be triggered by such changes, we have to ask ourselves, is this sane?

Table 5.3 only lists currently observed impacts. As Figure 5.10 shows, atmospheric carbon dioxide levels are still rising, implying ever greater radiative forcing. And, as we will see in Part III, most predictions of future world energy demand imply even greater emissions. In Part III we will also examine quantitatively and in detail what this could mean for the planet and particularly how these impacts might vary from region to region. Here we will simply summarize in a qualitative manner the likely consequences at three future dates if no serious attempt is made to reduce our emissions (Table 5.4).

The carbon cycle

Probably the most well understood climate control mechanism is based on the cycling of carbon dioxide through organic and inorganic systems. Rainwater dissolves carbon dioxide within the atmosphere to form carbonic acid, which reacts with rocks containing calcium silicate minerals to release calcium ions (Figure 5.13). These flow into the oceans and are used by various organisms to build their shells, the remnants of which build up over time on the ocean floor forming layers of carbonate rich sediments. This part of the process is termed the *biological pump*. Over millions of years these sediments are then

Table 5.4 Predicted consequences of climate change if no action is taken to curtail our use of fossil fuels. The level of confidence reflects uncertainties in the scientific understanding [data from IPC01]. Confidence levels are defined by the IPCC as very high (95% or greater), high (67–95%), medium (33–67%), low (5–33%) and very low (5% or less). Data from Watson, R. (ed.), Climate Change 2001 – Synthesis Report, third assessment report of the IPCC, ©IPCC

Parameter	2025	2050	2100
CO_2 concentration	405–460 ppm	445–640 ppm	540–970 ppm
Global mean temperature change from the year 1990	0.4–1.1 °C	0.8–2.6 °C	1.4–5.8 °C
Global mean sea-level rise from the year 1990	3–14 cm	5–32 cm	9–88 cm
Human health effects			
Heat stress and winter mortality	Increase in heat-related deaths and illness (high confidence). Decrease in winter deaths in some temperate regions (high confidence)	Thermal stress effects amplified (high confidence)	Thermal stress effects amplified (high confidence)
Vector- and water-borne diseases		Expansion of areas of potential transmission of malaria and dengue (medium to high confidence)	Further expansion of areas of potential transmission (medium to high confidence)
Floods and storms	Increase in deaths, injuries, and infections associated with extreme weather (medium confidence)	Greater increases in deaths, injuries and infections associated with extreme weather (medium confidence)	Greater increases in deaths, injuries and infections (medium confidence)
Nutrition	Poor are vulnerable to increased risk of hunger, but state of science very incomplete	Poor remain vulnerable to increased risk of hunger	Poor remain vulnerable to increased risk of hunger
Ecosystem effects			
Corals	Increase in frequency of coral bleaching and death (high confidence)	More extensive coral bleaching and death (high confidence)	More extensive coral bleaching and death of corals (high confidence). Reduced species biodiversity and fish yields from reefs (medium confidence)

Coastal wetlands and shorelines	Loss of some coastal wetlands to sea level rise (medium confidence). Increased erosion of shorelines (medium confidence)	Further loss of coastal wetlands (medium confidence). Further erosion of shorelines (medium confidence)	Further loss of coastal wetlands (medium confidence). Further erosion of shorelines (medium confidence)
Terrestrial ecosystems	Lengthening of growing season in mid- and high latitudes; shifts in ranges of plant and animal species (high confidence). Increase in net primary productivity of many mid- and high-latitude forests (medium confidence). Increase in frequency of ecosystem disturbance by fire and insect pests (high confidence)	Extinction of some endangered species; many others pushed closer to extinction (high confidence). Increase in net primary productivity may or may not continue. Increase in frequency of ecosystem disturbance by fire and insect pests (high confidence)	Loss of unique habitats and their endemic species (e.g., Cape region of South Africa and some cloud forests) (medium confidence). Increase in frequency of ecosystem disturbance by fire and insect pests (high confidence)
Ice environments	Retreat of glaciers, decreased sea-ice extent, thawing of some permafrost, longer ice free seasons on rivers and lakes (high confidence)	Extensive Arctic sea-ice reduction benefiting shipping but harming wildlife (e.g., seals, polar bears, walrus) (medium confidence). Ground subsidence leading to infrastructure damage (high confidence)	Substantial loss of ice volume from glaciers, particularly tropical glaciers (high confidence)
Agricultural effects Average crop yields	Cereal crop yields increase in many mid- and high-latitude regions (low to medium confidence). Cereal crop yields decrease in subtropical regions (low to medium confidence)	Mixed effects on cereal yields in mid-latitude regions. More pronounced cereal yield decreases in tropical and subtropical regions (low to medium confidence)	General reduction in cereal yields in most mid-latitude regions for a warming of more than a few °C (low to medium confidence)

(*Continued*)

Table 5.4   (*Continued*)

Parameter	2025	2050	2100
Extreme low and high temperatures	Reduced frost damage to some crops (high confidence). Increased heat stress damage to some crops (high confidence). Increased heat stress in livestock (high confidence)	Effects of changes in extreme temperatures amplified (high confidence)	Effects of changes in extreme temperatures amplified (high confidence)
Incomes and prices		Incomes of poor farmers in developing countries decreased (low to medium confidence)	Food prices increased relative to projections that exclude climate (low to medium confidence)
Water resource effects			
Water supply	Peak river flow shifts from spring toward winter in basins where snowfall is an important source of water (high confidence)	Water supply decreased in many water-stressed countries, increased in some other water stressed countries (high confidence)	Water supply effects amplified (high confidence)
Water quality	Water quality degraded by higher temperatures. Water quality changes modified by changes in water flow volume. Increase in saltwater intrusion into coastal aquifers due to sea-level rise (medium confidence)	Water quality degraded by higher temperatures (high confidence) Water quality changes modified by changes in water volume (high confidence)	Water quality effects amplified (high confidence)
Water demand	Water demand for irrigation will respond to changes in climate; higher temperatures will tend to increase demand (high confidence)	Water demand effects amplified (high confidence)	Water demand effects amplified (high confidence)

Extreme events	Increased flood damage due to more intense precipitation events (high confidence). Increased drought frequency (high confidence)	Further increase in flood damage (high confidence). Further increase in drought events and their impacts	Flood damage several-fold higher than 'no climate change' scenarios
Other market sector effects Energy	Decreased energy demand for heating buildings (high confidence). Increased energy demand for cooling buildings (high confidence)	Energy demand effects amplified (high confidence)	Energy demand effects amplified (high confidence)
Financial sector		Increased insurance prices and reduced insurance availability (high confidence)	Effects on financial sector amplified
Aggregate market effects	Net market sector losses in many developing countries (low confidence). Mixture of market gains and losses in developed countries (low confidence)	Losses in developing countries amplified (medium confidence). Gains diminished and losses amplified in developed countries (medium confidence)	Losses in developing countries amplified (medium confidence). Net market sector losses in developed countries from warming of more than a few °C (medium confidence)

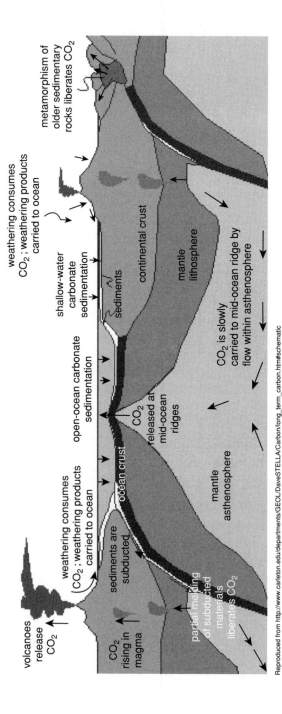

Reproduced from http://www.carleton.edu/departments/GEOL/DaveSTELLA/Carbon/long_term_carbon.htm#schematic

Figure 5.13 Schematic representation of the long-term, or geological, carbon cycle showing the flows of carbon that are important on timescales of more than 100 000 years. Carbon is added to the atmosphere through metamorphic degassing and volcanic activity on land and at mid-ocean ridges. Atmospheric carbon is used in the weathering of silicate minerals in a temperature-sensitive dissolution process; the products of this weathering are carried by rivers to the oceans. Carbonate sedimentation extracts carbon from the oceans and ties it up in the form of limestones. Pelagic limestones deposited in the deep ocean can be subducted and melted. Limestones deposited on continental crust are recycled much more slowly: if they are exposed and weathered, their remains may end up as pelagic carbonates; if they get caught up in a continental collision, they can be metamorphosed, liberating their CO_2 [CAR04]

slowly transported and subducted under the continental crust. At the high temperatures and pressures then encountered, silicate rocks are formed from these materials, which finally release carbon dioxide through slow leakage or violently via volcanic eruptions, thereby completing the cycle.

This is a negative feedback process, which works to suppress any fluctuations in temperature to create a global thermostat. An increase in temperature would increase evaporation, increasing rainfall and dissolving more carbon dioxide, thereby reducing temperatures. Conversely, a drop in temperature would decrease evaporation, reducing rainfall and the rate of carbon dioxide removal from the atmosphere, thereby eventually increasing temperatures.

There are other carbon-transporting processes that occur over a much shorter time frame. Carbon dioxide from the atmosphere dissolves in surface waters. Some of this carbon dioxide will remain as dissolved carbon dioxide but the rest will react to form other chemical compounds, including the living forms mentioned above. The un-reacted dissolved carbon dioxide is transported by ocean currents which, when they reach colder northern areas, will sink towards the ocean depths and spread through the ocean basins, only returning to the surface hundreds of years later. As colder water can hold more dissolved carbon dioxide than warm water, large quantities of carbon dioxide are locked up at great depths for long periods of time, thereby slowing the growth in atmospheric carbon dioxide – albeit temporarily. This process is termed the *solubility pump*.

Terrestrial life also pays an important role. At the scale of individual plants, carbon is extracted from the atmosphere during photosynthesis and returned during decay. On a regional scale, vegetation distribution is sensitive to temperature and rainfall, and differing types of vegetation store different amounts of carbon with forests storing more than grasslands. Natural changes in vegetation can therefore alter the amount of carbon stored and the atmospheric concentration of carbon dioxide, thereby forming a feedback loop. Manmade deforestation can have the same effect. Figure 5.14 shows the overall carbon cycle including the role of fossil fuels.

Whatever the evidence for anthropogenic climate change might be, what is clear is that there is a strong relationship between the composition of the atmosphere and the climate of the planet. For many the existence of such a relationship, regardless of its details, is enough to suggest that altering the composition of the atmosphere might be ill advised. Figure 5.15 compares the size of the various carbon stores and the implication for atmospheric concentration if they were fully exploited.

The natural carbon cycle is incapable of responding instantaneously to a sudden release of carbon dioxide; if it were, the concentration of carbon dioxide in the atmosphere would not be rising today. So slow is the carbon cycle that even if we were suddenly to reduce our emissions the concentration would continue to rise and the final equilibrium would not be at pre-industrial levels, but at a much higher value. The same is true of other climate related phenomena, including mean global temperature and sea level (Figure 5.16). Table 5.5 shows where our current emissions are being stored.

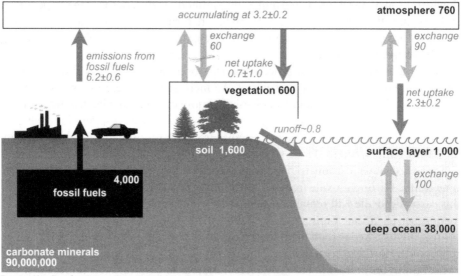

Figure 5.14   The world's overall carbon cycle (GtC) showing stores and flows [DTI00]

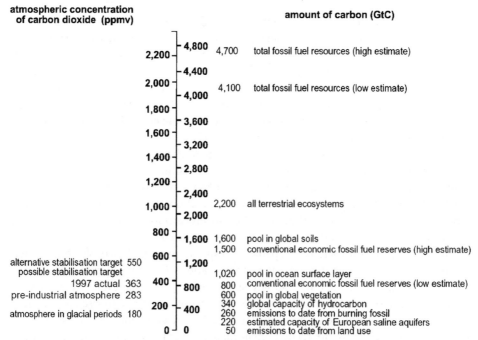

Figure 5.15   The size of various carbon stores and the impact on atmospheric concentration if they were instantaneously released [DTI00]

Table 5.5 Changes in the size of major carbon stores caused by anthropogenic emissions since the industrial revolution. Note how emissions and stores balance [adapted from IPCC00 and PEA03]

Emission or change in size of store	Gt of carbon
Total anthropogenic emissions from fuel burning and cement production 1850–1998	270 ± 30
Emissions resultant from land-use change	136 ± 55
Stored in the atmosphere	176 ± 10
Stored in oceans and forests $(1 + 2 - 3)$	230 ± 95

Other evidence and feedback processes

As we have seen, global mean temperatures have been rising. Sea levels have also been rising as the water in the oceans expands as it warms, and snow and ice cover has been lessening over a similar period. Levels of carbon dioxide are now rising much faster than they have done for at least one million years. Carbon dioxide is not however the only greenhouse gas. Surprisingly, water vapour is the most important greenhouse gas – simply because of the large quantity in the atmosphere. It is however one that humankind doesn't directly cause to be emitted, although it has a role within a feedback process and is therefore important. Methane (CH_4) is 23 times more powerful as a greenhouse gas per kilogram then carbon dioxide and is therefore described as having

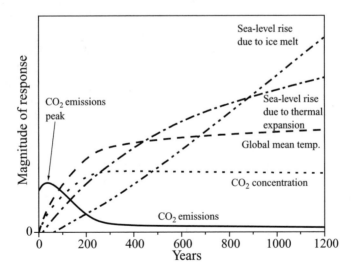

Figure 5.16 The response (or lack of it) of various phenomena to a reduction in anthropogenic carbon dioxide emissions [adapted from IPCC01 and PEA03]. Clearly, climate change will cause problems for future generations, whatever our ability to reduce emissions

a *global warming potential* [10] (GWP) of 23. It is released by both natural and man-made systems. Rice paddies, landfill sites, leaking gas mains and ruminant animals are all sources, as are waterlogged soils and marshes. It is also important as part of a feedback process that might amplify any small anthropogenic climate change: it is believed that the melting of areas of permafrost in northern Canada and other regions will release large quantities of the gas, further warming the planet and releasing yet more methane.

Most of the Arctic and nearly a quarter of all land in the northern hemisphere is underlaid by permafrost. The permafrost varies in depth but can extend to 1.5 km, its limit being set by heat rising from the lower regions of the earth. This geothermal heat means that permafrost is normally coldest at the surface and is seen to rise slowly in temperature with depth. This situation is now changing with the coldest regions far below the surface, suggesting that the surface of the planet is indeed warming. This is a clear sign of climate change. As we have seen, air temperature records fluctuate rapidly (even hourly); permafrost acts as a low-pass filter, removing these fluctuations, to expose any long-term trends.

The very top of the permafrost is called the *active layer* and can be 1 cm to 1 m deep; this freezes in winter, but thaws in summer to allow trees and smaller plants to grow. Because the temperature is still very low even in the active layer, dead plants do not fully decompose, but are covered over and eventually reach true permafrost depth where they sit in suspended animation. Grass that is still green has been found in permafrost dating back to the last ice age (over 100 000 years ago) [KOL05]. This biomass represents a vast store of carbon. The size of the store is unknown but might be as high as 450 GtC [KOL05].

The melting of permafrost soils and the subsequent production of methane form a *positive feedback* process. Negative feedback processes have also been suggested, such as the possible increase in cloud cover that a warmer world might cause through increased evaporation from the oceans. With more clouds, more of the sun's incident radiation would be reflected back into space, thereby cooling the planet; however clouds also reflect infrared radiation back down toward the planet, thereby warming it. Given the relative stability of the earth's climate over many millions of years, where rising temperatures have been reversed over long periods, climate researchers have concluded that both positive and negative feedbacks are critical to our thermal environment, and that a full understanding of them will be required before precise predictions of climate change can be made. The reduction of snow and ice cover in a warming world will also amplify the original temperature increase. The observed melting of polar sea ice is particularly worrying in this regard. Snow has a very high albedo (around 0.85), sea water a very low one (about 0.07). Therefore any reduction in the area of sea ice equates to the replacing of a highly reflective surface with one that absorbs most of the light (heat) from the sun: a clear positive feedback.

---

[10] A degree of caution is required here. Different gases last for different time periods in the atmosphere, so the GWP of a gas will vary depending on whether a 20, 100 or 200 year horizon is chosen. Also, GWPs are sometimes given on the basis of an identical number of molecules and sometimes on the basis of an equivalent mass of compound.

Ocean-centred processes provide both positive and negative feedbacks. Oceans provide most of the vapour in the atmosphere and they transport large quantities of heat to the atmosphere through latent heat. However, they also slow global warming by acting as a heat store for any initial warming. Most worrying, and difficult to predict, is their role in transporting heat around the planet in the form of ocean currents, such as the Gulf Stream. Such currents are believed to be sensitive to quite small changes in temperature and could either stop or change direction; changes which could have serious and very rapid effects on local and regional weather, fish stocks and on the global climate. Although the natural world and humankind are threatened by gradual temperature rises from global warming, it is these, non-linear effects that may prove the most damaging. They are also the most difficult to predict. Their importance is often underplayed. This is not due to their being less important than radiative forcing but simply a reflection of our current level of understanding. Two of the most talked about examples are the switching-off of the North Atlantic heat pump and the collapse of Antarctic ice sheets.

The North Atlantic heat pump refers to the observation that the Atlantic Ocean transports heat much further north than the Pacific Ocean. Ocean currents are mainly driven by two forces: winds and ocean density differences. The portion of ocean flow that is driven by density is called the *thermohaline circulation* (temperature and salinity together determine the density of ocean water). Global thermohaline circulation is sometimes described as the *great ocean conveyor belt* – with warm, less dense waters flowing in one direction at the sea surface and cold, dense waters flowing in the opposite direction in the deep ocean (Figure 5.17). The critical points of this conveyor belt are where surface waters sink into the deep ocean. This happens along the Antarctic shelf and at two sites in the northern North Atlantic (the Labrador Sea and the Nordic Seas). In these two North Atlantic sites, as the ocean loses its heat to the atmosphere, the surface waters become so cold (and therefore dense) that they sink into the deep ocean and then flow over the seafloor toward the equator. The dense waters that are exiting to the south in the deep ocean have to be replaced. This draws warm, surface currents farther north than they would normally flow and pumps additional heat to high northern latitudes [WHO04].

If conditions change in the North Atlantic Ocean such that surface waters can no longer become dense enough to sink, then the conveyor belt would slow or possibly stop altogether. The most likely agent of change is extra fresh water added to the ocean's sinking sites. If too much fresh water is added from melting ice and/or increased precipitation, then no matter how cold the surface waters become, they cannot become dense enough to sink.

Ice core data and deep-sea sediments provide evidence that the circulation has changed in the geologic past. The most prominent event occurred about 12 000 years ago, when the Atlantic heat pump ceased for a period of approximately 1000 years. It has also been speculated that changes in North Atlantic circulation may have contributed to a widespread cooling from 1300 AD to 1800 AD called The Little Ice Age [WHO04].

The rapid collapse of the Larson B ice shelf in Antarctica is a particularly worrying sign. In 2002, 3250 km^2 of 200 m thick ice disintegrated into smaller chunks and made

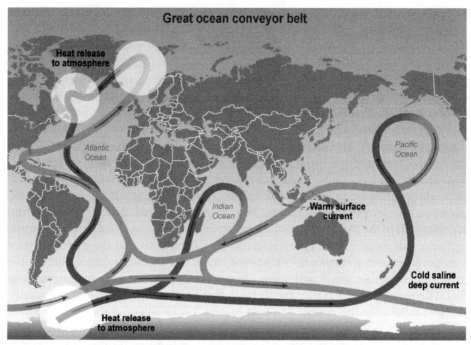

Figure 5.17 Major oceanic circulation patterns – the 'Great Ocean Conveyor Belt'. Light grey flows show warm surface currents; dark grey, cold, deep ocean ones. Warm, salty, surface waters lose their heat to the atmosphere at the points labelled 'heat release to atmosphere' and thereby sink [IPC01]

its way out to sea over a period of only a few months (Figure 5.18). The shelf sits in an area that is experiencing a five times greater temperature rise from global warming compared to the planetary average [NEW02]. Although this event will not raise sea levels directly, the shelf was holding back numerous glaciers. These are now flowing at three to eight times their previous rate, melting into the sea and raising sea levels. If the much larger glaciers in Western Antarctica suffer the same fate then sea levels could rise by more than a metre [NEW04].

Problem 5.3 The specific heat capacity of air and water are 1 kJ/kg and 4.2 kJ/kg respectively. Assuming the atmosphere is 40 km high with an average density of 0.2 kg/m^3 and the radius of the earth is 6370 km, estimate the depth of ocean that has the same thermal capacity as the whole of the atmosphere. This should give you some idea of how large a thermal store the whole ocean might be[11].

---

[11] The actual effectiveness of the whole ocean as a thermal store is complicated by the way the various layers of the ocean exchange heat and this exercise should only be taken as illustrative. A slightly more accurate answer is obtained by using the specific heat capacity of seawater, 3.8 kJ/kg.

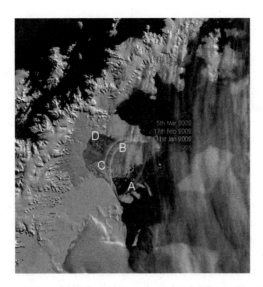

Figure 5.18 Collapse of the Larson B ice shelf in 2002 [ANT02]. The lines and letters show the position of the front of the shelf at various dates (A = 1995, B = 31st January 2002, C = 17th February 2002, D = 5th March 2002. Reproduced from http://www.antarctica.ac.uk/News_and_Information/Press_Releases/2002/20020319.html, British Antarctic Survey.

Although the existence of feedbacks makes the modelling of climate change more difficult, the situation is not quite as complex as it might seem. The feedbacks are driven via a change in temperature, not directly from the changing concentration of a particular greenhouse gas. This means that the feedback processes are generally blind to the original reason for the initial temperature rise. So, for example, the amount of increased evaporation from the oceans will not depend on whether the initial temperature rise was caused by elevated CO_2 levels or by increased emissions of methane. If the two possibilities lead to the same radiative forcing, then they will give rise to the same amplification by feedback processes. This means that most feedbacks can be separated from initial causes, and one can keep the discussion, and developing science of feedbacks, separate from the much more easily understood process of radiative forcing. So, if one or more scenarios exist for future anthropogenic emissions of greenhouse gases, then their impact via radiative forcing can be discussed in isolation and a predicted temperature rise obtained. A constant amount of feedback, or amplification, can then be added in later. This means that in rank ordering the effect of various emission levels of various gases, feedback processes can be ignored. Also, if in the future the science of feedback estimation improves, only the feedback constants need be changed: the rank ordering won't change.

Other greenhouse gases

As Table 5.1 showed, there are several other important anthropogenic greenhouse gases including ozone, chlorofluorocarbons and nitrous oxide.

Nitrous Oxide (N_2O), which has a GWP of 275, is also released by both natural and made-made processes; the most important being the increased use of fertilizers, nylon manufacture, deforestation and combustion of any fuel in air.

Although the majority of the concern about ozone (O_3) has been because of its destruction in the upper atmosphere over Antarctica at certain times of the year, creating the so-called ozone hole, it is also a greenhouse gas when found in the lower atmosphere. High-level ozone is critical as a filter of ultraviolet light, which would otherwise reach the surface and be harmful to life, hence the concern over the ozone hole. Naturally it is rare in the lower atmosphere, however there are now anthropogenic emissions, particularly from the action of sunlight on vehicle exhausts.

Like ozone, CFCs (Chlorofluorocarbons) are more frequently discussed in terms of the ozone hole (where they are the cause of ozone destruction) but they, and many of their more modern substitutes, are also greenhouse gases. Their GWP depends on the particular CFC in question, but values in the range 4500 to 7100 are typical.

Table 5.6 shows the GWP of the various anthropogenic greenhouse gases.

Although the GWP of each gas is important, the value only refers to a single molecule (or unit mass) of the gas compared with a single molecule (or unit mass) of carbon dioxide. The total number of molecules of each gas in the atmosphere at any time is also critical to its contribution to the greenhouse effect. Table 5.7 summarizes the role

Table 5.6 Global warming potentials and lifetimes of anthropogenic greenhouse gases, relative to 1kg of CO_2. GWPs calculated for different time horizons show the effects of atmospheric lifetimes of the different gases [data from GLO04]. There are many gases classified as chlorofluorocarbons, ethers etc. and only illustrative formulae and GWPs are given for these families; for compound specific values see [GLO04]. Data from University of Michigan, http://www.globalchange.umich.edu/globalchange1/current/lectures/samson/global_warming_potential/

GAS	Lifetime (Years)	Global warming potential (Time horizon in years)		
		20 yrs	100 yrs	500 yrs
Carbon Dioxide, CO_2	120	1	1	1
Methane, CH_4	12	62	23	7
Nitrous Oxide, N_2O	114	275	296	156
Chlorofluorocarbons				
CFC-11 etc.	55–550	4500–7100	3400–7100	1400–8500
Hydrofluorocarbons				
HFC-23 etc., CHF_3 etc.	0.5–260	40–9400	12–12000	4–10000
Fully fluorinated species				
SF_6, C_2F_6, etc.	2600–50000	5900–15100	5700–22200	8900–32400
Ethers and Halogenated Ethers				
CH_3OCH_3	0.015	1	1	$\ll 1$
HFE-125 etc., CF_3OCHF_2 etc.	0.22–150	99–12900	30–14900	9–9200

Table 5.7 Percentage contribution of the main anthropogenic greenhouse gases to the enhanced greenhouse effect using 100 year GWPs [data from IPC95]. Reproduced from Houghton, J.T. (ed.), Climate Change 1995: the Science of Climate Change, contribution of Working Group I to the second assessment report of the IPCC ©1996 Cambridge University Press

Gas	Main anthropogenic sources	Contribution (%)
CO_2	Energy use, deforestation and changing land use, cement production	65
CH_4	Energy production and use, enteric fermentation, rice paddies, wastes, landfills, biomass burning, domestic sewage	20
CFCs & HCFCs	Industrial: primarily refrigeration, aerosols, foam blowing, solvents	10
N_2O	Fertilized soils, land clearing, acid production, biomass burning, combustion of fossil fuels	5

of each gas by showing its contribution to the enhanced greenhouse effect, taking into account not only its GWP, but also its concentration in the atmosphere – carbon dioxide is seen to be the most important compound. As was mentioned earlier, the main anthropogenic source of carbon dioxide is energy use, however not all fuels or technologies release the same amount of carbon per unit energy. Table 5.8 shows that for electricity generation some technologies are ten times less polluting than others in terms of climate change.

There are other compounds that are *negative* greenhouse gases, most notably sulphate aerosols produced by the oxidation of sulphur dioxide. Such aerosols (which are toxic to plants) scatter incoming radiation in all directions including back into space, thereby reducing surface heating. Volcanic eruptions produce very large quantities, which persist in the atmosphere for several years. During this time they will have a negative radiative forcing – effectively cooling the planet. The 1991 eruption of Mt Pinatubo created

Table 5.8 Greenhouse gas emission intensities for power generation. Date from Boyle, G., Everett, B., and Ramage, J., Energy Systems and Sustainability, the Open University, Oxford University Press, 2003

Conventional coal	Advanced coal	Oil	Gas	Nuclear	Biomass	Photovoltaic	Hydroelectric	Wind
960–1300	800–850	690–870	460–1230[a]	9–100	37–166[a]	30–150	2–410	11–75

Note: (a) Natural gas and biomass fuel cycles were also analysed in cogeneration configurations, with the heat produced credited for displacing greenhouse gas emissions from gas heating systems. That approach reduced greenhouse gas emissions to 220 grams of carbon dioxide equivalent per kilowatt-hour for natural gas, and to minus 400 grams of carbon dioxide equivalent per kilowatt-hour for biomass [data from J. P. Holden and K. R. Smith in BOY03]

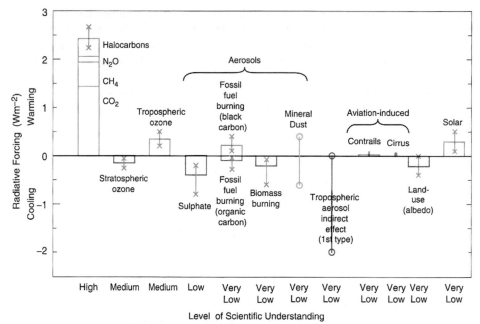

Watson, R. (ed.), Climate Change 2001 – Synthesis Report, third assessment report of the IPCC, © IPCC

Figure 5.19 Radiative forcing from various compounds and activities relative to the pre-industrial situation (1750) [IPC01]. Note the imprecision in some values as indicated by the vertical lines

about 25 million tonnes of sulphate aerosol and a radiative forcing of $-4\ \text{W/m}^2$ (note the minus sign) reducing over a two year period to $-1\ \text{W/m}^2$. Over this time, this negative forcing would have cancelled the positive forcing from all anthropogenic greenhouse gas emissions, and indeed, global temperatures fell in 1992 by about 0.4 °C. Figure 5.19 shows the radiative forcing from various compounds and activities. The situation is clearly a complex one with both positive and negative forcings.

Glancing back to Figure 5.2b, we see that much of the atmospheric spectrum is already completely opaque to infrared radiation, i.e. there is already 100 per cent absorption at these wavelengths. This means that increasing the concentration of greenhouse gases cannot lead to substantial increases in radiative forcing at these wavelengths. The only possibility for additional forcing is in the much smaller part of the spectrum that has less than 100 per cent absorption. Looking at the absorption spectra of the important greenhouse gases (Figure 5.2c), we see that some gases are more able to fulfil this requirement than others. Much of the carbon dioxide absorption spectrum is already saturated, so increasing carbon dioxide levels has a smaller forcing effect than one might expect. In fact there is a logarithmic relationship: if doubling the concentration of carbon dioxide in the atmosphere increases average global temperatures by 2 °C, doubling it again will only raise the temperature by a further 2 °C, not a further 4 °C. For other gases, the absorption spectra are not near saturation and there is a more powerful linear relationship between the amount of gas added to the atmosphere and the resultant radiative forcing. The following expressions [IPC05] can be used to roughly estimate the change in radiative forcing, δF (W/m^2), provided by a change in concentration of the major greenhouse gases.

CO$_2$	$\delta F = \alpha \ln(C/C_0)$	$\alpha = 5.35$
CH$_4$	$\delta F = \alpha(\sqrt{M} - \sqrt{M_0}) - (f(M, N_0) - f(M_0, N_0))$	$\alpha = 0.036$
N$_2$O	$\delta F = \alpha(\sqrt{N} - \sqrt{N_0}) - (f(M_0, N) - f(M_0, N_0))$	$\alpha = 0.12$
CFC-11[12]	$\delta F = \alpha(X - X_0)$	$\alpha = 0.25$
CFC-12	$\delta F = \alpha(X - X_0)$	$\alpha = 0.32$

where $f(M, N) = 0.47 \ln[1 + 2.01 \times 10^{-5}(MN)^{0.75} + 5.31 \times 10^{-15} M(MN)^{1.52}]$, C is CO$_2$ in ppm, M is CH$_4$ in ppb, N is N$_2$O in ppb, X is CFC in ppb and the subscript $_0$ indicates the previous (usually the pre-industrial) concentration.

Problem 5.4   Use the above equations to estimate the change in radiative forcing produced by the known change in concentration of carbon dioxide, methane and nitrous oxide since pre-industrial times.

The final change in global mean temperature, δT, reached by such changes in radiative forcing can then be estimated from [DEF05]:

$$\delta T = \varphi \delta F$$

where the constant of proportionality, φ, determines the sensitivity of the climate to changes in radiative forcing. Unfortunately it has proved difficult to identify the value of φ [DEF05] and in truth the presence of feedback processes is likely to mean the relationship is not strictly proportional. However, estimates of δT since pre-industrial times range from 1.5 to 4.5 K, and Problem 5.4 suggests that δF is around 2.4, therefore φ might be expected to be in the range 0.6 to 1.9. (See the book's web site for a simple climate model that uses these equations.)

We have yet to discuss in detail the likely effect of predicted temperature or sea level rises, or indeed, how large each of these will be. In order to do this, we need to have a good idea of the current and likely future emissions of greenhouses gases, the majority of which come from energy related activities – particularly the use of fossil fuels. These activities form the subject matter of Part II, and so the discussion of the impacts of a warmer world, and the response of the international community will be left until after this.

---

[12] The same expression can be used for all CFCs and CFC replacements, but with different values for α.

5.3 Other concerns

This book is largely motivated by how the way we derive our energy affects the global environment; however the concept of sustainability also includes concerns about local environments and their occupants.

Although climate change is probably the most important environmental concern in the long-term, other worries, such as accidents, can have a more immediate impact on individuals and communities. Some concerns are specific to a technology or fuel and will be discussed under the relevant technology or fuel; others are more general and therefore best considered together. Some of the resultant comparisons we will be forced to make are based on fairly accurate data, for example the emissions of various pollutants per GWh of generation; others, such as the death rate per GWh generated, much less so.

We will focus on the following four concerns: general pollution, acid rain, injury and fatality, and land-use. Then consider two frameworks for discussing some of these concerns: energy payback estimations and external cost calculations. Local air quality will be considered in Chapter 14 when discussing transport technologies.

General pollution

The term 'pollution' means different things to different people and in different contexts. Often the word is used to specifically identify the release of chemical compounds harmful to human health. Sometimes a much broader definition is used, giving rise to such terms as 'visual pollution' or 'noise pollution'. Greenhouse gases sit between these two extremes, being in general non-poisonous in the concentrations typically encountered, but having long-term consequences for humankind and other species. Our definition of pollution will reflect this diversity by staying somewhat fluid in order to echo the subject matter of the chapter in hand.

Table 5.9 lists a number of impacts, including pollution, to which various energy technologies can give rise and Table 5.10 attempts to quantify this pollution and compare it to the natural flow of relevant compounds and the emissions from non-energy industry related activities. Neither table is exhaustive, but both clearly show that energy transformation and use has a variety of impacts and that for a series of pollutants the energy industry is a major emitter. The natural baseline given in Table 5.9 represents emissions from non-anthropogenic sources and the *human disruption index* gives the ratio of the anthropogenic flow to the non-anthropogenic one. So for example, 160 million tonnes of methane is naturally released to the atmosphere annually and a further 2.3×160 million $= 368$ million tonnes is added from human activities.

This book concentrates on the impact energy production is having on global climate and we do not have the space to discuss each air-born and water-born pollutant in turn, nor their detailed impact on ecosystems. However, we shall return briefly to many of the pollutants and impacts in the coming chapters as we investigate the various energy

Table 5.9 Impacts and pollutants from various energy sources [data from BOY03, RAM97]

Source	Potential causes for concern
Oil	Global climate change, air pollution by vehicles, acid rain, oil spills, oil rig accidents
Natural gas	Global climate change, methane leakage from pipes, methane explosions, gas rig accidents
Coal	Global climate change, acid rain, environmental spoliation by open-cast mining, land subsidence due to deep mining, spoil heaps, ground water pollution, mining accidents, health effects on miners
Nuclear power	Radioactivity (routine release, risk of accident, waste disposal), misuse of fissile and other radioactive material by terrorists, proliferation of nuclear weapons, land pollution by mine tailings, health effects on uranium miners
Biomass	Effect on landscape and biodiversity, ground water pollution due to fertilizers, use of scarce water, competition with food production
Hydroelectricity	Displacement of populations, effect on rivers and ground water, dams (visual intrusion and risk of accident), seismic effects, downstream effects on agriculture, methane emissions from submerged biomass
Wind power	Visual intrusion in sensitive landscapes, noise, bird strikes, interference with telecommunications
Tidal power	Visual intrusion and destruction of wildlife habitat, reduced dispersal of effluents (these concerns apply mainly to tidal barrages, not tidal current turbines)
Geothermal energy	Release of polluting gases (SO_2, H_2S, etc.), ground water pollution by chemicals including heavy metals, seismic effects
Solar energy	Sequestration of large land areas (in the case of centralized plant), use of toxic materials in manufacture of some PV cells, visual intrusion in rural and urban environments

options open to us. Of course some forms of pollution from our use of hydrocarbons are more obvious, oil spills (Figure 5.20) being an obvious example.

Acid Rain

Unpolluted rainwater has a pH[13] of around 5.6, i.e. slightly acidic. This natural acidity arises mainly from the formation of carbonic acid (H_2CO_3) from carbon dioxide in the atmosphere and is part of the global carbon cycle:

$$CO_2(gas) + H_2O(liquid) \leftrightarrow H_2CO_3(aqueous).$$

Since carbonic acid is a weak acid, it partially dissociates:

$$CO_2(g) + H_2O(l) \leftrightarrow H^+(aq) + HCO_3^-(aq).$$

---

[13] The pH of a substance is a measure of its acidity/alkalinity and spans the range 0 (highly acidic) to 14 (highly alkaline). A pH of 7 indicates a substance is neither acidic nor alkaline.

Table 5.10 Global loading from various pollutants and human disruption [from BOY03 after HOL00]. Data from Boyle, G., Everett, B., and Ramage, J., Energy Systems and Sustainability, the Open University, Oxford University Press, 2003

Insult	Natural baseline (tonnes per year)	Human disruption index(a)	Share of human disruption caused by			
			Commercial energy supply	Traditional energy supply	Agriculture	Manufacturing, other
Lead emissions to atmosphere (b)	12000	18	41% (fossil fuel burning, including additives)	Negligible	Negligible	59% (metal processing, manufacturing, refuse burning)
Oil added to oceans	200 000	10	44% (petroleum harvesting, processing, and transport)	Negligible	Negligible	56% (disposal of oil wastes, including motor oil changes)
Cadmium emissions to atmosphere	1400	5.4	13% (fossil fuel burning)	5% (traditional fuel burning)	12% (agricultural manufacturing, refuse burning)	70% (metals processing)
Sulphur emissions to atmosphere	31 million (sulphur)	2.7	85% (fossil fuel burning)	0.5% (traditional fuel burning)	1% (agricultural burning)	13% (smelting refuse burning)
Methane flow to atmosphere	160 million	2.3	18% (fossil fuel harvesting and processing)	5% (traditional fuel burning)	65% (rice paddies, domestic animals, land clearing)	12% (landfills)
Nitrogen fixation (as nitrogen oxide and ammonium)(c)	140 million (nitrogen)	1.5	30% (fossil fuel burning)	2% (traditional fuel burning)	67% (fertilizer, agricultural burning)	1% (refuse burning)
Mercury emissions to atmosphere	2500	1.4	20% (fossil fuel burning)	1% (traditional fuel burning)	2% (agricultural burning)	77% (metals processing, manufacturing, refuse burning)

Di-nitrogen oxide flows to atmosphere	33 million	0.5	12% (fossil fuel burning)	8% (traditional fuel burning)	80% (fertilizer, land clearing, aquifer disruption)	Negligible
Particulate emissions to atmosphere	3100 million (d)	0.12	35% (fossil fuel burning)	10% (traditional fuel burning)	40% (agricultural burning)	15% (smelting, non-agricultural land clearing, refuse)
Non-methane hydrocarbon emissions to atmosphere	1 billion	0.12	35% (fossil fuel processing and burning)	5% (traditional fuel burning)	40% (agricultural burning)	30% (non-agricultural land clearing, refuse burning)
Carbon dioxide flows to atmosphere	150 billion (carbon)	0.05(e)	75% (fossil fuel burning)	3% (net deforestation for fuelwood)	15% (net deforestation for land clearing)	7% (net deforestation for lumber, cement manufacturing)

Notes: The magnitude of the insult is only one factor determining the size of the actual environmental impact. (a) The human disruption index is the ratio of human-generated flow to the natural (baseline) flow. (b) The automotive portion of anthropogenic lead emissions in the mid-1990s is assumed to be 50 per cent of global automotive emissions in the early 1990s. (c) Calculated from total nitrogen fixation minus that from di-nitrogen oxide (nitrous oxide). (d) Dry mass. (e) Although seemingly small, because of the long atmospheric lifetime and other characteristics of carbon dioxide, this slight imbalance in natural flows is causing a 0.5 per cent annual increase in the global atmospheric concentration of carbon dioxide

(a)   (b)

(c)   (d)

Figure 5.20   The sinking of the Exxon Valdez in Price William Sound, Alaska in March 1989 released 40 million litres (eleven million US gallons) of crude oil into the sea. The oil impacted 2080 km (1300 miles) of coastline, 200 miles heavily. At its peak 10 000 staff were involved in the clean-up. It has been estimated that 250 000 seabirds, 2800 sea otters, 300 harbour seals, 250 bald eagles and up to 22 killer whales, died [EXX04]. Clockwise: the vessel surrounded by a containment boom, cleaning one of the beaches, oiled cormorant, oiled and dead sea otter. Photos courtesy of the Exxon Valdez Oil Spill Trustee Council

The burning of fossil fuels and other industrial processes can lead to the formation of other acids and hence a further reduction in the pH of rainwater. As the pH scale is logarithmic, a change of one point on the scale implies a ten-fold change in the concentration of excess hydrogen ions, and so even a small change in pH can have a dramatic effect on the local chemistry, particularly the chemistry of living organisms.

The major culprits are the emission of sulphur dioxide (SO_2), which is a precursor of sulphuric acid, and nitrogen oxides (NO_x), which are precursors of nitric acid. The resultant acid 'rain' can either fall as genuine rain or be in the form of a *dry deposition* of fine particles, which later form acids when they dissolve in rain or river water. Chemically,

$$SO_2(g) + H_2O(l) \leftrightarrow H_2SO_3(aq) (\text{i.e. the production of sulphurous acid})$$

or

$$2SO_2(g) + O_2(g) \rightarrow 2SO_3(g) \text{ then}$$

$$SO_3(g) + H_2O(l) \rightarrow H_2SO_4 (aq) \text{ (i.e. the production of sulphuric acid)}$$

and

$$2NO_2(g) + H_2O(l) \rightarrow HNO_2(aq) + HNO_3(aq) (\text{i.e. the production of nitrous}$$

and nitric acid).

Sulphur dioxide and nitrogen oxides are no respecters of national boundaries and the acid rain which falls within an area is likely to have originated some distance away, possibly in another country. In Europe the hardest hit region is Scandinavia – from remote emissions in Germany, the UK and France. In North America, Canada has suffered greatly from emissions originating within the USA.

The impacts are diverse and include: damage to buildings[14]; the widespread death of plants, including trees; the removal of nutrients such as magnesium, calcium and potassium from soils; the release of toxic metals such as aluminium from soils; a reduction, often to zero, in reproduction and survival of fish, frogs aquatic insects and other species; and calcium deficiency in fish leading to deformed or dwarfed individuals.

Below a pH of 4.5 or 5, neither newly hatched fish nor aquatic plants are likely to survive and lakes will slowly 'die', leaving only some species of algae, moss and fungi. Figure 5.21 shows measured surface-water pH in the USA, and Figure 5.22 the effect acid rain can have on a forest. The scale of the problem depends not only on the amount of deposition but also on factors such as the type of soil and rock base onto which it falls. Limestone soils and rock bases tend to neutralise the acidity; granite and sandstone scarcely react with such acids and therefore offer little in the way of a neutralising effect.

As Table 5.9 shows, 85 per cent of sulphur emissions are from fossil fuel burning and thus acid rain is largely a problem caused by our reliance on such fuels. The amount of sulphur dioxide and nitrogen oxides released by the burning of fossil fuels depends on the fuel, the combustion technology and the emission control technology employed. The sulphur content of coal can range from 0.5 to 5 per cent and therefore the combustion of high sulphur coal (so called 'dirty coals') in power stations with little emission control can add greatly to the acid burden. Oil can also contain significant quantities of sulphur (0.2 to 2 per cent). Natural gas is processed to remove almost all the sulphur, yet if this processing is carried out with poor emission control, then large amounts of sulphur can be released in the process. Although it is possible to reduce sulphur dioxide emissions

---

[14] By reacting with the calcium carbonate to form soluble calcium hydrogen carbonate (calcium bicarbonate, $Ca(HCO_3)_2$): $CaCO_3$ + acid rain $\rightarrow Ca(HCO_3)_2(aq)$.

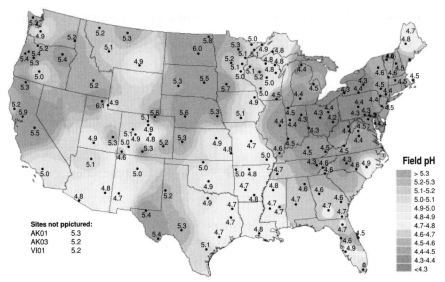

Reproduced courtesy of National Atmospheric Deposition Program (NRSP-3). 2006. NADP Griffith Dr., Champaign, IL 61820

Figure 5.21 Surface pH within the USA [NAT99][15] (For a colour reproduction of the figure, please see the colour section located towards the centre of the book)

Reproduced from Urban Options, photo by Paul Donahue: http://www.urbanoptions.org/RenewableEnergy/AcidRain.htm. Change caption to: Victims of acid rain - dead and dieing Red Spruce in Maine, USA.

Figure 5.22 Victims of acid rain – dead and dieing Red Spruce in Maine, USA (photo by Paul Donahue Urban Options [UOP07][16]).

---

[15] Source: National Atmospheric Deposition Program (NRSP-3), 2006. NADP Program Office, Illinois State Water Survey, 2204 Griffith Dr., Champaign, IL 61820.

[16] Source: T. Mabry, University of Texas.

Table 5.11 SO_2 emissions from various generating
technologies [data from BOY03]. Date from Boyle, G., Everett,
B., and Ramage, J., Energy Systems and Sustainability, the
Open University, Oxford University Press, 2003

Technology	SO_2 t/TWh
Hydro with reservoir	7
Diesel (0.25% S)	1285
Heavy oil (1.5% S) without SO_2 scrubbing	8013
Hydro run-of-river	1
Coal (1% S) without SO_2 scrubbing	5274
Coal (2%) with SO_2 scrubbing	104
Nuclear	3
Natural gas (delivery + 2000 km) combined cycle	314
Fuel cell (H_2 from gas reforming)	470
Biomass plantation	26
Sawmill wastes	26
Wind power	69
PV	24

by 90 per cent during the combustion of fossil fuels, the increased cost of doing so has limited the uptake of such technology.

For oil and gas the emissions of NO_x arise not from the fuel but from the high concentration (70 per cent) of nitrogen in the air used for combustion. In general, these emissions will be greater the higher the combustion temperature. Coal also contains significant amounts of nitrogen, increasing its NO_x emissions. Although the burning of biomass gives rise to little in the way of SO_2, it can release large quantities of NO_x and therefore form a significant source of acid precipitation.

The introduction of tighter legislation on emissions and the switch to natural gas has led to reductions in acid emissions from energy production in the developed world. Emissions in Africa and China are growing, suggesting that those regions may eventually suffer the problems of acid deposition found in northern Europe and North America.

Table 5.11 compares the SO_2 emissions of various energy technologies and points to the sustainable technologies discussed in Part IV having much lower emissions than their traditional counter-parts. Table 5.12 attempts the same analysis for NO_x (somewhat less successfully due to the greater variability caused by the effect of differing combustion possibilities.)

Injuries and fatalities

Clearly the idea of widespread injury or fatality from accidents clashes with our thoughts on sustainability. Having said that, it is also unlikely that energy production and supply is

Table 5.12   Range of NO_x emissions found in the literature from various generating technologies [adapted from data in BOY03]

Technology	NO_2 t/TWh (min. approx.)	NO_2 t/TWh (max. approx.)
Hydro with reservoir	150	150
Diesel (0.25% S)	310	12 000
Heavy oil (1.5% S) without SO_2 scrubbing	1300	2000
Hydro run-of-river	120	120
Coal (1% S) without SO_2 scrubbing	700	5000
Coal (2%) with SO_2 scrubbing	690	5000
Nuclear	150	150
Natural gas (delivery + 2000 km) combined cycle	77	1500
Biomass plantation	1100	2500
Sawmill wastes	69	1900
Wind power	77	310
PV	150	150

likely to be intrinsically safer than other industrial activities. This suggests that we should investigate the relative number of serious injuries and fatalities for each of the major energy technologies in relation to each other and with reference to industrial averages.

For the UK, [NAT02] reports an average of nine fatalities per annum for mining, quarrying and offshore production within energy related industries, and 366 serious injuries. The supply of electricity, gas and water add a further three fatalities and 215 serious injuries. These numbers are quite low given that there are around 240 workplace deaths in the UK per annum [HEA05]. The energy industry represents four per cent of the economic activity of the country [DTI03], and from the above we can see that the number of energy-related deaths is 12/240 = 0.05, or five per cent of the workplace total. Therefore the energy industry would not seem to be particularly dangerous. However, some energy technologies would appear to be more dangerous than others. Table 5.13 shows that oil and coal apparently result in a much higher fatality rate than,

Table 5.13   Deaths from power generation per GW_e year [data from W. D. Nordhaus in BOY03]. Date from Boyle, G., Everett, B., and Ramage, J., Energy Systems and Sustainability, the Open University, Oxford University Press, 2003

Fuel cycle	Occupational hazards per GW_eyr		Public (off-site) hazards per GW_eyr	
	Fatal	Non-fatal	Fatal	Non-fatal
Coal	0.2–4.3	63	2.1–7.0	2018
Oil	0.2–1.4	30	2.0–6.1	2000
Gas	0.1–1.0	15	0.2–0.4	15
Nuclear (LWR)	0.1–0.9	15	0.006–0.2	16

Figure 5.23 The Piper-Alpha oil platform in the North Sea, which exploded in July 1988 killing 167 (Image reproduced from the BBC News website http://newsbbc.co.uk/onthisday/hi/witness/july/6/newsid_3036000/3036510 (Accessed July 2006)

for example, nuclear power. One-off 'catastrophic' accidents can also alter the statistics in certain years, Figure 5.23 shows an example, and are not included in the table.

Reference [GIP01] suggests that there have been 20 industrial sector deaths from wind power and no injuries to the public. There have been approximately 200 recorded dam failures in the last 100 years outside of China, leading to an estimated 10 000 deaths. Within China, where a considerable number of dams have been built, reliable data is poor; however it is estimated [SUL95], that in 1975 alone a quarter of a million people may have died from a series of hydroelectric dam failures in the country.

Land use

Land used for energy production is land not available for other purposes, such as food production, other economic activities or recreation. For example, the construction of large dams during the twentieth century has flooded an area of half a million square kilometres (about the size of Spain) and led to the displacement of 30 to 60 million people [BOY03], although some of these projects were for irrigation and water supply purposes. Other technologies can also require substantial areas of land, for example, open cast mining and biomass growing. As later calculations will show, the un-concentrated nature of many sustainable energy resources implies very large areas of land will be required for energy production if such technologies are widely adopted. Table 5.14 shows that nuclear power has the lowest land area requirement per unit of delivered energy, although again, caution is required due to the numerous approximations and assumptions that go into such calculations. In the case of nuclear power, the land area required for long-term

Table 5.14 Land-area requirements of various generating technologies [adapted from data in BOY03]

Technology	km² per TWh (min. approx.)	km² per TWh (max. approx.)
Hydro with reservoir	0	200
Hydro run-of-river	1	5
Coal	4	10
Nuclear	0.5	5
Biomass plantation	533	2200
Sawmill wastes	1	3
Wind power	25	115
PV	30	45

storage and disposal can not be included because such technologies have yet to be developed; and although the land area required to do this is likely to be small, it may be required for thousands of years. Biomass technologies require the greatest land areas and could conflict with food production. Fossil fuel use only requires modest areas of land, but can affect much greater areas due to acid rain and climatic change. Wind power would appear to need substantial areas of land, however all but three per cent of this can normally be simultaneously used for other purposes such as agriculture. Photovoltaics (PV) would also seem to need large land areas, but this would be greatly reduced if roof-mounted systems were used.

Energy paybacks

Power stations, hydroelectric plants and oilrigs are not only producers of energy; they also use energy in their construction and operation. In the case of electricity production using heat, the efficiency will be very low. If we ignore this natural inefficiency by not including the primary energy supply – solar, wind, water or fossil fuel – but still count the energy required for construction, maintenance and provision of the fuel (what is commonly termed the *embodied energy*), we can estimate the *energy payback ratio* – the ratio of the energy produced during the plant's lifetime to that required to build, maintain and extract and transport the fuel to it. Accounting for the embodied energy is important because the energy used in the construction and maintenance cycle will typically be polluting in the various ways already discussed, even if the structure/machine constructed is a sustainable energy technology. An example being hydroelectric dam construction, where large quantities of fossil fuels will be used during the construction process. As we will see, the energy payback ratio can be surprisingly low. Table 5.15 shows this ratio for the major electricity technologies.

Hydropower clearly has the best ratio, with over 200 times as much energy being produced compared to the amount used in construction etc. All the fossil-fuel technologies have very low values, as does biomass (unless it is waste biomass). The value of 80 for

Table 5.15 Energy payback ratios for various electricity technologies (typical values for options available in North America [data from L. Gagnon in BOY03]. Data from Boyle, G., Everett, B., and Ramage, J., Energy Systems and Sustainability, the Open University, Oxford University Press, 2003

Technology	Energy output/ Energy input
Hydro with reservoir	205
Hydro run-of-river	267
Coal (1% S): without SO_2 scrubbing	7
Coal (2%) with SO_2 scrubbing	5
Nuclear	16
Natural gas (delivery + 2000 km): combined cycle	5
Fuel cell (H_2 from gas reforming)	3
Biomass plantation	5
Sawmill wastes	27
Wind power	80
PV	9

wind power is probably an over-estimate as it does not include the backup generation plant which needs to be constructed in order to provide guaranteed supply when there is insufficient wind.

External costs and sustainability

In general this book will not attempt to compare the costs of energy technologies. Such calculations are difficult and will naturally favour established technologies because of their pre-existing economies of scale. To make such comparisons fairly one would ideally like to establish a series of energy future scenarios and estimate their future costs. Unfortunately this is impossible as many of the technologies are in their infancy and it would be impossible to calculate their costs at the point in the future (possibly 30 years hence) when they have become widely adopted.

It is also true that in many countries much of the cost to the consumer of energy is a reflection of taxation policy, rather than a reflection of the true cost of the fuel. For example, in the UK and much of the world most of the cost of vehicle fuel is tax. This means that such an analysis would largely be a comparison of national tax policy rather than an analysis of energy costs. It is also possible that one of the central policies that will be used to tackle climate change in future will be some form of carbon tax – further complicating the analysis. What we can do however is to try and estimate each technology's *external cost*. This is the 'cost' of the pollution etc. that the technology creates. It includes for example, the costs of cleaning up oil spills, cleaning polluted buildings, the costs of adapting to climate change and the health cost of accidents and pollution. Many of these *externalities* represent problems or costs that will be born by

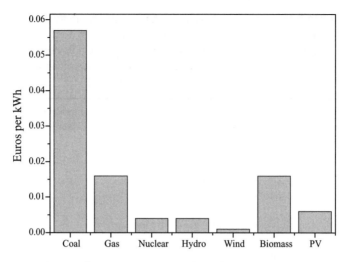

Figure 5.24 External costs of various technologies [data from European Commission ExternE Programme in BOY03]

later generations. Therefore they will, in general, need to be minimized in order not to conflict with our definition of sustainability.

Figure 5.24 shows the results from one such study. Other studies [GIP95, PEA03] have come up with different external costs – not surprisingly since, for example, we do not yet know the exact impact of climate change. However the ranking of the technologies would appear to be consistent across studies and suggests that in the UK the price of fossil fuel-based electricity needs to suffer a 50 per cent increase in cost if these externalities are to be included in the cost of the fuel. At present such externalities are mostly paid for by other forms of taxation, or health and building insurance – i.e. not at the point of energy use or purchase. Basic economic theory points to this being an inefficient use of money because it does not allow individuals or institutions to make rational decisions. For example, if fossil fuel-based electricity is much cheaper than wind power, most will opt to use fossil fuels, even though they are paying for the external cost of fossil fuels via other taxes and costs. This is because if they elect to switch to wind power their tax bill and other costs will not fall. However, if the money needed to clean up fossil fuel-based pollution is raised through the price of such fuels, then anyone electing to switch technologies might well be able to save money. Such savings go beyond those of the individual or business making the change, as the small reduction in pollution means a small reduction in external costs for everyone. This can lead to a more efficient use of money within the economy in general. The following problem attempts to illustrate this basic point.

Problem 5.5 A hypothetical country has two ice cream factories. They both produce a similar product, but one factory (factory A) does it at a lower cost but pollutes a nearby river. The cost of £2 per kg of ice cream sold, needed for cleaning up such pollution and the impact on the local fishing industry, is borne through taxation by the general population.

Factory B uses a different process that means it does not pollute its local river but this has doubled the cost of its ice cream. Factory A sells ice cream at £1 per kg and sells two million kilograms per year. Factory B sells ice cream at £2 per kg and because of its higher costs only sells half a million kilograms per annum.

Estimate the total money spent on ice cream in the country including tax under the current regime; and if Government intervention forces the external cost (the cost of cleaning up the river) to be born by the manufacturers – i.e. the externalities are *internalized*. This process can also be termed the principle of *the polluter pays*. Assume the total amount of ice cream sold does not change and that twice the amount of the cheaper product will be purchased relative to the more expensive one.

Although a simple economic analysis based on internalising external costs can indicate that it makes economic sense to make an environmentally sound choice, other analyses centred on a true cost of money can sometimes indicate the opposite. We will consider this difference in the next chapter.

5.4 Debating climate change and answering the sceptics

Within textbooks the topic of climate change is frequently discussed without presenting any opposing view, i.e. as though the science is well understood and the need for action almost self-evident. Yet in the press much is made of findings that do not seem to fit climate change theory, and of the views of 'climate sceptics'. Why this difference? Is it true that there are scientific results that disprove that humankind is altering the climate? Can the media be relied upon to publish a balanced viewpoint? Do they even see this as their role?

To answer these questions we need to have an understanding of the kind of questions science can answer and how scientists, politicians and the media phrase questions or make statements to add weight to their case. For an in-depth summary of the topic, with special reference to climate change see Andrew Dessler and Edward Parson's excellent book, The Science and Politics of Global Climate Change [DES06].

In order to see what climate scientists are trying to tell us, and what others might be doing to manipulate the results when debating climate change, we need to be able to dissect any question or statement into its component sub-questions or sub-statements and analyse each in turn. We can categorize statements, claims or questions as either *normative* or *positive*. A positive claim concerns the way things are, such as 'on average, men are taller than women'. Whereas a normative claim concerns how we think things should be, i.e. whether a thing is just, fair, right or wrong. An example being 'killing is wrong'. The first step in analysing any comment about climate change (or any other issue that involves a mix of science and political debate) is to ask whether the comment is normative or positive. Science can only address the latter and the former simply represents a viewpoint of the speaker.

For a statement to be positive does not require the answer to be known, just knowable at least in theory. So the statement 'the Western invasion of Iraq led to higher oil prices' is a positive statement open to analysis but might be very hard to prove.

For a statement to be normative does not require it to be controversial, for example 'all people should have access to clean drinking water' is, as a statement, unlikely to find many dissenters, but it is still normative. When trying to spot normative claims look for words such as 'should' or 'ought' or phrases that indicate a personal viewpoint is being expressed. Sometimes things can get a bit mixed, and it can be difficult to know if a positive or normative claim is being made. For example, the statement 'Germany should win the (Soccer) World Cup', might be positive, in that, it might be possible to construct a statistical model of team performance and show that the German team is the 'most likely' to win. However, the speaker may simply be expressing his, or her desire for Germany to win. The important thing is to be on guard against normative claims and to question whether the speaker is knowledgeable or qualified to answer any positive ones.

How science works

Many people have a highly romanticized view of science and of scientists, and imagine a solitary intellect battling for years on a problem then finally resolving the problem clearly and unequivocally. This is rarely if ever the case. Science is a collective activity and results are often published at early stages of an investigation to aid with further funding, to stake a claim on the ideas behind the work, to help other investigators, or because only a partial study was financed. A scientific case is finally proven by weight of evidence and over a long period of time. Initial theories are tested against an ever-growing number of experiments or data sets until it seems highly unlikely that the theory is not a true representation of the situation. This doesn't mean that theories cannot be shown to be incorrect at a later stage – sometimes they are – but that the likelihood of this is inversely proportional to the depth to which the theory has been probed by other, often competing, scientists. Well established theories based on numerical results are in fact very rarely overturned, they are much more often extended. Einstein's theory of relativity did not overturn Newton's work; it simply extended it towards newer more testing domains. Even today, Newton's methods are applied much more frequently by scientists and engineers than Einstein's, because Newton's equations describe the world in which we live accurately enough for almost all practical purposes. If a theory has qualitative predictive power that has been shown to work, then even if a newer theory comes along, the predictive power and results of the old theory are by definition still valid (as the old theory predicted the correct answers within the domain where it was tested). Apples didn't stop falling out of trees the day Einstein published his work.

This position of proved-by-weight-of-evidence is the situation we are now in with the science of climate change. An enormous number of peer-reviewed papers, which support the underlying theory, have been written by internationally respected scientists, and it seems almost impossible that the theory will change in any major way. This doesn't mean that there will not be surprises, for example the discovery of a new feedback process (see Section 5.2: other evidence and feedback processes) that might change the rate of warming. However, this doesn't mean that there have not been dis-senters. The number of dissidents has always been small, is becoming smaller, and their criticisms are becoming less credible as the weight of scientific evidence grows. They have had to reduce the scope of their attacks, with most now agreeing that the world is warming and that humankind is part of the reason. Most serious attacks on the science are about how fast the world will warm in future and the role of feedback processes.

How political debate and the media work

Policy debates in the public arena are not subject to the rules of science. A politician or a journalist is not required to produce a clear positive statement, carry out in-depth research over several years, make sure any results are mathematically consistent, subject themselves to peer review and await the results of further research by others. They can say almost whatever they wish, using a mix of positive and normative statements; use results that have never been confirmed by additional studies and will rarely be punished by their peers if history subsequently shows they were wrong. They are free to publicise the result of a single scientific study as though it formed definitive proof. Newspapers are particularly susceptible to this, and in many ways it is their job. Topical stories about political intrigue or the behaviour of sports stars often have only weak or mixed evidence behind them, and it may be impossible for newspapers ever to get to the bottom of a story before it is first published. However, applying this approach to scientific stories can cause problems. For example, a single study involving a small number of human subjects that links a lifestyle choice with cancer will be presented as big news, even though the scientists might well have said that their findings were only provisional, and that the data only shows a relatively weak statistical connection because of the small number of subjects. In scientific studies, data is frequently analysed with a requirement that it proves the hypothesis within a certain statistical accuracy. With difficult-to-gather or noisy data, or if the number of subjects or cases studied is small, the requirement is surprisingly low, possibly only a 90 or 95 per cent probability that the hypothesis is correct. This might seem reasonably high, but it implies on mathematical grounds alone that five to ten per cent of such single studies might be wrong. Fortunately, the probability that the hypothesis is correct will grow rapidly as the number of similar or supporting studies grows – particularly if they are carried out by other teams and are therefore less likely to be subject to bias. Unfortunately, subsequent studies indicating that the cancer-causing hypothesis was wrong will tend to get less attention in the media, as would studies that added weight to the original hypothesis. The public and politicians therefore tend to over emphasize the results of early studies, are not aware of the weight of evidence and tend to retain disproved theories as fact.

In theory, the media should be just as happy to report a result that 'proves' or 'dis-proves' climate change, as long as it is a new, and preferably exciting, story. Unfortunately anthropogenic climate change is now so well established as true, that only stories which are particularly apocalyptic or go against this view are seen as newsworthy. Politicians or the general public often feel even less of a need for balance. Usually, scientific results are not the trigger for their beliefs or actions, but simply supply supporting evidence for their own pre-existing policy commitments or beliefs. There are notable exceptions to this, but they are rare.

Both the press and politicians use a mix of positive and normative comments. It is unlikely that their audience will check to see if the results presented are correct and valid, or little more than urban myth. Most people, after all, would find it difficult to access and interpret the original scientific papers, let alone be aware of how much supportive evidence exists on one side or another, or the reputation of the scientists involved. Thus politicians are free to pick one or two scientific papers that support their view (however slightly) and ignore maybe several hundred papers that take a different view, and the public will be unable to gauge the inequality of the scale of the two opposing views – not even if one side has withdrawn their results or the results are many years out of date.

As an example of how we might analyse a statement about climate change, consider the following quote:

> 'There is little evidence that the planet is warming, and even if it were, reducing our emissions would be too costly.'

Such a comment achieves power by being: (a) short (much shorter than a scientific paper), (b) easy to remember and (c) might carry additional weight if uttered by a respected, or at least publicly known, individual. The subsequent use of such a quote by others is likely to rest on one or all of these three points. Whether it is true or not is unlikely to be much of a consideration.

The quote contains two statements. The first, about the evidence for climate change, is positive in nature; the second, about costs, is normative. The first is amenable to scientific analysis, although the use of the word 'little' makes the statement somewhat subjective as different individuals may require differing levels of proof. The statement does not provide any evidence to support its claim, so it is difficult to know if it is a considered judgement by an informed individual or little more than a random assertion. The second statement is not directly amenable to scientific analysis because the phrase 'too costly' has not been defined. The reader is not given the cost, or the possible cost to the economy of not tackling climate change. The speaker is setting out their political position, but no more. Many may consider it worth spending large sums of money on reducing any impacts, others not. Either way, we are not told how much it might cost and over what timeframe. This should make the reader immediately suspicious. By adding a positive claim to a normative one, weight is seemingly given to the normative statement and it can be easy to forget that it is still normative (even if the positive statement is true) and therefore represents a personal, moral or philosophical view.

Sometimes data is included in such a quote. Such data should be analysed by the reader immediately and any possible flaws identified using logic and simple mathematics.

(Encouraging students to do this is the reason this book contains so many in-text problems.) So, let us imagine the quote continued with:

'... – over €100 billion over the next 50 years for this country alone.'

€100 billion sounds an impressive number, and therefore possibly 'too costly', but in reality is it so impressive? If the country in question has a population of 100 million, then the cost per capita, per annum would be:

$$\frac{€100 \times 10^9}{100 \times 10^6 \times 50} = €20,$$

i.e. less than the cost of a tank of petrol (gasoline). Now, it is still possible that the author of the comment might consider this 'too much', the statement is after all normative, but it is just as likely that they never bothered to apply any form of quantitative analysis to it, and that many readers might take a different view. And, as mentioned above, the author of the comment has conveniently forgotten to mention the cost of not tackling global warming.

Scientific assessments

Given scientists who are mainly interested in doing good science, who cloak their results in mathematics and jargon and have little inclination to enter the policy arena, and a media only interested in 'new' news and politicians who want to push their own (or a voting block's) agenda, what source of information can the public rely upon?

The answer is to make extensive use of scientific assessments. These reports are normally commissioned by governments, government agencies or international bodies (such as the UN or the World Health Organization) and review all the available evidence – including the claims of those with fringe views. Often an organization will produce a pair of reports: an in-depth scientific analysis of the state of knowledge of the topic, and a summary for policy makers and the public. This latter will usually be the most useful and will be published on the web. Whenever you wish to get to the bottom of a scientific or environmental question, search the web for the appropriate scientific assessment. Such assessments were the route taken by the IPCC (see Section 18.1). Their assessments involve many hundreds of experts reviewing the evidence and producing a result with which it is very difficult to argue. Four scientific assessments have been carried out by the IPCC so far (1989, 1995, 2001, and 2007) and all have found that the climate is changing and that humankind is responsible – with the quantity and quality of the evidence growing with each report.

Major claims of climate change sceptics

We will now consider three frequently heard claims from climate change sceptics. These examples (and parts of the subsequent analysis) are taken from reference [DES06], but similar statements can be found in the press.

Claim 1

The earth is not warming

> *Evidence of recent warming rests solely on the surface thermometer record. Such data are contradicted by satellite measurements, which are far more reliable. Satellite measurements show a very small warming trend since measurements began in 1979 – about 0.06° C per decade, much too small to be noticeable [DES06].*

This claim consists of four positive statements, then a fifth which could be positive or normative depending on whether 'noticeable' is noticeable by a particular individual, or meant to imply 'measurable'.

The first statement is simply wrong: there are many other data sets that show the world is warming, including sea temperatures, ice cover, sea ice thickness and increased growing seasons. In addition, the basic science implies that the climate will warm as carbon dioxide concentration increases (and there is irrefutable evidence to show that this increase in concentration is occurring – see Figure 5.7), and computer models show that this warming trend cannot be accounted for unless anthropogenic emissions are included.

The third statement is also false: satellite measurements are not more reliable. Such remote systems are actually quite difficult to work with and calibration between satellites is a longstanding problem.

The second statement has a stronger basis. For a long time the satellite data (which shows the temperature of the lower atmosphere) appeared to show no warming – contradicting the rapid rise seen in surface data. This anomaly has subsequently been shown to have arisen from problems with calibration over time and between satellites and from incorrect assumptions in the analysis of the raw data [NEW06]. The range of predictions from surface data show a warming of 0.1 to 0.2 °C per decade, those from satellite data 0.06 to 0.26 °C per decade. So the two sources are in agreement and the satellite data is seen to offer no greater precision. Unfortunately, the supposed contradiction has become almost an urban myth and we can expect to find it recycled in the press for many years to come.

The use of 0.06 °C in the fourth statement is interesting: it represents the lowest estimate of the temperature trend discussed above, and there is no reason to believe that it more accurately reflects the reality of the situation than the highest figure, i.e. 0.26 °C per decade. The speaker is clearly manipulating the numbers for their own purposes. A more honest commentator might have chosen the arithmetic mean of the range, i.e. $(0.06 + 0.26)/2 = 0.16$ °C per decade.

The final normative point, that 0.06 °C per decade is too small to be noticed, is incorrect in that such a change would be 'measurable', also a small rise such as this would still be a concern given that greenhouse gas concentrations are still rising and we can therefore expect this number to rise. Any temperature rise found at current concentrations implies we have a serious problem, as we know the continuing expansion of the world economy will drive emissions far higher.

Claim 2

Humankind is not to blame

The earth may be warming, but human activities are not responsible [DES06].

Here we have a single positive statement. For it to be true, the warming must have another cause. One possible explanation often given is that the warming represents a recovery from the little ice age. Unfortunately for sceptics, it would appear that the little ice age (1550–1850) was a regional event centred on Europe, but the current measured warming is global. In addition, the recent warming would appear to be too fast to be consistent with post ice age warming. In essence, the sceptics are implying that the climate has some form of inherent natural variability not caused by an identified driving force: there is no evidence for this assertion.

This leads us to a second explanation often cited, namely that the driving force behind the current rise in temperatures is a change in solar output. There has indeed been an increase in the sun's output over the last 100 years. But the change has been small and cannot account for the scale of the warming, whereas the observed changes in greenhouse gas concentrations (about which there is no controversy) are of a scale sufficient to account for this warming. It is worth noting that the increased concentration of carbon dioxide in the atmosphere is known to be from fossil fuel use, not some other natural process. As discussed earlier, this is because the isotopic mix of carbon in today's vegetation is different to that in fossil fuels. By measuring the isotopic mix in atmospheric carbon dioxide we therefore know how much is natural and how much is anthropogenic.

Claim 3

No need to panic

Future warming will be small and easily manageable [DES06].

Here we have a positive claim followed by a normative one. The claim is usually based on the hope that the sensitivity of the climate to increasing concentrations of greenhouse gases is low, emissions will grow little and that all humankind, together with the planet's flora and other fauna, can adapt to such changes.

For the first claim to be true there needs to be one or more negative feedbacks which have not kicked in yet. One possibility might be the water vapour cycle. If the climate worked such that humidity drops as the planet warms then this would put a break on rising temperatures (remember water vapour is also a greenhouse gas – Table 5.1). Whether this is possible is as yet unknown, but the scientific consensus is the opposite, i.e. that as temperatures rise, evaporation will increase thereby increasing humidity, and that if additional rainfall does subsequently reduce the humidity (the iris hypothesis) the effect will be small and is unlikely to significantly change the expected rise in temperatures.

Another possibility is that our emission projections are far too pessimistic. Sceptics can therefore argue that emissions are unlikely to rise anything like as fast as predicted and therefore the rate of warming will be small. Looking at the world around you, the behaviour of your fellow citizens and the growth in developing world emissions, is this a credible stance? Is it really likely that we will all suddenly awake to the reality of climate change and mend our ways? This author thinks not. There is also a more fundamental reason why this argument is incorrect: to stop the predicted temperature rise, we need to not just stop the current growth in emissions, but to reduce our emissions. Even if emissions do not increase at all, the concentration of greenhouse gases in the atmosphere will continue to rise, as will the global mean temperature.

To what degree climate change is 'manageable' is less well understood than other aspects of the debate, and each citizen will have their own feelings about what is acceptable, but it is worth emphasising that the predictions from climate change impact modelling (see Section 17.4) make for sober reading.

5.5 The atmosphere

Evolution of the atmosphere

Today's atmosphere is distinctly different from that of the early Earth and we can divide its evolution into three periods based on atmospheric composition and chemistry. Although geological processes and inorganic chemistry have been important in the development of the atmosphere, it is clear that life itself has played an equally crucial role—suggesting to some that the planet can be seen as some kind of evolving super-organism: the Gaia hypothesis (see reference [LO00] for more details of this fascinating story).

Earth's original atmosphere (termed the *first atmosphere*) consisted primarily of helium and hydrogen. Much of this atmosphere was lost to space. However, around 4.4 billion years ago the Earth's crust had formed which allowed volcanoes to develop. These emitted large quantities of steam, carbon dioxide and ammonia and created an atmosphere (termed the *second atmosphere*)with 100 times more molecules of gas than today's, composed of water vapour, carbon dioxide, nitrogen (and possibly large amounts of hydrogen), but almost no oxygen. Via the greenhouse effect such large quantities of carbon dioxide kept the planet from freezing, and temperatures may have been as high as $70°C$, up to 2.7 billion years ago. Later as the atmosphere cooled, much of the carbon dioxide dissolved in the oceans.

Around 3.3 billion years ago cyanobacteria or similar organisms were responsible for producing large amounts of oxygen by photosynthesis and by 2.2 billion years ago the atmosphere had become oxygen rich and carbon dioxide poor. This oxygen reacted with ammonia to produce nitrogen. Nitrogen was also generated from the conversion of ammonia by bacteria. However today's high concentration of nitrogen is mainly due to the sunlight-powered photolysis (the break-up of molecules caused by photons) of ammonia from volcanoes.

Higher levels of oxygen allowed the ozone level to form—i.e. the presence of the ozone layer is of itself a sign of life—protecting life forms from high levels of ultraviolet radiation and creating Earth's *third atmosphere*, that which we know today, with 78% nitrogen, 20.9% oxygen, 0.93% argon, 0.037% carbon dioxide and many other trace gases, as well as 1-4% water vapour depending on the location and the time of year.

Today's atmosphere

Five distinct layers can be identified within the atmosphere, each defined by its temperature characteristics (Figure 5.25), its chemical composition and the way air moves within the layer. We can also define two regions based on how the ratio of the different gases changes with altitude (Figure 5.26). The *homosphere* extends from the surface to around 100 km. Here the composition (with the exception of water vapour) changes little with altitude because of vertical mixing. Above this we find the *heterosphere* were the composition changes with altitude because there is little mixing and because through the whole atmosphere the density of a gas falls exponentially with increasing altitude, but at a rate which depends on its molar mass. The composition is therefore gravitationally sorted in parts of the atmosphere where vertical mixing caused by convection is reduced.

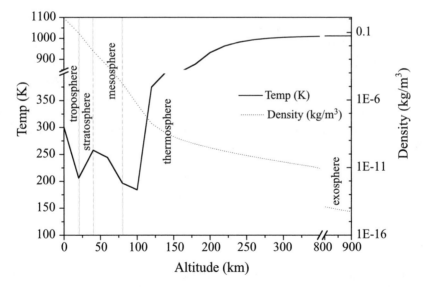

Figure 5.25 The change in temperature and density with altitude as one ascends through the layers of the atmosphere. Note that the temperature gradient changes sign several times [from data in ESA07]. (Note the two axis breaks.)

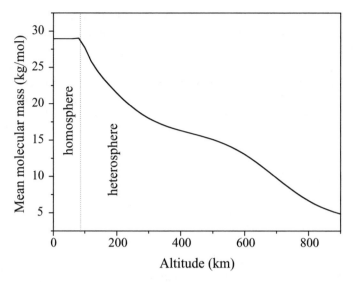

Figure 5.26 Change in average molecular weight of gases in the atmosphere as a function of altitude: a steady decline in mass is seen as the heavy species become less common once the boundary between the homosphere and the heterosphere is crossed [from data in ESA05].

The atmosphere, despite weighing over 5×10^{18} kg, is a much thinner layer than many realise. Although the atmosphere has no distinct end and extends over 1000 km from the surface of the planet, as can be seen from the drag it causes on satellites, three-quarters of the mass of the atmosphere lies below 11 km. One way of comprehending just how thin a layer this is, is to estimate how one might represent it on a $^1/_2$ metre diameter model globe. The Earth has a diameter of 12,740,000 m meaning the scale of the model would be 0.5:12,740,000 or 1:25,480,000 and thus 11 km would be represented by a thickness of 11,000/25,480,000 m or 0.43 mm—barely the depth of a layer of varnish. Alternatively, we can ask how thick a layer of water would have to be to have the same mass as the atmosphere. Water has a density of 1,000 kg/m^3 so 5×10^{18} kg of water would occupy 5×10^{15}m^3. The planet has a surface area of 510×10^{12} m^3, so the thickness would be $5 \times 10^{15}/510 \times 10^{12} = 9.8$ metres. It is therefore hardly surprising that our atmosphere is better viewed as a thin, delicate, protective blanket, which ensures our survival (Figure 5.27); and that our current activities have the potential to disrupt it.

We will now consider the five layers of the atmosphere in turn.

Troposphere

The troposphere extends from the Earth's surface to between 8 and 15 km (depending on the latitude and climatic conditions). As one ascends the temperature falls from around 14°C to −52°C (Figure 5.25). Almost all weather occurs in this layer and the air is vertically well mixed with no change in chemical composition with altitude.

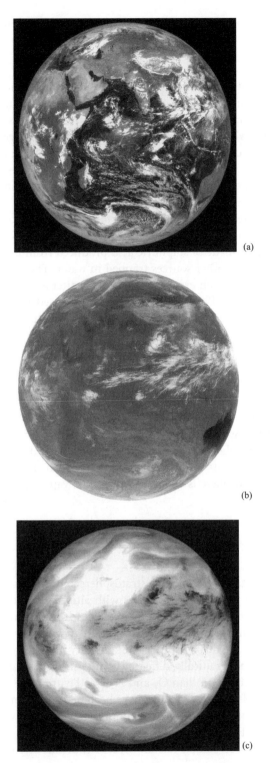

Figure 5.27 Earth as seen at (a) visible wavelengths, (b) broadband infrared wavelengths, and (c) around 6 μm i.e. within a water absorption band [IND02].

Stratosphere

The stratosphere extends from the top of the stratosphere to around 50 km. Because of low temperature of the upper reaches of the atmosphere water vapour finds it difficult to reach the stratosphere and consequently the air is drier. The stratosphere is warmed by the absorption of ultraviolet light (by in particular ozone and oxygen), with the upper regions being exposed to more absorbing and therefore being warmed more. This gives rise to a temperature inversion, with the temperature rising from $-52°C$ at the bottom to around $-3°C$ at the top, which stops most vertical mixing, maintains the inversion and forms a giant lid on the troposphere.

The lower stratosphere is particularly important to the study of climate change as it contains a layer of tiny liquid droplets typically less than 1 micrometer in diameter. These are primarily made of sulphuric acid produced from the release of sulphur dioxide by volcanoes and other sources trapped in water. These aerosols reflect light and infrared radiation and can exist for several years after an eruption. The eruption of Mount Pinatubo in the Philippines) in June 1991 raised the temperature of the stratosphere by over $1°C$ and reduced global surface temperatures by around $0.5°C$ for several years.

Mesosphere

The Mesosphere begins at the top of the stratosphere and extends to around 85 km. Here the temperature again falls with height to as low as $-93°C$. Temperatures are lower than in the stratosphere because of the lack of ozone at this height and therefore only relatively small amounts of solar radiation is absorbed. This then allows convection to occur. The mesosphere is where most meteors vaporise, so forming shooting stars.

Thermosphere

The thermosphere starts at the top of the stratosphere and extends to 600 km or more. As in the stratosphere the temperature rises with altitude, but this time because of the absorption of very high frequency ultraviolet light from the sun (which was easily dissociate oxygen and nitrogen molecules turning them into their atomic forms), but as the air is so rarefied, the thermosphere contains very little heat per unit volume. At this height the "air" has a radically different composition than at ground level with O and N being more common than O_2 and N_2 (see Figure 25.28), and by the top of the Thermosphere hydrogen and helium have become more common than nitrogen and the temperature can reach 2000 C depending on the intensity of solar ultraviolet radiation reaching the Earth.

The international space station has an orbit within the thermosphere at around 360 km.

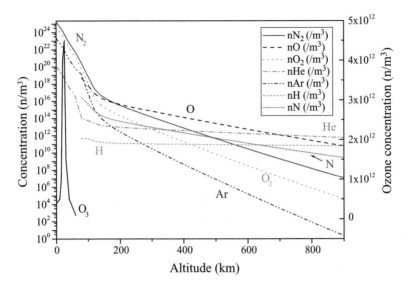

Figure 5.28 Changes in gas concentrations (in terms of molecules per m^3) as one ascends through the atmosphere. Note how in the lower atmosphere the ratio of gases stays almost constant, but high in the atmosphere the heavy species (e.g. oxygen and nitrogen) decline in abundance more rapidly than the lighter ones (e.g. hydrogen and helium) [from data in ESA07]. Also shown is the concentration of ozone—the position of the ozone layer is clearly visible.

Exosphere

The exosphere is the outermost layer of the atmosphere and extends from the top of the thermosphere to 64,000 km or so above the planet, or one-sixth of the way to the moon. Here the gases are extremely rarefied and the lighter ones can escape into space.

5.6 Student exercises

> *"There is broad agreement within the scientific community that amplification of the Earth's natural greenhouse effect by the buildup of various gases introduced by human activity has the potential to produce dramatic changes in climate. Only by taking action now can we ensure that future generations will not be put at risk."*
>
> —Statement by 49 Nobel Prize winners and 700 members of the National Academy of Sciences, 1990

1. Define *sustainable*. What characteristics might a sustainable energy technology have?

2. (Only for those with suitable computer skills.) Write a simple computer program or spreadsheet which randomly assigns a number of rainy days to each month in the last 300 years and calculates how many years one has to look back to find the same month (January, February, ...) that has as least as many rainy days. Plot the

number of years one had to look back against the month in question for the last 100 years.

3. Describe the natural greenhouse effect.

4. How do anthropogenic emissions enhance the natural greenhouse effect?

5. What evidence is there for the natural greenhouse effect?

6. What evidence is there for the anthropogenic greenhouse effect?

7. Why might changes in surface albedo change global mean air temperature? Outline which natural surfaces might change in albedo under the influence of climate change and whether they might provide a positive or negative feedback.

8. Describe the carbon cycle.

9. In what way does permafrost act as an indicator of climate change and how might it produce a positive feedback?

10. Outline the most common climate feedback processes detailed in the text.

11. What are the most important anthropogenic greenhouse gases, what are their global warming potentials and what are their contributions to climate change?

12. What is meant by *radiative forcing*?

13. What is acid rain?

14. Discuss some of concerns, other than climate change, that make energy provision typically unsustainable.

15. Expand the model you built for Problem 5.2 to include the additional radiative forcing from anthropogenic greenhouse gas emissions (see Problem 5.4). This is best done by adding four times the additional forcing to the Solar Constant, S (1370 W/m^2), in your atmospheric model. (The factor of 4 is needed because the solar constant is defined in terms of the radiation falling on a disk, not a sphere, at the distance the Earth is from the sun.) Use your new model to examine the likely temperature rise if carbon dioxide concentrations reach 550 ppm.

16. Use the model you built for Problem 5.2 and Exercise 15 to estimate the change in height of the emission level compared to pre-industrial times. (Hint: calculate the

change in emission level that gives the same temperature change as the additional radiative forcing (Exercise 15), and assuming the emission level was previously at 5.5 km. This can be done by estimating the temperature of the emission level from the lapse rate, its height (5.5 km) and the pre-industrial surface temperature, then seeing from what height one needs to descend (at the lapse rate) to give a surface temperature similar to the post-industrial value.)

6

Economics and the environment

"Economics is, at root, the study of incentives: how people get what they want, or need, especially when other people want or need the same thing."

–Freakonomics, Steven Dubner and Stephen Levitt

Several basic economic concepts are key to an understanding of why some environmental policies are adopted by individuals, companies and governments more willingly than others. This section introduces a number of these concepts and gives examples of how they influence environmental decisions. Not only does economics provide a way of understanding behaviour, it also provides a way of setting policy on a rational footing, and we will review three examples of this at the end of the chapter. Table 6.1 shows some of the economic topics we will be considering.

6.1 Key concepts

Supply and demand

The *law of demand* states that when the price of a commodity rises, the quantity demanded will fall. There are two reasons for this:

1. The income effect. People will feel poorer – i.e. they will not be able to afford to buy so much of the goods

2. The substitution effect. The goods will be more expensive relative to other goods. People will thus switch to an alternative product

Energy and Climate Change   David A. Coley
© 2008 John Wiley & Sons, Ltd

Table 6.1 A few of the interconnections between economics and the environment

Topic	Example	Economic concepts
The effect of a price change on behaviour	Increased fuel costs	Supply and demand curves, elasticity
The effect on prices of an increased supply of goods or services	Cheaper air flights	Demand curves, elasticity
Wealth of the nation	Relative cost of climate change	GDP
Cost	Relative cost of large and small-scale wind farms	Marginal vs. average costs
Environmental investments	How long will it take to return the cost of insulating your hot water tank	Payback periods, present value of money, discounting

The size of the income effect depends primarily on the proportion of income spent on the item. The more money spent on the item the more likely we will be forced to cut down on the amount we buy. The size of the substitution effect depends primarily on the number and closeness of alternatives.

We can represent such sensitivities with a demand curve. This relates the price and the quantity demanded. Each of us will have our own distinct demand curve for specific goods or services, but by averaging over the population a single curve can be obtained. This curve may then be used to influence pricing decisions.

We can repeat this analysis in reverse. If the price of a product or service rises then it will be more profitable to produce it. Therefore the *law of supply* is: when the price of a commodity rises the quantity supplied will also rise. The supply curve relates the amount producers would like to supply to the price. Supply and demand curves have opposite slopes, and cross at a point where supply and demand are in equilibrium (Figure 6.1).

Elasticity

Elasticity describes the responsiveness of demand to a price change – i.e. the slope (and direction) of the curves in Figure 6.1. This responsiveness will depend on the goods or service in question. For some goods even a small price change can greatly effect the demand (Figure 6.2, curve B), this is particularly true if a near-substitute item exists. For other goods, large changes in price only have relatively small changes in demand (curve A). Many environmentally degrading activities fall into this second category. For example, people are willing to absorb quite large increases in the cost of private road transport before they substantially reduce their consumption. This makes it difficult to use the increasing taxation of transport fuels as the only mechanism behind an environmental transport policy.

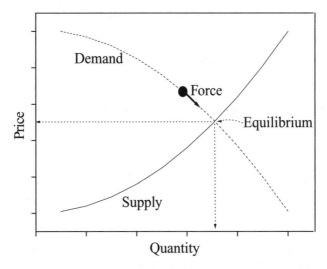

Figure 6.1 Demand and supply curves (for a single product or service) averaged over the population. The equilibrium is the point of intersection of the two curves. Any point way from the equilibrium will experience a 'force' attempting to return it to equilibrium. This will be achieved through either a change in price or a change in the rate of supply

Formally, the elasticity ε is defined as:

$$\varepsilon = \frac{proportionate\ (or\ percentage)\ change\ in\ quantity}{proportionate\ (or\ percentage)\ change\ in\ the\ determinant}$$

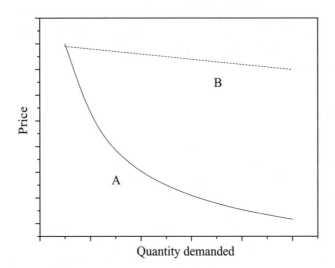

Figure 6.2 Two demand curves with different elasticities

Typically the *determinant* is price and the *quantity* is demand for a product or service, giving the *price elasticity of demand, $P\varepsilon_d$*:

$$P\varepsilon_d = \frac{proportionate\ (or\ percentage)\ change\ in\ demand}{proportionate\ (or\ percentage)\ change\ in\ price}$$

For example, if a 20 per cent rise in the price of electricity causes a 10 per cent fall in the quantity demanded, the price elasticity of demand for electricity will be:

$$-10\%/20\% = -0.5 \text{ (note the minus sign)}.$$

Problem 6.1   If the price elasticity of bus travel is –0.2, how many fewer journeys will be made on a bus route that currently attracts 10 000 passengers per annum if a price rise of 15 per cent occurs?

Other useful elasticities are: the price elasticity of supply, $P\varepsilon_s$; the income elasticity of demand, $Y\varepsilon_d$; and the cross-price elasticity of demand, $P\varepsilon_{ab}$:

$$P\varepsilon_s = \frac{proportionate\ (or\ percentage)\ change\ in\ quantity\ supplied}{proportionate\ (or\ percentage)\ change\ in\ price}$$

$$Y\varepsilon_d = \frac{proportionate\ (or\ percentage)\ change\ in\ demand}{proportionate\ (or\ percentage)\ change\ in\ income}$$

$$P\varepsilon_{ab} = \frac{proportionate\ (or\ percentage)\ change\ in\ demand\ for\ goods\ \mathbf{a}}{proportionate\ (or\ percentage)\ change\ in\ price\ of\ goods\ \mathbf{b}}$$

An example *of $P\varepsilon_{ab}$* might be the price responsiveness of the demand for public transport to a change in the price of private road tax.

As curve A of Figure 6.2 shows, the elasticity along the length of a demand curve is often not constant (i.e. the slope is not constant). In estimating the likely effect of a price change it is therefore important to know that the given elasticity is relevant to the current and future price.

The economy

We can use a simple model of the economy (Figure 6.3) to estimate the wealth of a nation, from which we can try to answer whether, for example, the environmental cost of pollution or switching to a new energy technology is affordable.

As Figure 6.3 indicates, money flows in a circular fashion from firms to households then back to firms. The speed of the cycling is termed the *velocity of circulation*. Any unit of money can flow around this cycle many times a year, and therefore it is important to distinguish between *money* and *income*. Money is a physical concept. If at any time there is £100 billion of money in the economy, all of which is paid out as income five times a year (i.e. it cycles around Figure 6.3 five times a year), then the *national income* is 5 × £100 billion = £500 billion.

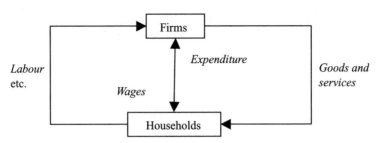

Figure 6.3 The circular flow of money

There are several methods used to measure national income. The most commonly cited in the media is the *gross domestic product* (GDP), which measures a nation's output. GDP can be arrived at in one of three ways:

1. Product method: add up the value of all goods and services traded in the country per annum, industry by industry

2. Income method: add up everyone's annual income from wages, profits, rents, interest etc.

3. Expenditure method: the total value of all sales in one year

In theory, all three methods should produce the same result.

GDP is only one of several measures of wealth that are needed for an economic analysis of an environmental question. Some of a country's GDP will be because of production by foreign companies within the country, the profits of which will leave the country and thus need to be removed from the GDP if the nation's wealth is to be accurately determined. In addition earnings by companies from production abroad need to be included. If this is done, the *gross national product* (GNP) will have been estimated.

As their names suggest, both GNP and GDP are *gross* measures and do not include depreciation of capital items (such as machinery and infrastructure – for example the aging of nuclear power stations). If this depreciation is subtracted from the GNP the *net national product* (NNP) is found. Table 6.2 shows the GDP of selected countries together with their per-capita GDP and their current rate of growth. Clearly there is great variety in national and personal income. In addition, some of the less wealthy countries have the highest rates of growth. This will affect their demand for new sources of energy and their ability to pay for the environmentally best technology.

Marginal and average costs

When considering the costs of changing the level of provision of a service or product (for example placing 31 wind turbines on a site rather than 30) it is very important to distinguish between *average* and *marginal* costs. The average cost of a product or service

Table 6.2 GDPs of several countries [data from CIA05]

Country	GDP (US$)	GDP per capita ($)	GDP growth % per annum
China	7.20 trillion	5 600	9.1
Japan	3.75 trillion	29 400	2.9
Singapore	121 billion	27 800	8.1
France	1.74 trillion	28 700	2.1
USA	11.8 trillion	40 100	4.4
Australia	612 billion	30 700	3.5
Canada	1.02 trillion	31 500	2.4
India	3.32 trillion	3 100	6.2
Mexico	1.01 trillion	9 600	4.1
Kenya	34.7 billion	1 100	2.2
UK	1.78 trillion	29 600	3.2
World	55.5 trillion	8 800	4.9
EU	11.7 trillion	26 900	2.4

is the total cost of the product or service divided by the number of units produced. For example the average unit cost of electricity supplied by a power company will be the total cost of running the plant (say €1m per annum) divided by the number of units generated (say 50×10^6 kWh), therefore:

$$\text{average cost per unit} = (€1 \times 10^6)/(25 \times 10^6)\,\text{kWh} = €0.04.$$

So the power company needs to sell its electricity for something in excess of €0.04 per kWh in order to make a profit. However, if the company decides to generate one more kWh it might well be able to sell it for much less than €0.04 and still make a profit. This is because the labour and other costs of running the plant will not have changed. The only additional cost will be fuel. This additional cost is called the *marginal* cost of the electricity.

Payback period and discounting

If an individual or company invests money in order to, for example, save electricity by installing energy efficient lighting, it is easy to calculate how long it will take for the investment to pay off and thereby whether such an investment is economically sensible. If each low-energy bulb costs £10 and saves £5 of electricity per year the *simple payback period* will be:

$$£10/£5 \text{ per annum} = 2 \text{ years.}$$

However, this is likely to be an underestimate of the true payback period. In order to spend the £5, the company (or individual) will have to either borrow the money, or decide not to invest it in some other activity. Thus, not only does the £10 have to be paid back, but also the lost interest (or the cost of the borrowing).

The true payback period can be estimated by using the *present value* (PV) of the investment and the technique of *discounting*. The present value of €1 (for example) is defined as the amount of money you would need to invest now in order to obtain €1 at some future date. For example, the PV of €10 if the investment is to be made for 2 years and the rate of interest is seven per cent (or 0.07) per annum is given by:

$$PV = \frac{\text{investment (or saving)}}{(1 + \text{interest rate})^{time}} = €10/(1 + 0.07)^2 = €8.73.$$

Consequentially it is not worth (economically speaking) investing now any more than €8.73 in order to save €10 in two years. The PV of €1 (or pound, dollar etc.) is termed the *discount factor* for a set number of years at a defined interest, or *discount*, rate.

The concept of discounting can have profound implications for environmental economics. Given an environmental problem, is it better to spend money now to solve the problem, or to wait until a later date and treat the problem then? The results of such an analysis can be surprising. If, for example, a radiation hazard from stored nuclear waste will cost €1000m to clean up in 50 years from now, and we assume an interest rate of five per cent, the PV is:

$$€1 \times 10^9/(1 + 0.05)^{50} = €87.2 \text{ million.}$$

Thus if the power company invests €87.2 m now it will have enough money to finance the clean-up. Alternatively, it implies that it is *not worth* the company spending more than €87.2m to stop the problem from occurring. With a higher interest rate and a greater time frame the results are even more surprising. If the problem will only occur in 100 years and the interest rate is assumed to be 10 per cent then the PV is:

$$€1 \times 10^9/(1 + 0.1)^{100} = €72566.$$

So, the power company might only be willing to invest €72 566 now in order to solve what will be a €1 billion problem for future generations.

Another relevant example is the cost of hydroelectricity. Large hydroelectric dams are very costly and take a long time to build when compared to gas turbine-based power stations; therefore considerable sums need to be borrowed for substantial periods. A PV analysis of their cost can lead to the conclusion that they can not compete with fossil fuels, despite the fact that their source of fuel is free. Those that are concerned by such calculations term the problem 'the tyranny of discounting'.

Discounted cash flow analysis

Typically the cost of a power station is not paid back at the end of its life with a single payment. In reality, a series of payments will be made to the lender (often a group of banks) over the length of the loan. Thus to compare the costs of different technologies we need to annuitize the capital cost (this is explained below), then add in the cost of fuel, labour and maintenance and divide the answer by the amount of energy generated, to give the cost per unit generated for each technology. In the following we will do this twice, firstly assuming no interest is paid, and without discounting, and secondly assuming interest is paid. We will then consider a more complex example.

Simple analysis

Taking a 500 MW$_e$ combined-cycle gas turbine as an example (note, the same analysis could be applied to any energy technology, not just electricity generation), we might expect the capital cost of the plant including associated buildings and connection to the grid to be €280m, the loan to be repaid over twenty years and the plant to run for 8000 hours per year. If the efficiency is 55 per cent, maintenance costs are approximately proportional to the amount of electricity generated, and assumed to be €0.005 per kWh, and the cost of gas to a power station is only €0.014 per kWh then:

1. gas consumption rate = 500 MW/0.55 = 909 MW

2. gas consumed each year = 909 × 8000 = 7 272 000 MWh

3. electricity generated each year = 500 × 8000 = 4 000 000 MWh

4. capital cost repayment (spread over twenty years and ignoring interest) = €280m/ 20 = €14m per annum

5. cost of gas = 7 272 000 × 1000 × 0.014 = €101 808 000 per annum

6. maintenance cost = 4 000 000 × 1000 × 0.005 = €20m

7. Therefore total cost per annum = €20m + €101 808 000 + €14m = €135 808 000

8. and the cost of electricity = €135 808 000/4 000 000 = €33.95 per MWh, or €0.034 per kWh.

The cost, or price, to the consumer would be higher as:

$$price = cost + profit.$$

The profit being necessary otherwise the electricity company would be better off simply investing the €280m elsewhere. However, ultimately the price will be set by the market, so this equation is more an aspiration than a given.

We will now repeat this analysis, but accounting for interest. To do this we need to adapt the discounting method discussed above to a series of annual repayments. If we borrow an initial sum, S, and pay it back over T years at an interest rate of r, then the annuitized amount, A, repaid to the bank each year will be given by:

$$A = \frac{rS}{1 - (1+r)^{-T}} \tag{6.1}$$

and the cost of the electricity generated will be given by:

$$\text{cost of electricity} = \frac{\text{annuitized capital cost} + \text{annual running cost}}{\text{annual electricity production}}. \tag{6.2}$$

For the power station in question, and assuming the interest rate is seven per cent per annum, we first apply Equation (6.1):

$$\text{repayment} = \frac{0.07 \times 280}{1 - (1 + 0.07)^{-20}} = \text{€}26.43 \text{ m per annum,}$$

and then Equation (6.2):

$$\text{cost of electricity} = \frac{\text{€}26.43\text{m} + \text{€}101\,808\,000 + \text{€}20\text{m}}{4000000}$$

$= \text{€}37.06$ per MWh, or $\text{€}0.037$ per kWh. This compares with the $\text{€}0.034$ per kWh we calculated earlier without discounting. The difference may seem small, but such differences can decide whether a technology is economic or not.

Problem 6.2   Produce a spreadsheet applying Equations (6.1) and (6.2) to any energy technology. By imagining sustainable energy technologies to have a capital cost of twice that of non-sustainable ones, the same maintenance costs, but no fuel costs, examine the relative cost per unit of energy of the two approaches. Run your spreadsheet for various fuel costs, interest rates and repayment periods.

Complex analysis

Figure 6.4 compares the costs and quantities of electricity generated from a nuclear and a gas-fired power station. Clearly the situation is a complex one, with large fluctuations in the costs to the generating company. In order to analyse such situations consistently, we need to extend the above discounted cash flow analysis by estimating the present value of each year's expenditure (or electricity generation) separately and summing

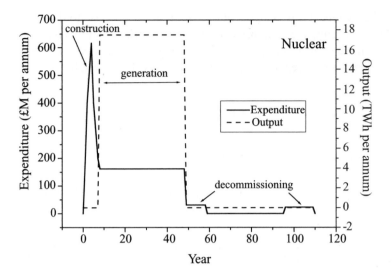

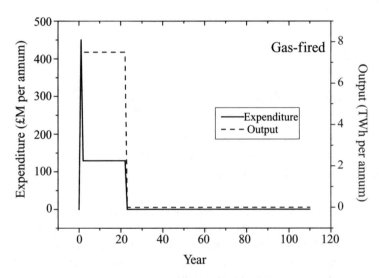

Figure 6.4 Possible costs (solid line) and output (dashed) of a 2 MW$_e$ nuclear facility (top) and a 1 MW$_e$ gas turbine power station (bottom). Note the complexity of the curves and the great delay between the start of the nuclear project and the start of energy generation; this has a great impact on the likely economic viability of a nuclear plant compared to a gas-fired plant; the lower operational and fuel costs of the nuclear plant (per GW); the longer lifetime of the nuclear plant; and the length of the nuclear decommissioning period

them over the whole period to give the net present value (NPV):

$$\text{NPV} = \frac{Z_1}{(1+r)^1} + \frac{Z_2}{(1+r)^2} + \frac{Z_3}{(1+r)^3} + \cdots + \frac{Z_n}{(1+r)^n} = \sum_{i=1}^{n} \frac{Z_i}{(1+r)^i} \quad (6.3)$$

where Z_i is the expenditure in any year, i, and n is the total number of years the project will run.

The cost of the electricity generated (or heat produced from a boiler) will then be given by:

$$\text{cost of electricity} = \frac{\begin{array}{c}\text{NPV of capital loan, fuel, operating, maintenance}\\\text{and decommissioning}\end{array}}{\text{NPV of electricity or heat}} \quad (6.4)$$

In general, the discount rate r may be different for the initial capital, the fuel, the decommissioning costs and the electricity produced, although frequently this complication can be ignored in the first analysis. (Note the NPV of capital is the NPV of the loan repayments, i.e. Equation (6.1) is applied first, and then the NPV estimated.) It may seem surprising that we need to use the NPV of the electricity, rather than just the total electricity produced, as we did in Equation (6.2), but this needs to be done to account correctly for situations where electricity production might not be constant throughout the life of the project (see Figure 6.4). Electricity produced early is more valuable as it can be sold to offset loans, or any profit simply invested. In the case where all discount rates are the same and the costs and the quantity of electricity generated are the same each year, Equation (6.4) simplifies to Equation (6.2).

We also need to account for inflation. This can be achieved by calculating a real interest (or discount) rate from the monetary interest (or discount) rate (i.e. the one quoted by a bank):

real interest rate = monetary interest rate – rate of inflation.

However, predicting future interest rates or the rate of inflation is not easy.

A sensible way to implement Equation (6.4) is via a spreadsheet, and it is recommended that students do so. This can then be used to study how sensitive the final cost of the electricity (or heat) is to assumptions about the various terms in Equation (6.4). For some technologies it is found that the cost is most sensitive to the price of the fuel, for others it is the initial capital cost and the length of the construction process.

Problem 6.3   (a) Produce a spreadsheet that applies Equation (6.4) to any energy technology. (b) Apply it to a 1 GW$_e$ nuclear station and a 1 GW$_e$ gas plant, assuming (gas plant figures in brackets): capital cost €1900 (€600) per kW; fuel, maintenance and operation €0.015 (€0.030) per kWh; operational lifetime 40 (20) years; construction time eight (one) years; decommissioning cost €3 billion in total evenly spread over 60 years,

(€3m total spread over two years). Assume all discount and interest rates are identical at six per cent. (c) Use your spreadsheet to investigate the sensitivity of the cost of electricity to a ±100% change in any of the input variables, including the construction time and the discount rate. Present the results as a pair of spider graphs – one for each technology. (Such a graph has the parameter variation in per cent on the abscissa and the cost per kWh of the electricity on the ordinate, with a separate curve plotted for each parameter.) Note: the results you produce are only suitable for general discussion and are unlikely to predict accurately the true cost of the technologies, as the data given is only approximate.

6.2 Environmental economics

In recent years there has been considerable interest in the impact of economic decisions on the environment. This interest is either in how economic activity can damage the environment, or in how economic instruments can be deployed to reduce such impacts.

The circular flow diagram shown in Figure 6.3 took no account of constraints imposed by environmental factors. It implied natural resources are limitless and ignores any waste disposal implications. A better diagram might be Figure 6.5. Here the environment is shown as having three economic functions: it is a source of amenity services, a resource supplier and a waste receptor. These functions are frequently interlinked, for example the amenity value, the fishing potential and the waste disposal possibilities of a coastline.

In this model, the economy and the environment are seen to be one. Everything that happens in the economy can have an impact on the environment, and vice versa. It cannot be overemphasized that the environment is not something external to the

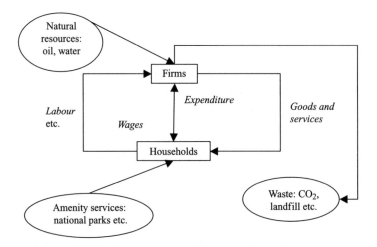

Figure 6.5 A model of the economy with the environment included

economy – the economy sits inside of the environment. Indeed, economic activity can be seen primarily as the conversion of natural resources into products. Without the environment there would be no oil, iron, water, food or air for the economy to access, or a location for waste disposal.

As the Earth is an approximately closed system, a finite set of resources is available for current and future growth. Even if resources are sufficient to permit growth, the extra production will 'drag through' more materials and energy in products, emissions from which the environment must ultimately assimilate. Such a philosophy of growth is clearly unsustainable.

This has led to a desire to decouple economic growth from environmental harm, thereby making growth sustainable. There are at least two ways this could be achieved:

1. Technological change

The increased use of clean technologies in the developed world has led to some pollution being decoupled from GDP (see Figure 6.6). Unfortunately, many developing countries have yet to reach this turning point for several key pollutants. In the case of greenhouse gases, the developed world is starting to decouple emissions from growth, but far too slowly to stop climate change. The developing world has yet to start the transition – mainly because of their greater growth rates, the greater cost of the new technologies and the need to apply discounting.

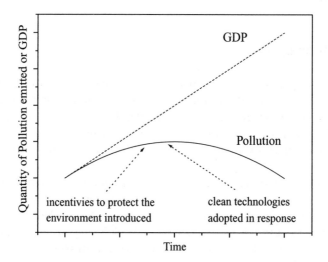

Figure 6.6 Breaking the link between GDP and pollution. The level of an emitted pollutant is seen to grow as a nation gains wealth and more use is made of the polluting technology. The introduction of new, cleaner technologies then allows growth to continue but with reduced emissions. An example being urban air pollution, which increased for decades but then reduced with the fitting of catalytic converters to cars

2. Conservation and recycling

The recycling of glass may, for example, save 25 per cent of the energy and raw material inputs; the recycling of aluminium, 95 per cent. Thus it is possible to expand the use of both of these materials without increasing energy demand (as long as an ever-greater amount is recycled). The same is true for conservation; by using more efficient machinery it is possible to increase mechanical output, without increasing fossil fuel use.

We have just seen that an increase in GDP can bring with it environmental costs. This has led economists to try and include such costs within the GNP measure. One example is the *Index of Sustainable Economic Welfare* (ISEW). This adjusts the GNP by including monies spent correcting any decline in natural resources, environmental and pollution damage:

> ISEW = GNP *minus*
>   depreciation of physical capital *minus*
>   defensive expenditures *minus*
>   depreciation of environmental capital *minus*
>   monetary value of residual pollution.

ISEW therefore accounts to some degree for the cost of pollution. Several studies have shown that ISEW is currently falling in the developed world even though GDP is growing – a worrying sign.

Cost benefit analysis

ISEW accounts for pollution at a national level; *cost benefit analysis* (or CBA) can be used to provide a monetary valuation of the perceived costs and benefits of an individual investment project or act. It allows, for example, the external costs (health damage, building damage etc.) of road transport to be internalized and thus given a financial value, allowing the true cost to be compared with the benefits (mobility, GDP growth etc.).

By comparing the marginal pollution cost (the environmental cost of one more unit of production) to the marginal net private benefit (the benefit to the firm in question) the true economic optimum level of pollution can be found. Note, the (economic) optimum level of pollution is not, in general, zero. This suggests we might not wish to implement policies that allow no anthropogenic carbon emissions.

A great deal of the above rests on our being able to attach monetary values to environmental effects. This is not easy. One way to estimate such costs is via *willingness to pay* information gathered from the public by questionnaire. Such surveys ask people to state how much they would be willing to pay to avoid an environmentally damaging event from occurring, or to solve a pre-existing environmental problem.

Economic solutions to environmental problems

Having carried out a CBA, we need to act to reduce the problem to the level suggested by the CBA, or lower. Three types of incentives to reduce the damaging effects of energy use are often discussed: environmental taxes (e.g. carbon taxes), tradable permits and regulation.

1. Environmental taxes

An environmental tax is a tax on a product or service that is detrimental to the environment. The idea is to increase the cost of the most polluting methods of production. If the externalities are fully internalized through such a method then, by definition, the marginal pollution cost will equal the marginal net benefit to the polluter, and the economic optimum level of pollution will have been reached. Taxation beyond this point will penalize the activity to a greater extent than is *economically* desirable.

Such taxes embody *the principle of polluter pays* and are common, with the U.K., for example, raising over eight per cent of its taxes in this way (mainly through fuel taxes).

2. Tradable permits

Here the polluter receives a permit to emit a specified amount of waste (e.g. CO_2 or SO_2). If they wish to pollute more than this they need to buy further permits (usually from competitors). Once more, a cost is placed on polluting and pressure is applied to find better means of production. Such permits find great favour with economists, because those who can save emissions at the lowest cost will do so first (and sell their permits). Thus the most environmental benefit is achieved at the minimum cost. This approach has been used successfully in the USA to reduce acid emissions from power stations and we will revisit the idea in Chapter 18 as a way of helping to solve the problem of climate change.

3. Regulation

Here the government sets fixed limits on emissions. Examples are exhaust emission regulations for cars and chemical emission regulations for power stations. Unlike environmental taxes, regulation can only work where an individual or firm can be linked with the pollution. Regulation is thought not to be the most economically effective means of reducing emissions because it requires a similar level of response from all polluters, some of which will need to spend more than others to solve the problem. By contrast, tradable permits encourage the polluter to make reductions below the regulation level and retrieve the cost by selling their permits to others who face prohibitive costs of reduction. However, because of this, tradable permits can increase pollution in some areas. Therefore they are most suitable for controlling national and global pollutants, rather than improving local air quality.

6.3 Student exercises

"Economics is the science of greed."

–F. V. Meyer

1. Supply and demand curves have different signs. Why?

2. Given a country with a GDP of €3 billion and €0.5 billion of money in circulation, what is the velocity of money in the country?

3. a. If spending €100 on insulation will save a household €25 per annum on heating bills, calculate the simple payback period of the insulation.

   b. If the embodied energy of the insulation (which lasts only a single year) is 1 kWh per cm of depth, and each cm applied will save 2 kWh/d, (where d is the depth in cm), what is the maximum sensible depth of insulation to the nearest cm? There are several answers to this question – discuss their differences.

4. Explain what is meant by the 'tyranny of discounting'.

5. Given a discount rate of seven per cent, what is the greatest sum it is economically worth spending now, to stop an environmentally damaging event costing €100 000 from occurring in 10 years time?

6. Describe the relationship between sustainable development and technological change.

7. Define ISEW. If the ISEW of a country is falling, yet GNP rising, what does this suggest about the country's environment?

8. Why are tradable permits possibly not always a guaranteed way of improving local air quality?

7

Combustion, inescapable inefficiencies and the generation of electricity

Electricity is the most important and most valuable source of energy humankind has. It is relatively easy to convert electricity into most other forms of energy, heat for example, but much more difficult to do the reverse. In this chapter we will see that these difficulties arise because of the nature of the physical processes involved and the complexity of the engineering required to generate and distribute useful amounts of electrical energy. Although fossil fuels are consumed via combustion in boilers, car engines and turbines, the chemistry involved is largely unaffected by the technology. We therefore initially consider combustion in general terms as a source of heat. We will then see how this heat can be used to drive a power station.

7.1 Combustion

Natural gas, or methane (CH_4) burns as

$$CH_4 + 2O_2 \rightarrow CO_2 + 2H_2O \tag{7.1}$$

Energy and Climate Change   David A. Coley
© 2008 John Wiley & Sons, Ltd

Table 7.1 Atomic masses of elements commonly involved in combustion

Element	Atomic mass
Carbon (C)	12
Hydrogen (H)	1
Oxygen (O)	16
Nitrogen (N)	14

and propane as

$$C_3H_8 + 5O_2 \rightarrow 3CO_2 + 4H_2O \tag{7.2}$$

Although the fuel burnt in these equations is different, the result is the same: a mix of carbon dioxide and water. These are termed the *products* of the reaction and are produced when any hydrocarbon is burnt. The hydrocarbon is termed the *fuel*. Due to impurities in the fuel, most notably sulphur, the practice of burning fuels in air rather than pure oxygen, and only realising partial combustion, the burning of fossil fuels also gives rise to other compounds. However, these additional products will be produced in much smaller quantities and the general point holds true: burning carbon-based fuels generates water and carbon dioxide.

Table 7.1 gives the relative atomic masses of the most common elements involved in combustion for energy production. From this we can see that one gram of methane is $1/(12 + (4 \times 1)) = 1/16$ mol[1] of methane and therefore its combustion will require $2/16$ mol of oxygen and produce $1/16$ mol of carbon dioxide and $2/16$ mol of water. Referring to Table 7.1 once more allows these molar quantities to be converted to masses:

$2/16 \times (2 \times 16) = 4$ g of oxygen

$1/16 \times (12 + (2 \times 16)) = 2.75$ g of carbon dioxide and

$2/16 \times ((2 \times 1) + 16) = 2.25$ g of water.

This will release 53.42 kJ of chemical energy in the form of heat.

Problem 7.1   Calculate the mass of carbon dioxide released from the combustion of 1 kg of propane.

A mixture of fuel and air that contains the theoretical minimum quantity of oxygen required for complete combustion of all the fuel is called a *stoichiometric mixture*. In

---

[1] One mol of an element contains Avogadro's number (6.02×10^{23}) of atoms. So 12 grams of carbon contains 6.02×10^{23} atoms of carbon. In a similar manner, one mol of a molecular substance contains 6.02×10^{23} molecules. (This simple picture is somewhat complicated by the existence of difference *isotopes* of the same element each with (usually) slightly different atomic masses, however for engineering level calculations this subtlety can be ignored.)

practice more air than this is used to ensure adequate mixing and combustion in a limited time. For hydrocarbons too little air will give rise to incomplete combustion and the formation of a mix of unburnt fuel, water, carbon monoxide (CO) and carbon dioxide. This will also produce less thermal energy.

Problem 7.2   Air is composed of 76.7 per cent nitrogen (N_2) and 23.3 per cent oxygen by mass. Estimate the mass of air required to form a stoichiometric mix with 1 kg of propane (C_3H_8).

After complete combustion no fuel is left. Therefore the heat produced must lie in the products of the reaction. Given *adiabatic* conditions, i.e. where heat is not lost to the outside world, most fuels (initially at room temperature) burn to produce products that reach temperatures of 2000 to 3000 K [EAS90]. In practice the mass of product will be greater than suggested by equations such as (7.1) and (7.2) because of the presence of excess amounts of air (and the nitrogen it contains) and therefore the temperature rise will be less (although the energy released will be the same – it is just that the energy is trying to elevate the temperature of a greater mass). This reduced temperature can be surprisingly useful as a way of controlling the temperature of a reaction so that the materials from which combustion chambers are formed do not fail.

7.2 Calorific values

As stated previously, the energy content of a fuel is given by its calorific value. If a fossil fuel, biofuel or hydrogen is burnt then the resultant water will be in the form of steam. As this steam cools it will condense to form liquid water. This act of condensation will also release internal, or heat energy. If we include this second source of energy in the measurement of the calorific value of a fuel we have the *gross*, or *higher*, calorific value. If not, and the water remains as a vapour, we have the *net*, or *lower*, calorific value. Table 7.2 lists the calorific values of common fuels.

7.3 Inescapable inefficiencies

The efficiency, η, of an energy conversion process is commonly defined as the ratio of useful energy out, to the total energy in, i.e. $\eta = E_{out}/E_{in}$. One might think that

Table 7.2   Net and gross caloric values of common fuels

Fuel	Net (MJ/kg)	Gross (MJ/kg)
Wood	10–13	13–16
Natural gas	48.16	53.42
Oil (gas oil)	42.8	45.6
Coal (anthracite)	28.95	29.65

maximising η is simply a matter of good engineering. Unfortunately no matter how good the engineering, if we wish to convert heat into work by using a 'heat engine', η will be far less than we would hope. By heat engine, a term more commonly shortened to simply 'engine', we mean a device for turning internal energy (heat) into mechanical energy, i.e. doing work. The internal combustion engine within a car and the steam turbine within a nuclear power station are both examples of heat engines. Although it is possible to turn mechanical energy into heat with 100 per cent efficiency, the reverse – turning heat into mechanical energy – can never take place without losses, however good our engineering.

As we have already said, heat is a form of energy; heat is not equivalent to temperature. The heat that can be stored, or given up by a body depends not only on its temperature, but also on the mass of the body and its thermal capacity per unit mass (termed the specific heat capacity), c_m.

The heat energy, Q, (or E_{th}) stored in a body of mass, m, which has been raised from temperature T_1 to T_2 is given by:

$$Q = mC_p\,(T_2 - T_1).\qquad(7.3)$$

Problem 7.3   1 kg of water at 80 °C is mixed with 0.5 kg of water at 50 °C. What is the final temperature of the mixture? Hint: Conservation of energy requires that the quantity of heat energy is not changed by this mixing.

Within a heat engine, as with any closed system, the total amount of energy must remain constant[2]. By closed we mean that neither matter nor energy can escape. So any change of internal energy within the engine must equal the net heat energy (Q) input, minus the net external work done, W. If the engine returns to the same state after each cycle (for example the cycle of compression, ignition and expansion of an internal combustion engine) then the energy stored will be zero, and

$$Q - W = 0.\qquad(7.4)$$

In reality, friction and other losses means that the system will not return to exactly the same state, nor can it be considered truly closed. However, such an idealized system is termed an *ideal heat engine* and forms a useful concept, which will allow us to estimate the maximum possible efficiency of *any* engine design and what this efficiency might depend on.

Figure 7.1 shows a schematic of an ideal heat engine. A source of heat at temperature T_{source} provides heat energy, Q_{source}, to an engine that does external work (for example in rotating the shaft that drives a car) and rejects any heat Q_{sink} not used, to a low temperature sink (typically the atmosphere, or the cooling water circuit in a power station).

---

[2] The first law of thermodynamics.

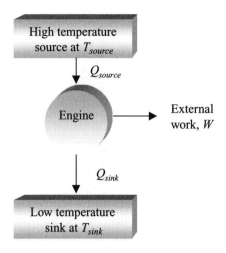

Figure 7.1 Schematic of an ideal heat engine

Rearranging Equation (7.4) shows that:

$$Q = W,$$

or in our case that

$$Q_{source} - Q_{sink} = W. \tag{7.5}$$

Clearly if we can make Q_{sink} as close to zero as possible then W will be maximized and we will have a very efficient machine where almost all the heat supplied is turned into work; but how close to zero can we make Q_{sink}?

As we stated at the beginning of this section, the efficiency, η, of any system is the ratio of energy (or work) out, to energy in, or in our case:

$$\eta = \frac{W}{Q_{source}}.$$

Substituting for W from Equation (7.5) gives

$$\eta = \frac{Q_{source} - Q_{sink}}{Q_{source}} = 1 - \frac{Q_{sink}}{Q_{source}}. \tag{7.6}$$

This is still as we might expect: if Q_{sink} is close to zero, $\eta = 1$ (i.e. 100 per cent efficiency).

In 1824 a French engineer by the name of Sadi Carnot proved that for an *ideal* heat engine (i.e. one with no losses),

$$\frac{Q_{sink}}{Q_{source}} = \frac{T_{sink}}{T_{source}},$$

and therefore for such an engine, by Equation (7.6)

$$\eta = 1 - \frac{T_{sink}}{T_{source}}. \tag{7.7}$$

This equation applies to ideal (or perfect) engines; therefore any practical (i.e. imperfect) engine will always have an efficiency less than this. So for a real engine:

$$\eta < 1 - \frac{T_{sink}}{T_{source}}. \tag{7.8}$$

This innocuous looking equation is full of ramifications for many of the systems studied in this book. In the case of an ideal heat engine, η is termed the *Carnot efficiency*, η_{carnot}. In Equations (7.7) and (7.8), both temperatures are measured in K, not °C and therefore η_{carnot} only equals 1 if T_{sink}, the sink temperature, is set to absolute zero (−273 °C, −459 °F) – an impossibility. Unfortunately we do not have readily available sinks at anything like this temperature. The air or oceans rarely fall below −30 °C. In addition, most materials fail at temperatures below 2000 °C, setting a maximum value for T_{source}. Therefore the maximum efficiency we could hope to reach is probably around:

$$1 - \frac{-30 + 273}{2000 + 273} = 0.89$$

or 89 per cent. (The use of 273 is required in order to convert celsius to kelvin.) More realistic temperatures, say 300 °C on the input side and 20 °C at the outlet would yield an efficiency of only 49 per cent. This is not a very impressive efficiency as it implies half of the energy provided by the burning of coal, oil or gas, or the heat provided by a solar thermal array or nuclear reactor, has to be wasted, no matter how good the engineering, if we try to use heat to carry out work. This still holds true if the purpose of this work is to generate electricity.

The Carnot efficiency of a system is the theoretical maximum efficiency of a reversible[3], theoretically perfect heat engine operating between two temperatures, regardless of the mechanics of the device. Real systems also suffer from genuine inefficiencies,

---

[3] A reversible process is one in which the system can be made to pass through the same states in the reverse order when the process is reversed.

such as friction, further reducing the overall efficiency, often to very low values. For example, the Carnot efficiency of an engine operating at temperatures typical of an internal combustion engine within a car is likely to be around 30 per cent. However a real car engine is typically only 20 per cent efficient, and the overall efficiency of the car is only 15 per cent if losses in mechanical linkages and the rolling resistance of the tyres etc. are taken into account, and only 7 per cent, if air resistance is included.

For some engines, for example gas turbines, there is the need to maintain minimum temperatures and pressures at the exhaust point in order for the engine to function, so the minimum sink temperature is often not the temperature of the surrounding environment, but the much higher exhaust temperature.

The good news is that all this rejected (or wasted) heat can still be used as heat, for example to warm houses, swimming pools etc. Unfortunately this is rarely done in most countries, and for mobile systems, such as cars, even this is not a possibility. For other systems that do not have a heat engine at their core, such as hydroelectric power stations, electric motors or photovoltaics, the concept of a Carnot efficiency does not apply and is not a limiting factor.

Problem 7.4 Estimate the Carnot efficiency of a power plant that uses steam at 300 °C and cooling water at 20 °C. Use the result to estimate the theoretical maximum quantity of electrical energy (in kWh) that could be generated from 1 kg of coal by such a plant.

Figure 7.2 shows an energy conversion diagram that highlights the underlying poor efficiency of converting thermal energy into mechanical energy. This is the weak link; not the conversion of chemical energy into thermal energy or the conversion of mechanical energy into electrical energy. This efficiency is so low that the heat loss from UK power stations exceeds the total UK domestic heating requirement [ATK86].

7.4 Heat pumps

The concept of a Carnot efficiency applies in reverse, i.e. to a refrigerator or heat pump. Figure 7.3 shows a schematic of such a device, which indicates that, counter to intuition, we can make heat flow from a cold to a hot reservoir, but only by carrying out additional work via the heat engine. For example a refrigerator extracts heat from the food stored inside it and uses it to heat the air within the kitchen.

A heat pump extracts heat from a large reservoir, typically the ground or the outside air and transfers it to a warm reservoir, typically the air inside a building. If more heat energy is transferred between the reservoirs than work is done in transferring the heat then the operation will have been worthwhile, and in money terms we will have gained heat or energy for free.

For an ideal heat pump the heat transferred to the hot reservoir will be the sum of the heat removed from the cold reservoir plus any work done by the engine (if the engine

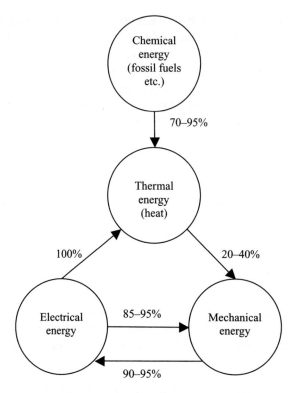

Figure 7.2 Energy conversion efficiencies [adapted from EAS90, p296]

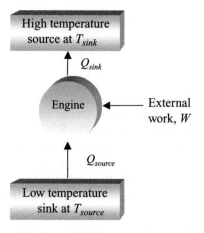

Figure 7.3 Schematic of a heat pump

sits inside the hot reservoir), i.e. energy is still conserved. Therefore:

$$Q_{sink} = Q_{source} + W. \qquad (7.9)$$

The 'efficiency' of the device will be the ratio of heat transferred to the hot reservoir, to the work input:

$$\eta = \frac{Q_{sink}}{W} = \frac{Q_{sink}}{Q_{sink} - Q_{source}} = \frac{1}{1 - (Q_{source}/Q_{sink})},$$

and for an ideal heat pump:

$$\eta_{Carnot} = \frac{1}{1 - (T_{source}/T_{sink})} = \frac{T_{sink}}{T_{sink} - T_{source}}. \qquad (7.10)$$

Hopefully more heat is gained than work is carried out and thus the 'efficiency' will be greater than 100 per cent.

Such an efficiency can only be greater than 100 per cent because we have conveniently forgotten to include the heat energy we are gaining for free from the cold reservoir. In order to avoid confusion with true efficiencies, which are always less than 100 per cent, the efficiency of a heat pump is usually termed its *coefficient of performance*, COP, or COP_{Carnot} in the ideal case.

We can now explore the likely COP of an ideal heat pump and discover how much energy we might gain for free. If $T_{sink} = 20\,°C$ (a typical air temperature for a building) and $T_{source} = 5\,°C$ (external air on a winter's day), then from Equation (7.10),

$$\eta_{Carnot} = \frac{T_{sink}}{T_{sink} - T_{source}} = \frac{20 + 273}{(20 + 273) - (5 + 273)} = \frac{293}{15} = 19.5,$$

i.e. for every watt of electrical power drawn by the heat pump nearly 20 watts of heat could be delivered to a building. Of course, real heat pumps are not that efficient and values of between three and five are more typical in practice (i.e. less than a quarter of the Carnot efficiency).

Problem 7.5 Calculate the maximum theoretical COP of a heat pump running between external air at −20 °C and a building at 20 °C. Repeat the calculation for an external air temperature of +10 °C. The result suggests that such heat such pumps are less likely to be practicable in colder environments, or during colder periods of the year – just where and when the demand for heat will be greatest.

7.5 Double Carnot efficiencies

As we have already seen in this chapter, the concept of a Carnot efficiency limits the efficiency of any heat engine-based power station, such as a coal, oil or gas burning one, or even a nuclear power plant. Typically only a third of the energy used will be turned into electricity; the majority of the rest will be rejected to the cold reservoir. This suggests that generating electricity from fossil fuels in order to provide heat to homes is not really a sensible idea, in that it might be better to burn the fuel within the house to generate heat directly. And, in general, this is indeed the case if a modern boiler is installed in the home.

The use of a heat pump can allow us to regain some of the loss that occurs within a power station. So, as long as the product of the efficiency of the power station and the *COP* of the heat pump is greater than the efficiency of a typical domestic gas, oil or coal fired heating system, electrical heating would be the better solution environmentally, in that fuel use would be minimized and overall emissions of carbon dioxide would be lower.

If a power station is only 35 per cent efficient and a domestic boiler 85 per cent efficient then the use of an electrical heat pump and electrical heating will make energetic sense when the *COP* of the heat pump is such that:

$$0.35 \times COP > 0.85,$$

i.e. when the *COP* > 0.85/0.35, or 2.4. Such a value is readily achievable in practice. As Problem 7.6 shows, if the minimization of carbon dioxide emissions is of concern, then the *COP* needs to be higher than this for this argument still to hold.

Problem 7.6   Contrast the carbon dioxide emissions, in terms of kg of CO_2 per kWh of delivered heat, from a coal burning power station and heat pump combination ($COP = 4$), to that of a domestic gas boiler.

Having studied the theoretical limitations of heat engines, we turn our attention to the generation of electricity by such engines.

7.6 The generation of electricity from heat

Electricity supplies around 15 per cent of the energy demand of most developed nations. It is therefore a very important energy source. However, electricity is also important because it allows us to run machines and power processes that would be impossible, or at least very difficult, using other energy sources. Although it is quite possible to replace a coal boiler with a gas-powered one, computers for example have to use electricity, as does the production of aluminium. Electricity is also important because it is easy to convert into kinetic energy or heat and is highly portable using wires – rather than the pressurized distribution systems required by gas or oil, or the trucks needed for the movement of coal. Such is electricity's importance that most primary sources of energy are converted to it, and therefore the generation of electricity often forms a common link by which to discuss nuclear, coal, oil, gas, hydro and alternative energy. In the case of fossil fuels and biomass, the primary source is only used to heat water and create pressurized steam. At the point of generation, the turbine/generator assembly, which fuel was used is largely irrelevant. Many of the same principles also apply to the generation of hydroelectricity. The generation of electricity directly without the need for a rotating generator is left until the introduction of solar power (Chapter 20) and fuel cells (Chapter 28). Figure 7.4 demonstrates the generation of electricity from heat and shows the scale of the losses at various points. We see that fuel is burnt to raise steam, which powers a steam turbine that turns the generator; the steam is then condensed using an external sink (in this case water from a nearby river).

We now need to discuss some of the fundamental aspects of electricity before moving on to talk about practicable generating machines.

Some fundamental concepts

Within a metallic conductor, such as a length of copper wire, some of the electrons from the atoms that make up the metal are not bound to their respective nuclei, but

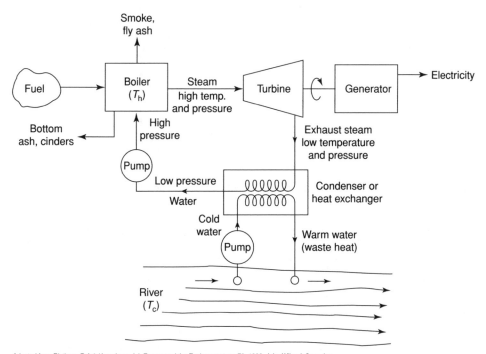

Adapted from Ristinen, R.A & Kraushaar, J.J, Energy and the Environment, pg 70, 1998, John Wiley & Sons, Inc

Figure 7.4(a)   Schematic diagram of a fuel burning power plant [RIS99, p70]

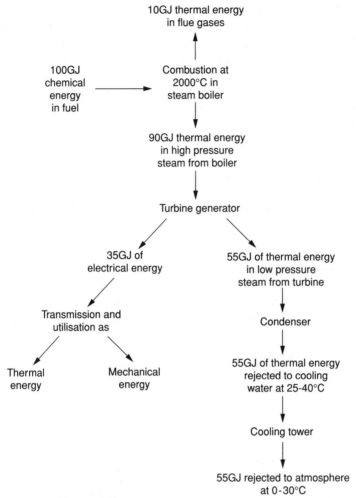

Figure 7.4(b)   Losses in the fuel burning power plant at various points in the process for 100 GJ of fuel [HEW00, p7]

free to travel around the bulk of the material. These *free* electrons randomly bounce about with speeds of the order of 1000 km/s. Because this motion is random, there is no net flow of energy along the wire. However, when for example a battery is connected across the ends of the wire, a small drift velocity is imparted to the electrons and their motion, although still largely random, becomes biased in one direction and they slowly drift along the wire. This drift velocity is small and often only of the order of 1 mm/s – much less than walking speed. It is this gentle flow that heats and lights our homes and powers our computers. The two most important aspects of this flow are described by the current and the voltage, the product of which gives the power flowing within an electrical circuit.

Current

If there are n charge carriers (electrons) per unit volume of wire then the current, i, flowing in the wire will be given by:

$$i = \upsilon n e A \quad \text{ampere (or amps)} \quad (7.11)$$

where A is the cross-sectional area of the wire, υ the drift velocity and e the charge on a single electron (1.60×10^{-19} coulomb).

Problem 7.7   In copper, there is approximately one free electron per atom, or about 8.5×10^{28} electrons per m^3. Estimate the drift velocity of electrons in a copper wire of diameter 1 mm, carrying a current of 1 ampere.

From this example we get an indication that the reason electricity can be so 'powerful' is not so much because individual electrons carry large amounts of energy about their person, but in part because there are simply so many of them in any conductor.

Problem 7.8   Estimate the volume of copper that contains the same number of free electrons as the planet does people. (The current population of Earth is 6.4 billion [CIA05].) The answer gives some idea of just how electron-dense conductors are.

Within a conductor the electrons are not entirely free to drift and there will be resistance in the form of collisions with the atoms that make up the conductor. During such a collision some of the energy of the electron will be imparted to the atom thereby increasing the conductor's internal, or vibrational, energy, i.e. increasing the temperature of the conductor. The electron is then re-accelerated by the *electric field* provided by the battery. It is only at such a point of transfer that the energy of the current is realized. This is why an electric bar fire gets hot or a light bulb glows, but the wiring to these items does not—the wiring has a much lower resistance. The situation is analogous to that of filling a car with petrol (gasoline). Although there is a large 'relocation' of chemical energy through the filling hose from storage tank to vehicle, none of this energy is realized until the fuel is burnt in the car's engine; there is therefore no energy flow. This is a general point. It is only at the point of transformation from one form of energy to another, or from one body to another, that anything of energetic interest happens.

Voltage

In order to establish this flow we need to create an electric field, or potential difference, between two points. An electric field between two points will cause charged particles to move between the points in much the same way as a gravitational field will cause a body with mass to move (fall) through the field. Two points in a gravitational field are described by the difference in the gravitational potential (height) between the points. By analogy, within an electric field points will generally be at different electric potentials. The difference, δV, in electric potential is described in terms of the work we need to do to force a charge, q, to move in a direction opposite to its natural inclination; so:

$$\delta V = -\frac{W}{q}. \tag{7.12}$$

If we have one *coulomb* of charge (the charge carried by approximately 6×10^{18} electrons [YOU92]) and we need to carry out one joule of work, then we need to have a potential difference of 1 J/coulomb, an amount termed 1 volt (V). A small battery typically has a potential difference of 1.5 volts between its ends. Mains electricity is supplied in Europe at approximately 230 volts; a power station might produce a potential difference of several hundred thousand volts.

Power

From Equation (7.12) we see that (ignoring the sign),

$$W = q\delta V. \tag{7.13}$$

Work has the units of energy, and returning to our wire of cross-sectional area A with n free electrons per unit volume, each with charge e, then Equation (7.13) suggests that if υneA amps per second pass any point in the wire then the rate of energy transfer (the power P) per second will be

$$P = \upsilon neAV$$

or

$$P = iV \text{ J/s or watts (W)} \tag{7.14}$$

i.e. the power flowing in a conductor is the product of the potential difference and the current.

Problem 7.9   An electric heater has a potential difference of 230 volts between its ends and a current of 4 amps flows through it. How much energy does the heater use in a year if left on continuously?

Having defined current, voltage and electrical power, we now look at how such power is typically generated and distributed to homes, offices and industry.

Electric generators

Although electricity can be generated directly from sunlight by using a photovoltaic cell (Chapter 21), or from a source of chemical energy such as a battery or fuel cell (Chapter 28), almost all the electric power in the world is produced by forcing conductors

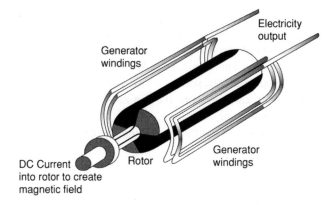

Figure 7.5 Schematic representation of an electric generator [BE05 with permission]

through magnetic fields. In a way that might seem almost circular, this magnetic field is in turn produced via electromagnets powered from some of the current produced by the generator. The coils of wire that form these electromagnets are termed the *field coils*, the rotating inner part of the generator is termed the *rotor* and the fixed shell the *stator*. The field coils can either be on the stator or the rotor. The main conductors that generate the electricity will then be on the other element. As the field coils carry much less current than the main conductors it is typical to place them on the rotor; the main conductors will then lie on the stator. This arrangement saves wear on the brushes that would otherwise have to be used to connect the conductors on the rotor to the wires carrying power away from the machine. Figure 7.5 shows the general arrangement and Figure 7.6 part of a real machine.

Such a generator produces alternating current (a.c.) and the drift velocity of the electrons in any circuit connected to the generator is repeatedly reversed – 100 times a

Figure 7.6 View of a typical 600 MW$_e$ generator set (note person for scale) [BE05 with permission]

second in Europe, 120[4] times a second in North America. The frequency (in cycles per second, or hertz, Hz) of the electric current is given by the product of the rotation speed of the shaft and the number of pairs of magnetic poles (field coils). Large power stations typically run at 3000 rev/min and therefore only require one pole pair to generate at 50 Hz. Many smaller generating sets, such as those using internal combustion engines, often run at slower speeds and therefore require more pole pairs. In either case the speed needs to be held constant or the frequency will change. This has implications for the design of generators such as wind turbines where the driving force can vary greatly second by second.

National and international electricity grids

The main reason for generating alternating current is that it is easily transformed to very high voltages for transmission via a national grid. By using high voltages long distance transmission losses are reduced because such *resistive* losses, (originating from collisions between the electrons and the atoms which make up the conductor) are proportional to the current flowing, not the potential difference. From Equation (7.14) we can see that we can carry 1 MW using a transmission cable at 10 volts and 100 000 amps; or at 100 000 volts and 10 amps. The latter will produce much lower losses. Even using high voltages, approximately 1.5 per cent of a nation's electricity can be consumed by transmission losses [DTI05]. Near the consumer these high voltages are transformed back down to more manageable values, then, if required, rectified into direct current (d.c. – where the drift velocity does not change direction) within electronic goods (Figure 7.7). However, the use of a.c. is not without problems. It means that all generators connected to a grid must use the same generating frequency and be in phase (shuffling the electrons in the same direction at the same time). This can be difficult to organize, especially if electricity is to be exported across national boundaries. With the development of solid-state high voltage rectifiers it is now possible to produce high voltage d.c. for transmission, removing this difficulty: France and the UK are connected by such a d.c. link which is used to import and export electricity (Figure 7.8).

Figure 7.9 shows the layout of the UK national grid and the distribution of power stations. The grid came into being in the 1930s and enabled the most efficient power stations to be used more intensively and cut over-capacity from 70 per cent to 15 per cent. It also meant power stations could be built closer to sources of fuel. The grid consists of more than 807 000 km (500 000 miles) of overhead lines and underground cables and delivers electricity to over 27 million locations [ELE04]. The network is separated into high voltage (400 or 275 kV) grids, plus 14 regional distribution networks operating at 132 kV or less.

---

[4] This is twice the *frequency* of the current as there are two reversals per cycle, or wavelength.

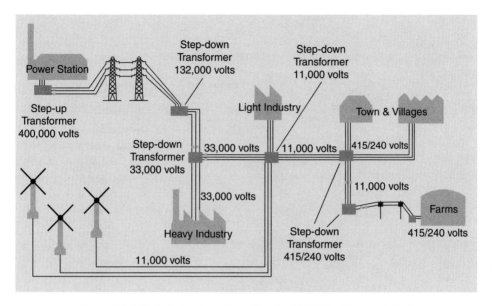

Figure 7.7 Typical structure of a national grid [BE05 with permission]

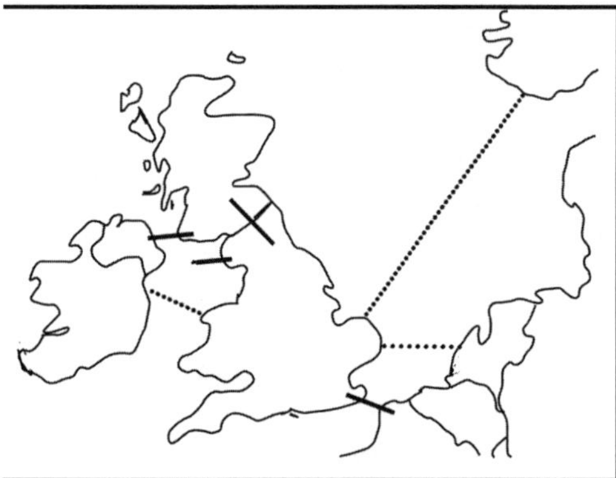

Connector	Capacity (MW)	Length (km)
Britned (Netherlands)	4,320	200
Northsea (Norway)	1,320	730
Scottish	1,600	
Moyle (Northern Ireland)	500	55
French	2000	70
Isle of Man	40	100

Solid line = existing; dotted = planned

Data from Electricity Association, http://www.energynetworks.org/spring/indexpages/enaindex_default.asp (accessed August 2006)

Figure 7.8 UK interconnectors [ELE04]

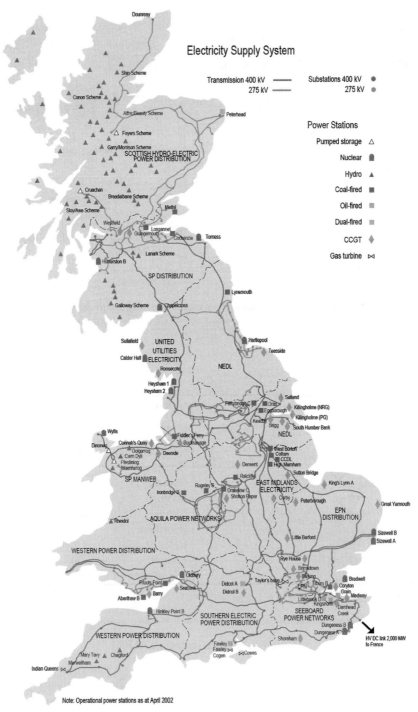

Note: Operational power stations as at April 2002

Reproduced from Power station effects - http://www.powerstationeffects.co.uk/pdf/UK-PowerPlants-NatGrid-DistAreas.pdf

Figure 7.9 UK power stations and the national grid [ELE04]

7.7 Student exercises

"Humanity is acquiring all the right technology for all the wrong reasons."

–R. Buckminster Fuller

1. What are the main products of burning fossil fuels?

2. What is a stoichiometric mixture?

3. Why is the efficiency of any power station that uses heat to make electricity likely to be very low? What are the likely physical and engineering limits to greatly improving such efficiencies?

4. How can using electricity derived from fossil fuels to heat a home be more efficient (in terms of mass of carbon dioxide emitted) than using the same fossil fuels directly to heat the home?

5. Outline with one or more sketches the form of a power station and a national grid. Include relevant temperatures, voltages, etc.

6. Why do the interconnectors joining countries often use d.c. rather than a.c.?

7. What is the maximum theoretical efficiency of a steam engine which uses reservoirs at 150 °C and 20 °C?

8. An engine with a Carnot efficiency of 40% uses a low temperature sink at 15 °C. How much does the temperature of the high temperature source need to rise to increase the efficiency to (i) 50% or (ii) 99%?

9. Why does the national grid use such high voltages to transmit electricity long distances?

10. For different types of power plant, what are the main considerations in choosing a suitable location?

11. If 240 V is applied to a 500 ohm resistance, what current is flowing and how much energy is dissipated? If only 12 V were available, what would the current have to be to maintain the same power

PART II

Unsustainable energy technologies

"No young American should be held hostage to America's dependence on oil"

–John Kerry

In the previous chapters we have seen how dependent the development of modern society has been on the availability of a seemingly limitless supply of cheap energy. We have also seen how the emissions of greenhouse gases from the use of this energy now pose a serious threat to mankind and the planet's ecosystem. In Part II we will be investigating each of the main current energy technologies upon which we are reliant, namely: coal, oil, gas, nuclear power and hydroelectricity (Figure II.1). In addition we will investigate non-traditional sources of fossil carbon (oil shale, tar sands and gas hydrates) and review some of the special problems that arise from our use of fossil fuels to power transportation systems.

This review of energy technologies and fuels will be carried out with little regard to how they combine to form the energy infrastructure of a nation. In the final chapter we investigate this by studying energy provision within the UK[1] in detail and consider how this has changed over time. This will allow us to ask whether it is possible for a nation or an individual to emit large quantities of greenhouse gasses, or trade and amass wealth from supplying fossil fuels to others, and still consider their actions to be moral.

Figure II.2 shows how the world uses its fossil fuels, from which we see that we can separate use into four main sectors and that these sectors use a very different mix of fuels, with most oil being used in transport or heating and most coal being used for electricity production.

---

[1] The UK has been chosen mainly because it is an industrially developed nation that is both a fossil fuel user and producer; the majority of the analysis would be applicable to many other nations.

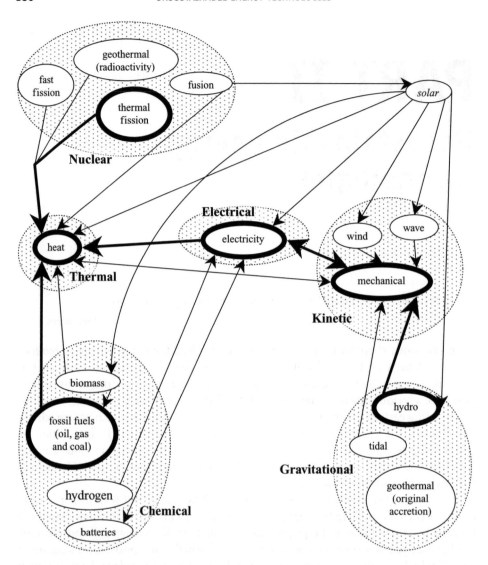

Figure II.1 Energy flows, fuels and technologies studied in this book. Those discussed in Part II are highlighted

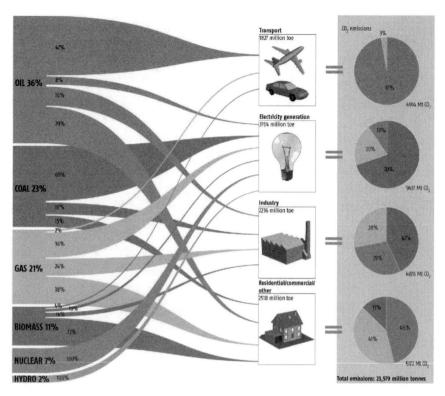

Figure II.2 World energy use (including biomass) and the sectorial split of fuel use and carbon emissions [NEW05a]

8
Coal

"Among the scenes which are deeply impressed on my mind, none exceed in sublimity the primeval forests undefaced by the hand of man. No one can stand in these solitudes unmoved, and not feel that there is more in man than the mere breath of his body."

–Charles Darwin

If it were possible to walk through the world's forests during the carboniferous period (350 million years ago) the most striking thing one would notice, except for the huge trees, would be the silence. Apart from sounds made by a buzzing insect, or a reptile crawling thorough the lush undergrowth or maybe the splash of an amphibian, all would be quiet. This was a time before mammals and before animals had learnt to call, shout and scream their presence. It was also a time when much of the world's vast coal reserves started to be laid down.

The dying vegetation in these swamps of giant trees and ferns accumulated and was initially decomposed by the action of aerobic bacteria and then, as it was covered over, anaerobically. This produced peat, which was buried and heated by further acculimation slowly to produce concentrated carbon in the form of coal.

The biological *peatification* stage took 10 000 to 15 000 years, with anaerobic processes increasing the carbon fraction within the material, for example the conversion of cellulose within the plant to peat:

$$12C_6H_{10}O_5 \rightarrow C_{62}H_{72}O_{24} + 20H_2O + 8CO_2 + 2CH_4.$$

Like oil, coal comes in various forms, but unlike oil, its calorific value varies greatly between these forms. Anthracite is the oldest coal and is chronologically followed by bituminous coal and finally by lignite. In general, the older the coal the higher the carbon content and the more valuable the resource is. The quality and usefulness of any coal depends not only on the organic remains from which it solidified, but also on

Energy and Climate Change   David A. Coley
© 2008 John Wiley & Sons, Ltd

Table 8.1 Coal rankings (data from [McV84])

Name	Carbon (%)	Hydrogen (%)	Calorific value (MJ/kg)
Anthracite	95−98	2.9–3.8	>32.5
Bituminous	82.5−92	4.2–5.6	26.7 − >32.5
Sub-bituminous	78−82.5	5.2–5.6	19.3 − 26.7
Lignites	73−78	5.2–5.6	<19.3

the inorganic fraction added by water seepage from the surrounding mineral deposits. This second fraction can be up to 30 per cent of the coal by weight and contains sulphur, arsenic, cadmium, mercury and radioactive material – all unwanted for energy generation and potentially harmful in any waste stream.

Coal is most usefully ranked according to its percentage carbon content (the higher the better). Table 8.1 gives the percentage carbon content of coals together with their calorific values, the latter varying by a factor of almost two.

8.1 History

As we saw in Chapter 4, although some use had been made of fossil fuels before the sixteenth century the adoption had been marginal. This changed when the use of biomass by industry and in the home reached a level such that it was no longer sustainable, with wood or peat having to be collected from an ever-widening area. England was the first country to make the transition to coal due to the increasing cost of wood and charcoal during the sixteenth and seventeenth centuries [HAR74], and almost all of the country's coalfields were opened between 1540 and 1640. By 1650 output was two million tonnes per annum and 10 million tonnes by 1790.

Because coal contains large quantities of impurities it could not be used directly in iron production – a severe limitation. However in much the same way as charcoal is produced, the heating of coal in the absence of air produces pure carbon, and in 1709 Abraham Darby produced pig iron with this new fuel. This removed the biomass restriction on iron production and truly opened the door to mass production.

Coal could also be used to produce gas, which in turn could be used in lighting installations. As early as 1812 a company had been formed to provide London with a centralized gas supply from this source [SMI94, p160]. However, the most important breakthrough was the invention of the steam engine, which, with the use of seemingly limitless quantities of fuel, freed the whole of industry from the cost and location difficulties of biomass and waterpower. Today we see oil as the most prominent fossil fuel, but it was coal, not oil, that powered the industrial revolution and proved critical to the transformation of society. Today its main role is as the power source for giant turbines that supply 39 per cent of the world's electricity [WOR05].

Figure 8.1   Opencast mine in Germany [RHC04]. Note person for scale

8.2 Extraction

The traditional image of coal mining is one of deep underground mines and winding gear at the head of the pit, but this approach is steadily being replaced by more economical opencast operations (Figure 8.1). The two technologies operate at very different depths. Underground mines can reach depths of over 1000 metres, whereas opencast extractions rarely operate at more than a tenth of this depth.

Coal occurs in stratified layers, or seams, separated by sedimentary rocks. The average deposit is at 100 metres with a seam thickness of one to three metres, although 30 metre thick seams do occur. If a deep mining approach is used only 50 to 60 per cent of the coal can be recovered, opencast mines can return around 90 per cent and this is the preferred approach in the former Soviet Union, the USA and Australia.

Although much coal is exported around the world, a great deal is used locally by the producer countries themselves. Transportation can be by road, rail, ship or slurry pipeline. Such pipelines contain a mix of pulverized coal and water in equal quantities travelling distances of up to 400 km. At the final destination the coal is separated from the carrying water by centrifuges.

Problem 8.1   The Arizona–Nevada slurry line consists of a 46 cm (18 inch) diameter pipe and has the capacity to carry around 2.6 Mt of crushed coal a year suspended in a slurry of 50 per cent water [MAN05]. Estimate the speed of flow in the pipe in metres per second. How does this compare with walking speed?

As a high proportion of the coal mined is used for electricity generation, one sensible approach is to locate power stations as close as possible to mines. The difficult transportation of millions of tons of coal is then replaced by the simpler transportation of electricity via high-voltage power cables.

8.3 The combustion of coal

Almost all coal used today in the developed world is burnt within boilers, i.e. with the purpose of heating water. Coal boilers vary in design and size – from a few kilowatts in the home to over 600 MW in a power station, but all include:

- a combustion chamber
- a feed system to place the coal into the chamber
- a system to supply air
- a flue/chimney, and
- a collection mechanism for the ash

Within the boiler the coal burns as:

$$C + O_2 \rightarrow CO_2,$$

$$H_2 + 1/2O_2 \rightarrow H_2O,$$

$$S + O_2 \rightarrow SO_2, \text{ and}$$

$$N \text{ (both fuel-bound and atmospheric)} + O_2 \rightarrow NO_x.$$

For anthracite this leads to the production of 98 kg of carbon dioxide per GJ of energy released.

Problem 8.2   The world consumption of coal is approximately 2600 MT_{oe} per annum [BP04]. Estimate the mass of carbon dioxide and water produced annually by this consumption.

Small boilers up to one megawatt are gravity fed and the ash is removed by hand. Larger boilers up to around two megawatt use an Archimedean screw or similar mechanism to deliver the coal slowly but constantly. Both have combustion chambers directly below a heat exchanger through which water is passed, or an exchanger transversed by pipes carrying the reaction products (the exhaust gases), Figure 8.2.

The bigger the boiler, the bigger the combustion chamber and the greater the quantity of coal held in it. Above around eight megawatts, stoking (i.e. vibrating the coal to allow ash to fall out of the chamber) becomes impossible and pulverized coal is employed and injected almost as a vapour into the chamber.

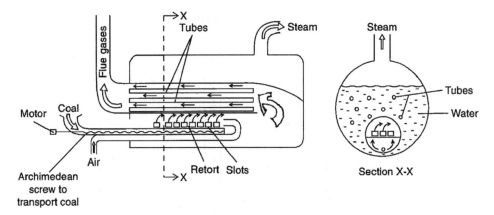

Figure 8.2 Boiler with screw feed

8.4 Technologies for use

Until 1950, most coal was burnt on open fires. With the advent of electric and gas-fired heating such usage is declining rapidly in the developed world, which is just as well given the air pollution such uncontrolled burning causes. Coal has also been used to power vehicles via steam engines, produce chemical feedstocks and as a source of liquid and gaseous fuels; however currently its most important application is in electricity generation. The amount of pollution depends on the technology used. The four main pollutants of concern are carbon dioxide, particulates, sulphur dioxide and nitrogen dioxide, the latter two producing acid rain (Chapter 5). In the controlled environment of a power station, such emissions can be reduced by various measures. Most of the inorganic sulphur can be removed by washing the coal prior to combustion. The remaining inorganic sulphur together with the organic sulphur locked up in the coal itself can be removed from the combustion gases by passing them over a chemical scrubbing agent such as lime or sodium. Nitrogen oxide formation can be inhibited by lowering the combustion temperature, restricting the air intake, recirculating the flue gases and injecting water.

The inhalation of fine particulates is now considered a major health risk. These particles, which are also produced in large numbers by diesel vehicles, can travel considerable distances if not removed at source. Four technologies are used to carry out this removal. Within electrostatic precipitators, the flue gases travel between high-voltage electrodes and precipitate onto one of the electrodes. This can remove 99.9 per cent of the particulate matter by weight. Mechanical collectors use centrifugal or other forces to separate out the particles. Wet scrubbers use water to wash the particles from the exhaust stream. Finally, large fabric bags can be used to filter the exhaust gases.

Reducing carbon dioxide emissions is harder, but possible. Advanced techniques remove carbon dioxide directly from the exhaust stream and are discussed in Chapter 29. A more straightforward approach arises from the fact that it is only the final heat or electricity that we are after, rather than the burning of a fixed quantity of coal. This

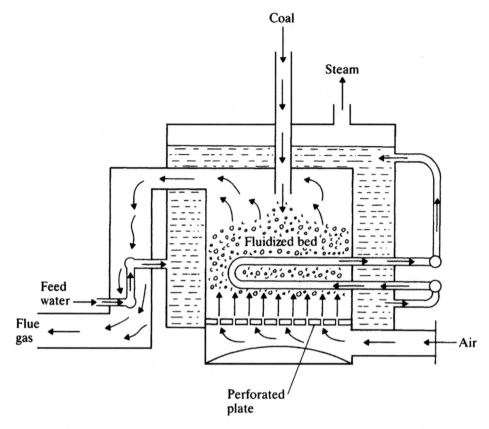

Figure 8.3   Schematic of a pressurized fluid bed plant

means that improving the efficiency of either the burning or the electricity generating plant effectively leads to a reduction in carbon emissions per joule of energy produced. The leading technology to achieve this is the fluidized combined-cycle power plant (Figure 8.3). In such a plant the coal is held in the form of a fluidized bed, mixed with limestone (to absorb sulphur) and burnt in suspension in the presence of a rising stream of injected air. Because of the efficiency of combustion provided by the fluid-like behaviour of the bed, temperatures are lower, thereby reducing the production of NO_x.

Other technologies rely on the ability of coal to be converted into other fuel products. However, even those states with large coal reserves have failed to fully pursue these technologies. The main reason is the relatively low cost of oil. When cost has not been the main consideration, for example when South Africa faced a trade embargo, or when Germany became cut off from its traditional oil supplies during World War II, commercial production has been initiated. South Africa was using 30 million tonnes of coal per annum for liquid fuel production (mainly gasoline, diesel and kerosene) in the mid 1980s, which illustrates that it would in theory be possible to switch the world's economy from naturally occurring oils to coal if needed, although this would be at considerable cost. The four current technologies – indirect liquefaction, pyrolysis, solvent extraction and catalytic liquefaction – require the coal to be heated, and thus tend to be energy inefficient. They also require inputs in the form of catalysts or other products and can lead to concentrated residues of hazardous material.

Indirect liquefaction

Coal can be gasified by heating it in the presence of steam to give a mix of carbon monoxide and hydrogen (*synthesis gas*):

$$\text{heat} + C + H_2O \rightarrow CO + H_2,$$

from which liquid petroleum products may be formed by the use of purification (to remove trace compounds), suitable catalysts, high temperatures and high pressures. This is the process used by the SASOL plant in South Africa.

If the synthesis gas is mixed with more steam, a mix of carbon dioxide, hydrogen and heat is produced:

$$CO + H_2O \rightarrow CO_2 + H_2 + \text{heat}.$$

The original carbon monoxide and this hydrogen can then be converted to methane:

$$CO + 3H_2 \rightarrow CH_4 + H_2O + \text{heat}.$$

Pyrolysis

By heating coal in the absence of air and hydrogenating the products with hydrogen already removed from additional coal, various liquid products can be obtained.

Solvent extraction

Here finely crushed coal is treated with hot hydrogenated solvents to produce a range of products including diesel.

Direct hydrogenation

This uses high temperature hydrogen to react directly with the coal in the presence of catalysts and high pressures. By definition this, just like pyrolysis and solvent extraction, needs additional hydrogen, whereas indirect liquefaction does not.

Figure 8.4 The Drax coal burning power station (U.K.) [DRA05]

8.5 Example applications

The Drax power station, UK

Drax power station in North Yorkshire, UK (Figure 8.4) generates 82.8 PJ (23 000 GWh) of electricity per annum [DRA00]. At the core of the plant are six 660 MW$_e$ coal-fired generating units which produce power from high temperature steam. These burn 8.5 Mt of crushed coal per annum and 37 kt of oil (for start-up and combustion stabilisation) with 361 kt of limestone. Water is converted to steam at 568 °C and 156 bar, by pipes running through the boiler, and used to drive steam turbines. The resultant low-temperature steam is then condensed with the aid of 160 million litres (per day) of water from a local river before it is re-introduced to the boiler. About 50 per cent of the cooling water is lost through the cooling towers and the rest is returned to the river at a slightly elevated temperature. (Some of the heat in the cooling towers is used to warm a nearby set of glasshouses.)

The plant creates 585 kt of gypsum, which is sold to make plasterboard and other plaster products and 1.3 Mt of ash, of which one per cent escapes to the atmosphere. Of the collected ash, 940 kt is sold to the construction industry and the remaining 360 kt goes to landfill.

Electricity is produced at 23.5 kV and is stepped up by transformers to 400 kV for distribution via the national grid. Some 19 Mt of carbon dioxide is produced annually by the plant together with 31 kt of sulphur dioxide, 59 kt of oxides of nitrogen and 16 kt of hydrogen chloride.

Eggborough power station, East Yorkshire, UK

Eggborough is a 2 GW$_e$ coal-fired power station supplied by rail (see Figure 8.5, the numbers in the following paragraphs refer to the figure). Coal is delivered in 1100 t trainloads [BE00] in bottom opening wagons. These are unloaded without stopping as the train passes through the hopper house (1) at approximately 1 km per hour. The coal travels by conveyor to the junction tower (2) where it is screened before more conveyors take it to the boiler house. The Redler coal feeder (6) controls the flow of coal into the Foster Wheeler tube ball mill (7) where it is crushed to a powder (resulting in a dust of particles of less than 0.1 mm in diameter). The pulverized coal is then entrained in a stream of hot air from the primary air fan (8) and sent to the burners (9) where it combusts in a similar manner to a gas jet to produce heat for steam generation. The hot air used by the primary fans is pre-heated in the mill air heaters (10). The forced draught fan (11) draws warm air from the top of the boiler house (12) and passes it through the main air heater (13) to the burner windbox (14) to provide a secondary source of combustion air.

Ash left by the combustion process falls to the bottom of the boiler (15) and is sluiced away to ash pits. The dust is carried by the flue gases to precipitators (16) where it is extracted by high voltage electrodes and blown into storage bunkers. Additional fans (17) then take the cleaned exhaust gases to the main chimney (18).

The heat realized by the combustion process is absorbed by the 50 km (31.5 miles) of tubing which line the boiler walls. Here the water is converted to high-pressure steam. The steam is then further heated by a superheater (19) and passes via control valves into the high-pressure steam turbine (20) and is discharged through nozzles onto the turbine blades. After passing through the high pressure turbine the still hot steam is returned to the boiler for reheating (21) and passed to the intermediate pressure turbine (22) and then to the three low-pressure turbines (23). Coupled to the turbine shaft is the rotor (24) of the generator, which rotates at 3000 revolutions per minute. The electrical rotor is enclosed in a water-cooled stator (25) and the interior of the generator is cooled by hydrogen (which is in turn cooled by water). Electricity is produced in the stator copper conductor bars by the rotation of the magnetic fields created by the rotor electro-magnets. 500 MW$_e$ (per generator) is generated at 22.5 kV and then transformed (26) to 400 kV for export via the National Grid.

The spent steam is turned back into water in the condenser (28) and re-used by the boiler. The water is pumped by the condensate extraction pump (29), heated by two low pressure direct contact feed heaters (30) and pumped by lift pumps (31) to the deaerator (32) – a large heated storage tank, which, as its name suggests, removes any

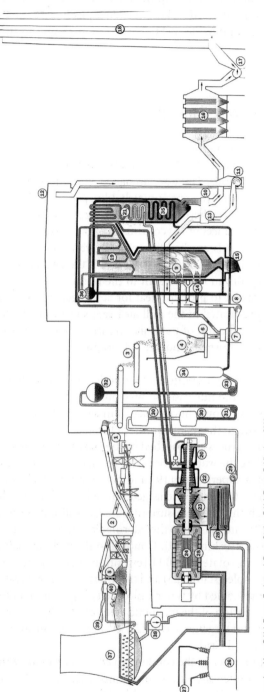

Figure 8.5 Schematic of Eggborough power station [BE00]. Key:

1 Track hopper house
2 Junction tower
3 Shuttle conveyors
4 Coal bunkers
5 Bucket wheel machine
6 Redler coal feeder
7 Foster Wheeler tube ball mill
8 Primary air fan
9 Furnace burners
10 Burners

11 Forced draught fan
12 Air intake to FD fan
13 Main air heater
14 Burner windbox
15 Furnace ash box
16 Electrostatic precipitators
17 Induced draught fan
18 Chimney
19 Superheater
20 High pressure turbine

21 Reheater
22 Intermediate pressure turbine
23 Low pressure turbine
24 Generator rotor
25 Water cooled stator
26 Generator transformer
27 Transmission tower
28 Condenser
29 Condensate extraction pump
30 Low pressure feed heaters

31 Lift pump to deaerator
32 Deaerator
33 Boiler feed pumps
34 High pressure feed heaters
35 Economizer
36 Boiler steam drum
37 Cooling tower
38 Circulating water pumps
39 River Aire
40 Make-up pumps

Table 8.2 Key parameters of Eggborough power station [BE00]. Adapted from British Energy, Eggborough Power Station, British Energy Publications, 2000

Object	Specification
Civil engineering	
Site area	380 acres
Turbine hall	210 × 60 × 35 m high
Boiler house	220 × 55 × 60 m high
Cooling towers	8 of height 115 m and base diameter 115 m
Chimney	198 m high, 18 m diameter
Boiler	
Type	Single furnace
Super heated steam	169 bar, 568 °C
Reheated steam	44 bar, 568 °C
Furnace	29.2 × 9.45 × 29.6 m high
Fuel	800 t/hour (coal)
Coal diameter	75 % < 76 μm
Generators	
Rating	2 GW (4 × 500 MW)
Terminal voltage	22 kV
Terminal current	15.44 kA
Stator mass	211 t
Condenser	
Circulation water	15.13 m³/s
Total cold water circulation of the station	61 m³/s

gases from the water. The water then falls under gravity to the boiler feed pump (33) and back to the boiler.

8.6 Global resource

Figures 8.6 to 8.9 analyse the regional reserves, production and consumption situation in detail. Country specific data is contained in Appendix 1; data for subsequent years may be found on the book's web site.

Here the term *resource* is defined as all the un-extracted material or energy, whereas the term *reserve* is commonly used to identify those resources, which with current or expected technology can be recovered. The *reserves to production ratio* (termed the R/P ratio) is the ratio between the reserve and the current annual production and therefore gives the number of years until all reserves are exhausted.

The present global reserve is 984 Gt, which at current extraction rates will last 192 years (R/P = 192). To allow easy comparisons between fossil fuels to be made, it is common to express consumption and production of all fossil fuels in terms of tonnes of oil equivalent, t_{oe}, where 1 t_{oe} = 1.3 t of coal (anthracite) or 42 GJ. Unlike oil, the resource is widely spread, suggesting supplies are more secure against geopolitical tensions. Again

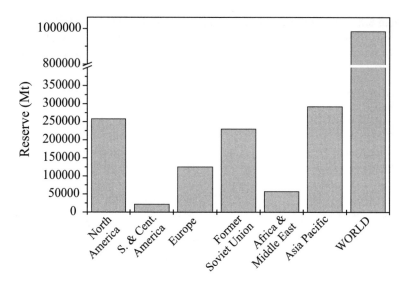

Figure 8.6 World coal reserves [data from BP04]. Regional definitions are given in Appendix 1

unlike oil, regional (and country-centred) production and consumption are fairly balanced, with the main market being the country of origin. In the UK, production and consumption had been balanced (75.7 and 70.7 Mt_{eo} respectively, 1981) but production is now only half of consumption (19.6 and 40.3 Mt_{eo} respectively, 2001) with the coal mining industry having rapidly declined.

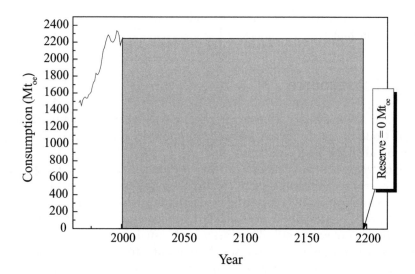

Figure 8.7 Time until total reserve depletion (estimated by assuming production at today's rate) [data from BP04]

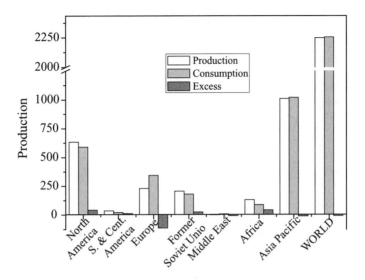

Figure 8.8 Annual regional coal production and consumption [data from BP04]. Total world production = 106 EJ (5.13Gt, 2519 Mt$_{oe}$)

Figure 8.7 shows the current and historic consumption of coal, together with a projection of how long the reserve should last. Assuming production matches consumption, the area under such a curve, from the date of first extraction until all the coal is removed should equal the total original reserve available to humankind. This is often termed Q_{∞}. Given knowledge of the R/P ratio, we can estimate the date, t_{∞}, when all reserves

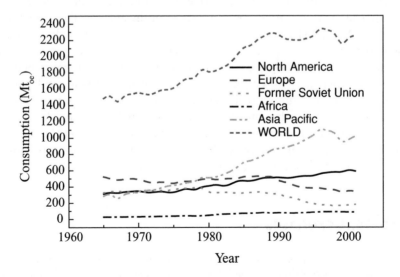

Figure 8.9 Trends in consumption for major coal using regions [data from BP04]

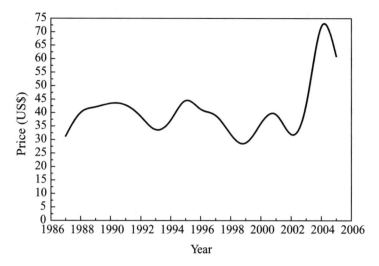

Figure 8.10   Trends in coal prices (based on North Western European prices) [data from BP04]

should be exhausted. As R/P = 192 then for coal $t_\infty = 2196$. Although Q_∞ should be a fixed quantity, it is liable to revision as the definition of *available* changes as new methods of extraction become economic and new coalfields are discovered. Fossil fuel reserves were created over several hundreds of millions of years; Figure 8.7 shows that by comparison we are using this resource at a phenomenal rate. Many have argued that it is an injustice to future generations to squander such a legacy in so short a time and is in complete contradiction to the idea of sustainability.

The total world resource (rather than reserve) is unknown but is many times the reserve, particularly if low-grade coal is included in the analysis. For some very low-grade hydrocarbons, such as tar sands and shale oil, Q_∞ is currently close to zero.

Despite competition from more modern alternatives for power generation, world coal consumption is rising (Figures 8.8 and 8.9), with the dominant growth being in the USA and Asia (where growth is rapid in several countries). Of note are: Japan, which produces very little coal but is a major user; Australia, a major exporter; and France, which having few reserves has opted to generate the majority of its electricity from nuclear sources. Consumption in both Europe and the former Soviet Union is falling.

In the recent past the fluctuations in the world coal price have been small and there seems to be no marked progression in price (Figure 8.10).

8.7 Student exercises

"There's a better scientific consensus on this than on any issue I know – except maybe Newton's second law of dynamics," said D. James Baker, administrator of the U.S. National Oceanic and Atmospheric Administration. "Man has reached the point where his impact

on the climate can be as significant as nature's." Warrick, Joby. "Consensus Emerges Earth Is Warming – Now What?"

–Washington Post 12 Nov. 1997

1. Outline the history of coal use.

2. What are the distinguishing features of opencast and underground mines?

3. What are the main pollutants from coal combustion and how are power stations designed to reduce these?

4. Describe the main technologies for converting coal into other hydrocarbons.

5. Using the data in Appendix 1, or that from the book's web site, plot the ratio of production (Mt_{oe}) to consumption (Mt_{oe}) for all the countries listed. (A spreadsheet loaded using cut and paste from the data on the book's web site would be ideal for this.) Which countries are the major importers and exporters, and approximately how much coal is traded internationally?

6. How did the proportion of coal used in electrical generation in the UK change from 1980 to 2004? (See Chapter 15 for data.)

9
Oil

"Moses dragged us through the desert for 40 years to bring us to the one place in the Middle East where there was no oil."

–Golda Meir

Unlike coal, oil has an aquatic origin and is believed to be formed from marine plants. Material that accumulated on the ocean floor decomposed through the action of bacteria and was slowly covered by sediments, then heated and pressurized by surrounding material. Being liquid or gaseous, the resultant hydrocarbons were free to travel and accumulate under the force of gravity but also restrained by the geologic formations around them. The oldest reserves are found in sedimentary rocks over one billion years old, i.e. over twice the age of coal and before plants existed on land. Above an oil deposit will be found a reservoir of natural gas (sometimes no oil is present at all) and sometimes additional gas can be dissolved within the oil itself. The oil and gas are not contained within a void but within the pores of the rock itself. Thus an oil field is less an underground reservoir and more a bath sponge of oil and gas. This has profound implications for the extraction techniques that are required and the percentage of oil that can easily be extracted. Above the oil bearing rock lie impervious rocks that trap the oil and gas and above this may lie many metres of land or sea. Figure 9.1 illustrates a typical formation.

The oil recovered from a well is called crude oil and has very little use until it is separated into its constituent fractions and impurities removed by refining. Both the raw crude and the final liquid products can be measured in either volume units (barrels, bl; millions of barrels, Mbl) or mass units (tonnes), or of course joules. A barrel of oil is 159 litres, or 42 US gallons or 34.97 Imperial gallons. Standard conversions are: one barrel weighs 0.136 tonnes and contains 5.694 GJ of energy.

Problem 9.1 Norway's oil fields produce 3.4 Mbl of oil every day [BP04]. If this were burnt in oil fired power stations with a conversion efficiency of 30 per cent, how much

Energy and Climate Change   David A. Coley
© 2008 John Wiley & Sons, Ltd

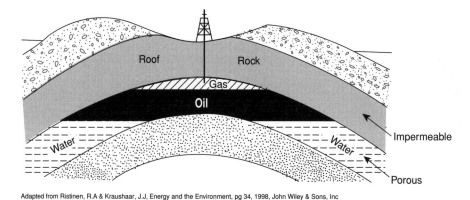

Adapted from Ristinen, R.A & Kraushaar, J.J, Energy and the Environment, pg 34, 1998, John Wiley & Sons, Inc

Figure 9.1 A typical oil and natural gas deposit. Gas is found above the oil and the oil above water, all interspersed in a porous formation [RIS99, p34]

electricity would be produced and what would be the combined generating capacity of these power stations?

Problem 9.2   A person can provide around 6 kJ per minute of useful work [SMI94, p86]. Use the answer to Problem 9.1 to estimate how many people would be needed to provide as much work as could be achieved with Norway's oil output. Compare the answer to the population of Norway (4.6 million [CIA05]).

9.1 Extraction

Oil extraction and use has a long history. Four thousand years ago oil from natural seepages was used for a variety of purposes, from caulking boats to preserving mummies [SHE97]. By the seventh century AD distillation was being used to produce light oils for use in incendiary weapons.

Use was modest until the modern oil era was started by 'Colonel' Edwin L. Drake in 1859 in Titusville, Pennsylvania, USA, when he drilled to just over 20 metres and pumped the first large quantities of oil to the surface. By 1862, Pennsylvanian production had risen to three million barrels per annum from 75 different wells. In 1909 US production stood at half a million barrels per day. During the 1920s and 1930s large fields were discovered and developed in the Middle East.

Until 1973 the price of crude oil was controlled by the large European and American oil producing companies and prices had been stable since the end of the nineteenth century (Figure 9.2). However, in 1973 the dominant force became the producing nations in the form of the Organization of Petroleum Exporting Countries (OPEC, Table 9.1), which had come into existence in 1960 to represent the 12 countries, which together are responsible for almost half of world oil production. OPEC quadrupled the price of oil, which caused an 'oil crisis' in the developed world (often represented by the members of the Organization for Economic Cooperation and Development or OECD –

Table 9.1   The OPEC and OECD members as of 2001

OPEC	OECD		
Iran	Austria	Italy	United Kingdom
Iraq	Belgium	Luxembourg	Australia
Kuwait	Czech Republic	Netherlands	Canada
Qatar	Denmark	Norway	Japan
Saudi Arabia	Finland	Poland	Mexico
Algeria	France	Portugal	New Zealand
Libya	Germany	Slovakia	South Korea
Nigeria	Greece	Spain	USA
Indonesia	Hungary	Sweden	
Venezuela	Iceland	Switzerland	
United Arab Emirates (Abu Dubai, Ras-al-Khaimah and Sharjah)	Republic of Ireland	Turkey	

see Table 9.1). The price almost tripled again in 1979, triggering another oil crisis. The power of OPEC can in part be explained by the fact that they are responsible for 40 per cent of oil production, but also by what they do with this oil. Production by OECD members is approximately 1 Gt per annum and consumption just over 2 Gt, a production/consumption ratio of 0.5. On the other hand, OPEC members consume only 230 Mt for a production of nearly 1.5 Gt, a production/consumption ratio of 6.5. So, although the OECD members produce a large amount of oil, they also use a lot; whereas OPEC members use little, implying that the majority of traded oil is from OPEC countries. This imbalance has the potential to cause severe consequences if geopolitical tensions were to arise between OPEC and OECD member states. This

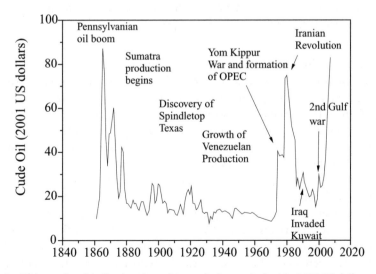

Figure 9.2   History of world oil prices in real terms (price equivalent in 2001 US dollars per barrel) [data from BP04]

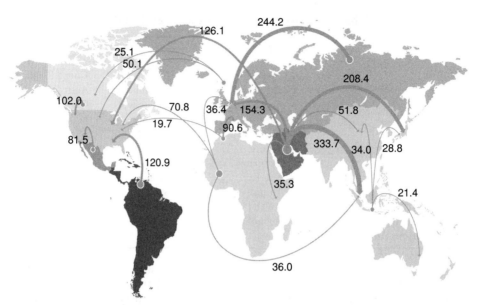

Figure 9.3   Major oil trade movements [BP plc, *BP Statistical Review of World Energy*, BP, 2004]

analysis is confirmed by a consideration of Figure 9.3, which shows the major trade movements of oil. Since 1986 the price of crude oil has returned to more *normal* values and has until recently been priced in real terms at much the same level as it was in 1880 – however at the time of writing (2007), it is once more on the rise.

Problem 9.3   Identify all the movements shown in Figure 9.3 as either originating in OPEC countries or not. This should show that the vast majority of traded oil is from OPEC countries, with one exception. Who is this exception?

Crude oil extraction is a difficult and costly business. Only about one in five exploratory drillings is likely to strike oil and, if the resource is offshore, drilling will be extremely expensive, as will be the cost of constructing, positioning and maintaining the oil platform in the area if oil is discovered. There will be additional costs if the platform is eventually returned to land for decommissioning. The deeper the water and the further from shore, the greater the cost. Wells are now being sited in over 2500 metres of water and 300 km offshore – a very harsh environment.

Only about 15 to 20 per cent of the oil within a well can be recovered by relying on natural pressure differences or by pumping (*primary extraction*). Another 15 per cent can be removed if either water or gas is pumped into the well (*secondary extraction*). The reason for these low returns is, as was mentioned above, the fact that the oil does not sit in large voids, but is within the rock or sand. This means that viscosity and surface tension must be overcome. The viscosity can be reduced by injecting carbon dioxide gas, which dissolves in the oil, or by heating the oil with either steam injection or by

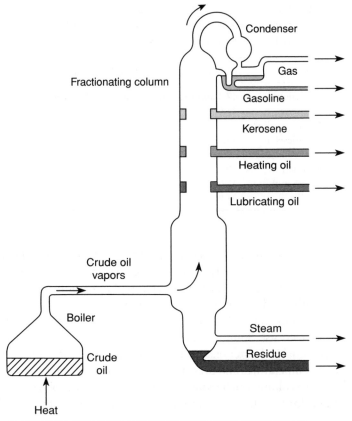

Adapted from Ristinen, R.A & Kraushaar, J.J, Energy and the Environment, 1998, John Wiley & Sons, Inc

Figure 9.4 Schematic of a fractional distillation column [RIS99]

burning some of the oil in place after the injection of air or oxygen. Surface tension can be reduced and the oil allowed to flow out of the rock pores in which it is trapped by the use of detergents. As can be imagined such *tertiary*, or *advanced extraction* methods are not energy neutral and can have environmental consequences; however, a further 20 per cent of the resource might be recoverable in this way.

The recovered oil is transferred to the refineries by either pipelines or ocean-going tankers. The latter can be over 450 metres in length, take 3 km to stop and carry 0.56 Mt of oil. Several collisions have occurred between such tankers and the sea floor resulting in large-scale loss of oil (Figure 5.20). Such spillage can have severe consequences for the environment and be costly to contain and disperse.

Crude oil is refined by fractional distillation to create useful products, a process which uses around five per cent of the chemical energy of the crude. The basic principle is a simple one, although the scale on which it is carried out in order to satisfy our insatiable appetite for oil is truly impressive. Figure 9.4 shows a schematic of a fractional distillation column. Crude oil is heated (to approximately 345 °C) at the base of the column and

Table 9.2 Net calorific values and densities of common liquid fuels. Note how much smaller the range of values is compared to those of coals [OSB85]

Product	MJ/kg	Density at 15 °C (kg/l)
LPG (liquid)	47.3 approx.	0.58 (butane), 0.51(propane)
Kerosene	43.6	0.79
Motor gasoline	44.8	0.80
Gas/diesel oil	43.3	0.84
Heavy fuel oil	40.2	0.97
Bitumen	40.2	
Bioethanol	37.0	
Biodiesel	37.0	

rises as a vapour. The various fractions (which are a mix of compounds) in the vapour condense at various heights as the temperature reduces with height. Less volatile oils, such as heating oil, condense near the bottom of the column, more volatile compounds such as gasoline, only condense in the upper, cooler regions.

The proportions of the various fractions produced do not correspond to the proportions that we as a society use. Some of the less useful heavy oils can be converted to more useful, and more valuable, lighter oils such as gasoline by either *thermal cracking* of the molecules at high temperatures and pressures, or by the use of catalysts. Conversely, the gases at the top of the column can be converted into liquid fuels by *polymerization*. Table 9.2 gives a list of calorific values and other physical properties for the most important fractions.

Not only is 'oil' not a single compound, the individual fuels obtained by distillation are a mix of compounds. For example, petrol (gasoline) is a mix of hydrocarbons of between six and twelve carbon atoms (i.e. C_6H_{14} to $C_{12}H_{26}$) with an average of eight (i.e. octane C_8H_{18}).

9.2 The combustion of oil

Unlike coal, oils are used to power a wide variety of combustion systems: from simple domestic boilers to aircraft turbines. Using octane as an example (a common constituent of petrol (gasoline)), oil burns as:

$$2C_8H_{18} + 25O_2 + \text{contaminates} \rightarrow 16CO_2 + 18H_2O \text{ plus minor products.} \quad (9.1)$$

From which we see that burning oil produces approximately 74 kg of CO_2 per GJ of energy released.

Problem 9.4 The world consumes 152 EJ (3637 Mt) of oil per annum [BP04]. Estimate the mass of carbon dioxide and water produced annually by this consumption.

9.3 Technologies for use

Oil is an expensive fuel, which means it is second choice to natural gas and to coal for the generation of electricity. However it is highly flammable compared to coal and easier to store than gaseous hydrocarbons, which makes it ideal for systems that have to be switched on and off regularly and are mobile: hence its adoption for many transport systems, including aeroplanes, cars and ships. It is also used as a heating fuel. In mobile systems it is usually burnt in internal combustion engines where the required power is likely to fluctuate and capital cost is important, or in turbines where power levels are likely to be constant for long periods and initial capital costs of less importance. Other uses include solvents, lubricants, candles, roofing materials and road construction.

As Table 9.2 demonstrates, oils have similar calorific values, so the choice of which to use for a particular application usually depends on other physical properties, most notably the viscosity. High viscosity oils are hard to pump and difficult to atomize to effect efficient mixing with air to allow high-speed combustion. (Preheating the oil (to around 100 °C) will greatly reduce its viscosity.) Small domestic boilers run on kerosene, industrial boilers on light fuel oil and very large boilers on heavy fuel oil. Aircraft turbines are fed a form of kerosene or gasoline, internal combustion engines gas-oil (diesel) or gasoline (petrol).

9.4 Example application: the motor car

Although estimates vary, there are probably around 600 million cars in use worldwide [PHY05]. Within developed states they account for in excess of 25 per cent of greenhouse gas emissions and primary energy use. They also emit a large quantity of particulates and other air pollutants. The vast majority of cars are powered by either diesel or petrol (gasoline) engines. Chemical energy contained in the fuel is converted first to heat energy and then to rotational motion. This rotation is translated via mechanical linkages to the wheels to provide movement. Figure 9.5 shows the four phases of a single cylinder (typically there are four cylinders) of a petrol engine. Fuel and air mixed in a ratio of one to 15 (by mass) [PUM05] within the cylinder is first compressed (a), then a high-voltage spark is released (b), igniting the fuel which burns (or explodes). The temperature in the cylinder reaches over 1000 °C and the expanding gases thrust the piston downwards and cause the camshaft, to which it is connected, to rotate. The reaction creates mainly carbon dioxide and water with a small amount of other, toxic, gases. A valve then opens (c) to exhaust the combusted gases, and finally a further valve opens (d) to re-fuel the cylinder. The whole process takes around 0.034 seconds.

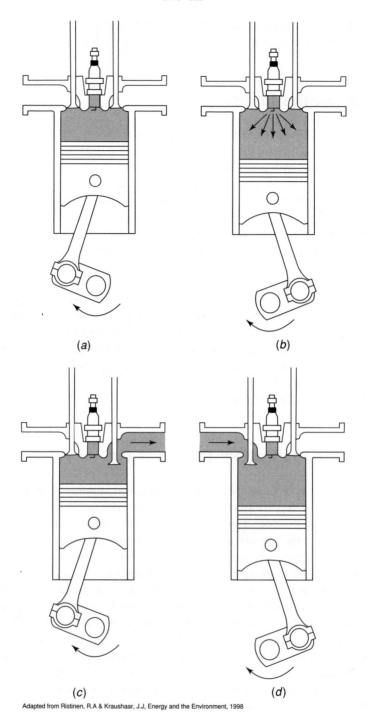

Figure 9.5   The four strokes of a four-stoke (or cycle) spark ignited internal combustion engine: (a) compression, (b) combustion, (c) exhaust, (d) fuel-air intake [from D. W. Devins in RIS99]

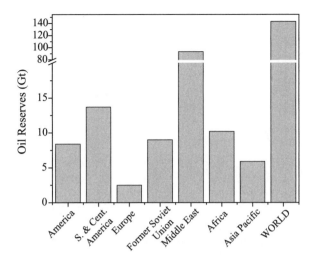

Figure 9.6 World oil reserves [data from BP04]. Regional definitions are given in Appendix 1

Problem 9.5 Cars are usually run with a fuel/air ratio of one to 15, how close to the stoichiometric ratio is this?

A typical car engine delivers 75 kW of power and is 20 per cent efficient in itself. However, when the frictional losses in the drive train and air resistance are included, the whole automobile efficiency is reduced to only seven per cent. Ultimately we are rarely interested in moving the vehicle itself, only the occupants and any goods carried. For single occupancy, this further reduces the 'efficiency' to 0.7 per cent in terms of body mass moved[1]. In Chapter 28 this figure will be compared to those for other transport systems. Efficiency can be advanced by reducing the mass of the vehicle, improving its aerodynamics, encouraging the fuel-air mixture to completely combust and increasing occupancy levels, thereby reducing the number of vehicles in use at any one time.

Problem 9.6 A car engine typically operates at 3000 revolutions per minute. For two revolutions, each cylinder will fire only once. Assuming a four-cylinder engine and an efficiency of 14 km/l (40 miles/gallon), estimate how much fuel is combusted in a cylinder during each firing if the car is travelling at 70 km per hour (44 miles per hour).

Problem 9.7 From your own experience, estimate what percentage of the hours in the year a typical car is in use. Estimate the total (new) value of all vehicles in use today in your country and comment on whether it seems sensible for society to have made such a

---

[1] i.e. a 750 kg car carrying a 75 kg person implies 75/750 or only one tenth of the energy is being used to transport the person. (This is a very approximate estimate; a more accurate figure could be arrived at with knowledge of the wind resistance of a person compared to that of a car.)

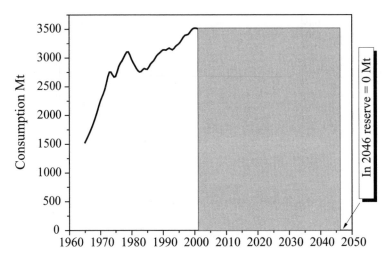

Figure 9.7   Time until total oil resource depletion [data from BP plc, *BP Statistical Review of World Energy*, BP, 2004.]

large capital investment in goods that are so rarely used. Can you think of something the engines in these vehicles could be doing when the vehicles themselves are not in use?

9.5 Global resource

Figures 9.6 to 9.9 analyse the regional resource, production and consumption situation in detail. Country specific data is contained in Appendix 1; data for subsequent years can be found on the book's web site.

The present global reserve of oil is 6580 EJ (157 Gt), which at current extraction rates would last 41 years (R/P = 41), suggesting $t_\infty = 2046$. Unlike coal, the resource is not widely spread, with two-thirds lying in the Middle East. Saudi Arabia alone has 25 per cent of the world's reserves and Iraq nearly 11 per cent. This suggests supplies might be insecure with respect to geopolitical tensions. Again unlike coal, regional (and country-centred) production and consumption are highly unbalanced, with the main market often being export rather than internal use.

Once more, caution is required with respect to the meaning of the word *reserves*. Because this refers to the proven reserves, i.e. those that are both known and can be recovered in the future under *existing* economic and operating conditions, any data are liable to revision as more deposits are found, technologies advance and economic conditions change. For example, oil reserves grew by 72 per cent between 1980 and 2003, despite large-scale withdrawals from the same reserves (currently running at 2.5 per cent of the reserve per annum). Thus, although the oil, gas and coal will run out at some point in the future if we carry on using them, they are likely to last much longer than many of the published figures suggest. The term resource, rather than reserve, is more correctly used when referring to *all* the coal, gas or oil that at present lies unused. This

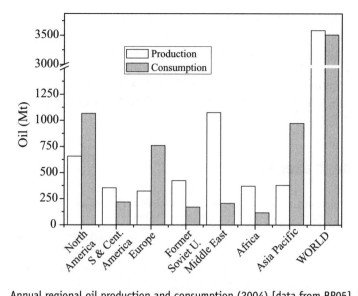

Figure 9.8 Annual regional oil production and consumption (2004) [data from BP05]. Total world production = 155 EJ (3697 Mt)

distinction might ultimately prove to be academic, as we may have to leave much of the resource untouched if climate change is to be avoided.

Despite the switch to gas for heating and electricity generation in much of the world, oil consumption has risen over the last 10 years, with the dominant growth being in the Asia Pacific region, which has overtaken all areas except North America. With the

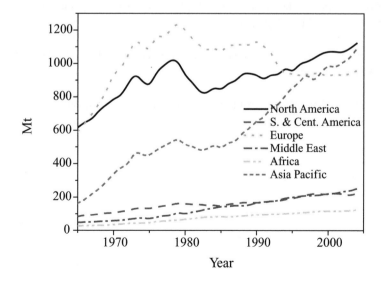

Figure 9.9 Regional trends in oil consumption [data from BP05]

exception of the former Soviet Union, all regions show a growing trend only modified by recessions. The USA accounts for 25.5 per cent of world consumption [BP04] despite only having 4.6 per cent of the world population; conversely, China uses 6.6 per cent for 20 per cent of the world population [BP04, CIA05].

As Figure 9.2 indicates, the price of crude in 2000 was similar to that at the end of the nineteenth century. This figure also shows that the price is sensitive to both the discovery of new fields and major world events.

9.6 Student exercises

"The earth we abuse and the living things we kill will, in the end, take their revenge; for in exploiting their presence we are diminishing our future."

–Marya Mannes, American Writer (1904–1990)

1. If OECD countries produce only a third less oil than OPEC countries, why does the majority of the power to influence oil prices lie with OPEC?

2. Describe the geological structures in which oil is found and how it is extracted. (Include a description of primary, secondary and tertiary methods.)

3. How, and why is crude oil distilled?

4. Outline the workings of an internal combustion engine. Estimate the Carnot efficiency of such an engine and compare this to typical efficiencies.

5. If a typical super-tanker carries 300,000 t of oil, how many tanker deliveries per annum are needed to supply Japan?

6. Why does the USA have such a large per capita consumption of oil compared with the rest of the world?

7. For many decades, oil consumption has continuously risen, yet reserves have risen not fallen. Why?

10

Gas

"There is a fine line between passion and gas."

—On Late Night with Conan O'Brien, 2005

Natural gas is a highly valued fuel. It burns cleanly, is easy to transport by pipeline and can be used as a replacement for oil in many systems from generating electricity to domestic heating. Such gas is often found above oil deposits, from which it derives, but it is also found remote from any oil, in which case its origin is usually lower-lying coal deposits that have been degassed by high temperatures. Natural gas connected with oil deposits is termed *associated gas* or *condensate*, that which is not is termed *non-associated*. When extracted it is usually around 80 to 90 per cent methane (CH_4) with small amounts of ethane, butane, propane, nitrogen, hydrogen sulphide and carbon dioxide; the mix depending on its location and history. Before retail distribution, this natural composition is standardized to 95 per cent methane, with small amounts of ethane, propane and butane. The resultant product is distributed to customers via pipelines, or in liquid form (liquid natural gas, LNG) via road tankers to in-situ tanks, or in bottled form.

Coal gas, rather than natural gas, is made by the industrial processing of coal but has become an unpopular fuel since large natural gas deposits have been developed.

If after reading the previous chapter on oil you thought the 'barrel' an odd and inconvenient unit, then you are unlikely to be impressed with the plethora of possibilities for measuring quantities of gaseous fuels. Natural gas is commonly measured in cubic metres, cubic feet, British Thermal Units (BTU), tonnes of oil equivalent and kWh, i.e. just about everything but joules. Table 10.1 gives the calorific values of various gaseous fuels.

10.1 Extraction

Natural gas was used as early as the sixth century BC in China and Japan for lighting and was distributed via bamboo pipelines [RIS99, p46]. Modern usage started in 1821

Energy and Climate Change   David A. Coley
© 2008 John Wiley & Sons, Ltd

Table 10.1 Calorific values of common gaseous fuels (for reference, 1 British thermal Unit (BTU) = 1055 J and 1 therm = 105.5 MJ, density of LNG = 512–575 kg/m^3)

Gas	Chemical formula	Gross Calorific Value (MJ/kg)
Methane	CH_4	55.6
Propane	C_3H_8	50.4
Butane	C_4H_{10}	49.5

in New York on a small scale and then expanded in 1883 in Pittsburgh, USA, with the creation of the first modern natural gas pipeline (with a length of 22 km). Before this, natural gas was generally regarded as an inconvenience and if produced alongside oil simply burnt in a flare. Coal gas however, which could be produced near the point of use, was extensively exploited.

Natural gas is extracted by drilling in much the same way as oil, with offshore recovery requiring heavy investment. After extraction, delivery can be via direct pipeline (Figure 10.1) or sea-going gas tanker. Three quarters of traded natural gas is distributed

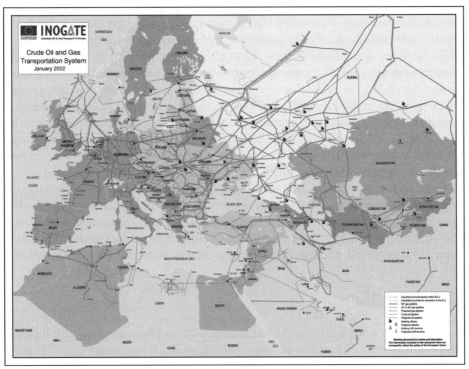

Reproduced from www.inogate.org/en/resources/map_gas

Figure 10.1 The large-scale gas and oil pipelines of Europe and nearby states (2002) [EU05]

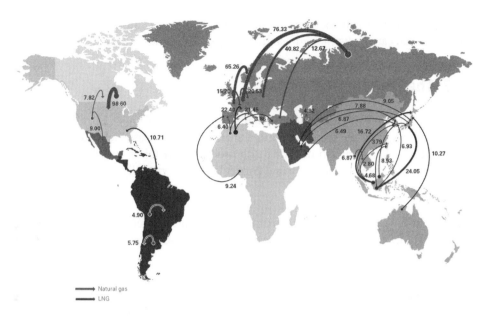

Figure 10.2 Major trade movements in gas [BP plc, *Putting Energy in the Spotlight*, BP, 2005]

via pipeline rather than tanker because the use of ships requires the gas to be liquefied. Figure 10.2 shows the major trade movements.

Problem 10.1 Compare the energy being transported by a large sea-going gas tanker (capacity 120 000 m³ of liquefied gas) with the annual energy used for domestic lighting in a town with a population of 100 000.

Up until the middle of the twentieth century, almost all natural gas was produced in the USA, with marginal additions from Russia and Venezuela. From 1959 large West European fields were discovered in Holland and the North Sea – where gas originates from deep coal layers topped by porous sandstone and sealed, or capped, by impermeable rocks. Since then, Eastern Europe, and in particular Russia, has expanded production greatly.

The nineteenth and first half of the twentieth century was the age of coal gas (rather than natural gas). Coal gas was used not only for heating but also, in an age before the widespread use of electricity, for lighting – including both domestic and street lighting. Given the size of the world's coal reserves, it is possible that large-scale production might emerge once more if natural gas reserves start to reach exhaustion.

Coal gas is produced by gasification. The coal is heated in the presence of air, or air and steam, or oxygen; and a mix of carbon monoxide, nitrogen (if air is used) and hydrogen is produced. The mix can either be burnt on an industrial scale in suitable boilers, or used to run a gas turbine for power generation. If the mix is methanized (see Chapter 8: Coal) methane gas is produced with an identical calorific value to natural gas and can be distributed via the same pipe network.

10.2 The combustion of gas

As we saw in Chapter 7, methane burns as $CH_4 + 2O_2 \rightarrow CO_2 + 2H_2O$ plus other minor products from the combination of contaminates with the fuel and nitrogen in the air; with 56 kg of carbon dioxide being produced per GJ of energy released in a typical large-scale boiler. Unlike oil and particularly coal, virtually no sulphur dioxide is produced.

Problem 10.2   The world consumption of gas is approximately 2.3 Gt_{oe} per annum [BP04]. Estimate the mass of carbon dioxide and water produced annually by this consumption.

Almost all natural gas combustion takes place either in boilers of various sizes, or in gas turbines used for electricity generation. Being a gas, the calorific value per unit volume is low and therefore it is rarely used in mobile systems and never in aircraft. Gas boilers are very similar to their coal and oil-fired cousins, and duel-fuel burners that can utilize either gas or oil are common. Although natural gas is the most common fuel, propane is often used for domestic heating and other applications in locations not connected to the gas distribution network. Propane and butane can be classified as liquefied petroleum gases (or LPGs). This term covers a range of 'oils' that are liquid at ambient temperature and moderate pressure. This makes them suitable for transportation in pressurized cylinders. LPG is slowly becoming a popular fuel for automotive use as it can be burnt within internal combustion engines and has lower emissions of various air pollutants than gasoline (petrol) or diesel.

10.3 Technologies for use

As stated above, natural gas is used for heating in both the domestic and industrial sectors, as a fuel to drive power stations and increasingly (when liquefied) in road vehicles. The reason for its rapid expansion in all of these sectors is two-fold: firstly cost and secondly environmental considerations. Despite the need to build vast networks of pipes crossing whole continents, it has proved to be extremely economical, and is cheaper than oil as a fuel. It is also a very clean fuel, containing fewer contaminates than either coal or oil and producing correspondingly fewer contaminants when burnt in various technologies. When used as the heat source in power stations it has one further advantage: it fits naturally with *combined cycle generation technologies*, with the fuel being first burnt in a turbine to produce electricity and then the waste heat from this process being used to raise steam for a conventional steam generator, thereby increasing the overall efficiency of the station.

However, its use is not without problems. Being a carbon-based fuel, it still produces carbon dioxide when burnt, although 25 per cent less than oil per unit of heat. Because it is burnt in air, atmospheric nitrogen is converted to oxides that contribute to acid rain and other problems. Most importantly, and unlike oil, it is in itself a greenhouse gas, with a global warming potential 23 times that of carbon dioxide (per unit mass). This means that leakage during extraction, processing or distribution can negate any environmental gains from its naturally lower carbon emissions.

Problem 10.3 Fuel switching is often seen as a way to reduce greenhouse gas emissions. Burning natural gas in a domestic boiler to produce 1 kWh of heat typically produces about 0.201 kg of CO_2; oil would produce about 0.266 kg of CO_2 for the same heat. Being a gas, natural gas escapes far more easily during transportation to the home than oil, offsetting some of this saving. Estimate the maximum percentage loss of gas allowable to still show savings in greenhouse gas emissions.

10.4 Example application: the domestic boiler

Gas-fired central heating has become popular in locations where natural gas is distributed via pipelines. There are two basic designs: boilers with hot water storage cylinders and combination (or 'combi') boilers, which do not need hot water cylinders. In addition both types come in traditional forms which only access the net calorific value of the fuel and *condensing* ones that access the gross calorific value by condensing the water in flue gases. In theory a condensing version should have an improved efficiency of:

$$100 \left(1 - \frac{Net C_v}{Gross C_v} \right) = 100 \left(1 - \frac{48.16}{53.42} \right) = 9.8 \text{ per cent.}$$

However this value is rarely realized. In part this is because condensation occurs when the flue gas is used to warm the water returning from the radiators. If this water returns hot then it will be less able to cool the flue gases to the point where condensation occurs.

Figure 10.3 shows the major components of a modern gas-fired condensing combi boiler and Figure 10.4 the flow of water within the device. Note the presence and finned form of the heat exchanger (Chapter 19). Table 10.2 lists the basic operational data for the design.

Table 10.2 Operational data for the boiler shown in Figure 10.3

Heating output (min/max)	16.0/22.0 kW
Hot water output (min/max)	7.9/22.3 kW
Central heating flow temperature	40–80 °C
Hot water flow rate (for 35 °C temp. rise)	9.6 l/min

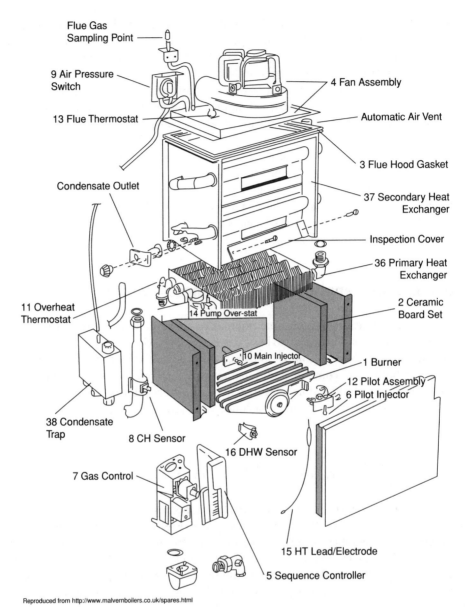

Flue Gas Sampling Point

9 Air Pressure Switch

4 Fan Assembly

13 Flue Thermostat

Automatic Air Vent

3 Flue Hood Gasket

Condensate Outlet

37 Secondary Heat Exchanger

Inspection Cover

36 Primary Heat Exchanger

11 Overheat Thermostat

14 Pump Over-stat

2 Ceramic Board Set

10 Main Injector

1 Burner

12 Pilot Assembly

6 Pilot Injector

38 Condensate Trap

8 CH Sensor

16 DHW Sensor

7 Gas Control

15 HT Lead/Electrode

5 Sequence Controller

Figure 10.3   Schematic of a Malvern condensing combi [MAL04]

10.5 Global resource

We will now analyse the regional resource, production and consumption situation in detail. Country specific data can by found in Appendix 1; regularly updated information may be found on the book's web site.

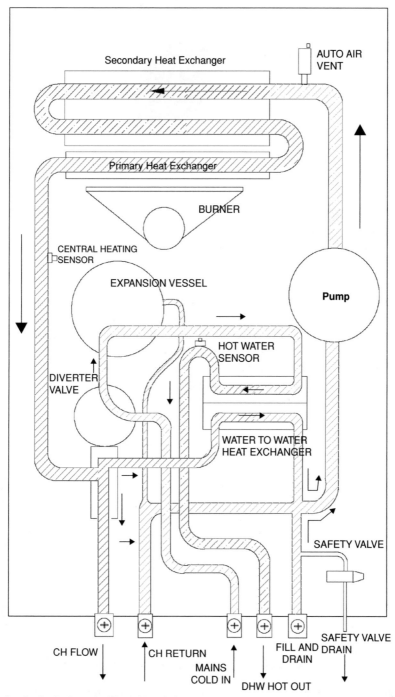

Figure 10.4 Water paths within the boiler shown in Figure 10.3 [MAL04]

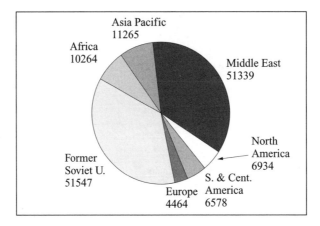

Figure 10.5 Regional natural gas reserves (trillion cubic metres) [data from BP05]. Regional definitions are given in Appendix 1

The current global proven reserve is 142 400 Mt_{oe}, which at present extraction rates, would last 62 years. The resource is more widely spread than oil (Figure 10.5), but less so than coal. The most notable difference compared to oil being the very large reserves in the Russian Federation. This suggests supplies are reasonably secure against global geopolitical tensions, but might be affected by tensions within both the Russian Federation and the other states of the former Soviet Union. The majority is used for

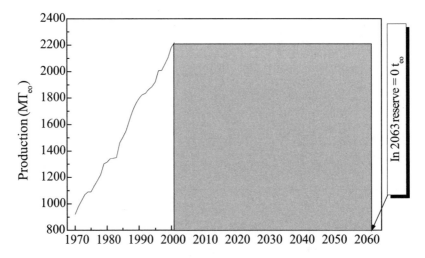

Figure 10.6 The rising consumption of natural gas and the time until total reserve depletion [data from BP04]. Unless additional reserves are found, t_∞ is likely to be before 2060 as consumption is rising rapidly

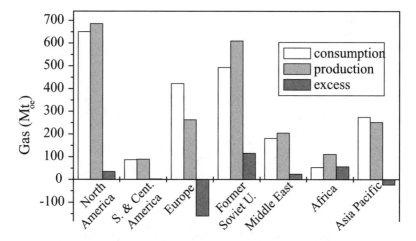

Figure 10.7 Regional annual natural gas production and consumption together with their excess production [data from BP04]. Note: the USA itself has a large negative excess; the sign of the excess is reversed in the graph by the excess production of Canada. World annual production is 99 EJ (2357 Mt$_{oe}$)

power generation and space heating. Again, unlike oil, regional (and country-centred) production and consumption are fairly balanced.

Because of the many technological and environmental advantages offered by natural gas, world consumption is rising (Figure 10.6). This growth is wide-spread and includes Asia, the former Soviet Union, Europe and Australia, but growth has been much slower in North America since 1976 – possibly in part because of its historically large-scale use of the fuel. Figure 10.6 shows the estimated time until resource depletion, assuming

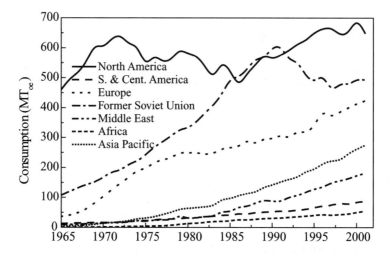

Figure 10.8 Regional trends in natural gas consumption [data from BP04]

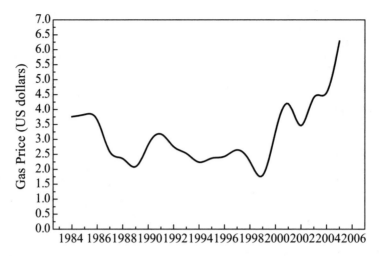

Figure 10.9 Trends in natural gas price per million BTU (based on European Union prices that include insurance and freight) [data from BP04]

consumption stays at the current rate. Looking at the slope of the consumption curve this seems most unlikely and therefore the reserve will only satisfy us for a much shorter time. However, we are likely to find some new deposits and develop more effective extraction technologies. Countries of note are: Japan, which has no substantial reserves but imports from Indonesia, Malaysia and the Middle East an amount similar to UK annual consumption; Canada which exports more than it uses to the USA; and the Russian Federation which holds over one third of global reserves, against the USA's three per cent.

Figures 10.7 and 10.8 analyse the world's production and consumption situation in detail. The price of natural gas is proving to be variable (Figure 10.9).

10.6 Student exercises

> "The fact that some geniuses were laughed at does not imply that all who are laughed at are geniuses. They laughed at Columbus, they laughed at Fulton, they laughed at the Wright brothers. But they also laughed at Bozo the Clown."
>
> –Carl Sagan

1. What is natural gas, and how is it formed?

2. What advantages and disadvantages does natural gas have over oil?

3. What is the difference between a condensing and a non-condensing boiler?

4. Use the data in Appendix 1 to list, in order, the world's ten largest producers and consumers of natural gas.

11

Non-conventional hydrocarbons

"Overdrafts on aquifers are one reason some of our geologist colleagues are convinced that water shortages will bring the human population explosion to a halt. There are substitutes for oil; there is no substitute for fresh water."

–Paul R. Ehrlich

Alongside the conventional well developed fossil fuels covered in the last three chapters, the earth contains large reserves of other hydrocarbons, namely: oil shale, tar sands and methane hydrate. These have been less well developed as resources for a combination of reasons centred on issues of low concentration, difficulty of access and the need for additional processing. However as our fossil fuel reserves dwindle, these other compounds are likely to become more attractive. The size of these reserves is less well known than those of oil, coal and gas, but they are substantial and would in theory be able to ward off any energy crisis.

11.1 Oil shale

As its name suggests, oil shale is an inorganic non-porous rock that contains a high proportion of organic matter. The waxy carbon-based solid it contains is called *kerogen*, typically at a concentration of four parts in 100. It was formed in a similar way to oil deposits, but was never subjected to the high temperatures required for the final transformation to crude oil. Although some oil shales can be directly ignited and burnt as a very low-grade solid fuel, interest in the substance is focused on the extraction and transformation of the kerogen into oil by *retorting* it (i.e. heating it to 350–400 °C in a closed vessel). This typically produces around 27 litres (six gallons) of oil per tonne of shale, although the highest yielding rock can give eight times this.

Energy and Climate Change   David A. Coley
© 2008 John Wiley & Sons, Ltd

The shale can either be mined and the retorting carried out on the surface, or the retorting can be done in-situ. The latter requires holes to be bored into the shale and hot gases injected. One of the main advantages of in-situ retorting is that it eliminates the need to dispose of the waste rock, as it is never removed. Due to the low concentration of kerogen in the original rock, such spoil would be much greater per kilogram of hydrocarbon recovered than, for example, in coal mining.

Oil recovered from the oil shale has less carbon and hydrogen than normal crude and more nitrogen, oxygen and sulphur. This implies that oil refineries would need adapting before the range of oil products to which we are used could be produced. Tight emission controls would also be needed when the resultant fuel was burnt if SO_x and NO_x emissions were to be acceptable.

Global resource

There is currently no commercial exploitation of oil shale, although there are research projects, most notably the Stuart Project in Queensland, Australia, that are refining the technologies involved in commercial extraction and processing. During the latter part of the nineteenth century and the first part of the twentieth, commercial production did exist near Edinburgh, Scotland, but with the discovery of large deposits of liquid oil around the world, production became uneconomic.

Probably the most impressive thing about oil shale is the size of the deposits. The largest known reserves, the Green River deposits, lie on the borders of Wyoming, Colorado and Utah in the USA. These are believed to contain 270 billion tonnes of shale.

Problem 11.1   Assume the Green River deposits can be mined and retorted to produce 100 litres (20 gallons) of oil per tonne of shale. How much oil might be produced in total? Compare the result to current world annual oil use.

The above problem illustrates that there are reserves of non-traditional hydrocarbons that could avert the much heralded oil crisis, although the costs of the oil produced are likely to be much higher.

11.2 Tar sands

Tar sands (also known as oil sands) are deposits of sand, or in some cases porous carbonate rocks, bound by a heavy bituminous material at a concentration of 10 to 15 parts in 100. Tar sand is easier to process than oil shale and can also be extracted from its surrounding rocks in-situ or remotely. The former method requires steam to be injected via boreholes, thereby reducing the viscosity of the bitumen to the point where it can be pumped to the surface. Whereas remote recovery requires it to be mined then mixed with steam and water, the bitumen subsequently floating to the surface and the sand falling under gravity to the bottom of the processing vessel. In either case, heating

Image courtesy of Syncrude Canada

Figure 11.1 Some of the world's largest diggers and haul trucks mining tar sand in Athabasca, Canada. The door and ladders give some idea of the scale of the machines and the amount of sand that needs to be moved due to the dilute nature of the resource [Syncrude Canada Ltd, http://www.syncrude.ca/users/folder.asp?FolderID=5703 (accessed August 2006)]

the freed bitumen to above 500 °C creates a product similar to crude oil, which can then be distilled to produce kerosene and other oils.

Global resource

The largest reserves of tar sands are in the Athabasca area of northern Alberta, Canada. Substantial deposits also exist in the former Soviet Union and Venezuela. The total resource exceeds the world's entire endowment of conventional crude oil [BOY03, p281]. The Athabasca deposits are commercially exploited and around 64 million litres (400 000 barrels) of oil are produced daily from half a million tonnes of opencast-mined rock by the world's largest diggers (Figure 11.1).

11.3 Methane hydrate

Gas hydrates occur abundantly in arctic regions and in marine sediments. A gas hydrate (or clathrate) is a crystalline solid formed of gas molecules, usually methane, each

Figure 11.2 Methane hydrate (white) as laminae within sediment (black) [USG05]

surrounded by a cage of water molecules. The result has the appearance of water ice (Figure 11.2). It occurs at low temperatures and at depths greater than 300 metres where it helps to cement loose sediments into a bonded layer several hundreds of metres thick. Because the gas is held within a crystal structure, the packing density of the gas is much higher than it would be were it truly gaseous – around 3000 times that which it would have at surface temperatures and pressures. This layer can also form a cap over a trap of gaseous methane.

Methane hydrates form a large carbon resource. Their detailed distribution is not fully known but they are believed to be very widely dispersed, with a geographical spread greater than other fossil fuels (Figure 11.3). This would make their adoption ideal from a geopolitical standpoint.

Several schemes have been proposed for methane hydrate exploitation, although technological and economic barriers have meant that these have never been developed beyond an experimental stage. The Japan National Oil Company has been one of the pioneering forces and has experimental wells in the Mackenzie Delta in Northern Canada. One concept yet to be explored is the possibility of sequestering carbon dioxide as a hydrate (which would be thermodynamically more stable than methane hydrate) into existing seafloor clathrates thereby releasing methane [CGH04]. Such a facility would be particularly attractive, acting as a source of methane and as a sink for the resultant carbon dioxide produced by its combustion. (See Chapter 29 for more about carbon sequestration technologies.)

One concern about gas hydrates is their potential to form a geohazard. If a large quantity of hydrate was destabilized by drilling for oil or gas at depth, or by a natural geological event, a large and rapid conversion to gaseous methane might take place. This poses a hazard to oil and gas rigs, but it could also make the sea floor unstable. In

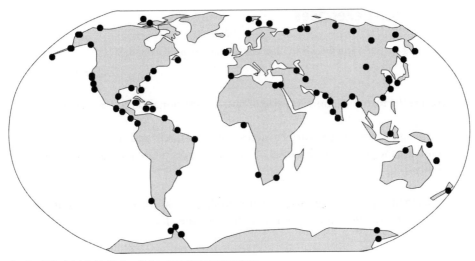

Data from US Geological Society, http://woodshole.er.usgs.gov/project-pages/hydrates

Figure 11.3 Locations of methane hydrates [USG05]

an extreme case it might lead to underwater slope failure (i.e. rapid slippage of inclined parts of the sea floor) and the creation of tsunamis [CGH04].

A second concern to us here is the role of methane as a greenhouse gas (see Chapter 5). Increasing global air temperatures will eventually lead to elevated ocean temperatures. This could destabilize methane hydrates causing the release of enormous quantities of methane thereby adding to radiative forcing, higher temperatures and the release of yet more methane: a classic positive feedback cycle. Indeed, hydrate dissociation might have played a part in climate regulation in the past. Cold periods led to extensive glaciations and thereby a reduction in sea levels of 300 metres or more. The resultant reduction in water pressure on the ocean floor may have caused hydrate dissociation thereby warming the planet via increased global warming, an increase in sea level and ultimately a reduction in methane emissions: i.e. a negative feedback cycle.

Global resource

The size of the methane hydrate resource is not well known but is believed to be in excess of twice the amount of carbon in all recognized traditional fossil fuel reserves [USG05]. Other sources [ORN04] put the figure higher at 1.43×10^6 EJ of methane – re-emphasising the view that we are unlikely to face a future shortage of carbon.

Problem 11.2 Estimate how many years the reserves of coal, tar sands, oil shale and gas hydrates would last if they were used to supply all of the world's primary energy.

11.4 Student exercises

"Science is a wonderful thing if one does not have to earn one's living at it."

–Albert Einstein

1. Describe the form, location and size of non-conventional carbon resources.

2. Describe the processing required to access these resources and any advantages or disadvantages the geographic location of these resources might have for world political stability.

3. Produce a histogram comparing the world's oil, gas, coal and non-conventional fossil fuel reserves (*resources* for the non-conventional fuels).

12
Nuclear power

"There is not the slightest indication that nuclear energy will ever be obtainable. It would mean that the atom would have to be shattered at will."

–Albert Einstein, 1932

In 1957 the first commercial nuclear power station started production of electricity at Shippingport, Pennsylvania, USA. The development of the technology had been surprisingly rapid. Only 19 years had passed between the first practical demonstration of the physics and the commercialization of a new generation technology. This was an amazingly short period of time and worth comparing with the length of time many of the other alternative renewable technologies have taken to reach commercialization.

In 1938/39 it had been noted that uranium nuclei would split into two, smaller, nuclei in the presence of an additional neutron and that the process released a large quantity of energy. The first experimental nuclear reactor was built in 1942 where this process, called *fission*, could be sustained – creating 200 watts of heat.

After Shippingport, development was equally rapid. A large number of reactors were built in many places around the globe and there are now 442 reactors in 31 countries supplying 10 per cent of the world's total electricity demand [IAEA05]. However few countries have switched the majority of their generation to nuclear fission (France being a notable exception) and orders for new reactors have declined – effectively to zero in North America, the UK and most of Europe. Before analysing the reason for this and discussing the details of the particular technologies involved, we need first to see how fission takes place and how the energy released is converted into electricity.

12.1 Physical basis

When chemical reactions occur, atoms break the bonds that hold them and then form other alternative collections of atoms. As discussed in Chapter 2 this can either release

Energy and Climate Change   David A. Coley
© 2008 John Wiley & Sons, Ltd

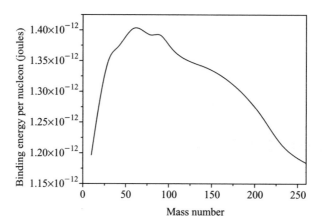

Figure 12.1 Binding energy per nucleon for various numbers of nucleons [adapted from data in *Encyclopaedia Britannica*, Encyclopaedia Britannica (UK) Ltd, 2002]

or require a net quantity of energy, depending on whether the new substance formed requires less or more energy to hold it together. In an analogous fashion, the nuclei that lie at the heart of atoms can undergo reactions, which will release energy if the newly formed nuclei require less energy to hold them together. The energy required to hold, or bind, a nucleus together is called the binding energy. As Figure 12.1 shows, some nuclei require a greater binding energy than others.

Each atom in the universe is formed from a number of protons, neutrons and electrons. Every element has a unique number of protons, but the number of neutrons can vary. Each variant of an element with a different number of neutrons is called an *isotope* of the original element. On paper we distinguish between isotopes of the same element with an addition to the element's chemical symbol or name which gives the sum of the number of neutrons and protons within the isotope (i.e. the total number of nucleons). For example, the majority of natural uranium contains 92 protons and 146 neutrons and is therefore referred to as uranium-238 (as $92 + 146 = 238$), U238 or ^{238}U, but a small percentage of natural uranium contains 143 neutrons (but still 92 protons – otherwise it would not be uranium) and is referred to as uranium-235.

In Figure 12.1 we see that the curve that represents the energy required to disintegrate a nucleus peaks at about 60 nucleons. This suggests that any nucleus that lies to the right of the peak can form a lower energy state by disintegrating into two lighter components. For example, if a slow moving neutron strikes a uranium-235 nucleus it will be absorbed briefly but the resultant nucleus will rapidly undergo fission to create two nuclei of roughly equal mass and a further two or three neutrons plus a quantity of gamma radiation. For example:

$$^{235}U + n \rightarrow {}^{92}Kr + {}^{141}Ba + 3n + \gamma. \tag{12.1}$$

As predicted by Figure 12.1, the mass of barium (Ba) and krypton (Kr) plus the three neutrons is less than the original uranium and neutron by 3.57×10^{-28} kilograms.

Applying $E = mc^2$ to this missing mass we see that:

$$E = 3.57 \times 10^{-28} \times (3 \times 10^8)^2 = 3.2 \times 10^{-11} \text{ J is released per fission.}$$

This energy is mostly carried away in the form of kinetic energy by the barium and krypton (comically termed by the nuclear industry the *fish* and *chips*). Within the fuel rod of a nuclear reactor these will almost instantly smash into surrounding atoms and thereby convert this kinetic energy into heat. Any radioactive products produced will also undergo radioactive decay, producing yet more heat – even if the reactor has been shut down and fission stopped. The additional neutrons produced (see Equation (12.1)) are free to cause further fissions and thereby even more neutrons.

Problem 12.1   By considering the energy released by a single fission, estimate the energy released by the fission of one kilogram of uranium-235. How much oil would have to be burnt to release an equivalent amount of energy?

It is interesting to compare the energy produced by nuclear and chemical reactions with that available to traditional societies that might rely on gravitational systems such as waterpower. Dropping a kilogram of uranium (or anything else) from a height of 100 metres will release:

$$mgh = 1 \times 9.81 \times 100 = 981 \text{ J.}$$

One kilogram of oil has a calorific value of 35 MJ and as implied above, the fission of uranium-235 releases 8.2×10^{13} J per kg. From this analysis we see that the ratio between the energy released in gravitational and chemical reactions is of the order 1:36 000 and that between chemical and nuclear reactions is 1:2.3 $\times 10^6$. This, at least in part, explains how during the industrial revolution the use of fossil fuels allowed economies to expand so greatly. Initially it was thought that the same jump might be fired by a move from chemical to nuclear power, with electricity produced by nuclear fission being 'too cheap to meter'.

12.2 Technologies for use

Unlike coal, oil or gas, there is only one commercial use for nuclear fission: the generation of electricity for supply to consumers via a national grid. Smaller scale uses have been made of the process, for example, powering nuclear submarines. However, virtually all of the world's installed capacity is in the form of nuclear reactors connected to steam turbines generating electricity for sale. Such reactors are similar in many ways to any fossil fuel plant: nuclear fission is used simply as a heat source which is used to raise steam, which in turn is used to run a turbine generator, usually producing about one gigawatt of electricity. Differences in the systems actually installed in various parts of the world lie primarily in the design of the core of the reactor and the fluid used to remove heat from it.

Before describing the various possibilities for reactor design, it is necessary to introduce some common principles of fuel and reactor design as well as some basic reactor physics.

Basic reactor design

As an illustration of the basic components we will sketch out the design and operation of an advanced gas cooled reactor (AGR) (Figure 12.2) as the majority of structures and processes are shared between this design and other reactors. Figure 12.3 shows the advanced gas cooled reactor at Dungeness, UK.

Within an AGR, pellets of uranium oxide are held in stainless steel cans stacked in fuel assemblies and inserted into vertical channels within a massive block of graphite. The whole assembly is contained in a concrete pressure vessel filled with high-pressure carbon dioxide gas, which acts as the coolant. For an AGR, this outer vessel also acts as the biological shield containing the radiation produced by the reactor. Heat removed from the fuel cans by the coolant is then used to raise steam for electricity generation. The reactor/generator assembly is a heat engine, and therefore the concept of Carnot efficiency applies and typical overall efficiencies are similar to fossil-fuel power stations. The final element of the design is the control rods used to control the neutron population, and thereby the reaction rate. These are made from good neutron absorbers, typically a steel/cadmium or steel/boron mix. Like the fuel assemblies, these slide vertically into channels.

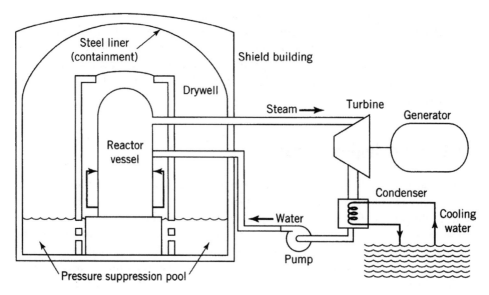

Figure 12.2 A basic reactor [A.V. Nero in Ristinen, R.A. and Kraushaar, J.J., *Energy and the Environment*, Wiley, 1999, p182]

Figure 12.3 The Dungeness-B AGR (UK) [BE05]

In essence, designing, building and operating a nuclear reactor is about creating a finely tuned environment where the fuel is consumed at a steady rate. As Equation (12.1) shows, each fission requires exactly one neutron to initiate it but produces two or three neutrons. If each of these were to react with a ^{235}U nucleus they would, in total, produce nine neutrons and an escalating chain reaction would have begun. On the other hand, if on average less than one of the three resultant neutrons were to cause a subsequent reaction, the number of reactions occurring each second would steadily diminish. The required balance of on average one resultant neutron causing one additional fission is ensured by a mix of reactor geometry and materials and the use of neutron absorbing structures (in the form of control rods) that can be inserted into the reactor to mop up excess neutrons.

In addition to the need for the reactor to absorb excess neutrons, the fission of uranium-235 itself requires the initialising neutrons to be moving relatively slowly (about 2 km/s) (see Figure 12.4), whereas the neutrons created by the reaction will be far more energetic and travelling at around 20 000 km/s. A *moderator*, usually water or graphite, is used to slow the neutrons to a suitable speed with the fast neutrons simply bouncing around the moderator until they have lost most of their energy. Such slow moving neutrons are termed *thermal* neutrons and a reactor designed around them a *thermal reactor*.

We can get a good idea about the properties required of a moderator by comparing what happens when a low mass object (a neutron) strikes a stationary high or low mass object. In the case of a head-on collision with a high mass object such as a large nucleus, the neutron suffers only a small loss of energy but a reversal of its momentum, i.e. it bounces off. In the case of a collision with an equally low-mass object the neutron will lose half of its energy. In general the fractional loss in energy will be:

$$1/(A+1) \tag{12.2}$$

if an object of unit mass collides with a heavier mass of A units.

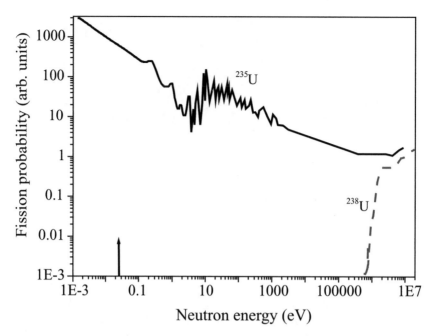

Figure 12.4 Probability of a neutron causing fission against neutron energy: note both scales are logarithmic [RIS99, p178]. The arrow marks the energy of thermal neutrons (4×10^{-21}J or 0.025 eV). For uranium-238, the probability only becomes appreciable above 1.6×10^{-13} J (1 MeV)

We can therefore conclude that low mass, common (i.e. cheap) nuclei that can exist in a high-density form at the required temperatures and pressures would be ideal. A gas would have a disadvantage, as the lower density would imply fewer, energy-losing collisions. This rules out hydrogen or helium, the lightest elements. However hydrogen nuclei can be found in abundance as H_2O, making water an ideal combined moderator and coolant. Unfortunately, water also has the tendency to absorb neutrons and this is one of the reasons why many reactors use enriched, rather than natural uranium.

The fissile isotope ^{235}U is not mined in its pure form, but as uranium oxide (U_3O_8). The majority (99.3 per cent) of this uranium is non-fissile isotope ^{238}U with only 0.7 per cent being ^{235}U. Commonly the fuel is enriched until it contains around three per cent ^{235}U.

Although uranium-238, which makes up most of the uranium within the reactor, cannot directly undergo fission (except by interaction with very high energy neutrons), it can absorb a neutron to become uranium 239 and then undergo beta-decay[1] to

---

[1] A form of radioactive decay in which a neutron loses a beta particle, i.e. an electron. This will increase the atomic number of the atom by one by turning the neutron into a proton. The atom's atomic mass number stays the same because the total number of protons and neutrons remains the same.

produce ^{239}Np, which in turn, beta-decays to Plutonium 239:

$$^{238}U + n \rightarrow {}^{239}U$$

$$^{239}U \rightarrow {}^{239}Np + e$$

$$^{239}Np \rightarrow {}^{239}Pu + e.$$

Plutonium-239 is fissionable by neutrons and contributes about one third of the energy produced by a reactor.

In addition to the extra heat produced by the plutonium, many of the fission products are radioactive and thereby heat producing. This means that even if the control rods are fully inserted to mop up neutrons and 'stop' the reactor, large quantities of heat will still go on being produced (about 10 per cent of the full power rating of the machine). Thus, even when stopped, the reactor will still need cooling for some time. As we will see later, this fact can have severe implications for safety.

Reactor physics

The number of additional neutrons going on to produce fission from each fission event is termed the multiplication factor, k. In the steady-state of power production, k must equal 1. Any greater and the rate of fission within the core will spiral out of control; any less and the fission rate will decline and the reactor eventually stop producing useful quantities of heat. The speed at which the control rods can respond (by being lowered or raised, into or out of the reactor at approximately 0.5 m/s) needs to match the rate at which the neutron production grows or declines if k is not exactly equal to 1. The term *prompt* is used to describe those neutrons which are released immediately by the fission event described by Equation (12.1), and the term *delayed* to describe those that arise from the subsequent decay of radioactive products. Neutron lifetime in a reactor is very short, averaging 0.005 seconds; however the *effective* lifetime as measured by time between the originating fission event and the absorption of the resultant neutron will be very different for prompt and delayed neutrons. Prompt neutrons have effective lifetimes of only 0.0001 to 0.001 seconds, delayed neutrons 0.6 to 80 seconds – a period dominated by the radioactive lifetimes of the element that produced them. Because of the extremely short period between generations of prompt neutrons, if $k > 1$ for such neutrons the neutron density would multiply many times in only a fraction of a second.

Figure 12.5 shows how neutrons are processed within a thermal reactor.

Problem 12.2   If $k = 1.005$ find the number of prompt neutrons within a reactor after one second and after 10 seconds, assuming initially only one neutron is present.

Clearly no control system based on slowly lowering or raising control rods will be able to effectively control such a reaction. This means that the control system must rely

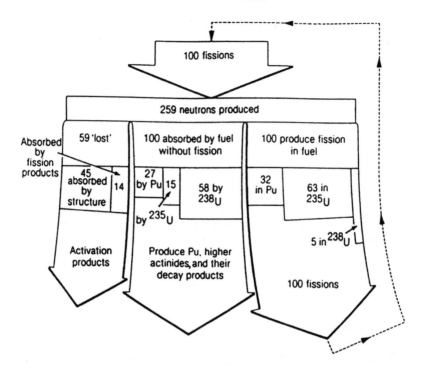

Figure 12.5 Creation, reaction and absorption of neutrons in a thermal reactor [Hewitt, G.F. and Collier, J.G., *Introduction to Nuclear Power* (2nd edition), Taylor and Francis, 2000, p30]

on controlling the population of delayed neutrons – which form only approximately 0.7 per cent of the total. The number of prompt neutrons is essentially maintained by the fundamental design of the reactor and ensures that $k < 1$. The additional delayed neutrons push k up over unity and the control rods control these neutrons, the lifetimes of which are of the same order of magnitude as the response time of the control rods.

We will now look at several thermal reactor designs in detail. An alternative technology, the fast breeder reactor, will be covered in Chapter 27. Table 12.1 gives a comparative overview of their characteristics.

Problem 12.3   Moderators are used to slow the fast reaction neutrons to thermal energies. This means reducing their kinetic energy from around 1 MeV to about 1 eV. Convert this energy to joules and estimate the speed of travel of the neutrons before and after this reduction. (A neutron has a mass of 1.6749×10^{-27} kg.)

Magnox reactors

A natural uranium graphite-moderated (Magnox) reactor uses carbon dioxide as a coolant, at 2000 kPa (20 atmospheres, or 20 times atmospheric pressure; for comparison a racing bicycle tyre is typically at 750 kPa or 7.5 atmospheres). This is circulated through

Table 12.1 Fundamental characteristics of various reactor designs. All values are typical/approximate. Note the much smaller size of the fast-breeder design (Chapter 27) and its much greater volumetric power density [adapted from Hewitt, G.F. and Collier, J.G., *Introduction to Nuclear Power* (2nd edition), Taylor and Francis, 2000, p39 and other sources]

Type	Thermal power (MW$_{th}$)	Coolant	Moderator	Core volume (m^3)	Volumetric power density (MW/m^3)	Fuel rating (MW/tonne)	Exit coolant temp. (°C)
Magnox	225–1875	CO_2	Graphite	449–2166	0.5–0.87	2.2–3.15	400
AGR	1500	CO_2	Graphite	550	2.5	11.2	650
CANDU	3425	Heavy water	Heavy water	280	12.2	26.4	293
PWR	3800	Water	Water	40	95	38.8	332
BWR	3800	Water	Water	75	51	24.6	290
RBMK	3140	Water	Graphite	765	4.1	15.4	–
Fast (PFR)	1000	Liquid sodium	None	1.5	400	150	–

the moderator that forms the reactor core. The moderator itself is made of graphite bricks punctured by channels for the coolant to flow through and the fuel to be placed in. The fuel is natural uranium bars within magnesium alloy (Magnox) cans – an alloy that absorbs few neutrons – allowing natural rather than enriched uranium to be used. The reactor core is typically 8 m high by 14 m in diameter, the coolant leaves the core at 400 °C and the overall efficiency is around 31 per cent.

Magnox reactors have been built in France, Italy, UK and Japan and some have operated for over 40 years – a testament to the simplicity and reliability of their design. However, they have a low power output per unit volume of core or fuel. This implies the technology has a high capital cost.

AGRs

Advanced gas cooled reactors (AGRs) were built on the experience gained from Magnox power stations and have greater volumetric power densities and operating temperatures. Again, carbon dioxide is used as the coolant but now at 4000 kPa (41 atmospheres) and an outlet temperature of 650 °C. These higher temperatures and pressures mean stainless steel tubes have to be used to contain the fuel. Because stainless steel is a significant neutron absorber, this in turn leads to the need to use enriched uranium. The fuel, which is in the form of hollow uranium oxide pellets, is enriched to 2.3 per cent ^{235}U. Bundles of 36 fuel tubes are grouped together to form fuel assemblies (Figure 12.6). The higher temperatures imply that the Carnot efficiency of the reactor is higher, and the realized efficiency can be as high as 42 per cent. This higher power density allows for a capital-effective design. These higher temperatures initially led to problems particularly from the reaction of the coolant with the graphite moderator to form carbon monoxide and thereby cause corrosion.

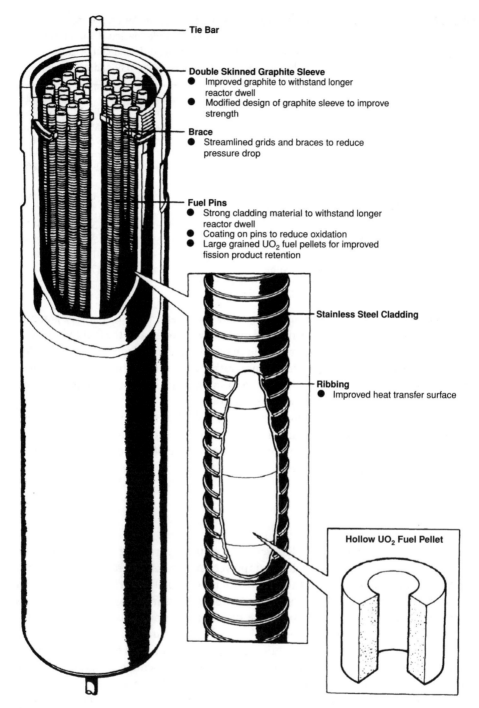

Figure 12.6 AGR fuel bundle [Hewitt, G.F. and Collier, J.G., *Introduction to Nuclear Power* (2nd edition), Taylor and Francis, 2000, p42]

PWRs

Pressurized water reactors (PWRs) use, as the name suggests, pressurized water (at 1500 kPa or 150 atmospheres) as the coolant – the high pressures ensuring that the water does not boil. Light, i.e. ordinary, water (where the hydrogen nuclei consist of just a proton) is used. This is a good neutron absorber, meaning that the fuel needs to be highly enriched (3.2 per cent ^{235}U). The fuel is in the form of uranium oxide pellets contained in 4 m tubes of zirconium alloy mounted in bundles of 17 rows of 17 tubes. As with most reactors, the hot pressurized water exiting the reactor is not used directly as a source of steam for electricity generation, but to raise steam within separate external steam generators. The overall efficiency is 32 per cent.

With over 250 commercial PWR reactors currently operating, the design has proved to be the most popular for power station use. The main reasons for this being the low capital cost of construction and the large amount of heat generated per unit mass of fuel.

Some have criticized the design on safety grounds because the high power density, even when the reactor is shut down (from heat produced by radioactive decay of fission products), means that the core has to be kept covered with water at all times. If the water were to flash to steam, due to a loss of pressurization, the resultant steam would not be able to cool the reactor due to steam's much lower thermal capacity per unit volume.

BWRs

Boiling water reactors (BWRs), at least on the surface, probably represent the simplest solution to the problem of how to use a large mass of fissile material to raise steam in order to generate electricity (Figure 12.7). The design dispenses with the need for external steam generators and boils approximately 10 per cent of the coolant water entering the reactor for distribution to steam turbines, which subsequently return it to the coolant loop; power densities half that of PWRs are achieved.

Ordinary water is used both as the coolant and moderator. The fuel assembly sits in a bath of water that boils from the heat produced by the fuel. Steam is produced at 290 °C and 70 atmospheres and passes out of the top of the reactor to the turbine hall. The core of the reactor consists of around 46 000 zirconium alloy fuel rods each about four metres long and 1 cm in diameter. The fuel is 155 000 kg (342 000 lb) of enriched uranium oxide in pellet form, with each pellet having a diameter of 10 mm (0.4 inch). 170 control rods inserted from the bottom control the reaction, which has a power density of 54 kW per litre of core. Initially the fuel is 2.8 per cent uranium 235, but after one year this has fallen to 0.8 per cent and the fuel is replaced.

Potentially, there are problems with the design. Any contamination from the turbines will end up in the reactor; conversely, radioactive products will pass through the turbines – possibly meaning higher radiation doses for operators.

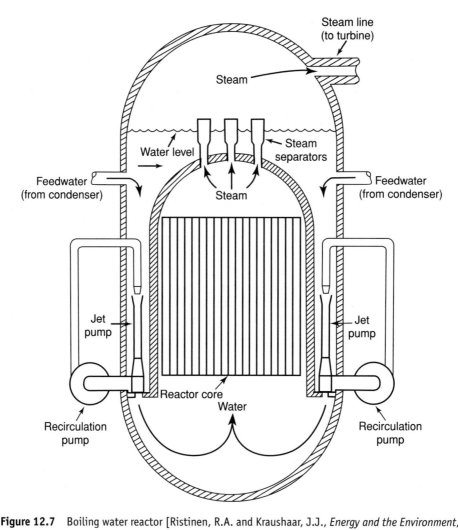

Figure 12.7 Boiling water reactor [Ristinen, R.A. and Kraushaar, J.J., *Energy and the Environment*, Wiley, 1999, p183]

CANDUs

Canadian deuterium-uranium reactors (CANDUs) return to the idea of using natural, unenriched uranium as the fuel. This is achieved by using heavy water as the coolant. Heavy water (or deuterium oxide, D_2O) is chemically identical to H_2O but each hydrogen nuclei contains both a proton and a neutron. D_2O makes up 0.016 per cent of ordinary water, from which it is extracted. Heavy water is used because it absorbs fewer neutrons than normal water, enabling natural rather than enriched uranium to be used. The massive thick-walled pressure vessel found in BWRs and PWRs is replaced by a bath (or calandria) filled with heavy water held at a low pressure and modest temperature.

The fuel bundles are again constructed of zirconium alloy cans but they lie horizontally rather than vertically within the calandria. Each tube is pressurized (to 9700 kPa, or 95 atmospheres) and filled with heavy water coolant/moderator. The fuel is in the form of uranium oxide pellets, with 12 bundles within each pressure tube. The volumetric power density is only one tenth of that of a PWR.

Because natural uranium is used, fuel costs are lower. Unfortunately the use of costly heavy water, about two per cent of which is lost every year, offsets this.

RBMKs

The boiling water, graphite-moderated direct-cycle reactor is unique to the former USSR, and from its name you should by now be able to guess most of its fundamental characteristics. It was an RBMK that was involved in the nuclear accident at Chernobyl (Section 12.3) in 1986.

The reactor is built from graphite blocks punctured by vertical channels containing zirconium alloy pressure tubes; in a 1 GW reactor there are 1663 channels. Uranium oxide pellets are used at an enrichment of 2.4 per cent. Boiling light water is used as the coolant at 7900 kPa (78 atmospheres) and about 14 per cent is extracted as steam for direct use within the steam turbines. The maximum graphite temperature is 700 °C.

Table 12.2 lists the national breakdown of reactor type and the fraction of national electricity they produce.

Since 1956 a total of 19 nuclear power stations, comprising 41 reactors, have been constructed in the UK. Of these, 16 stations, comprising 35 reactors, are currently (2006) fully operational, and three stations, each with two reactors, have been closed down and are being decommissioned. The first generation stations were Magnox reactors; these were followed by a series of AGRs that were commissioned between 1976 and 1988. A PWR station (the UK's first, Sizewell B) was commissioned in February 1995 [DTI05a].

Nuclear power stations in the UK are operated by four companies: Nuclear Electric Ltd (NEL), Scottish Nuclear Ltd (SNL), Magnox Electric plc (Magnox) and British Nuclear Fuels plc (BNFL). NEL and SNL are wholly owned subsidiaries of British Energy plc, which was privatized in July 1996. Magnox and BNFL are wholly owned by the Government [DTI05a].

NEL operates five AGRs and one PWR (total capacity of 7.2 GW_e). SNL operates two AGR stations, each with two reactors (2.4 GW_e). Magnox currently operates six Magnox stations (2.9 GW_e). BNFL operates two Magnox stations (0.4 GW_e). The UK nuclear power stations currently generate about 25 per cent of the UK power demand. Additionally, 17 TWh of nuclear electricity is imported each year from France via the cross-channel interconnector (2.0 GW_e) (see Chapter 7) [DTI05a].

12.3 Environmental concerns

Environmental problems associated with nuclear power include the generation of large amounts of waste material from the mining of the uranium ore; the release of radioactive

Table 12.2 Nuclear power reactors in operation and under construction in the world (as of June 2004). Note: The total includes the following data on Taiwan, China: 6 units, 4884 MW$_e$ in operation; 2 units, 2600 MW$_e$ under construction; 37.37 TWh of nuclear electricity generation, representing 21.5 per cent of the total electricity generated in 2003; 137 years of total operating experience [data from IAEA04]

Country	Reactors in operation		Reactors under construction		Nuclear electricity supplied in 2003		Total operating experience to June 2004
	No of units	Total MW$_e$	No of units	Total MW$_e$	TWh	% of total	Reactor years
Argentina	2	935	1	692	7.03	8.59	51
Armenia	1	376			1.82	35.5	36
Belgium	7	5 760			44.61	55.5	195
Brazil	2	1 901			13.34	3.65	26
Bulgaria	4	2 722			16.04	37.7	131
Canada	17	12 113			70.29	12.5	495
China	9	6 587	2	2 000	41.59	2.18	43
Czech Republic	6	3 548			25.87	31.1	77
Finland	4	2 656			21.82	27.3	101
France	59	63 363			420.7	77.7	1 375
Germany	18	20 643			157.4	28.1	657
Hungary	4	1 755			11.01	32.7	76
India	14	2 550	8	3 622	16.37	3.3	230
Iran			2	2 111			0
Japan	54	45 464	2	2 371	230.8	25.0	1 150
Korea, Dem. People's Rep. of			1	1 040			0
Korea, Republic of	19	15 850	1	960	123.3	40.0	230
Lithuania	2	2 370			14.3	79.9	37
Mexico	2	1 310			10.51	5.23	24
Netherlands	1	449			3.8	4.48	59
Pakistan	2	425			1.81	2.37	36
Romania	1	655	1	655	4.54	9.33	8
Russian Federation	30	20 793	3	2 825	138.4	16.5	776
Slovakia	6	2 442			17.86	57.3	103
Slovenia	1	656			4.96	40.4	22
South Africa	2	1 800			12.66	6.05	39
Spain	9	7 584			59.36	23.6	223
Sweden	11	9 451			65.5	49.6	316
Switzerland	5	3 200			25.93	39.7	146
Ukraine	13	11 207	4	3 800	76.7	45.9	286
United Kingdom	35	12 052			85.31	23.7	1 343
United States Of America	104	98 298			763.7	19.9	2 923
Total	442	363 819	27	22 676	2524.74		11 364

radon gas from mines; production of radioactive materials from the enrichment process; and leakage of radioactive gases (krypton and xenon) and of radioactive tritium (^{3}H). However the two greatest concerns (for most people) are the possibility of a large-scale nuclear accident and the question of what to do with the spent fuel rods. The only option for the latter is either reprocessing (a costly and difficult task) or storage for a *considerable* time. How long 'considerable' is can be gleaned from an examination of the half-life[2] of the elements involved, some being many thousands of years. The radioactivity within the waste will initially reduce rapidly as those elements with short half-lives decay, leaving the long lived (and therefore less radioactive) elements to be stored for perhaps 100 000 years in some form of yet undecided storage facility. This length of time is so long it is not clear how it might be achieved given the possibility of war and geological disaster that could occur during such a time period. For this reason, as it leaves an environmental, technical and economic challenge to later generations, many would conclude that nuclear power clashes with our definition of sustainability:

'Meeting the needs of the present generation without compromising the ability of future generations to meet their own needs' [BRU87]

There have been few major nuclear accidents and only one disaster: this within a background of over 11 000 cumulative years of total operating experience. An example of a serious incident being Three-Mile Island, Pennsylvania, USA, where the reactor suffered a loss of coolant in 1979 and was destroyed by the residual heat in the core even though the reactor had been shut down. However, there was no major release of radioactive material. There have also been various small-scale leaks of radioactive material both from power stations and from ancillary facilities.

In 1986 a reactor at the Chernobyl power station in the Ukraine was destroyed in an explosion (Figure 12.8). A very large quantity of radioactive material was released and distributed across much of Europe. Many tens of thousands, if not hundreds of thousands, of people were displaced by the accident and both soil and farm animals contaminated. Many site personnel and resource workers received a fatal dose of radioactivity and it has been estimated [RIS99, p190] that about 47 000 excess deaths may occur over the next 50 years from the isotopes released (however this number is highly controversial). The Chernobyl reactor, the remains of which are now entombed within concrete, was a boiling water reactor, designed to produce both electricity and plutonium-239 for nuclear weapons. This meant that the fuel rods had to be extractable from the reactor so the plutonium could be harvested whilst the reactor was still running, thus leaving a large void above the reactor and no steel or concrete containment vessel. The design was also potentially unstable at low power levels.

In order to carry out a series of legitimate experiments, several safety systems were disconnected including the reserve cooling system. This was in breach of safety rules at the plant. A surge of power within the core could not be contained as the control rods had

---

[2] The half-life of a radioactive substance is the time taken for its radioactivity to drop by one half.

Figure 12.8   The Chernobyl reactor after the explosion [Ukranian Web, http://www.ukrainianweb.com/chernobyl_ukraine.htm, (accessed June 2006)]

been withdrawn. This caused a rapid rise in temperature and pressure and enormous steam pressures in the coolant circuit. The control rods could not be fully lowered to shut the reactor down and the uranium fuel channels ruptured. Steam explosions followed, blowing up the pressure vessel, the reactor core and the reactor hall. Hot radioactive debris was thrown into the air and ignited, and a massive escape of highly radioactive gas occurred.

Although the accident at Chernobyl can in part be blamed on the reactor design, it was mostly a case of poor operating practice and human error. Such problems are much harder to predict than failings in pumps, valves and other engineered items. Estimates have been made of the probability of a serious accident causing loss of life for various reactor designs (Table 12.3). In all cases the probabilities are small, especially when compared on a unit of electricity basis with values from other technologies such as oil and coal.

Two of the central difficulties with estimating the probability of any accident, not just one connected with nuclear power, are human error and *common mode failure*. As

Table 12.3 Estimated probabilities of various accidents and fatalities for nuclear power stations

Event	Estimated number of deaths		Probable number of events per million reactor-years	Predicted number of deaths per thousand reactor-years
	Immediate	*Delayed*		
Meltdown without major breach of containment	0	0–10	10–100	0–1
Meltdown with breach of containment, average conditions	1–10	1000–10000	0.1–10	0.1–100
Meltdown with breach of containment, worst conditions	1000–10 000	10 000–100 000	0.001–0.01	0.001–1

an example of common mode failure, imagine you have a meat pie and a cream bun for lunch everyday (probably not the healthiest option). You are told that there is a one in 100 chance of a pie not containing any meat and a one in 200 chance of a cream bun not containing any cream. What is the probability of not having any meat or cream for lunch? One might make the estimate:

$$\frac{1}{100} \times \frac{1}{200} = \frac{1}{20\ 000}.$$

So, at one in 20 000, it would seem very unlikely you would suffer such a poor lunch. However, if you look closer into why such mistakes are made at the bakery, you might find problems tend to occur when new staff are being trained, or when quality control is temporarily unavailable. This might indicate that such failures, despite originating in differing processes, have a common link and tend to occur together, making the probability of a no-meat and no-cream lunch closer to one in 200 (simply the lower of the two probabilities), i.e. at least once a year – a very different conclusion. The same logic holds for safety systems. A valve might have a small chance of failure, and the chances of the backup valve and the backup to the backup valve all failing at the same time might be astronomically small under most circumstances, but there may be unplanned events that can affect all three valves at the same time, for example, physical damage if they were co-located. And, as Chernobyl (Figure 12.8) has taught us, valves and wires aren't the only thing that can fail.

A common mode failure presented itself at Brown's Ferry nuclear power station, Alabama, USA, in 1975 when a small fire caused by some maintenance work (the use of a candle to locate an air leak) took hold within a wall cavity damaging control systems linkages. Unfortunately primary and backup control systems passed through the same cavity, causing 2000 cables to be affected, a potentially serious situation, the plant to be shut down for a year and €7.8m (£5.6m, US$10m) of costs incured.

12.4 Waste

Much of the pollution and the wastes from energy production, distribution and use are short lived. Some pollution however has a much longer time constant in the atmosphere, ground or ecosystem: silting of hydroelectric reservoirs, greenhouse gases and nuclear waste being prime examples. Given that some nuclear wastes will remain radioactive and a danger to the environment for thousands of years, such products certainly breach our definition of sustainable, in that they represent a problem for future generations. The implication being that fission is inherently unsustainable. There is one major difference between nuclear wastes and many other forms of pollution – they are highly contained – suggesting that the problem may not be that great and all that is required of future generations is some form of safe storage facility. However, it could be argued that storing anything for 1000 years safe from water ingress, earthquakes and other unknowns might not be so simple.

Waste products arise at various stages of the nuclear power cycle and include: spoil from uranium mining; products of the enrichment process; spent fuel, which even if reprocessed to extract useful uranium and plutonium, leaves highly radioactive fission products; reprocessing plant wastes, including liquid wastes and the metal cladding from fuel assemblies; and the reactor itself, which must be decommissioned at the end of its useful life.

Most of the waste produced by nuclear power stations is only weakly radioactive and is termed *low-level waste* (although there can be very large quantities of it). The smaller amount of highly radioactive (or high-level) waste, mostly from the fuel itself, has proved the greatest challenge. Given enough time, the radioactivity of all waste products will fall to background levels found in the original uranium ore or other natural substances such as granite. By definition, highly radioactive substances cannot remain so for long – as they are being transformed very rapidly. However, very safe storage will still be needed in this initial period. During this period the waste will also be producing large quantities of heat (from radioactive decay), which will need to be removed.

For biological systems, such as the human body, the physical half-life of radioactive containation is not the most important time-constant. Such products entering the body will be processed and eventually excreted. The *biological half-life* is the time taken for the amount of radioactivity in the body after ingestion or inhalation to fall to half. These two half-lives can take very different values. For example, strontium-90 and caesium-137 both have similar physical half-lives, yet strontium-90 has a biological half-life of

30 years and caesium-137 only around 70 days. This difference is explained by the fact that strontium tends to accumulate and become trapped within bones. This means that a full understanding of the danger of varying waste streams depends not only on a consideration of the physical properties of the waste, but also how any leaks to atmosphere or into natural water courses will be processed by the biological systems they pollute.

Figure 12.9 outlines the disposal routes for high-level wastes, with and without processing. Note that both streams end in a final solid product that will need surface storage in a controlled environment (partially to remove heat) for around 50 years, then storage under the sea or within old mines or specifically created caverns for maybe 1000 years. In addition, unprocessed fuels need to be stored, typically in cooling ponds (Figure 12.10) for at least 10 years. Because few communities want a nuclear storage facility near them, and because of the cost and technological problems of long-term storage, the world nuclear industry has yet to complete this storage cycle.

Low-level wastes should be much easier to deal with. However there will be far greater quantities. A wide variety of materials need to be treated, from rubber gloves – which can simply be stored in steel drums and buried – to vast quantities of concrete that need dedicated underground storage facilities. Sweden has such a facility at Forsmark, and Finland has one at the Olkiluoto power plant.

It is expected that fission reactors will not be fully dismantled for many years after they stop producing power, and that they will be left for levels of radioactivity to fall to more easily manageable levels. Decommissioning might be in stages: de-fuelling (approx. three years); removal of non-radioactive materials and plant (approx. 5–10 years); dismantling the reactor and returning the site to green field conditions (approx. 100 years after shut down). An alternative would be to mothball the plant for 100 years, then complete all three stages more rapidly. As few reactors have been decommissioned and long term storage not completed, it is not known how much this process might add to the cost of nuclear power.

12.5 World resource

The estimated known uranium reserves amount to 3.12 Mt or 1400 EJ if used in thermal reactors[3]. With present usage, this uranium stock will only last around 70 years and if nuclear power programmes are expanded the supply will last even less time. As is the case with fossil fuels, there are currently speculative reserves. These have been estimated to equal around 4000 EJ. By comparison, current day primary fuel use is around 409 EJ per annum, so we can see that a policy aimed at removing the threat of climate change based solely on the expansion of thermal fission is not likely to be successful for very long.

---

[3] This is thermal energy; assuming a generating efficiency of 35 per cent implies $1.4 \times 10^{21} \times 0.35 = 490$ EJ of electricity.

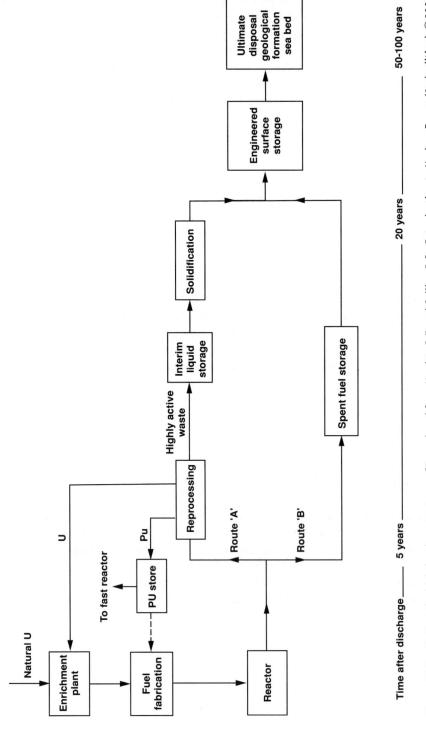

Figure 12.9   Options for high-level waste storage [Reproduced from Hewitt, G.F. and Collier, J.G., Introduction to Nuclear Power (2nd edition) ©2000 Taylor and Francis]

Figure 12.10 Nuclear waste storage ponds at Sellafield, UK, used to dissipate heat [Reproduced from www.uic.com.au/opinion6.html, The Royal College of Physicians]

Uranium ore (an oxide, U_3O_8) is not found in an undiluted form, but mixed with surrounding material. The less diluted, the easier and cheaper the extraction. Therefore the size of the estimated reserve is really just a reflection of the price one is willing to pay for the refined product. Historically, uranium has traded at around $20 per kg and accounts for less than one per cent of the cost of electricity production for a nuclear station. Therefore, it would be possible to extract uranium at greater cost without overly affecting the cost of nuclear generated electricity. There is also an argument that if we wish to tackle the threat of global warming then it would seem sensible to subsidize the cost of non-carbon fuels. Such subsidies would then allow even more costly uranium resources to be accessed. At $260 per kg, Ristinen [RIS99, p183] estimates the reserve to be in excess of 28 Mt, or 1.26×10^{22} J.

There is a much larger source of uranium: the oceans. Seawater contains uranium at a concentration of three parts per billion. The surface area of the planet is 510×10^{12} m^2, 71 per cent of which is water. Assuming an average ocean depth of 3.7 km, the total seawater volume is 1.34×10^{18} m^3. At three parts per billion (by mass), this equates to 4000 Gt of uranium (density 19050 kg/m^3), or 1760×10^{21} J. This would be enough for almost 100 000 years at the current rate of use. Unfortunately, filtering the whole of the world's oceans may well be unfeasible, and the extraction of uranium from seawater is at present uneconomic. It has also been suggested that the extraction of uranium from highly dilute sources, land- or sea-based, might well also use more energy than it returns. There is however an alternative to thermal fission, which has the potential to unlock much more of the energy content of natural uranium than thermal reactors do. These *fast* reactors use the much larger fraction of uranium, that is ^{238}U rather than ^{235}U, and produce 50 times more energy per kilogram of uranium. They have been developed to the point of commercialization in several countries. Such reactors are covered in Part IV along with nuclear fusion.

Fission has others problems though, that need to be overcome if its share of the electricity generating market is to increase: accidents have occurred; the public is fearful of the technology; and no acceptable solution has been found to the problem of nuclear waste. But above and beyond all these hurdles is the simple fact that in a world of cheap fossil fuels, and almost as cheap wind power, nuclear power is uneconomic without large government subsidies. Unless this position changes, nuclear expansion will be limited.

12.6 Example applications

We will review two facilities, one an advanced gas cooled reactor, the other a pressurized water design.

Heysham 2, Lancaster, UK

Heysham 2 is a 1.3 GW$_e$ AGR (Figure 12.11), which uses carbon dioxide as a coolant within a pair of graphite moderated reactor cores to give a thermal efficiency of 40 per cent [BEP00a].

In the following paragraphs all numbers refer to Figure 12.12.

Figure 12.11   Heysham 2 [Reproduced from British Energy, http://www.industcards.com/nuclear-uk.htm]

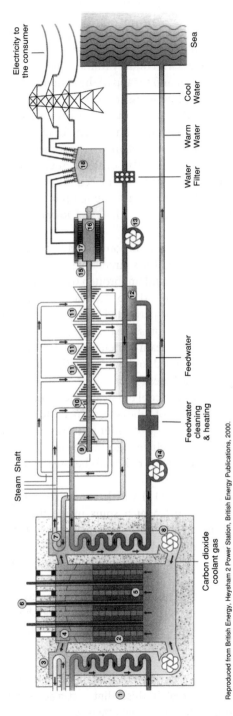

Figure 12.12 Schematic of Heysham 2 – an AGR [BEP00a]. See text for an explanation of the numbered items

Reproduced from British Energy, Heysham 2 Power Station, British Energy Publications, 2000.

The graphite core/moderator (2) is penetrated vertically by 332 fuel channels. Each reactor is encased within a concrete vessel (3) with walls over six metres thick pre-stressed by heavy steel tendons wound helically from top to bottom. The principal role of the pressure vessel is as a container for the steel liner that surrounds the core, the boilers and the coolant. It also plays a secondary role as a barrier against radiation from the reactor. The top of the steel liner and the top of the containment vessel are breached by a series of standpipes (4) that allow access to each fuel channel.

As discussed above, the fuel within an AGR is in the form of small pellets of uranium dioxide contained in a stainless steel can with 36 cans built into a graphite sleeve to form a single fuel element (5). Eight such elements are joined together vertically by a steel tie pin which passes through the centre of the elements to form a fuel stringer topped by an end-cap, or plug. One stringer is placed into each of the 332 standpipes with the plug forming a seal. The core is also punctured by 89 additional channels that hold the boron steel control rods (6). These are lowered and raised by electric motors.

Heat is extracted from the circulating carbon dioxide coolant by 12 boiler circuits grouped into four quadrants (7). Eight gas circulators (8), at the base of the reactor, pump the coolant through the boilers and up through the fuel channels (where the heat is generated). The coolant, now at around 635 °C, passes around the top of the boiler and back down to the gas circulators. Superheated high-pressure steam (at 541 °C) from the boilers is pumped first to the high-pressure turbine (9) where it is directed through nozzles onto the turbine blades. The steam is then returned to the reactor for reheating in the lower boilers before being passed to the intermediate pressure turbine (10) and finally to the three low-pressure turbines (11).

The condenser (12) uses seawater to cool the steam and change its state back to water. The condenser consists of thousands of tubes over which the steam passes, containing water pumped (13) at a rate of up to two million litres per minute from Heysham Harbour. The now warmer water is returned to the sea and the boiler water is stored temporarily in the deaerator (where air is removed) before the feed pump (14) returns it to the boilers.

The turbine drives the generator (15) which comprises a large hydrogen-cooled electromagnet, called the rotor (16), which spins inside the stator (17) – a water cooled electrical winding where the electricity is generated at 23 500 volts. A generator transformer (18) raises this to 400 kV for distribution via the national grid. Together, the two units produce 1.32 GW$_e$. Table 12.4 lists the key parameters of the design.

Used fuel elements are stored in cooling ponds for 90 days to allow for the decay of short-lived isotopes. The spent fuel is then packed into special shielded flasks and taken by rail to Sellafield in Cumbria.

Problem 12.4   Heysham 2 converts only 40 per cent of the energy generated in the reactor into electricity. How wasteful is this? Hint: Estimate the Carnot efficiency of Heysham 2 then compare this with the realized efficiency.

Table 12.4 Heysham 2: key parameters [data from British Energy, *Heysham 2 Power Station*, British Energy Publications, 2000]

General		Core	
Reactor type	Advanced gas cooled	*Moderator*	Graphite
Number of reactors	2	*Height*	8.31 m
Output	1344 MW (gross), 1230 MW (net)	*Diameter*	9.46 m
Thermal efficiency	40%	*Weight*	1808 tonnes
Reactor		*Fuel channels*	332
Thermal output	1596 MW (each)	*Control rod elements*	89
Coolant	High pressure carbon dioxide	*Fuel elements per channel*	8
Reactor inlet temp.	298 °C	*Fuel*	Enriched uranium in cylindrical hollow ceramic pellets of uranium dioxide clad in steel pins
Mean channel gas inlet temp.	336 °C	*Fuel form*	36 pin cluster in graphite sleeve
Mean channel gas outlet temp.	635 °C	*Length of element*	1.04 m
Total gas mass at outlet	4269 kg/sec	*Diameter of element*	190 mm
Pressure vessel		*Mass of uranium per reactor*	113.5 tonnes
Material	Pre-stressed concrete lined with steel	*Design refuelling rate per reactor*	2.2 channels per week
Internal diameter	20.25 m	**Boilers**	
Internal height	21.87 m	*Superheated steam temp.*	541 °C
External diameter	31.8 m	*Reheated outlet steam temp.*	539 °C
External height	39.8 m	*Feed temp.*	156 °C
Design gas pressure	45.7 bar (663 psi)	**Turbines**	
Condenser		*Speed*	3000 rpm
Exhaust pressure	40.5 m bar with 13 °C inlet	*Inlet steam*	583 °C at 159.6 bar
Cooling water	23 m³/s	*Feed water temp.*	156 °C
Surface area	30 565 m²	**Generator**	
		Output	660 MW, 19.62 kA, 23.5 kV
		Efficiency	97%

Problem 12.5   Estimate the volumetric power density (in W/litre) of Heysham 2's core. How quickly would a litre of boiling water need to cool to room temperature to realize an equivalent power density?

Sizewell B, Suffolk, UK

In the following paragraphs all numbers refer to Figure 12.13.

Sizewell B is a pressurized water reactor, or PWR – a design that has proved popular in the USA in particular. Unlike an AGR, the PWR uses water as both the coolant and the moderator. Sizewell B consists of a single reactor centred on a pressure vessel (1) through which water (2) is pumped. The core of the reactor consists of 193 fuel elements (3) held vertically [BEP00b]. Each element contains 264 fuel rods each 3.85 metres long. The reaction rate is controlled by 53 neutron absorbing control rods made of silver-indium-cadmium alloy. The rods are raised and held in position by electromagnets. If electrical power is lost, or the reactor needs to be closed down quickly the rods fall vertically into the core – rapidly reducing the neutron flux.

The core is surrounded by a 13.6 m high steel pressure vessel approximately 20 cm thick, which is connected to four steam generators (4) and four pumps (5) and a pressurizer (6). The pressurizer consists of a partly water filled vertical tank heated to a temperature even higher than that of the water in the rest of this primary pressure circuit. This arrangement ensures the high pressures (in excess of 150 bar) within the whole circuit, needed to keep the water in the reactor liquid at over 300 °C and therefore capable of moderating and cooling the reactor effectively.

The primary circuit is held within a 70 m high pre-stressed concrete reactor building (7), which acts as a radioactive shield and containment vessel. The steam generators (4), which consist of bundles of 690 metal tubes, use the hot water from the reactor to heat more water held in a secondary circuit. Because the pressure is lower in the secondary circuit the water boils – raising steam for the two turbines (9) mechanically joined to two generators (10).

As is the case with Heysham 2, seawater is used to cool the outlet water allowing it to condense back to water and be reused. Three million one hundred thousand litres (690 000 gallons) of water are required each minute. This is drawn by a pump (14) through filters (13) to remove fish and debris – with the fish being returned to the sea unharmed – through to the condenser (15). The cooling water, now approximately 10 °C warmer, is then returned to the sea.

Every 18 months one third of the fuel is replaced. This requires all fuel assemblies to be removed and taken to a cooling pond where the central third of the fuel is removed. The remaining fuel is then moved inwards and new fuel added to the outside of each assembly before it is placed back inside the reactor.

Problem 12.6   Compare the Carnot efficiency of the two reactor technologies given as example applications.

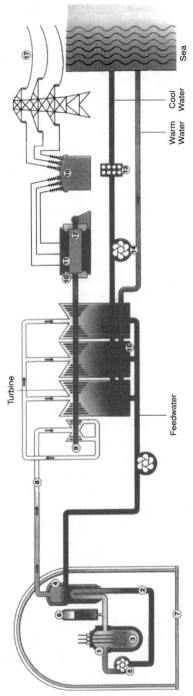

Reproduced from British Energy, Heysham 2 Power Station, British Energy Publications, 2000.

Figure 12.13   Schematic of Sizewell B, a PWR [BEP00b]. See text for an explanation of the numbered items

12.7 Is nuclear power the solution to global warming?

Nuclear power is sometimes touted as the answer to climate change. After all, thermal nuclear reactors have been producing electricity for nearly 50 years; they generate 10 per cent of the world's electricity [IAEA05] and emit very little carbon dioxide. So why try and develop alternative technologies such as wind and solar power when we have a mature technology in place? There are several arguments against expanding the role of nuclear power based on concerns about: the proliferation of reactors capable of generating reactor products suitable for nuclear weapon development; safety; the likelihood of terrorist attack against a plant or storage facility; terrorists obtaining plutonium and polluting water supplies with it; and cost. In addition to this it is also true that by itself nuclear power cannot 'solve' climate change. This is because electricity generation is only one source of carbon emissions and the fastest growing energy sector, namely transport, cannot make use of nuclear power unless alternative energy storage and drive systems are developed. As we will see in Chapter 28 this is unlikely to be easy. However, it is true that an expansion of nuclear capacity would reduce the growth in carbon emissions, and so the question becomes, is there a better, safer, or cheaper way to reduce carbon emissions than the worldwide adoption of nuclear power? If not, then an expansion of nuclear power would seem a reasonable policy.

Although many western-style democracies are in favour of expanding nuclear capacity, few welcome an expansion to other states. Part of this position might stem from little more than thinly veiled xenophobia and a belief that some nations cannot be trusted. However, there are genuine questions about security, the use of civil nuclear sites as part of a nuclear weapons program, accountability, and track records of how non-radioactive pollution has been dealt with in some regions. There is also the question of leaving poorer nations with the costs of decommissioning, of waste storage and a reliance on other nations for the initial technology and maintenance.

Leaving such arguments aside, it is worth exploring the costs and logistics of a policy centred on reducing global carbon emissions by building large numbers of nuclear power plants. The following analysis is based on that found in [LEG90]. If we are to reduce carbon emissions rapidly, it would seem sensible to replace the generating technology that has the highest carbon emissions (per unit of electricity generated) first. Table A4.1 shows this to be coal. So we will assume that the world coal-fired capacity is to be replaced by nuclear plants. We also assume that little opposition to nuclear power is met with and that the project can be completed within 35 years. The following is also assumed:

1. Nuclear power is cheap – only costing US$1000 per kW of installed capacity (cf. US$3000 in the USA today). This will allow a cost of US$0.05 per kWh to be realized (cf. US$0.13 per kWh in the USA today). This would seem sensible as standardization and mass plant production would reduce costs.

2. Plants can be built in only six years.

3. All problems with waste and safety have been solved.

4. Nuclear power stations take no energy and emit no greenhouse gases in their construction.

Given that the world-wide coal-fired installed generating capacity is in excess of 5000 GW$_e$ we will need in the order of 5000 one gigawatt nuclear plants; with around half located in the developing world. Figure 12.14 illustrates the approximate regional distribution required.

Problem 12.7   If 5000 nuclear plants are to be built in 30 years, how often would a new plant need to come on-line?

Is such an expansion likely or even possible? The cost would be considerable: US$5300 billion in total, or US$144 billion per annum. It is probably impossible for the world to consider financing this [LEG90, p300], especially if governments were unwilling to fund it, leaving the private sector to be relied upon. Unfortunately, even if this massive project were completed, projections of future oil and gas demand (Chapter 16) suggest total carbon emissions would continue to grow.

A very similar analysis could be carried out on many other potential technologies that might replace fossil-fuel electricity generation. This suggests that we are unlikely to see a single technology suddenly emerge and solve global climate change – a point that will be expanded upon in the final chapter. It also underlines the need for energy efficiency to be at the centre of any energy strategy aimed at ensuring atmospheric carbon concentrations stay at a safe level. Indeed, our primary objective must be to reduce energy use where possible. Without this it is unlikely that the development of new sustainable technologies can even keep up with increasing energy demand.

Estimates [GEL87, KRA88, KEE89, JAC89] suggest that energy efficiency is almost seven times more cost effective (in terms of kilograms of carbon not emitted) than nuclear power. Given the low cost of building and operating gas-fired plants, this might well suggest that we would be better off building gas-fired power stations rather than building nuclear ones and using the savings to promote energy efficiency, particularly in the developing world where efficiency is low and thus the savings would be greatest per dollar spent. When we consider the massive methane resources mentioned in Chapter 11, this might also seem a reasonable course from a resource perspective[4].

Of course, no matter how energy efficient we are, some generation will still be required. Could this be the role for nuclear power? Even if it is better to spend our money now on efficiency, could we reach a point where any further efficiency gains are either impossible or not cost effective compared to nuclear power? Such a point is a long way off, but it might be reached. However, nuclear power will then have to be judged both

---

[4] Although it is not clear yet what these might cost to access.

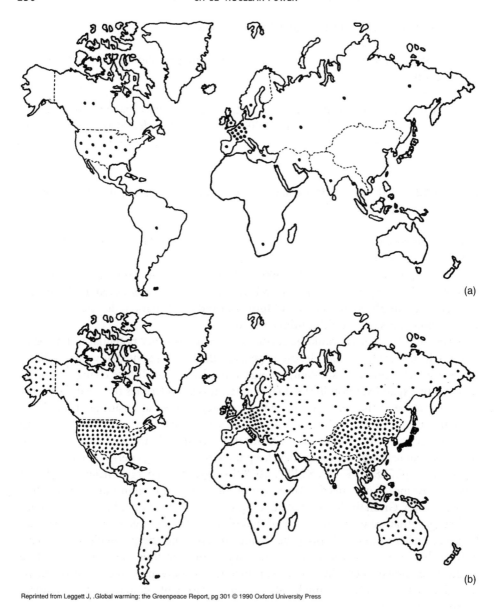

Figure 12.14 Approximate regional distribution of nuclear power stations, (a) current, and (b) if nuclear power replaced all coal-fired generation. Each dot represents ten 1GW$_e$ power stations [Leggett, J., *Global Warming: the Greenpeace Report*, Oxford University Press, 1990, p301]

economically and environmentally against other sustainable generating technologies, the cost of which would appear to be falling rapidly already. It is not clear yet whether nuclear power would be competitive against mature sustainable energy technologies, nor even if it were, whether it would be the public's first choice.

12.8 Student exercises

"The last people you'd want deciding what to do with nuclear waste are politicians."

–Ian Campbell, Australian Federal Environment Minister

1. Compare and contrast the design and operational parameters of PWRs, BWRs, AGRs and CANDUs.

2. Explain the role of the moderator within a reactor.

3. Without reference to nuclear power, describe your personal feelings and think analytically about the number of people who might die in an accident and the probability of the accident happening. For example, why does the press concern itself so much with rare accidents where relatively large numbers are killed, but ignores regular accidents that kill few per accident, but many over a year?

4. Leaving aside issues of safety and waste disposal, explain why nuclear power might not be a solution to climate change.

5. How much coal is required to produce the same amount of heat as 10 kg of U-235 (Assume complete fissioning of the uranium.)

6. If a substance has a half life of 50 days, how long does it take for its radioactivity to fall to (i) 1%, (ii) 0.00001% of its original value?

7. Given that U-235 has a half-life of 710 million years, how long will it take for its radioactivity to fall to 1% of its original value? What does this say about its level of radioactivity?

8. Why might a government find it difficult to privatise its nuclear industry?

13
Hydropower

"The drive toward complex technical achievement offers a clue to why the U.S. is good at space gadgetry and bad at slum problems."

–John Kenneth Galbraith

Many readers might have expected to find hydropower (the generation of electricity from the movement of non-tidal waters) under alternative energy sources in Part IV; after all, is it not really just a form of solar power driven by the global cycle of evaporation and precipitation; and therefore a true renewable responsible for no greenhouse gas emissions? There are however several reasons for taking a different view: large-scale hydro systems involving the construction of dams and the flooding of valleys can have substantial environmental and social consequences; the technology can not really be considered new or alternative as such systems currently provide around 16 per cent of the world's electricity; the reservoirs at the centre of such schemes are prone to silting up, implying that the resource is time-limited; the flooding of valleys can lead to large methane emissions as the vegetation decomposes (meaning such schemes are a source of greenhouse gases); and there are other possibilities for using water to generate electricity with less environmental impact that are genuinely new (such systems are discussed in Part IV).

13.1 History

The use of falling or moving water to provide energy has a long history in many parts of the world. The earliest use was probably as a simple way of raising water from a stream or river to adjacent fields for irrigation and was in use in the Middle East by 3000 BC. By 200 BC the technology had taken a leap with the invention of the vertical-axis mill for grinding corn (Figure 13.1a), and then the horizontal axis geared system of

Energy and Climate Change   David A. Coley
© 2008 John Wiley & Sons, Ltd

(a)

(b)

Figure 13.1 (a) Sixteenth-century horizontal corn-grinding waterwheel with no gears [A. Ramelli in SMI94, p104]; (b) an eighteenth-century horizontal axis overshot wheel with extensive geared drive chain [SMI99, p121]

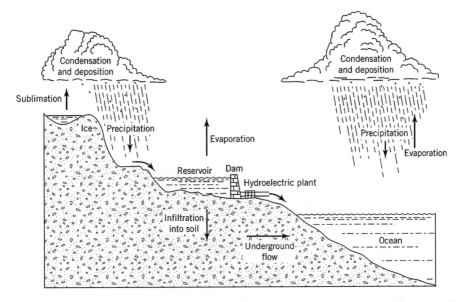

Figure 13.2   The interruption by a dam/reservoir/generator system of the return of water to the oceans, thereby converting potential energy into electric current [Ristinen, R.A. and Kraushaar, J.J., *Energy and the Environment*, Wiley, 1999, p128]

Figure 13.1b. This final design formed the basis of all arrangements for the next 2000 years. With only a few alterations, it proved so successful that it became the main source of mechanical power in much of the world and was used for pumping, iron working and paper manufacture.

The introduction of steam power during the industrial revolution reduced the need for waterpower for several reasons, the key factors being the greater power offered by steam and the severing of the link between the source location (the river) and the requirement location (the factory). The re-emergence of waterpower during the twentieth century was due to the simultaneous solution of these two problems. The first, by the invention of more efficient systems with far greater hydraulic heads (traditional systems had heads no greater than the water wheel diameter), and the latter by the introduction of electricity distribution grids.

13.2 Technologies for use

There are three technologies commonly in use for converting the potential and kinetic energy of water into electrical power: the reactive water turbine; the propeller; and the impulse turbine, or wheel. Before discussing the differences between these three approaches and indicating where each is used, we need to look in general at how water might be manipulated to provide substantial amounts of potential and kinetic energy.

The source of power lies in the hydrologic cycle (Figure 3.6). The natural evaporation of water results in a large mass being continuously elevated to great heights, gaining

Figure 13.3 Natural energy loss in a river system in Iceland (photo by author). In Chapter 24, we will see how such sites could be used to generate electricity for remote communities

potential energy as it rises. Subsequent precipitation leads to run-off that gathers in streams, rivers and lakes and returns to the sea losing any potential energy gained. Hydropower generating stations interrupt this cycle briefly (Figure 13.2) to extract some of the potential energy and convert it to kinetic energy within a rotating turbine or wheel. This is energy that would otherwise be lost to friction within the river system (Figure 13.3). It is only this frictional loss that is being extracted: the water finally ends up in the same place and at the same elevation – the sea – whether or not it passes through a hydroelectric power station. The hydrologic cycle itself is driven by the sun and therefore hydro electricity can reasonably be seen as solar power, along with wind power, wave power and biomass.

Common to the majority of hydropower stations is the idea that a large store of water is held at a considerable height then dropped down and through a turbine, or wheel. The greater the height (or head) of the water, the greater the potential energy that can be converted into rotational energy and then into electricity. The potential energy, E_{pot} (in joules), lost per second will be given by:

$$E_{pot} = mgh \tag{13.1}$$

where m is the mass of water (kg) passing through the system per second, h the hydraulic head (metres) and g the gravitational acceleration (9.8 m/s^2). Often the quantity of water involved is more easily described in volumetric units, in which case Equation (13.1) becomes:

$$E_{pot} = q\rho gh \tag{13.2}$$

here q is the volume of water (m^3) flowing through the system per second and ρ the density of water (1000 kg/m^3).

This loss of potential energy must equal the gain in kinetic energy of the water as it reaches the bottom of the hill. Allowing for frictional losses in pipes gives a PE to KE conversion efficiency of around 90 per cent. Water turbines are also highly efficient, with an extraction efficiency of greater than 90 per cent. The final conversion from rotational energy to electrical energy within the generator is typically even more efficient at around 95 per cent. This gives an overall efficiency of:

$$0.9 \times 0.9 \times 0.95 = 0.77, \text{ or } 77 \text{ per cent.}$$

This high overall efficiency should be compared with the typical efficiency of fossil fuel power stations of only 33 per cent. The much greater efficiency arises from the restrictive limits implied by the Carnot efficiency not being applicable as no heat engine is involved.

To increase the amount of electricity generated Equation (13.2) implies that either the volume of water passing through the generating station must increase, or the head must be raised. By building a sizable dam at the mouth of a valley a large catchment area can be created ensuring a large and constant supply of water. If the dam is tall enough, electricity can be generated at the base of the dam as the head is measured from the top of the reservoir. For shallower reservoirs, a long pipe, or penstock, can be used to take the water to a lower level, greatly increasing the head. Figures 13.4b and 13.4c contrast these two approaches. The geography of the environment will usually define which approach is implemented. That used in Figure 13.4b will usually make more sense if a large pre-existing river valley is to be damned. Without the dam the water would only drop through the very small height given by the incline of the streambed, so the dam effectively raises the water to a substantial height (often over 100 metres). This process can be repeated many times along the river. The approach of Figure 13.4c is often used to create or enlarge a mountain lake where most of the head will be given by the lie of the land, and the dam provides little additional height but does ensure a constant supply of water. Figure 13.5 shows examples of both approaches.

As previously mentioned, there are three common approaches for converting the potential energy of the water in the dam into rotational energy used to drive the generator: the water turbine, the propeller and the wheel.

The Francis turbine

In truth, all three systems we will be discussing can be considered to be turbines, however, one – the Francis turbine – looks and operates in a very similar way to the gas turbines found in other power stations. The central rotating part of the turbine is called the *runner* and is shown in Figure 13.6. This is housed at the tail end of the penstock, which looks like a snail shell and has controllable guide veins on its inner surface. Water flows around the circumference of the turbine and is directed by the guide veins on to the runner blades. Because the water exerts continuous pressure on the runner blades these

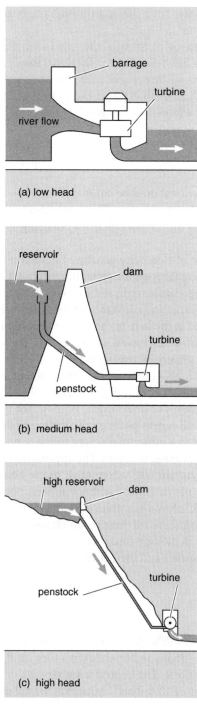

Figure 13.4 Types of hydroelectric installation [BOY96, p197]

(a)

(b)

Figure 13.5 (a) The Hoover dam, Colorado River, USA – a typical large river dam (note the very tall dam and the location of the generating equipment at the foot of the structure). Copyright Bureau of Reclamation [USB05]. (b) A high mountain dam (note the long penstock and the relatively low dam) [UNE05]

react continuously to this force, giving rise to the generic name *reaction* turbine. As the water strikes the rotating blades it loses most of its momentum and then flows out of the central tube. The system produces power for two reasons. Firstly the water is forced to change direction and lose its momentum, and secondly, there is a large pressure drop across the machine. Clearly if water is to exit the machine, which it has to if new water is to enter from above, this water must have a velocity. In turn, this must mean that not all the energy was extracted. The volume of water entering the machine each second must equal that exiting, but by flaring of the exit pipe to a larger diameter the exiting water is brought almost to a standstill. This has the effect of increasing the pressure drop across the machine thereby generating yet more power.

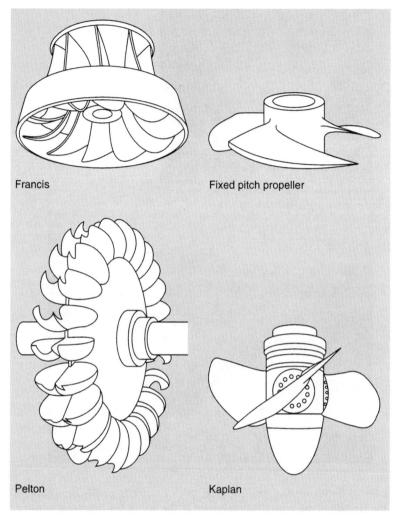

Francis

Fixed pitch propeller

Pelton

Kaplan

Figure 13.6 Types of turbine [BOY96, p199]

Francis turbines are most commonly used for power stations supplied by medium heads. Low heads imply that large volumes of water must be supplied and large turbines constructed. The low water speed of low-head systems also means that the blades must be more twisted or propeller like, suggesting that ultimately a propeller might be a better solution. Conversely, at very high heads the water speed will be considerable, and if most of the energy is to be extracted from this motion, the runner speed must be almost equal to the water speed, creating various engineering challenges. The following two systems have been designed to replace the Francis turbine when low or high heads are encountered.

The propeller

Like the Francis turbine, a propeller system is totally submerged and is also classed as a reaction machine. The supply side has the maximum possible area – the whole area of the blades – thereby making it suitable for use at low heads. As it is easy to change the pitch of the blades, the same revolution speed can be maintained for differing flow rates. The design is also particularly efficient in situations where demand varies.

In order for the propeller to interact with as much water as possible, the blade speed is greater than that of the water (often by a factor of two) just like a wind turbine; again suggesting that the machine will work well at low heads.

The Pelton wheel

For very high heads (greater than 250 metres) the most efficient design proves to be little more than a large number of buckets or cups sitting on the circumference of a wheel, a design reminiscent of the water wheels of the pre-industrial era. One or more jets of water from the penstock, and controlled by a needle valve, are fired at the wheel and strike each cup as the wheel rotates. The kinetic energy of the water is converted into rotational energy in a series of rapid pulses – hence the generic name *impulse turbine*.

As the wheel gains all of its rotational energy from the kinetic energy of the water, and none due to a pressure drop, it is easy to analyse the situation and arrive at a relationship between the head and the power generated. As discussed above, the potential energy of the water given by its head must equal the kinetic energy of the flowing water. So,

$$E_{pot} = E_{kinetic}$$

or

$$q\rho g h = \tfrac{1}{2}q\rho v^2 \qquad\qquad (13.3)$$

therefore, $gh = \frac{v^2}{2}$,
or

$$v = \sqrt{2gh}.$$

At the level of accuracy in which we are interested, g ($= 9.8$ ms^{-2}) can sensibly be replaced by the value 10. So the speed, v, of the water striking the wheel will be:

$$\boxed{v = \sqrt{20h}}. \qquad\qquad (13.4)$$

A surprisingly simple expression.

For a flow of q cubic metres per second, $q\rho g h$ of potential energy will be converted into kinetic energy every second. As the density of water is 1 kg per litre, or 1000 kg/m^3

this implies a power of

$$q \times 1000 \times g \times h \text{ (Joules per second)} \qquad (13.5)$$

or

$$10qh \text{ (kW)}$$

will be developed.

If the water velocity is given by Equation (13.4) then the flow, q, will be given by:

$$q = a\sqrt{20h}$$

where a is the cross-sectional area of the jet, or jets (in m^3). The power, P, is given by:

$$P = 10a(\sqrt{20h})\, h$$

or

$$P = 45ah^{3/2}. \qquad (13.6)$$

This indicates that doubling the head will increase the available power by $2^{2/3}$, i.e. nearly triple it. Whereas, Equation (13.5) implies that doubling the flow only doubles the power. This has obvious consequences for the siting and sizing of dams. We will meet a similar expression with similar consequences when we discuss wind power in Chapter 22.

Problem 13.1   Use Equation (13.6) to compare the theoretical power of a hydro station with a 20 m head to one with a 200 m head.

13.3 Example application: Itaipu hydroelectric station

Sited on the border between Brazil and Paraguay, the Itaipu hydroelectric station (Figure 13.7) is the largest power station in the world. Currently it is rated at 12.6 GW (due to expand to 14 GW)) and typically supplies 93 000 GWh annually, thereby meeting 24 per cent of Brazil's and 95 per cent of Paraguay's electricity requirements (all data from [ITA04]).

In analysing the site we start by looking at the water supply then move on to the engineering specifics. The dam sits on the Paraná River, which is itself fed by four other major rivers. This gives it a massive catchment area of 820 000 km^2. Taking the square root of this number, we see that this is equivalent to a square of side 286 kilometres. On to this area approximately 1400 mm of rain falls annually, although the majority of this is lost to evaporation. On average 12 000 cubic metres per second reaches the dam although the figure can range from 6000 to 40 000 on a seasonal basis. The 170 kilometre long reservoir has a volume of 19×10^9 cubic metres and covers an area of 1460 km^2 when full. The head is dependent on how full the reservoir is and varies between 84 and

Figure 13.7   The Itaipu dam [ITA05]

128 metres. The dam is truly massive and involved the excavation of over 55 million cubic metres of rock and soil. This was replaced by 15 million cubic metres of rock, 12 million cubic metres of concrete and required half a million tonnes of steel. This is eight and a half times the volume excavated for the Channel tunnel that runs between England and France and 15 times the volume of concrete used in that project. In total, and including the rock and earth banks to the side of the main concrete dam, the dam spans 7.7 km. The dam feeds 18 penstocks each with a diameter of 10.5 metres and each connected to its own Francis turbine (Figure 13.8).

Problem 13.2   If all the rain that falls on Itaipu's catchment area were to pass through the power station's turbines, how much electricity would be generated? Compare this figure with total world electricity consumption.

The turbines are connected to 18 generators in an arrangement which, if it were not for the scale of the objects, would look like a drawing from a car maintenance manual. Each turbine weighs 3360 tons and each generator over 3200 tons.

The powerhouse in which the generators sit is on a similar scale to the dam: one kilometre long and over 100 metres high. The generators produce 18 000 volts at 50 or 60 Hz (nine of each – as the two countries use different frequencies) with a rated power of around 800 MW each. A bank of transformers increases the voltage to 500 000 volts for distribution via the national grids (Figure 13.9). During the last 10 years, the power station has been able to generate for an average of 80 per cent of the year.

(a)

(b)

Figure 13.8   Itaipu turbine runner being lowered into place [ITA05]

Figure 13.9 Itaipu sub-station hall [ITA05]

The total cost of the project was 12 billion US dollars or 1000 US dollars per megawatt of installed capacity. By dividing the installed capacity by the flooded area we can estimate the *reservoir utilization index*. For Itaipu this has a value of 9.3 MW/km^2, which compares favourably with other reservoirs in the region where the index can be as low as 0.25 MW/km^2. In order to construct the reservoir and dam, 65 000 people needed to be resettled. Because of silting up of the reservoir, the useful life of the site is likely to be around 200 years.

Problem 13.3   Estimate how much coal would have to be burnt annually to supply an equivalent amount of power to that produced by Itaipu. How much carbon dioxide would this produce?

13.4 Environmental impacts

Large-scale hydroelectricity has many benefits: it is often cheap, uses no fossil fuels, produces no direct carbon emissions, and the reservoir may create a recreational facility. However, there are a series of negative impacts, some of which are considerable, and have led to major protests about the construction of new dams in various sensitive regions. Amongst the possible impacts are:

- the need to resettle large numbers of people

- the loss of important archaeological remains

- loss of habitat

- loss of rare species

- major impacts on river wildlife and humans on the downstream side of the dam

- methane production from rotting vegetation in the flooded area

- loss of human life from dam failures, and

- amplification of interstate tensions from diverting water resources

Not all of these will apply to every project, and several of them apply in some form or other to many energy related or economic projects. However it is probably true to say that the impacts are on a much larger scale. For example, the building of a new road might entail the compulsory purchase of land and the resettlement of a small number of people, but nothing like the one and a half million people [CNN05] that the construction of the Three Gorges hydroelectric scheme in China will require. This project will submerge over 100 towns and it is worth asking whether this would even be contemplated in the developed world.

Although some water is lost by evaporation from the reservoir surface, the major water loss comes from the diversion of water for agricultural purposes that the dam makes possible. This can lead to very low down stream flows and great problems for river wildlife in the dammed river.

If the construction of a hydroelectric scheme requires the flooding of a large area of forest, a substantial amount of biomass will be trapped beneath the surface. This will tend to decay anaerobically and produce methane, whereas the original forest would have decayed aerobically to produce carbon dioxide. As we saw in Chapter 5, methane is 23 times more powerful as a greenhouse gas than carbon dioxide. Such a reservoir can therefore be viewed as a source of anthropogenic greenhouse gases. Additional vegetation entering the reservoir from the sides or further upstream will add to this problem. The amount of methane produced will depend on the climate of the region, the extent of the original vegetation and whether some of this was cleared prior to flooding. Reliable figures are as yet unavailable, but there have been suggestions that some hydroelectric schemes might have an almost equivalent impact on climate change per kilowatt-hour generated to that of conventional coal-fired generators. However, such claims remain controversial.

Problem 13.4 Assuming vegetation decomposing within a reservoir does so anaerobically, estimate the mass of methane that must be given off per kWh of generation for a hydroelectric power station to have the same climate change impact as coal-fired generation.

A more direct impact that calls for consideration is the possibility of dam failure from either poor engineering or because of an earthquake. Again, the danger of this will depend on constructional details and the geology of the region and seems to be decreasing with time, however failures have been known to cause great loss of life.

Finally there is the question of how long a hydro scheme is likely to last. Many reservoirs show a tendency to silt up – that created by the Hoover dam in Arizona and Nevada, USA (constructed in 1936) has now lost half of its original storage capacity. This suggests that electricity from many hydro schemes should not be considered as wholly renewable in the long term. It should be noted that many of the impacts discussed above are greatly reduced for the small-scale schemes that will be covered in Chapter 24.

Before we leave hydropower it is worth introducing a novel use for the technology: pumped storage.

13.5 Pumped storage

Unlike chemical energy, electrical energy is very difficult to store. This means that it is usually created on demand, which as demand varies throughout the day and throughout the year, causes great problems for generating companies. Fortunately water turbines and generating sets work efficiently in reverse, allowing water to be pumped up to high-level reservoirs when demand is low. When demand rises, the upper reservoir can be drained back through the turbine into a lower reservoir. The overall efficiency of this process can be in excess of 70 per cent and maximum generation can be reached in only a few seconds. The original electrical energy that has effectively been stored is usually from a source that finds it hard to match the rapid fluctuations in demand, nuclear or coal being obvious examples, or one where the efficiency drops rapidly when it is not running at full power. Another up-and-coming example is wind power, where there is little control over when the electricity is generated. If the upper reservoir also has a substantial catchment area, then the facility will also be a net generator of electricity. If such a scheme is used to integrate a time-varying sustainable energy system into a national grid, we need to question whether the resultant energy system is truly sustainable, given that we have classified hydropower as unsustainable? If the dam is chosen carefully the answer is probably yes, but clearly there is a concern here.

As an example, the facility at Ludington, USA, takes water from Lake Michigan at night, pumps it to a height of 100 metres and releases it back during the day. It has a generation capacity of 1.8 GW_e fed by an upper reservoir of 6.4 km^2. A four kilometre long net in Lake Michigan is used to prevent fish from entering the turbines.

13.6 Global resource

We can return to the analysis first attempted in Chapter 3 to estimate the upper limit of the resource. About a quarter of the sun's incident energy is involved in the hydrologic cycle. This evaporates water from the surface and carries it aloft where it condenses and falls as rain and snow. The latent heat of evaporation (i.e. the energy required to change water from a liquid to a vapour) is high – 2453 kJ/kg at 20 °C. All this energy will be returned when the vapour condenses to a liquid at high altitude, during cloud formation and precipitation, and accounts for over 99 per cent of the energy in the cycle. However the small remaining fraction still represents a large amount of potential energy stored in the form of floating clouds. Only precipitation that falls over land is of use to us for hydropower and this is estimated to be about 1×10^{17} kg annually. Given a mean planetary height above sea level of 850 metres and assuming half the precipitation evaporates from land rather than making it to the sea, then the potential energy resource equals:

$$\tfrac{1}{2}\, mgh = \tfrac{1}{2} \times 1 \times 10^{17} \times 9.81 \times 850 = 4.17 \times 10^{20} = 417 \text{ EJ per annum.}$$

This is slightly more than the global primary energy use. More realistic assumptions about how much of this could be captured suggest quartering this figure to 100 EJ: still a phenomenal amount. As has been mentioned when discussing fossil fuels, it is unlikely that it would ever be possible to access all of this resource, partly for reasons of geography, partly because of the separation of resource from point of need (although expanding national grid systems are reducing this problem), but also for economic reasons. However, given the increasing level of concern about the environmental and sociological problems of large-scale hydropower, much of this is only likely to be accessible if a switch to small-scale hydro is made (Chapter 24). The picture is further confused by the fact that many hydro stations will not go on supplying electricity indefinitely as their respective dams will silt up. Note how we have changed our emphasis in the above discussion as we have started to consider the resource *per annum*, whereas for the other fuels we have not questioned how quickly we might extract the resource. For most renewables, there are limits to how quickly we can extract it (in this case no quicker than the annual rain fall replenishes the rivers), and conversely, we can not bank the resource over a long period by slowing our rate of extraction – last year's rains are lost.

The following graphs indicate the global distribution of hydropower and recent trends in development. Updated, country-specific, data can found on the book's web site.

The current global consumption of hydroelectricity is 2631 TWh. Treating this consumption as a resource, Figure 13.10 shows that the resource is more widely spread than oil, and more equally spread than coal. Supplies are reasonably secure against global geopolitical tensions, but might be affected by tensions between states that share rivers or lakes.

Because of the many economic and environmental advantages offered by hydroelectricity, world consumption is slowly rising (up by a factor of 2.8 since 1965, Figure 13.11), but is growing much faster in regions such as South America and Pacific Asia. Growth is slow in Africa and the Middle East, and modest in the former Soviet Union.

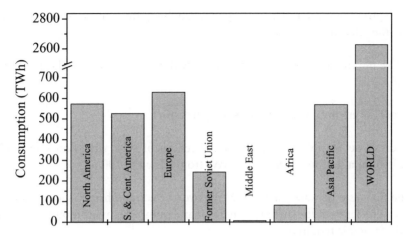

Figure 13.10 Regional hydroelectricity consumption [data from BP04]

13.7 Student exercises

"Physics is like sex: Sure, it may have practical results, but that is not the reason we do it."

–Richard Feynman

1. Describe the characteristics of the three common approaches for the conversion of potential energy of water into rotational energy.

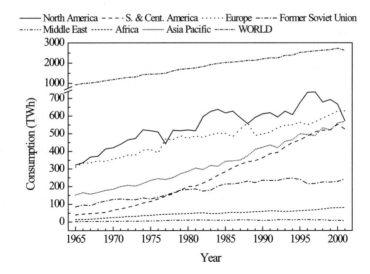

Figure 13.11 Regional changes in the consumption of hydroelectricity since 1965 [data from BP04]

2. Which gives the greater increase in power, doubling the flow or doubling the head? (Give the relevant equations.)

3. Outline the character of the power station with the largest installed capacity in the world.

4. Discuss the environmental impacts of hydropower.

5. If hydropower stations typically have efficiencies of 80%, what flow rate would be required for a station to generate 1 GW_e from a head of 100 m?

6. If the flow rate of water in a 150 MW_e hydro-electric scheme is 40 m^3/s and the overall efficiency of the turbine, generator and pumped storage motor is 71%, what head of water is required?

7. The mean height of the feeder reservoir in the Dinorwig pumped storage hydroelectric scheme is 568 m. At rated load the overall efficiency is 86 per cent. If the plant operates for 5 hours, delivering 1750 MW_e, what mass of water has passed through the turbines? What has been the flow rate? [From SHE97.]

8. Estimate the hydroelectric potential of an area or location, chosen from an atlas. Use the following technique, having chosen location X [from SHE97].

   a. What is the lowest altitude of X?

   b. What area of X lies more than 300 m above the lowest level?

   c. What is the annual rainfall on the high parts of X?

   d. If all of the rainfall ran to the lowest level, what amount of potential energy per year in MW would be given up by the moving water?

   e. What factors would prevent all of the rainfall being converted to electricity?

   f. Estimate the fraction of the rainfall potential energy that might be convertible to electricity.

   g. If your selected location X already contains a hydroelectric power station, compare your estimate of its potential capacity with the station rating. Comment on any large differences.

14

Transport and air quality

"I think cars today are almost the exact equivalent of the great Gothic cathedrals. I mean the supreme creation of an era, conceived with passion by unknown artists, and consumed in image if not in usage by a whole population which appropriates them as a purely magical object."

–Roland Barthes' The New Citroën (1957)

In this book we have been looking at where we get the energy we need on a fuel-by-fuel basis. This has proved a reasonable approach, as it is the characteristics of the fuel that define the technology, and to a large extent the final use of the energy. This will also allow us to follow a consistent path in Part IV, where we will be looking at alternative energy systems, resource by resource. However, energy use in transport has characteristics that make it distinct. Many of the environmental problems caused by transport systems, and several of the difficulties in solving these energy-related issues, come not just from the fuel used, but because it is consumed within mobile systems. This presents particular difficulties for any new technology that we might try to use to break our reliance on oil.

We will also need to question why we need transport. Or at least why we need so much. For other sectors, conservation usually implies using energy more efficiently – but still obtaining the same end result. With transport we need to consider more efficient vehicles, but also look at restructuring our lives and our societies so less transportation is required.

For these reasons transport is being considered as a separate topic. This chapter considers energy demand in our current transport systems and the problems this causes; in Chapter 28 we will look at how transport systems might change in the future so as to reduce these impacts.

Although there are a lot of shared problems, the precise details of transport systems and their degree of use are country specific. We don't have space to list the details for each country, nor the necessary data, so will concentrate on data from a single representative

Energy and Climate Change   David A. Coley

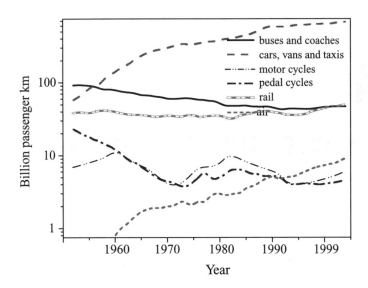

Figure 14.1   Growth in UK transport, 1952–2003 [data from UK Department for Transport, *Transport statistics Great Britain 2004*, HMSO, London, 2005]

industrialized country – the UK – but the analysis and problems identified are typical of most developed nations.

14.1 Present day problems

In much of the world there has been enormous growth in transport over the last 50 years (Figure 14.1). In the UK we now travel around 30 kilometres (18 miles) per person per day; 50 years ago it was around nine kilometres (six miles). Clearly travelling this far each day means not relying on walking or cycling but on some form of mechanized transport. The majority of this growth has been in road transport, most notably the car, and is continuing with growth rates of greater than one per cent per annum in many industrialized countries (Figure 14.2).

Although trains can easily be powered by an external energy source (electricity) cars have, since their invention, been inextricably linked to fossil fuels and to oil in particular. Given that the motor car meets the majority of our transport needs, we can easily make an estimate of how much oil we must be using each year to simply move ourselves around, and how much carbon dioxide this must be producing. At 30 kilometres per person per day, a UK population of 60 million and an assumed fuel efficiency of 10 km/l, the UK will be using very approximately

$$60 \times 10^6 \times 30/10 \times 365 = 6.57 \times 10^{10} \text{ litres, or } 5.5 \times 10^{10} \text{ kg}$$

of oil for transport per annum. Oil has a carbon dioxide emission factor of 74 kg/GJ and a calorific value of around 42.8 MJ/kg; therefore this analysis suggests that the

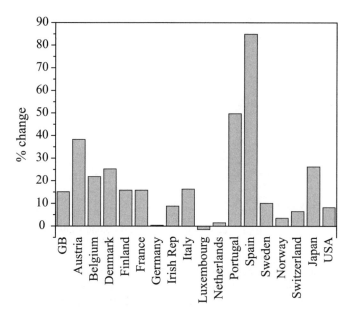

Figure 14.2 Total growth in road transport in a selection of industrialized countries (1991–2001). Note: growth in the USA is likely greater than indicated as the 2001 figure for freight has been assumed to be the same as in 1991 [data from UK Department for Transport, *Transport statistics Great Britain 2004*, HMSO, London, 2005]

UK produces approximately 1.7×10^{11} kg of carbon dioxide from our desire to travel. Chapter 15 gives the total mass of carbon dioxide produced by the UK as 5.58×10^{11} kg, therefore we can estimate that 29 per cent of the carbon dioxide we produce each year is from transport (the true figure is 26 per cent, see Figure 14.3). This result would be very approximately true for any industrialized country.

We seem to love our cars. In 1952 only 14 per cent of UK households had a car, now the number is 74 per cent, with 22 per cent having two cars and five per cent having three or more. This growth is even greater than it might at first appear. Since 1952 the number of UK households has grown from 14.5 million to 24.4 million due to factors such as people getting married later, increasing divorce rates and people living longer, i.e. there are now more households with cars than there were households in total in 1952.

The use of such a large number of internal combustion engines burning such a vast amount of oil has led not only to the production of a great amount of carbon dioxide but also to other pollutants more directly harmful to human health. Importantly, much of this production has been in close proximity to areas of high population density – city centres. For most of the developed world (and growing amounts of the developing world) problems in local air quality are dominated by questions about emissions from road vehicles. One good way of visualising that there might be a problem with vehicle emissions is to imagine the sight of a busy road full of cars but with the cars stripped of their body work etc., leaving just the running engine and the exhaust pipe sitting stationary in the road. Given a choice, would we naturally choose to run such a large number of polluting sources all day right in the centre of our towns?

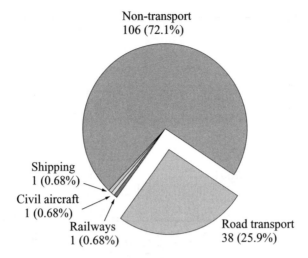

Figure 14.3 Carbon emissions from the UK in 2002 (MtC). Note how road transport dominates transport emissions and also makes up over one quarter of all emissions [data from UK Department for Transport, *Transport statistics Great Britain 2004*, HMSO, London, 2005]

Our examination of present day transport systems will revolve around three topics: energy use, health impacts and why we travel. The first two of these are amenable to changes in technology. The third only to alterations to the way we design our cities, our working practices and the way we lead our lives.

Energy use and carbon emissions

Having seen that 26 per cent of UK carbon dioxide emissions arise from transportation, it would seem sensible to ask how these emissions are shared between the various common modes: rail, road, air transport and shipping. Figure 14.4 shows this modal split, and indicates again that from an energy or climate change perspective, problems deriving from transport boil down to problems of road transport and in turn to cars.

This result should not stop us considering the climate impacts of other modes, particular that from aviation – which is now the fastest growing mode in the developed world.

Unfortunately, it has proved to be very difficult to judge the climate change impact of air travel. Aeroplanes emit pollutants, from ground level to 10 000 metres or so; sunlight, temperature and chemical regimes vary greatly across this range of altitudes, as they also do with latitude, season and time of day. This makes it very difficult to define a global warming potential for aviation emissions because the reactions, and rates of reactions for the pollutants will also vary greatly. For example, the radiative forcing from ozone produced by NO_x emissions is not simply linearly proportional to the amount of NO_x emitted, but also depends on location and season. In addition, highflying aircraft produce *contrails* (cloud-like vapour trails). Contrails reflect incoming sunlight (thereby

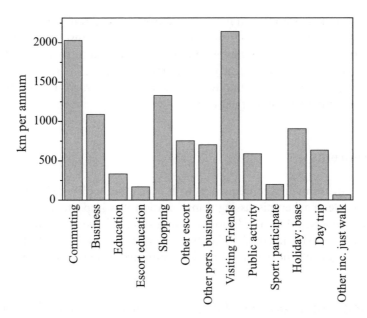

Figure 14.4 UK trip distances per person per year by main purpose for all modes. (The main purpose is the mode used for the longest part of the journey; escort means a driver or passenger who is not in the vehicle for the purpose of the journey, e.g. a parent driving a child to school.) [Data from DFT05]. Note, for example, "Visiting Friends" > "Commuting"

cooling the planet) and also reflect outgoing infrared radiation (thereby warming the planet). Unfortunately the warming dominates (especially at night). Contrails can also trigger the formation of clouds, greatly increasing the effect. It has been estimated that on average 0.1 per cent of the world's sky (and 2 per cent of Europe's and the USA's sky) is covered by contrails at any one time [SAU98].

To partially solve the problem of not being able to define a global warming potential for aviation, the *radiative forcing index* (RFI), has been introduced. This is defined as the ratio of total radiative forcing to that from CO_2 emissions alone. It is the sum of all forcings, direct emissions (CO_2, soot, etc.) and indirect atmospheric responses (including methane, ozone, sulphate and contrails). The RFI for the global aviation fleet is thought to be around 2.8, implying that air travel is almost three times as damaging as a simple consideration of carbon emissions would imply.

For an in-depth analysis of the future climate change impacts of aviation see reference [IPCC99].

Why we travel

Reducing these emissions will either require us to switch to non-carbon based fuels for cars, or for us to change the way we lead our lives. Contrary to popular belief, commuting or business travel does not account for the majority of road traffic, recreational travel does (Figure 14.4). This is an observation of which any new transportation policy, such as electric trams, or any urban planning initiative will need to be aware. The correct siting of out-of-town shopping centres and of recreational facilities, and ensuring that they are integrated into the public transportation system might be as important as encouraging commuters out of their cars.

Health concerns

Most of the environmental consequences with which this book is concerned are global in nature; however when discussing transportation it is important not to forget the direct health concerns that arise from the combustion of large quantities of petroleum within an urban environment. Another major health concern is road traffic accidents. On average, 10 people die per day in the UK from such accidents. Yet another concern is noise.

Problem 14.1   What percentage of UK citizens die in road traffic accidents rather than from all other causes? (The UK has a population of around 60 million.)

In the following we will expand upon these concerns by discussing each pollutant in turn and then summarize the total and relative emissions in a table.

14.2 Air quality and health

The various traffic pollutants created arise mainly from a mix of inefficiencies in the combustion process, or because the fuel/air mix contains impurities. The production of many of them can be controlled by either improving the design of engines, or post-processing the exhaust gases to remove them with the aid of catalytic converters.

Carbon monoxide (CO)

Approximately 50 per cent of UK carbon monoxide emissions are from transport (47 per cent from road transport alone). The production of carbon monoxide represents an inefficiency in the combustion process and arises from a failure to fully react the carbon in the fuel with air to create carbon dioxide. Once in the atmosphere, carbon monoxide is converted to carbon dioxide by reacting with hydroxyl radicals. This in turn reduces the general atmospheric concentration of radicals, thus increasing the concentration of both methane and low-level ozone that would normally react with such radicals. Carbon monoxide reacts with haemoglobin in the blood stream to form carboxyhaemoglobin, thereby reducing the blood's oxygen carrying capacity. Although fatal in very high concentrations, low concentrations can affect the functioning of the central nervous system, slowing mental function, causing headaches, drowsiness and disruption to vision. The UK guideline maximum concentration, above which an *air quality management zone* would need to be established, and policies introduced to reduce concentrations, is a mean of 10 mg/m^3 over any eight-hour period. Concentrations are likely to be highest in urban areas in winter.

Nitrogen oxides (NO$_x$)

Nearly 43 per cent of UK emissions of nitrogen oxides are from transport (37 per cent from road transport alone). Such oxides are initially released in the form of nitrous oxide, formed from burning fuel at high temperatures using the mix of nitrogen and oxygen found in air. The nitrous oxide is oxidized to form nitrogen oxide which then forms nitrous and nitric acid leading to acidification of the environment (see Chapter 5). Nitrogen oxides cause respiratory problems by inflaming the epithelium (surface) of the airways, predisposing individuals to infections and bronchitis. The UK guideline maximum concentration of NO$_2$ is 200 µg/m^3 (not to be exceeded more than 18 times a year) and an annual mean of 40 µg/m^3.

Nitrous oxide (N$_2$O)

Approximately 10 per cent of UK emissions are from transport (the majority are from natural sources such as emissions from soils). Apart from its role in the production of nitrogen oxides, nitrous oxide is a greenhouse gas.

Tropospheric (low-level) ozone (O$_3$)

Unlike high-level ozone, which protects us from ultraviolet radiation from the sun, low-level ozone is a pollutant. It is a secondary pollutant formed by a series of chemical and photochemical reactions involving oxygen, nitrogen dioxide and radicals derived from hydrocarbons. It irritates mucous membranes, reduces lung function and increases the susceptibility of individuals to infections. Asthmatics are particularly sensitive.

Methane (CH$_4$)

There are no health effects from the concentrations found in the roadside environment, and only one per cent of emissions are from transport. However, it is a greenhouse gas and emissions would increase if natural gas became a popular replacement for liquid transportation fuels.

Volatile organic compounds (VOCs)

Nearly 15 per cent of UK VOC emissions are transport related, including that from fuel evaporation (14 per cent from road transport alone). VOCs cover a wide variety of compounds, which if absorbed by the lungs can cause carcinogenic metabolites; others such as benzene are believed to be directly carcinogenic, still others such as aldehydes are toxic. Being carcinogenic, there are essentially no safe levels. Although catalytic converters can reduce hydrocarbon emissions by 90 per cent, the use of unleaded petrol (gasoline) can lead to increased emissions from the higher levels of aromatic compounds (including benzene) in the fuel. For benzene the UK guideline maximum concentrations are an annual mean of 16.25 $\mu g/m^3$ (reducing to 5 $\mu g/m^3$ by 2010).

Particulates

Particles with a diameter of less than 10 μm (0.01 mm), and referred to as PM_{10}, are termed *inhalable*; those with a diameter of less than 2.5 μm are termed *respirable* and can penetrate deep inside the lung and remain there. Such particles are believed to increase mortality via various routes, and are considered to be the most harmful of vehicle-based emissions. Some 24 per cent of UK emissions of PM_{10} are from transport (23 per cent from road transport alone). The UK guideline maximum concentrations are 50 $\mu g/m^3$ (not to be exceeded more than 35 times in any year) and 40 $\mu g/m^3$ (annual mean). Diesel-fuelled vehicles produce far greater quantities of particulates than their petrol equivalents.

Other pollutants

Environmental levels of platinum, palladium and rhodium have been increasing since the introduction of catalytic converters. The effects these might have on human health are as yet unknown.

Table 14.1 lists the quantity of major air pollutants released by road transport in the UK each year and their relative contribution to overall emissions.

Three-way catalytic converters

One of the most important developments in improving urban air quality has been the widespread use of catalytic converters. Three beneficial chemical reactions occur in an

Table 14.1 Major air pollutants and their relative contributions to total UK emissions (2004) [data from DFRA06]

	Thousand tonnes	Percentage
Nitrogen oxides		
All transport	692	42.6
Non transport	929	57.3
All emissions	1621	100
Carbon monoxide		
All transport	1436	49.0
Non transport	1494	51.0
All emissions	2930	100
Volatile organic compounds		
All transport	150.8	14.8
Non transport	873.2	85.2
All emissions	1024	100
Lead (tonnes)		
Road transport	1	1
Other sources	133	99
All emissions	134	100
Particulates (PM_{10})		
All transport	37	24.4
Non transport	117	75.6
All emissions	154	100

active three-way catalytic converter. These are the oxidation of hydrocarbons (HC) and carbon monoxide (CO), and the reduction of NO:

$$2HC + 5/2O_2 \rightarrow 2CO_2 + H_2O$$

$$CO + 1/2O_2 \rightarrow CO_2$$

$$NO + CO \rightarrow CO_2 + 1/2N_2$$

$$10NO + 4HC \rightarrow 4CO_2 + 2H_2O + 5N_2$$

$$NO + H_2 \rightarrow 1/2N_2 + H_2O$$

The result being that HC, CO and NO are simultaneously converted to benign CO_2, N_2 and H_2O.

Structurally, the converter sits between the engine and the exhaust pipe and consists of two honeycombs of catalysts. The honeycomb structure ensures that the gases have maximum opportunity to interact with the catalysts. The first honeycomb consists of platinum and rhodium and reduces NO. The second consists of platinum and palladium and oxidizes the unburned hydrocarbons and CO.

The number of cars fitted with three-way catalytic converters is growing, but a high fraction still do not benefit from the technology. Unfortunately, catalytic converters can reduce the efficiency of the combustion process, and thereby increase carbon dioxide emissions. Such converters are also unsuitable for use in diesel vehicles.

The use of diesel as a fuel is common in buses and other public transport systems and therefore it is used in great quantities within cities, creating a potential problem with air quality in such locations. Its greater efficiency as a fuel though does mean its CO_2 emissions per km driven are typically lower than equivalent petrol (gasoline) vehicles. Its widespread adoption in private vehicles will therefore be beneficial in the fight against climate change, but worsen local air quality. This is an example of an environmental dichotomy. We have a pair of options, neither of which is environmentally benign and, because they have different environmental effects, it is very difficult to compare the relative magnitudes of the impacts. In this case the dichotomy is deepened because the impacts affect differing groups. Local air quality is largely 'democratic', in that those affected either own cars themselves or derive wealth from the use of transport within the economy in which they live. However, the majority of greenhouse gas emissions are from more developed countries, whereas the majority of the future victims of climate change are likely to live in less developed economies. Climate change can therefore be seen as highly 'undemocratic', thus reducing the pressure for change.

This idea of an environmental dichotomy can also be used to discuss the benefits of other technologies, for example nuclear power. Nuclear power has very low carbon dioxide emissions, but the potential for releasing radioactive pollution on a local or even regional scale.

Impacts

The above analysis suggests not only that transportation is a major source of air-borne pollutants but also that the majority of transport emissions are from personal road transport. The rate of production of a pollutant depends on both how the vehicle has been maintained, and on how it is being driven. PM_{10} emissions will be greatest when

Figure 14.5 Photochemical pollution in Los Angeles, USA [EPA05]

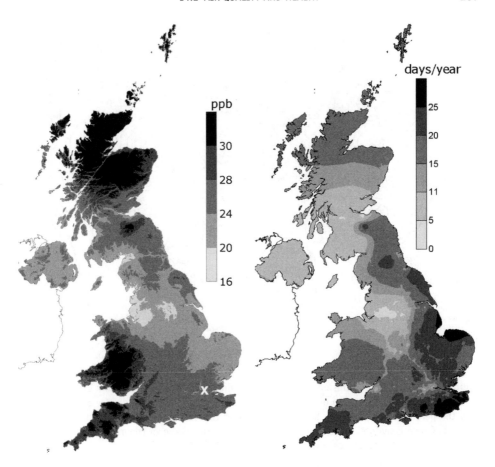

Figure 14.6 Ozone concentration in the UK. The left image shows annual mean concentrations, the right the number of days when the eight-hour running mean concentration exceeded 50 parts per billion (ppb), both 2001. The darker the shade the higher the concentrations/number of days exceeded. Concentrations are often lowest in the large cities (London is marked with an 'X') and highest in rural areas. Transnational ozone can also be seen migrating from continental Europe (to the right/East of the UK) giving rise to short-lived high concentrations in the southeast [Reproduced courtesy of the Centre for Ecology and Hydrology, NERC]

the engine is cold, as will those from incomplete combustion. This is just when catalytic converters tend to be performing sub-optimally as they too need to warm up. Thus converters will be less effective in reducing emissions during short journeys, or the initial part of any journey, a time when the vehicle is likely to be within a densely populated environment. On the other hand, nitrogen oxide production increases with engine temperature and therefore with vehicle speed.

Although it is known that polluted cities have a higher mortality rate, it is not clear how much of this is directly contributable to vehicle emissions. It has been suggested

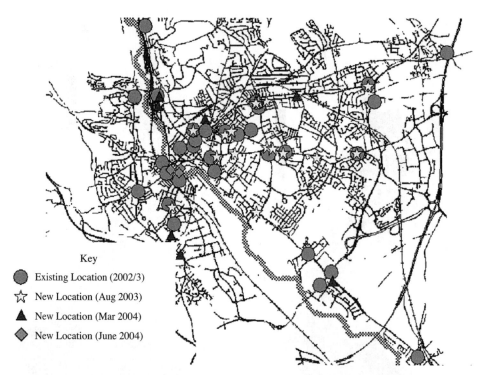

Key

● Existing Location (2002/3)

☆ New Location (Aug 2003)

▲ New Location (Mar 2004)

◆ New Location (June 2004)

Figure 14.7 Location of diffusion tubes to monitor nitrogen dioxide levels within Exeter [MIT05]. (This map, which covers an area of around 7 by 6 km, is reproduced from Ordnance Survey material with the permission of Ordnance Survey on behalf of the Controller of Her Majesty's Stationery Office © Crown copyright. Unauthorized reproduction infringes Crown copyright and may lead to prosecution or civil proceedings. Exeter City Council 100025458, 2004)

[RC94] that PM_{10} emissions alone might be responsible for 10 000 premature deaths per annum in the UK. This is a staggering figure.

Problem 14.2 Using the above death rate, estimate the percentage of UK deaths that might be attributable to PM_{10} emissions. (UK population = 60 million.)

The weather plays an important role in how these emissions can lead to high concentrations within the urban environment. Anticyclonic (high pressure) weather implies lower wind speeds thereby reducing dispersion. Such weather can also reduce mixing between vertical layers in the atmosphere particularly in winter. If these conditions are aided by the geography of a location (for example a large city sitting in a bowl), high concentrations of pollutants can become trapped. Figure 14.5 shows this occurring in Los Angeles (USA).

Like acid rain, ozone is a regional as well as a local pollutant. Because nitric oxide (NO) from vehicle exhausts acts as a scavenger molecule mopping up ozone, concentrations in urban environments are often lower than those in rural areas. Ozone created

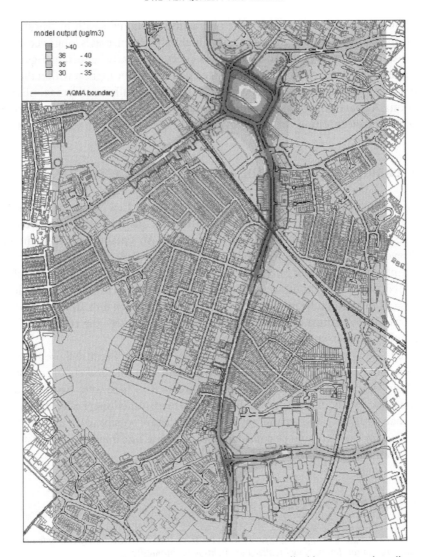

Figure 14.8   Detailed plot of predicted annual average nitrogen dioxide concentrations (in μgm^{-3}) for 2005 at 1.5 m elevation, for main roads in part of Exeter [MIT05]. The large roundabout lies at the three diamonds marked on Figure 14.7. Dark areas have an annual average nitrogen dioxide concentration in excess of 40 μgm^{-3} – the guideline maximum value. This suggests that an air quality management area (AQMA) will have to be declared even in this modest sized town and action taken to reduce concentration. (This map is reproduced from Ordnance Survey material with the permission of Ordnance Survey on behalf of the Controller of Her Majesty's Stationery Office © Crown copyright. Unauthorized reproduction infringes Crown copyright and may lead to prosecution or civil proceedings. Exeter City Council 100025458, 2004)

within the urban environment from the interaction of sunlight with hydrocarbons is free to disperse outwards, and once away from high vehicle densities is long lived and can drift across countries or even continents. This effect can clearly be seen in Figure 14.6. One might expect levels of such a traffic-related pollutant to be highest within the larger conurbations; however the opposite is true, with rural areas suffering the maximum concentrations. A clear example being London (population nine million) which experiences relatively low concentrations of low-level ozone, despite a very high vehicle density.

14.3 Example application: air quality in Exeter, UK

The UK requires local authorities to carry out periodic reviews of air quality in their areas, and to assess present and predicted future air quality against the prescribed objectives. If there were likely to be any exceedences, Air Quality Management Areas would have to be declared, followed up by focussed action plans addressing the problem, for example restricting traffic flows. In the case of Exeter (a small town – population 111 000 [NOS05] – in the southwest of England) the assessment involved both collecting data from 41 diffusion tube monitors (which measure nitrogen dioxide) and modelling concentrations by using a diffusion model. The example shows the effort that is now being applied, even in small towns, to this issue.

Figure 14.7 shows the network of diffusion tubes in Exeter. Because these are limited in number, and cannot account for the likely growth in traffic and changes in vehicle technology, modelling is also carried out (Figure 14.8). There were two conclusions from the work. Firstly, exceedence was likely for 15 roads/junctions (i.e. a concentration greater than $40\,\mu g/m^3$), and secondly such exceedences are possible even in small towns. This second conclusion could be considered surprising as one might think that it is only in large cities that air quality would be a problem.

14.4 Student exercises

At first I thought I was fighting to save the rubber trees; then I thought I was fighting to save the Amazon rainforest. Now I realize I am fighting for humanity.

–Chico Mendes

1. How has the need for transport grown in the developed world over the last 50 years? (Use the UK as an example.)

2. Using the data in Figure 14.4, apportion all travel to the categories *business/education* or *personal*. Which is the greater fraction?

3. List the major air pollutants from transport systems and describe their health impacts.

4. Describe (with chemical formulae) how catalytic converters work.

15

Figures and philosophy: an analysis of a nation's energy supply

"[W]hen we look at the graphs of rising ocean temperatures, rising carbon dioxide in the atmosphere and so on, we know that they are climbing far more steeply than can be accounted for by the natural oscillation of the weather... What people (must) do is to change their behaviour and their attitudes... If we do care about our grandchildren then we have to do something, and we have to demand that our governments do something."

–Sir David Attenborough, naturalist[3]

In the previous chapters we have seen how the majority of the world's energy is supplied. We have done this on a fuel-by-fuel basis with little consideration of how these various fuels might make up an individual nation's energy supply. Reducing our analysis to the level of the nation state is important because energy policy – which fuels to use, how much tax to charge, etc. – is decided at the national level. We will work by example and analyse the current energy supply for a particular country and how this has changed over the last two decades. We have chosen the UK because it is both a major energy user and supplier, it uses a mix of fuels, including nuclear and hydro, it represents a fairly typical developed country and its government has indicated that it would like to move toward a low-carbon economy [DTI03]. As Figure 15.1 shows, a nation's energy supply forms a complex picture.

Our analysis will cover the following subjects: the economy, energy production, energy consumption, prices, fuel poverty, energy efficiency and carbon emissions. In Chapter 30 we will see how the current UK situation might be altered to provide the much desired low-carbon economy. Our coverage will be brief. This is deliberate: we wish only to present the overall situation over the last 30 or so years to give an idea of the

Energy and Climate Change   David A. Coley
© 2008 John Wiley & Sons, Ltd

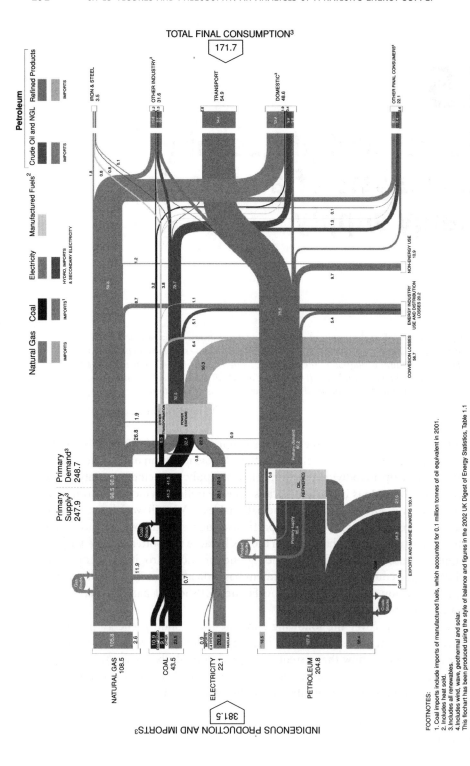

Figure 15.1 UK energy supply in 2001 (Mt$_{oe}$) [DTI01b] [Reproduced from http://www.dti.gov.uk/files/file11248.pdf (A larger version can be found on the book's website.)]

position we are in, the direction we are heading and the scale of the changes in energy supply that might be necessary to arrive at a low-carbon economy. This will require the majority of our energy to be provided from sustainable sources. It is important to get some idea of the historic rate of growth in the energy sector because we will need to size future energy sources based on the amount of energy required at the time they are introduced, not the amount of energy we currently consume. As we work through the possible alternative energy technologies in Part IV it is worth returning to the numbers given in this chapter, or similar numbers for your own country, to try and gauge the scale of each renewable that may be called for to substantially cut carbon emissions. Unless otherwise indicated, all data are from [DTI04 or DTI05d]. Data for other countries can be found in Appendix 1, or on the book's web site.

In addition to the question of the scale of our energy requirement, there is the need to consider several philosophical and moral points connected with energy purchasing and trade at the international level.

15.1 The economy

The population of the UK is 60.1 million (2003) and the country covers a land area of 244 820 km^2 (95 633 square miles) [CIA04]. The energy industry represents 4.3 percent of GDP. (GDP, or gross domestic product, is the sum of all wages paid, or money spent, in a country within a year and is a common way of representing the wealth of a nation – see Chapter 6.) The UK's energy industry employs 165 000 people directly (Figure 15.2) and many more indirectly, for example in oil rig design and manufacture. In 2002, the UK had a trade surplus in fuels of £6 billion. This compares with a GDP of £1780 billion.

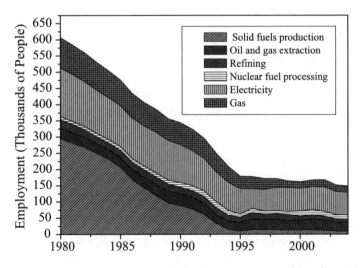

Figure 15.2 Employment in UK energy industries, 1980 to 2004 [data from DTI05d]

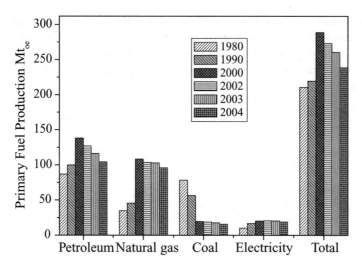

Figure 15.3 UK production of primary fuels, 1980 to 2004 [data from DTI05d]. In addition, 2.7 Mt$_{oe}$ of alternative (e.g. wind) and waste-derived (from the burning of municipal and other wastes) energy is produced. Primary electricity is the energy used to make the electricity, rather than the amount of electricity generated. So for example, it includes the calorific value of oil burnt in a power station. For hydro and nuclear power this conversion is meaningless, and it is usual to simply apply the national figure for fossil-fuelled plant (i.e. to assume an efficiency of 25–35 per cent). Here only electricity from non-fossil sources is included, as much of UK oil, coal and gas production is used in the electricity industry, meaning double accounting would occur

15.2 Production

The UK produces 11.6 EJ (275.5 Mt$_{oe}$) of primary energy per annum (Figure 15.3). Since 1970, oil and gas have come to dominate this production, mainly at the expense of coal.

15.3 Consumption

For most countries, for any particular fuel, consumption does not equal production because some fuels are produced in excess and exported, whereas others are imported. There are commonly two ways of presenting consumption data, one gives the total primary energy consumed and ignores the conversion inefficiencies of producing electricity, the other accounts for it by subtracting this loss to give the *final* energy consumption. The UK consumes 9.7 EJ per annum (230 Mt$_{oe}$) of primary energy and 6.6 EJ (157 Mt$_{oe}$) of final energy (Figure 15.4). As Figure 15.4 shows, since 1970 coal use has dropped and gas and nuclear production have greatly increased, as has total energy consumption.

It is common to disaggregate final energy consumption by sector: domestic, industry, transport and service industries (Figure 15.5). We can see from Figure 15.5 that since

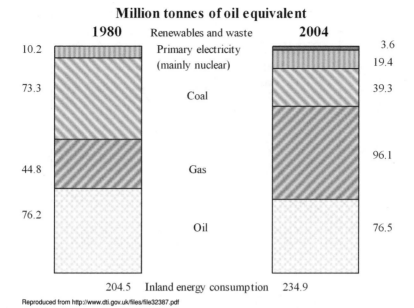

Figure 15.4   UK primary energy consumption, 1980 and 2004 [data from DTI05d]

1970 energy use in industry has declined by 44 per cent as the country has moved away from heavy industry toward a service economy. These savings have been offset by a 30 per cent growth in energy use in the domestic sector and an even greater, almost monotonic, 94 per cent growth of energy consumption in the transport sector.

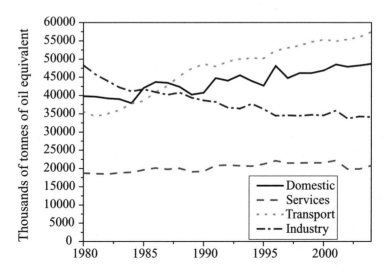

Figure 15.5   UK final energy consumption, 1980 to 2004 [data from DTI05d]

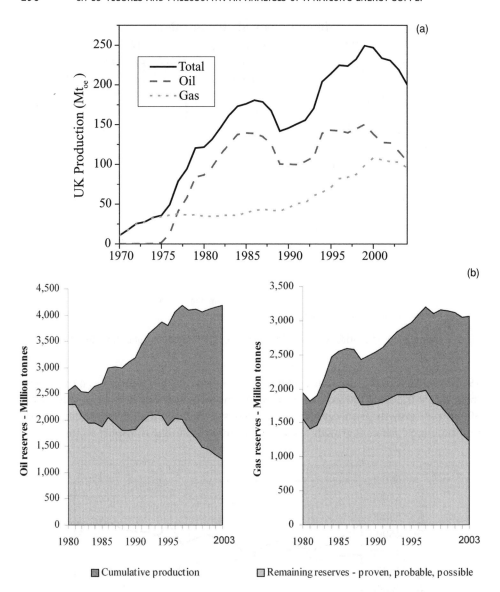

Figure 15.6 (a) UK oil and gas production, together with (b) remaining reserves [data from UK Department of Trade and Industry, *UK Energy in Brief*, the Stationary Office, London, 2005.]

15.4 Oil and gas production

The UK is not only a consumer of fossil fuels; it is also a producer and exporter. Oil and gas production has grown substantially since 1970 to 5.3 J (127 Mt$_{oe}$) of oil per annum and 4.3 J (103 Mt$_{oe}$) of gas. Meanwhile, oil and gas reserves have remained almost static (Figure 15.6).

Because the UK produces more oil and gas than it consumes, it has the opportunity to earn export revenue by selling the excess, currently £14.2 billion per annum. We now stumble upon an interesting philosophical point which was mentioned in the introduction: given that emissions from the use of fossil fuels are altering our climate and that this alteration will have severe consequences for future generations including large-scale loss of life, is this an ethical trade? This is a timely observation as the current administration (2007) believes it has put in place an ethical foreign policy covering, amongst other things, restrictions on arms trading. For the substances responsible for climate change, who is the polluter: the supplier of the fuel or the user? It is worth pausing for a moment to consider this in your own mind. Who is at fault here, the oil and gas producers, who emit relatively little carbon, or us, the users who account for the majority of emissions? This is a general point, but for fossil fuel exporting countries it is particularly pertinent as some of our wages and tax revenues are paid for by this trade. There are possible analogies with previous export industries such as the fur and slave trades, where there was a chain of responsibility from supply to use, and where others (who might not have agreed with the situation) still gained indirectly from the supply of goods and from wealth generation. The developed world has become rich through using fossil fuels and at the expense of those nations most likely to suffer from climate change. The question is, with whom does the responsibility lie – the world, the nation or the individual? As we will see in Part III, we will probably need to reduce global carbon emissions per capita by 60 to 90 per cent in the developed world in order to attain a sustainable position. This suggests the problem is less our level of fossil fuel use, but that we use fossil fuels at all. Returning to the analogy with the slave trade, would the plea 'I will only use slaves until a viable economic alternative is devised' have been a reasonable moral position in the nineteenth century? Most would now think not, yet tomorrow most of us will still use a car, or bus, heat and light our homes – all with fossil fuels. How will we be judged by later generations? Not just as a society, but as individuals?

Having mused on the personal moral position in which those of us in fossil fuel exporting, or using, countries find ourselves, we will return to our analysis of energy use in the UK by considering briefly how much of each fuel is used by each sector. Then we will consider how much energy costs us in our daily lives before estimating the total emissions of carbon dioxide that this usage implies, and how much impact alternative energy sources are having. Figures 15.7 to 15.9 show how coal, petroleum and gas consumption has changed since 1970. Coal consumption has fallen dramatically (61 per cent in total) across all sectors, with use almost ceasing in industrial and domestic settings. Much of this change can be accounted for by an increased reliance on gas, particularly in the electricity generating and domestic sectors, although the expansion would seem to be universal, totalling 85 per cent. Petroleum use has changed little since 1980 (up 0.39 per cent by 2003). However this hides an increase in oil used for transport (up 57 per cent over the same period) and a fall in the use of oil for heating and generation. Electricity use has grown across all sectors in a less dramatic way, but there have been substantial changes in how it is provided with the 'dash-for-gas' mentioned above (Figures 15.10 to 15.13).

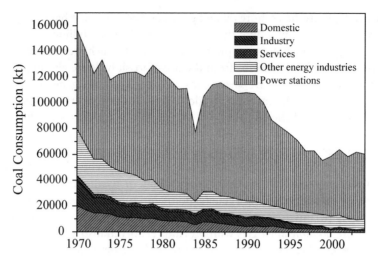

Figure 15.7   UK coal consumption, 1970 to 2004 [data from DTI05d]

15.5 Prices

One way of curbing the use of any commodity or service is to control its price, and indeed, governments raise a great deal of money by taxing fuels. It might therefore be seen as surprising that fuel prices have changed little in the last 30 years. In fact, given the increase in efficiency, reliability and cost of many delivery systems including cars and domestic heating systems, the cost of the services provided by the use of energy

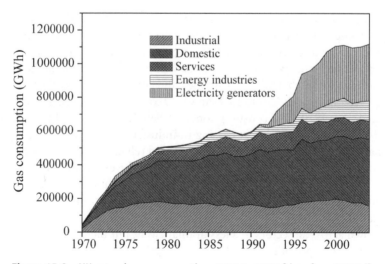

Figure 15.8   UK natural gas consumption, 1970 to 2005 [data from DTI05d]

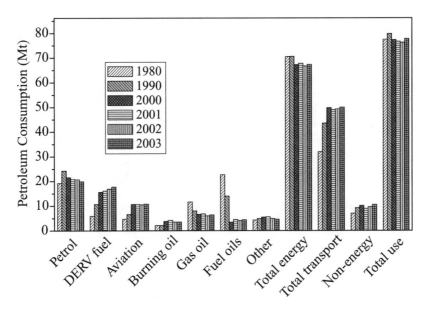

Figure 15.9 UK petroleum consumption, 1970 to 2003 [data from DTI04]

has fallen. Although it might appear that petroleum for your car is ever increasing in cost, the real cost of driving a kilometre has decreased substantially and the number of hours the average worker needs to work to earn the money to do so has fallen even more. This latter point is an important one. The prices presented in Figures 15.14 to

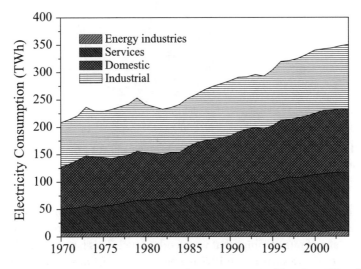

Figure 15.10 UK electricity consumption, 1970 to 2004 [data from DTI05d]

1990

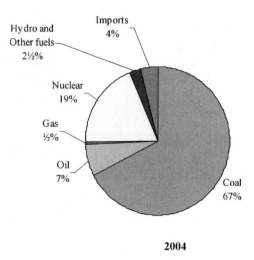

2004

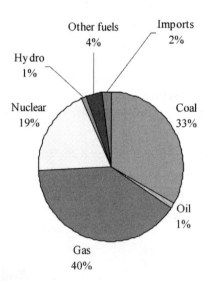

Figure 15.11 UK electricity production by fuel type, 1990 and 2004 (note the very large change in the relative importance of coal and gas) [data from DTI05d]

15.16 have only been adjusted for inflation. During the period shown, inflation led to a rise in the cost of living, however UK wages increased by far more. This is reflected in the percentage of our earnings that we spend on fuel. This has steadily decreased – a fact that may come as a surprise. In 1970 6.3 per cent of UK household expenditure was on vehicle fuel, now (2006) it is only 2.9 per cent (or €556/£398/US$720 per annum).

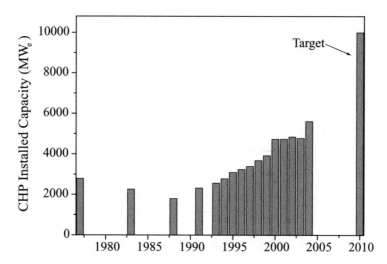

Figure 15.12 UK Electricity production by combined heat and power (see Chapter 19 for a description of this technology), 1977 to 2010 [data from DTI05d]

Problem 15.1 (a) Using the data in Figure 15.16, find the annual average and total reduction in the price of petrol (gasoline) if improvements in the fuel efficiency of vehicles are included. Assume vehicles in 1980 used 10 l/km and 2003 vehicles 17 l/km. (b) Estimate the cost of petrol at the pump to make the 2003 price equivalent to that in 1980 when efficiency is included.

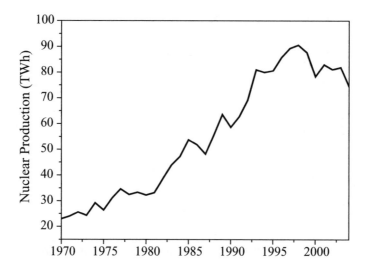

Figure 15.13 Nuclear power production in the UK, 1970 to 2004 [data from DTI05d]

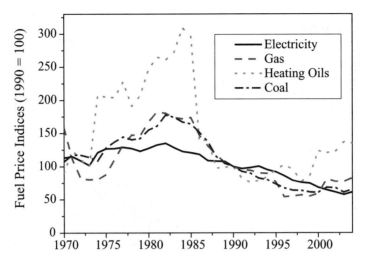

Figure 15.14 UK Fuel price indices for the industrial sector, 1980 to 2004 [data from DTI05d]

15.6 Fuel poverty

There have been other, non-economic, benefits from the growth in energy use and from its reduction in price. Fuel poverty (the inability of a household to afford the energy it needs) has fallen. This is in part because of increasing wealth and levels of employment, but also because of changes in the way the Government transfers money to the less well off and because of schemes to promote energy efficiency in the home. In 1996 it was

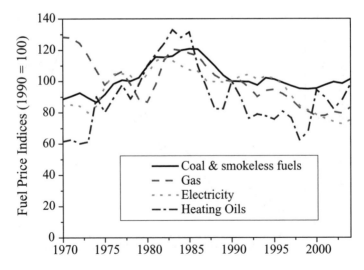

Figure 15.15 UK Fuel price index for the domestic sector, 1980 to 2004 [data from DTI05d]

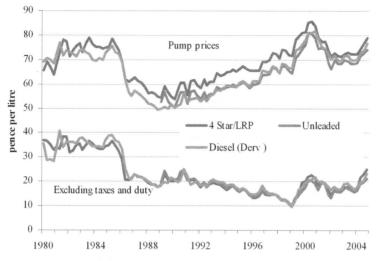

Figure 15.16   Pump prices of transport fuel, 1980 to 2005 [data from DTI05d]. (Prices inflation adjusted to 2000 prices) [Reproduced from http://www.dti.gov.uk/files/file32387.pdf]

estimated that 5.5 million UK households were in fuel poverty, the majority of these from vulnerable groups such as the elderly and those with disabilities or long-term illness or children. By 2002 this had been reduced to 3 million.

15.7 Carbon emissions

In 2003, the main sources of carbon dioxide emissions were power stations (30 per cent), industry (23 per cent), transport (22 per cent) and the domestic sector (16 per cent) [DTI04a]. An estimated 152.5 million tonnes of carbon were emitted as carbon dioxide from the UK during this year. Between 1990 and 2003, emissions fell by 7.5 per cent, despite small increases in emissions between 1999 and 2001 and again in 2003 [DTI04a]. These increases were due to greater coal consumption in power stations because higher gas prices increased the commercial attractiveness of coal [DTI04a]. In 2000 coal was also used by power stations to make up for a shortfall during maintenance and repair of gas and nuclear stations.

Figures 15.17 to 15.20 show that emissions have been falling since 1970; the UK is likely to match its Kyoto commitment (Chapter 18); emissions from transport are rising, whereas those from generation are falling; and that carbon intensity is improving.

15.8 Sustainable energy in the UK: the current state of play

UK primary energy use has grown by 13 per cent since 1965, despite positive shifts in energy conservation and the switch to more energy efficient fuels, with UK carbon

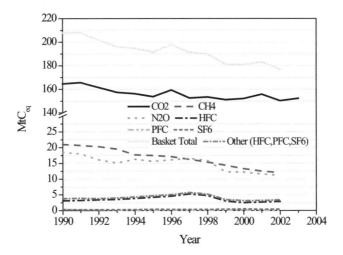

Figure 15.17   UK greenhouse gas emissions adjusted to take account of the GWP of each gas [data from DTI05b]

emissions now totalling 153 MtC per annum, or 2.52 tC per person, against a world average of around 1 tC per person. So, what attempts have been made to reduce these emissions by finding alternative means of producing the energy we demand or increasing the efficiency with which it is used?

Energy consumption per unit of delivered output, termed the *energy intensity*, gives an indication of energy efficiency. Figure 15.21 shows how energy intensity has changed over the last 30 years. This change has only in part been driven by improving efficiency in the way that term is normally used (see Chapter 19); much is a result of changes in

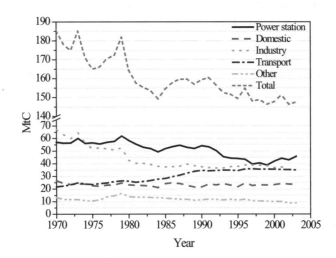

Figure 15.18   Change in UK carbon dioxide emissions by source [data from DTI05b]

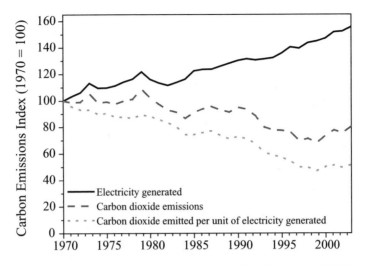

Figure 15.19   Carbon emissions from power stations [data from DTI05b]

industrial processes (e.g. a reduction in the size of energy intensive industries such as steel production) and the way in which companies operate. The trend has been largely towards reduced energy intensity across most sectors.

The UK has started to use more sustainable sources of energy (Figure 15.22), but it must be said that progress has been slow. Currently the UK produces 134 PJ (3.2 Mt$_{oe}$) of sustainable energy, if hydropower is considered sustainable, whereas UK total energy demand is 9660 PJ (230 Mt$_{oe}$).

The UK Government's Energy White Paper, *Our energy future – creating a low carbon economy* [DTI03a], published in February 2003, suggested that the UK should make a

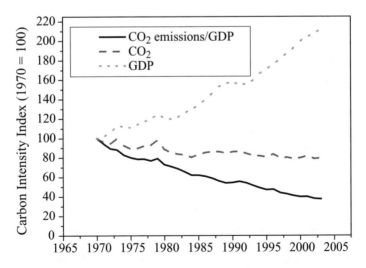

Figure 15.20   UK carbon intensity [data from DTI05b]

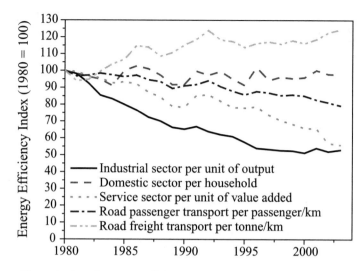

Figure 15.21   UK energy efficiency 1980 to 2004 [data from DTI05d]

substantial shift in the way the country provides its energy, with more renewable energy, much greater levels of energy efficiency and finally a move to hydrogen as the major transportation fuel. Although criticized for lacking detail and stringent targets, the title of the paper can be seen as a major breakthrough, with the government accepting

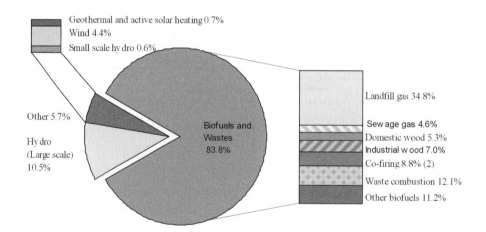

Total renewables used= 3.81 million tonnes of oil equivalent

Figure 15.22   UK renewable energy production in 2004 [data from DTI05d]. (1) Excludes all passive use of solar energy and all non-biodegradable wastes. (2) Biomass co-fired with fossil fuels in power stations. (3) 'Other biofuels' include farm waste, poultry litter, meat and bone, and short rotation coppice [Reproduced from http://www.dti.gov.uk/files/file32387.pdf]

that the future will not be fuelled by carbon alone. In Part IV we will investigate some of the new technologies that the world's nations might deploy in this move away from fossil fuels. In chapter 30 we will review the White Paper in more detail.

15.9 Student exercises

"It doesn't matter how beautiful your theory is, it doesn't matter how smart you are. If it doesn't agree with experiment, it's wrong."

–Richard Feynman

1. Outline how employment in energy industries, energy production and energy consumption has changed in the UK (or your nation) since 1980.

2. Are there any ethical questions about a country trading in fossil fuels? What are your thoughts about your own use of fossil fuels?

3. How have UK (or your own nation's) fuel prices changed since 1980?

4. What is energy intensity and how has it changed in the UK (or your country) since 1980?

5. Compare the proportions of prime fuels used for electricity generation in the UK between 1980 and 2004. What are the present trends?

PART III
Climate change: predictions and policies

"The greatest of evils and the worst of crimes is poverty."

–Bernard Shaw in the preface to Major Barbara, 1906

In Part II we looked at each of the major energy-producing technologies and fuels deployed today. In this part we will consider the amount of energy the world will use in future decades and then estimate the greenhouse gas emissions that this implies. We need to do this for two reasons: firstly, we are interested in meeting future energy consumption from sustainable resources, not just matching today's level of use (although that would be a start), and secondly, any predictions of the future impacts of climate change need to be based on future levels of greenhouse gas emissions.

Having estimated future emissions we will see what this implies for a series of impact indicators, namely: temperature, sea level, flora, water resources and food supply. We will be particularly interested in how variable these impacts are across the planet, i.e. is climate change likely to be 'democratic', in that we will all suffer equally (or the worst polluters suffer the most), or will some of those with the lowest emissions and the least amount of money to adapt suffer the greatest changes in temperature, rainfall etc.? What is clear is that the impacts will be dramatic, with the World Health Organization predicting that climate change is killing at least 150 000 people per annum (mainly from increases in malaria (which is acutely sensitive to climatic changes), heatstroke, dehydration and malnutrition), and that this number will at least double in the next two decades [GUA03]. At the same time we will see how the computer models that are used to study impacts also form one of the strongest strands of evidence proving that anthropogenic climate change is real.

The final chapter will discuss what success the world community has had in trying to reduce worldwide emissions and so reduce the level of future impacts. We will see that the answer is not a lot, and that this lack of progress arises from a combination of vested interest, difficulty in assigning responsibility and absence of a clearly ethical framework in which the discussion can take place. One possible framework – contraction and convergence – is then introduced as a way of solving this problem.

PART III

Climate change: predictions and policies

16

Future world energy use and carbon emissions

"Prediction is very difficult, especially about the future."

–Niels Bohr

In previous chapters we have studied the fossil, hydro and nuclear powered systems that were at the centre of our economies during the twentieth century. But what of the twenty-first century? Given a growing world population and the threat of climate change, can we afford simply to assume we can expand our use of these technologies as and when required regardless of any consequences; or are there potentially less harmful alternative futures?

In this chapter we will be looking at the world's energy demand and how it is currently supplied, then studying predictions of future demand and what this might mean for carbon emissions. In most of this book we only consider *commercial energy*, not more traditional sources, such as locally collected non-traded biomass and tractive power provided by animals. However, in this chapter we will also consider non-commercial fuels at several points. This is important because as the developing world evolves, animal power will reduce in importance as the use of agricultural machinery expands. If the total energy supplied by new machinery is only equal to that of the pre-existing system, then no extra energy is being supplied. However, the use of commercial fuel will have expanded. Equally, if car use doubles but fuel efficiency also doubles, no increase in commercially traded energy has occurred.

As has been outlined previously we have been using the term sustainable rather than alternative or renewable energy because technologies like hydropower are not truly renewable, although the resource is. But what about biomass burning? This is truly renewable, if carried out with care, but is hardly alternative, as fire was one of our first

Energy and Climate Change   David A. Coley
© 2008 John Wiley & Sons, Ltd

major inventions. The large-scale burning of domestic waste to raise steam and generate electricity is certainly alternative, in that it is relatively new, but is it a green or sustainable source of power? At one level it looks to be renewable – there seems to be no end to the wastes the modern world produces – but at a higher level, this cycle is not closed and therefore not renewable as such burning counts as resource depletion. In addition, many would argue we should be recycling waste rather than burning it. It is important that students are cognisant of these issues and the existence of alternative definitions and classifications of resources and technologies that might be used in other texts and the media.

As far as world energy use is concerned, the two definitions that are likely to cause most confusion are whether hydropower is counted under renewable/alternative energy and whether non-commercial fuels such as wood collected by the end user in the developing world is counted at all. That such concerns are more than just pedantic musing is brought home by the realization that hydropower produces more electricity than any other fuel except for coal and that 95 per cent of non-hydropower renewable energy (and 77.5 per cent of all renewable energy) used in the world today is simply local biomass burning. By switching hydro in and out of the renewable category and forgetting where much of the world gets much of its energy, newspapers and other sources can paint a wide variety of pictures of both the present and the future.

16.1 The world's future use of energy

Humankind currently uses 409 EJ (9741 Mt_{oe}) of commercially traded primary energy each year. Primary energy encompasses not only the final use of the energy but energy used in energy transformation industries such as electricity generation, e.g. the energy contained in the coal that was burnt to produce the electricity. Primary energy is used in preference to final energy when discussing changes in humankind's energy requirements because changes in efficiency, in for example power station boilers and generators, will show up as reductions in primary energy demands, whereas such technological improvements do not lead to a reduction in energy demand from customers.

Figure 16.1 shows how the world's primary energy demand is shared amongst fuels both in energy terms and as a percentage share of the total.

Our current level of primary energy use releases 29 Gt of carbon dioxide into the atmosphere each year, or 4.4 t per person. In terms of just carbon mass this is 7.9 Gt of carbon (GtC), or 1.2 t per person. As Figure 16.1 shows, this level will increase during the twenty-first century unless the fractional share of each fuel changes.

The predicted growth in energy use and therefore in carbon emissions is substantial and greatly amplifies any concerns we may have over whether our current emissions may be altering the world's climate. There have been many attempts at accurately predicting future energy requirements and the resultant carbon emissions, with a whole raft of possible scenarios being suggested and examined in detail. We will only have time and space to mention two, that of the Energy Information Administration, or EIA, (termed

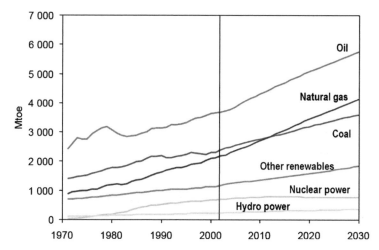

Figure 16.1 Historic division of world primary energy amongst fuels including non-commercial fuels (included under and dominating 'other renewables') and (right of vertical line) one prediction of future use (the reference scenario used later) [data from EIA04] [Adapted from data in Energy Information Administration, International Energy Outlook 2004, EIA, 2004]

the *International Energy Outlook, or IEO*[1]), which concentrates on energy demand and fuel type, and those of the Intergovernmental Panel on Climate Change (IPCC) which concentrate on greenhouse gas emissions from all sources including land use change and extend further into the future. Such predictions are not easy to make, and require knowledge of future world population in maybe 100 years time, and how much the economic gap between the developed and developing nations will change.

EIA predictions

Each year the EIA publishes an international energy outlook, which predicts the world's future energy requirements (details can be found on the book's website). We will only have space to consider what the EIA terms its *reference scenario*. In this the EIA predicts energy use to rise by 60 per cent (between 1990 and 2020) to 650 EJ. Consumption is not expected to rise uniformly. The greatest growth will be in the developing world, which by the end of the period will have almost equivalent use to that of the developed world (Figure 16.2).

The EIA expects oil to continue its role as the world's favourite fuel with its 40 per cent share remaining constant (Figure 16.1). The reason for oil's continuing dominance

---

[1] The International Energy Agency also publishes predictions in their annual *World Energy Outlook*, or WEO. Both the IEO and the WEO are available on the web and readers might like to study the current versions – see the book's website.

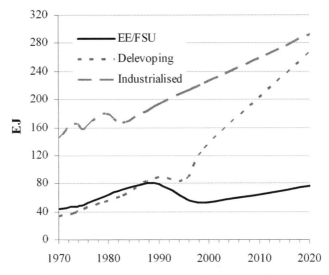

Figure 16.2 Historic and future world energy consumption by region (historic to 1999 and reference scenario after). EE/FSU = former Soviet Union and Eastern Europe [data from EIA02] [Adapted from data in Energy Information Administration, International Energy Outlook 2002, EIA, 2002]

will be because of its use in transportation systems. This is a rapidly growing sector that is treated in detail in Chapters 14 and 28. Transport use is predicted to accelerate, with oil consumption expected to grow at 2.2 per cent per annum even if electricity generation continues to move from oil to natural gas, hydro and alternative fuels.

In the EIA 2002 reference scenario, the fastest growing source of energy is anticipated to be natural gas, with consumption doubling (1999–2020) to 177EJ (162 trillion cubic feet, 4200 Mt$_{oe}$). In 1999 natural gas use surpassed coal use for the first time and is expected to represent 28 per cent of the energy market by 2020. World coal use is expected to continue to grow (at 1.7 per cent per annum), but its share of the market will slowly decline. Much of the growth is predicted to be in China where coal dominates the electricity sector. The situation with nuclear power is difficult to predict. Many of the world's fission reactors are now quite old, and in the past nuclear generation has been forecast to decline. However, improvements in technology and extended plant lifetimes mean that nuclear capacity will probably increase from 350 GW in 2000 to 363 GW in 2010 before falling to 359 GW in 2020. The developing nations of Asia, where over half the reactors currently under construction are sited (eight in China, four in South Korea, two in India and two in Taiwan), are expected to see the greatest expansion in nuclear power.

In the EIA reference scenario, renewables growth is modest. Although a 53 per cent increase is expected (1999–2020) this actually represents a slight decrease in market share from nine to eight per cent; most of the growth coming from large-scale hydro projects especially in China, India, Malaysia or other developing nations.

A large increase in electricity consumption is expected, from 13 TkWh (47 EJ) to 22 TkWh (79 EJ) (1999–2020). Again, vigorous growth is foreseen in Asia – at 4.5 per cent per annum – as demand for air conditioners, refrigerators and space heating grows.

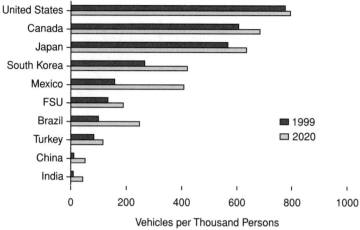

Vehicles per Thousand Persons

Adapted from data in Energy Information Administration, International Energy Outlook 2002, EIA, 2002.

Figure 16.3 Car ownership in selected countries (historic and reference scenario). FSU = former Soviet Union. Note the relatively small change in the developed world and the large change in the developing world [data from EIA02]

The requirement for energy to run transportation systems is likely to grow at 1.7 per cent per annum in the developed world and 3.8 per cent per annum in the developing world. The variation in these values is a reflection mainly of differences in access to personal transport at the present time, with near market saturation anticipated in car ownership throughout the developed world by 2020. Although these figures imply more than a 100 per cent increase in car ownership in the developing world, the combination of a very low current level of ownership, together with a rising population, mean that the inequality between developed and developing world is likely to be maintained. The USA is expected to reach almost 800 vehicles per 1000 persons by 2020, whereas China, despite a five-fold increase in ownership over the same period, is only predicted to reach 52 vehicles per 1000 persons (Figure 16.3).

All this growth in energy use has obvious implications for carbon dioxide emissions. The EIA 2002 reference scenario implies that world carbon emissions from energy use will rise from 7.9 GtC_{eq} (1999) to 9.9 GtC_{eq} (2020). Plotting this trend (Figure 16.4), we see that this implies not only an increase, but also an accelerating one. The majority of this increase will be from the developing world.

The one piece of good news is that the EIA projections suggest improving *energy intensity* (the ratio between energy consumption to gross domestic product, i.e. the amount of energy required to provide one monetary unit of economic activity). This points to improving energy efficiency. An improvement of 1.3 per cent per annum is expected in the developed countries and by 1.2 per cent per annum in the developing world. The greatest improvement is liable to be in Eastern Europe and the former Soviet Union, where energy intensity has been traditionally poor (i.e. high) (Figure 16.5).

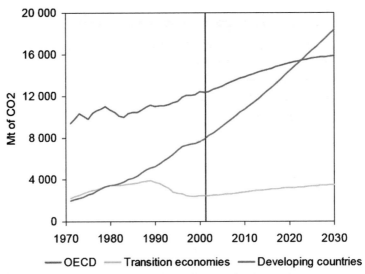

Figure 16.4 Accelerating carbon dioxide emissions as predicted (from 1999) by the IEO 2002 and 2004 reference scenario [data from EIA04]

Such changes in energy intensity are predicted to give rise to improvements in *carbon intensity* (the ratio of carbon emitted to gross domestic product). These should arise in part from fuel switching, for example from coal to gas. Figure 16.6 shows current and predicted carbon intensities for selected countries.

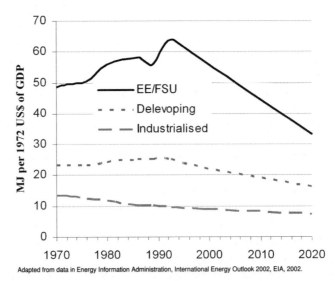

Figure 16.5 Energy intensity (historic and reference scenario). (EE/FSU = Eastern Europe and former Soviet Union.) Note the poor historic performance of EE and FSU compared to both the developed and developing world [data from EIA02]

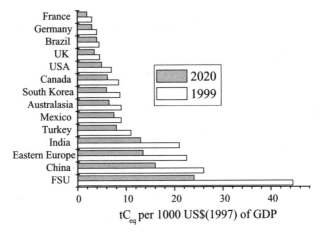

Figure 16.6 Carbon intensities: historic and predicted (reference scenario) [data from EIA02]

IPCC projections

Unlike the EIA predictions, those of the Intergovernmental Panel on Climate Change reach further into the future, to 2100, and are hence more speculative. They are also explicitly about carbon and greenhouse gas emissions.

The IPCC was jointly established by the World Meteorological Organization and the United Nations Environment Programme. Its aim is to assess the scientific, technical and socio-economic information relevant to an understanding of anthropogenic climate change. The many reports, papers, methodologies and guidelines produced have become an invaluable resource for both scientists and policy makers.

The IPCC has commissioned a special report on emissions scenarios, SRES [IPCC00], which predicts future emissions of greenhouse gases. None of the possible scenarios considered by the IPCC explicitly assume the world's governments take a strong stand to arrest climate change by dramatically reducing emissions in the near term: therefore they do not include all possibilities. However, SRES was developed to represent the broad range of possibilities found in the literature.

SRES contains 40 alternative scenarios. Luckily these can be collected into four story lines (or families of scenarios), each of which represents different demographic, social, economic, technological and environmental developments. Within each story line a range of scenarios was developed using various modelling approaches to examine the scope of possible emissions. Of the 40 scenarios, 13 explicitly explore variations in energy consumption and provision.

SRES considers the following anthropogenic emissions: carbon dioxide (CO_2); methane (CH_4); nitrous oxide (N_2O); hydrofluorocarbons (HFCs); perfluorocarbons (PFCs); sulphur hexafluoride (SF_6); hydrochloro-fluorocarbons (HCFCs); chlorofluorocarbons (CFCs), the aerosol precursor; and the chemically active gases sulphur dioxide (SO_2), carbon monoxide (CO), nitrogen oxides (NO_x), and non-methane volatile

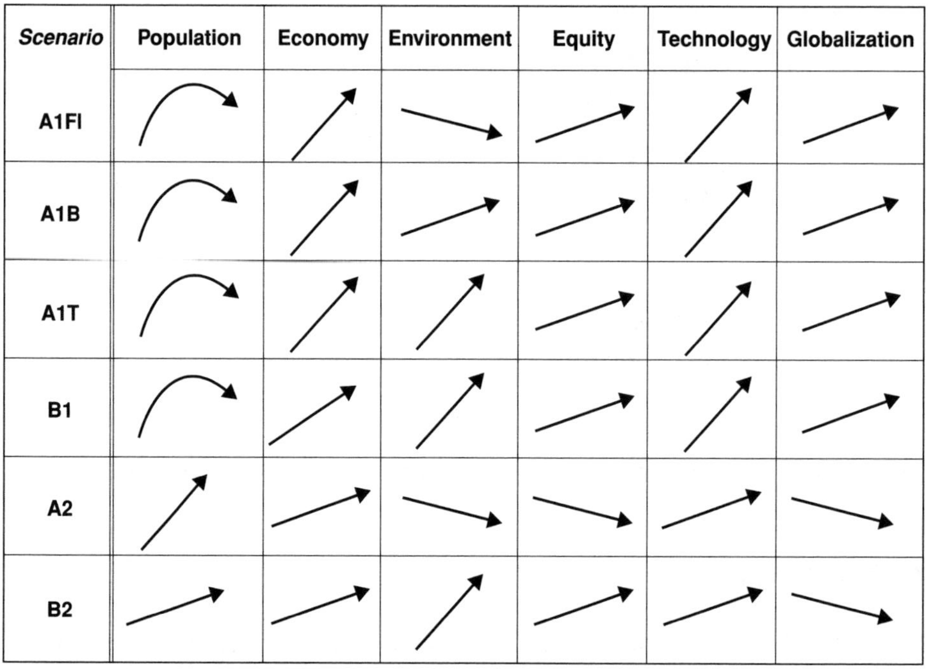
Scenario
Population
Economy
Environment
Equity
Technology
Globalization
A1FI
A1B
A1T
B1
A2
B2

Reproduced from IPCC, Technical Summary: Mitigation, IPCC, 2000

Figure 16.7 Qualitative direction of the SRES scenarios [SRES00]

organic compounds (NMVOCs). But does not include any feedback effects that might lead to additional emissions from the biosphere, or from ourselves as the climate changes. The storylines are called, rather prosaically, A1, A2, B1 and B2, and are outlined below [IPCC01a] (as just mentioned, none represent a world pursuing policies designed explicitly to tackle climate change):

- The A1 story line and scenario family describes a future world of very rapid economic growth, global population that peaks in mid-century and declines thereafter, and the rapid introduction of new and more efficient technologies. Major underlying themes are convergence amongst regions, capacity building, and increased cultural and social interactions, with a substantial reduction in regional differences in per capita income. The A1 scenario family develops into three groups, which describe alternative directions of technological change in the energy system. The three A1 groups are distinguished by their technological emphasis: fossil fuel intensive (A1F1), non-fossil fuel energy sources (A1T), or a balance across all sources (A1B).

- The A2 story line and scenario family describes a very heterogeneous world. The underlying theme is self-reliance and preservation of local identities. Fertility patterns across regions converge very slowly, which results in continuously increasing global

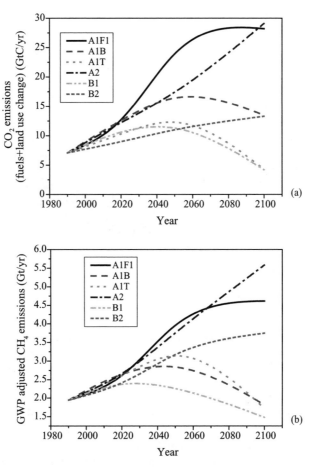

Figure 16.8 Future carbon dioxide (a) and methane (b) emissions for the SRES scenarios [data from IPCC01]. (The methane data has been adjusted to equivalent carbon dioxide using a GWP of 23 then multiplied by 0.273 to convert to carbon)

population. Economic development is primarily regionally orientated and per capita economic growth and technological change are more fragmented and slower than in the other story lines.

• The B1 story line and scenario family describes a convergent world with the same global population that peaks in mid-century and declines thereafter, as in the A1 story line; but with rapid changes in economic structures towards a service and information economy, with reductions in material intensity, and the introduction of clean and resource efficient technologies. The emphasis is on global solutions to economic, social, and environmental sustainability, including improved equity, but without additional climate incentives.

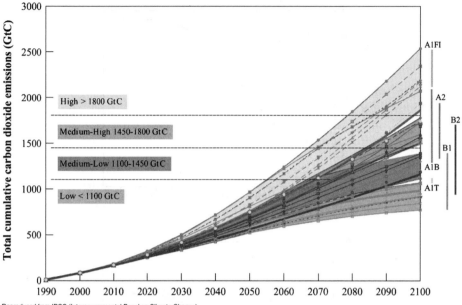

Figure 16.9 Cumulative emissions of SRES scenarios [IPCC00]. The figure shows individual alternative scenarios within the main family groups and the range of predictions for each family (For a colour reproduction of the figure, please see the colour section located towards the centre of the book)

- The B2 story line and scenario family describes a world in which the emphasis is on local solutions to economic, social, and environmental sustainability. It is a world with continuously increasing global population at a rate lower than A2, intermediate levels of economic development, and less rapid and more diverse technological change than in the B1 and A1 story lines. While the scenario is also orientated towards environmental protection and social equity, it focuses on local and regional levels.

The A1 scenario family is based on low population growth (8.7 billion in 2050 and declining to 7 billion in 2100 – the current world population is 6.4 billion (2005)). Both low infertility and low mortality are assumed. The B2 scenarios assume medium population growth (10.4 billion by 2100). The A2 scenarios are based on high population growth (15 billion by 2010).

All the scenarios describe futures that are generally more affluent than today. The lowest see world GDP rising 10-fold by 2100, the highest assumes a 28-fold increase. Within each story line, some of the scenarios assume emissions of greenhouse gases continually increase, others that they will increase for a time then fall. Figure 16.7 summarizes the scenarios.

Hopefully this description gives some flavour of both the number of possibilities that the IPCC have considered and the range of possible futures we could face. As Figure 16.8 indicates, all the scenarios point to a world in which emissions rise substantially (at

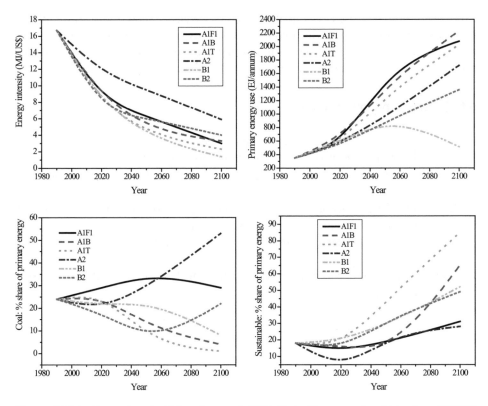

Figure 16.10 Energy-based driving forces of SRES illustrative scenarios [data from IPCC00] [Adapted from data in Emissions scenarios: summary for policy makers, IPCC, 2000, www.ipcc.ch/pub/syreng.htm]

least in the short to medium term) from their current baseline – a baseline that already appears to be affecting our climate.

Although all the lines in Figure 16.8 show increased emissions, the range of possibilities is quite diverse: with cumulative carbon emissions in 2100 of 770 GtC to 2540 GtC. We know from other work by the IPCC [IPCC01a] that it is the cumulative emission between now and when an equilibrium concentration occurs that is of importance, rather than the time-series of emissions. By way of comparison, the total anthropogenic carbon emission during the 100 years between 1800 and 1900 was 12.2 GtC and that between 1900 and 2000 was 302 GtC. Focusing on the cumulative emissions allows us to simplify Figure 16.8 to give Figure 16.9, where fewer of the scenario paths cross and we can clearly distinguish between the various story lines. A1F1 (continuing high reliance on fossil fuels) is seen to result in much higher cumulative emissions than A1T (a switch to alternative energy sources).

Figure 16.10 shows some of the energy based driving forces within the scenarios that are particularly important to us. It seems that the world faces a range of futures, with the possibility that energy intensity (efficiency) might improve by a factor of eight by

2100; primary energy use rises initially under all scenarios, then decreases slightly or increases by a factor of 3.3; coal use may increase slightly, or almost disappear completely; and alternative, non-carbon, energy sources (including non-commercial) may halve in importance or become the norm. The question is, which future will we choose?

16.2 Student exercises

"Tell a man that there are 300 billion stars in the universe, and he'll believe you. . . . Tell him that a bench has wet paint upon it and he'll have to touch it to be sure."

–Raimond Verwei (also credited to Albert Einstein)

1. By drawing distinctions between the developed, transition (Eastern European and former Soviet Union states) and the developing world, discuss how the EIA views future levels of car ownership, car use and carbon intensity.

2. Produce a quantitative and qualitative analysis of the illustrative IPCC scenarios discussed in this chapter.

17

The impact of a warmer world

Computers are useless. They can only give you answers.

–Pablo Picasso

The Intergovernmental Panel on Climate Change (IPCC) concluded in 1985 that the 'balance of evidence suggests a discernible human influence on global climate'. There are at least five sources of evidence for this: (1) our understanding of climate physics and the relationship between atmospheric composition and radiative balance, (2) data from ice cores and other sources that suggest a historic relationship between atmospheric composition and temperature, (3) measurements of increasing levels of carbon dioxide in the atmosphere (i.e. our emissions are not being processed by the planet at the same rate as we are producing them), (4) current and past records of temperature and sea level, and (5) computer models that indicate the world is warming and will continue to warm as we further load the atmosphere with greenhouse gases. This chapter will concentrate on the last of these and review what such models suggest our world will look like in the future. There are several forms of model, each distinguishable by the detail to which it treats the various physical processes involved.

In the following sections we will be relying heavily on results from the UK Meteorological Office's Hadley Centre, one of the world's leading centres for climate modelling. Before we study what our future climate will look like, we need to see how well these models represent our current climate and that of the recent past (for which we have very good records). This is one of the acid tests of a climate model – given knowledge of solar output and atmospheric composition and how these have changed in the past, how closely does the model predict the observed changes in global mean temperature? As Figure 17.1 would seem to suggest, this mean has been rising for some time. The decade 1990–2000 was some 0.6 °C warmer than the period 1860–1920. Nine of the warmest years since detailed records began have occurred since 1980. But is this just

Energy and Climate Change   David A. Coley
© 2008 John Wiley & Sons, Ltd

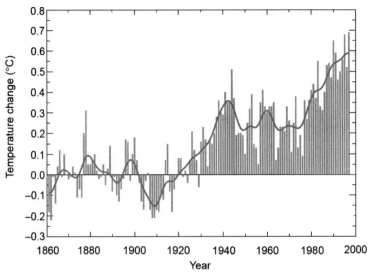

Figure 17.1 Global mean temperature since 1860 (line-air, bars-sea). © Crown copyright 2005, published by the Met Office Hadley Centre [MET97]

natural variability in which we see a trend only because we are looking at a very short time scale? This is the type of question climate modelling can help to answer.

17.1 Climate models

There are many factors that affect the accuracy of the predictions from any climate model. The model itself will be an ensemble of equations representing different physical processes such as changes in atmospheric pressure and solar radiation. The climate models used today are far too complex to be handled analytically and must be run on computers, often some of the fastest computers in the world. The accuracy of any model will be limited by the detail in which it represents processes and the capacity of the computer. Our knowledge of all the important processes is incomplete so the model will naturally be an approximation, with some processes more approximated than others (Figure 17.2). Typically the model divides the planet's atmosphere and ocean into a large number of *cells* which collectively form a three dimensional grid. In order to obtain answers within reasonable times with today's computers there is a limit to the number of cells that can be included. Another limitation is the accuracy of the input data, particularly the concentration of various greenhouse gases at future dates. We do not yet fully understand the carbon cycle, or the cycles of other greenhouse gases. As we saw in the last chapter, for anthropogenic emissions it is also difficult to state the likely rate of emission at any future date – and the further into the future we work, the greater the difficulty. Population and economic growth clearly play an important

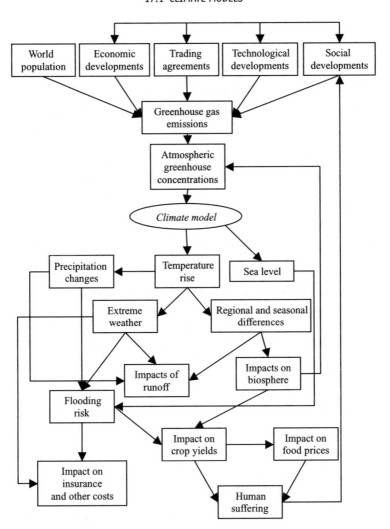

Figure 17.2 A few of the inputs and outputs of a climate/impact model, and some of their relationships. Errors within any of these can affect the results and it is therefore common to run the model with a range of possibilities at each stage

role here, but so does how seriously the world's policy makers take the threat of global climate change and hence alter energy policy. Often a model is run using a series of scenarios, each representing a different economic and energy future; this allows us to obtain a range of answers and get an idea of how sensitive the climate might be to future emission scenarios.

Climate models typically have a resolution of tens or hundreds of kilometres in the horizontal plane. Many important elements of the climate work on much smaller scales: clouds, hills and coastlines being three examples; this limits their accuracy. We can use these limitations to form a hierarchy of models, based on special resolution and complexity:

1. **One-dimensional radiative–convective atmospheric models.** Globally averaged in the horizontal plane, but with many layers vertically in the atmosphere.

2. **One-dimensional up-welling–diffusion ocean models.** Here the atmosphere is a single well-mixed box, which exchanges heat with the ocean and the land. The ocean is considered to be a one-dimensional column representing a horizontal average over the real ocean. Sinking water in the Polar Regions and the subsequent up-welling are treated separately.

3. **One-dimensional energy balance models.** Here the atmosphere is averaged vertically and in the east–west direction. But variation, particularly in heat transfer within the oceans and the atmosphere, in the north–south direction is modelled.

4. **Two-dimensional atmosphere and ocean models.** Such models can represent variation in both latitude and height (depth), but not in longitude.

5. **Three-dimensional atmosphere and ocean general circulation models.** These models divide the atmosphere and oceans into three-dimensional cells with a resolution of four degrees or less and 20 or more layers in the vertical. They simulate winds, ocean currents and other important processes and features. Coupled atmosphere–ocean models link the ocean to the atmosphere and include feedbacks from water vapour, clouds, and seasonal variations in snow and ice cover. Because the atmosphere is linked to the ocean, the heat uptake of the oceans, and the time delay this causes to climate change, is included. Figure 17.3 shows the major processes incorporated into such a model. The Hadley Centre model is a coupled ocean–atmosphere general circulation model with the form shown in Figure 17.4.

The simpler models discussed above allow basic relationships between the components of the climate system to be investigated and connections studied in isolation. Their speed also means that they can be run many times at low cost on modest computers. The three-dimensional models are the only true representation of the climate and therefore give a much better understanding of the important processes. However, they are very slow to run, even on super-computers and, because the results represent planet-wide variability, they can be more difficult to draw general conclusions from.

17.2 Natural variability and model reliability

We have accurate temperature records for approximately the last 140 years. If we simulate the climate over the last 1000 years using a climate model we should be able to get an idea of how variable the climate is during any 140-year period. We can then compare this to the observed variability. By taking the range of natural variability as the maximum range observed in the simulated results for any 140-year period, we can do this in a quantitative manner. Figure 17.5 does this and compares the maximum range of natural variability with the observed temperature trend since 1860. The conclusion would seem to be that

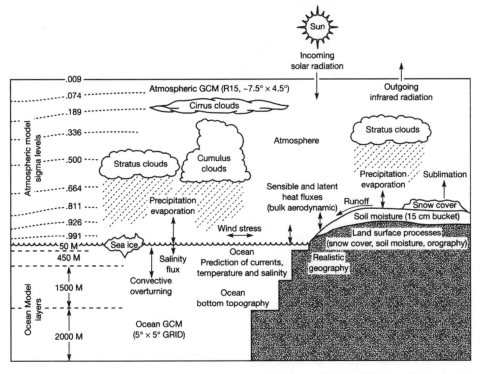

Figure 17.3 Major processes considered inside a coupled atmospheric–ocean general circulation model [W.M. Washington and G.A Meehl, in HAR00, p89] [Reproduced from Washington, W.M., and Meehl, G.A., Climate Sensitivity Due to Increased CO2 – Experiments with a Coupled Atmospheric and Ocean General Circulation Model, Climate Dynamics, no. 4, p1–38, 1989, pg 89 in Harvey, L.D.D, Climate and Global Environmental Change, p89, Prentice Hall, 2000]

we are experiencing a climate change outside the natural level of variability. This is a very important result.

How can we be sure that the reason we are now outside of the maximum range of natural variability is because of anthropogenic emissions? This is another question climate models can help to answer. By running the model with and without anthropogenic emissions we should be able to see if such emissions are the probable cause of our current high temperatures. The Hadley Centre model would appear to only represent the current climate if anthropogenic emissions of greenhouse gases and sulphate aerosols are added to the list of natural climate altering effects, such as changes in solar output and volcanic activity. This result is particularly impressive because not only does the model predict the global mean temperature reasonably well, it also captures much of the variability across the globe. Figure 17.6 shows regional variability in temperature between the beginning and end of the 20th century, as observed (a) and as simulated by the Hadley model (b) if anthropogenic emissions are included. There is broad general agreement, and features such as the greater warming at higher latitudes occur in both images. Much of this fingerprint of change arises from our emissions

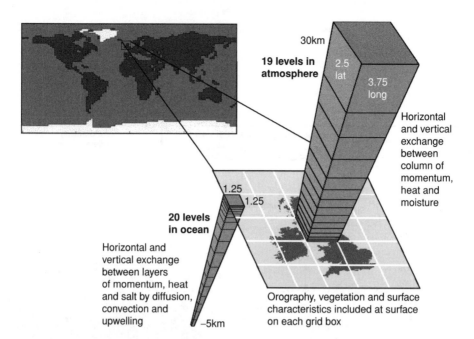

Figure 17.4 Basic design of the Hadley climate model. The world's oceans and atmosphere are sub-divided into a large number of cells. Heat, mass and mometum are allowed to flow between cells as time is stepped forward and the Earth rotates around the sun. As time progresses the solar output, greenhouse gas concentrations and other parameters can be altered in line with historical evidence or predictions. Horizontal divisions are in degrees of longitude or latitude; note how the vertical divisions increase in size away from the surface to save computing time. © Crown copyright 2005, published by the Met Office Hadley Centre

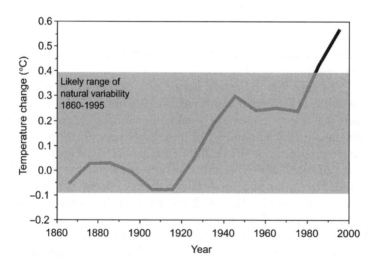

Figure 17.5 Observed change in global mean temperature (in the form of a decadal average (line)) compared with the likely maximum range of natural variability (the shaded region). © Crown copyright 2005, published by the Met Office Hadley Centre. [MET97]

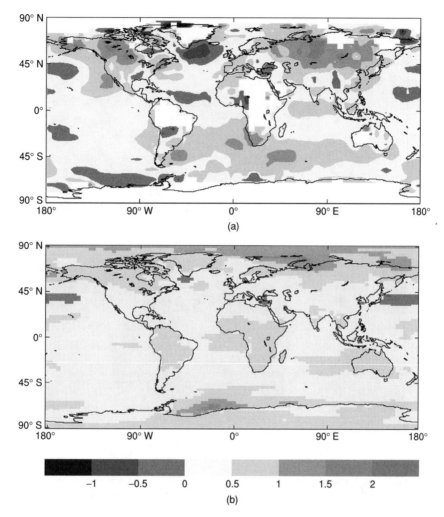

Figure 17.6 Regional variability in temperature (°C) between the start and finish of the twentieth century: (a) as observed and (b) as simulated by the Hadley model. Although hard to see without the aid of statistical analysis, there is broad general agreement about the form of the regional variability between the two images. © Crown copyright 2005, published by the Met Office Hadley Centre [MET97] (For a colour reproduction of the figure, please see the colour section located towards the centre of the book)

of aerosols – which have strong regional effects on climate. This visual agreement is confirmed and strengthened by statistical analysis of the results, and allows us to conclude that the observed regional alterations to climate are most probably caused by humankind.

Modern climate models are three-dimensional. This allows us to repeat the exercise described above, but vertically. Figure 17.7 shows the difference in temperature as observed (a) and as modelled (b) between 1961 and 1996. In both images we see a cooling at higher altitudes (>10 kilometres) and a warming at lower ones. Anthropogenic

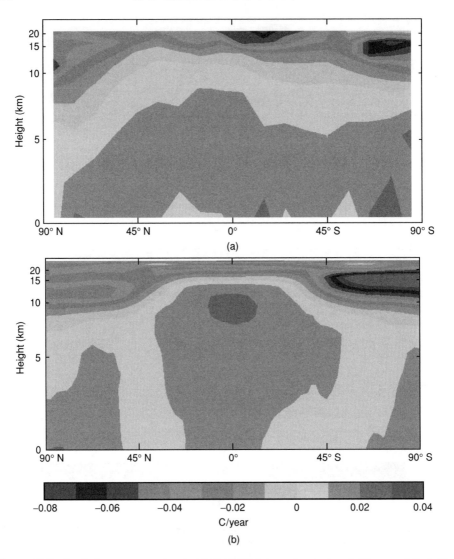

Figure 17.7 Vertical temperature trends 1961–1996: (a) as observed and (b) as modelled. The cooling of the stratosphere (i.e. above approximately 12 km) and warming of the troposphere (i.e. below approximately 12 km) can only be explained by human influence. © Crown copyright 2005, published by the Met Office Hadley Centre [MET97] (For a colour reproduction of the figure, please see the colour section located towards the centre of the book)

emissions of greenhouse gases are the only known mechanisms that can account for this change.

Given this evidence, it would seem reasonable to again conclude that much of the observed recent global temperature increase is due to human activities, i.e. that anthropogenic climate change is a reality.

In the following sections we will look at what future temperature rises will mean for both the natural world and for humankind. We will only have time and space to look at some of the impacts and will concentrate on changes in temperature, sea-level rise, flora, food supply and water resources. There are, however, likely also to be impacts on disease and health, fauna, storm damage, insurance premiums and possibly on geopolitical stability – see Table 5.4 for a list of other possible effects. Before examining the likely scale of these impacts we need to ask by how much will our climate change?

17.3 Future climate change

The accuracy of our climatic predictions is limited by the accuracy of our models, which is in turn limited by the factors discussed earlier in this chapter. In addition, we cannot yet account for possible non-linear effects such as changes in the strength of ocean currents (which could have a dramatic effect on regional climates). Ignoring these possibilities, Figure 17.8 shows the predicted global mean rises in temperature and precipitation. As we saw in the previous section, these match the observed rises in temperature seen in recent decades and this result suggests a continuation of this trend to a much warmer world. Figure 17.8 assumes we continue to expand our emissions of greenhouse gases and represents a 'business as usual' scenario[1]. Figure 17.9 shows the regional variability of this rise.

The results seem to suggest a mean temperature rise of 2.5 °C by 2100 but with much regional variability: there would be temperature rises of up to 4 °C in heavily populated regions as soon as 2050. Such a temperature rise would have dramatic consequences for any region. By way of comparison, the average mean temperature difference between London[2], England and Nice[3] in the South of France is less than 5 °C [LON04, NIC04]. An even greater mean temperature rise (3 °C) is suggested if the cooling effects of sulphate aerosols are not included. As we have only an approximate knowledge of what future emissions of these aerosols are likely to be it is probably sensible to ignore this cooling.

We will now turn our attention to discussing what impacts such climatic changes may have.

17.4 Impacts

Water resources

Some parts of the world will see a reduction in resource availability, others an increase. The Comprehensive Assessment of the Freshwater Resources of the World (CAFRW) estimated that in 1997 around one third of the world's population lived in countries

---

[1] IPCC reference scenario, in which atmospheric carbon dioxide concentration rises at approximately one per cent per annum.
[2] 10.5 °C
[3] 15.3 °C

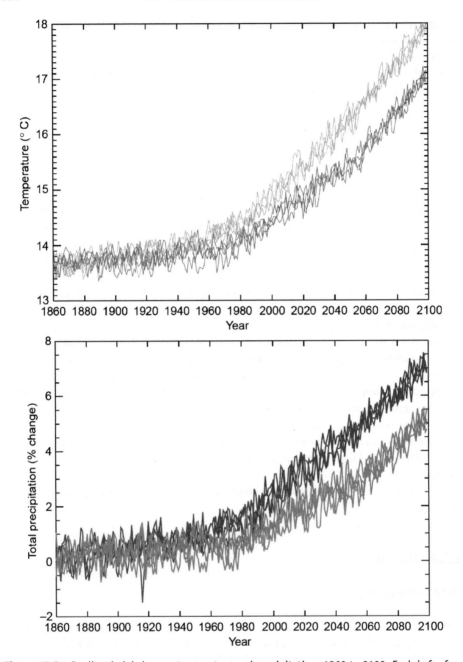

Figure 17.8   Predicted global mean temperature and precipitation, 1860 to 2100. Each is for four runs of the model – hence the noise in the results. The lower curves in each case include sulphate aerosols, the higher curves do not. © Crown copyright 2005, published by the Met Office Hadley Centre [MET97]

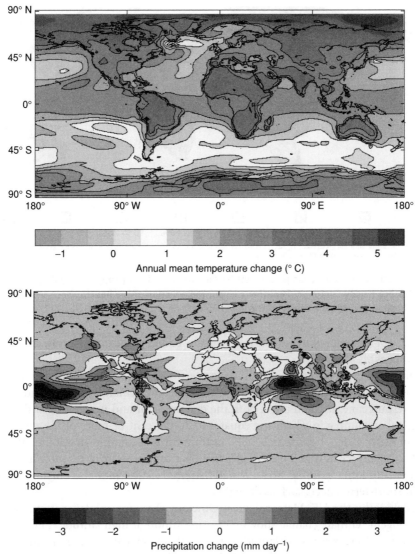

Figure 17.9 Predicted regional variability in temperature and precipitation for the 2050s as given by the Hadley model and an annual increase of one per cent in atmospheric carbon dioxide (i.e. less than its current rate). © Crown copyright 2005, published by the Met Office Hadley Centre [MET97] (For a colour reproduction of the figure, please see the colour section located towards the centre of the book)

experiencing moderate to high water stress, and forecast that by 2025 as much as two thirds of a much larger world population could be under stress conditions simply due to the rise in population, i.e. without additional factors such as climate change [MET97]. The basic measure of stress used by CAFRW is the ratio of the annual water withdrawal

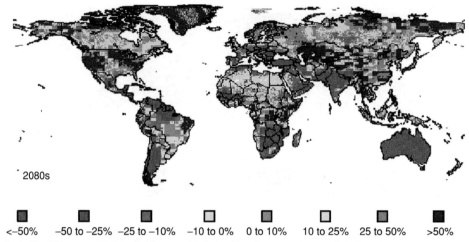

2080s

■   ■   ■   □   ■   □   ■   ■
<−50%   −50 to −25%   −25 to −10%   −10 to 0%   0 to 10%   10 to 25%   25 to 50%   >50%

Reprinted from UK Department of the Environment, Transport and the Regions, Climate Change and its Impacts – a Global Perspective, the Met. Office, 1997 ©
Crown copyright 2005, Published by the Met Office Hadley Centre

Figure 17.10 Percentage change in runoff by the 2080s. © Crown copyright 2005, published by the Met Office Hadley Centre [MET97] (For a colour reproduction of the figure, please see the colour section located towards the centre of the book)

to the size of the renewable water resource in any country (including inflow from other nation states).

Withdrawals will change with population, economic development and changes in irrigation practices. Climate change might also affect the demand for water, and the size of the resource: i.e. it could affect both the denominator and numerator of the ratio. In general, a country in which use equates to 20 per cent or more of the resource is likely to suffer water stress, which will constrain future economic development of the country. A country in which use equates to 40 per cent or more of resource is deemed to potentially suffer extreme stress. A water stressed nation may also encounter more serious short-term effects such as droughts.

Modelling by the UK Meteorological Office of changing rainfall patterns due to anthropogenic climate change shows increasing runoff at high latitudes and near the equator, but decreases at mid-latitudes. The predicted global increase in runoff is 2.9 per cent by the 2020s, 4 per cent by the 2050s and 6.5 per cent by the 2080s. From Figure 17.10, which shows these changes in detail for the 2080s, we see that there are great regional variations, with some areas experiencing runoff increases or decreases of more than 50 per cent.

Some of the countries experiencing the largest decreases in runoff already suffer from water stress, including much of Africa and the Indian sub-continent. Not only is total precipitation predicted to change, but timings will also change. Higher temperatures will decrease winter snowfall, increase winter rainfall and reduce the size of the spring/summer snowmelt and the subsequent spring/summer runoff. The reduced snow-pack may also affect precipitation-reliant industries such as skiing. Monsoon size and timings may also change.

In 1990 around two billion people in 15 countries were using at least 20 per cent of their potential resource. Accounting only for population growth this would rise to five billion by 2025, and six billion by 2055. Climate change is likely to elevate these numbers further. Not all countries show an increase in stress, for example China and the USA both show an improving situation. But again, some of the poorest parts of the world have deteriorating prospects. Similar rises are expected in populations suffering extreme water stress. In 1990, some 450 million people experienced extreme water stress. If climate change is included in the estimates then we might expect 2.4 billion people to have experienced extreme water stress by 2025, 3.1 billion by 2050 and 3.6 billion by 2085.

Food supply

Changes in climate will cause changes in crop yields. This will have an effect on total world food production, which in turn will affect world food prices and the number of people at risk of hunger and malnutrition. Figure 17.11 shows the Hadley Centre predictions of changes in world crop yields for the 2020s, the 2050s and the 2080s.

As with the supply of water, there are winners and losers. Much of the North will show increasing yields and the South decreasing yields. This is because of the fertilising effects of the increased atmospheric carbon dioxide concentrations but the positive or negative effect of high temperatures. Some countries that initially showed gains might well lose these as temperatures rise further.

Predicting the aftermath of this on the potential for hunger is difficult. It requires the linking of climate models with yield and economic models, which in turn require assumptions about population growth, farming practices and the liberalization, or otherwise, of trade agreements. Results that do not allow for climate change indicate world cereal production will grow from 1800 Mt (1990) to 3500 Mt (2050) and that this will match world requirements. Food prices are estimated to rise slightly, but the risk of hunger to decrease. The situation changes if the effects of climate change are included: world cereal production will grow more slowly compared to the no climate change reference scenario, the shortfall being 15 Mt by 2020s and 105 Mt (two per cent) by the 2080s; food prices will increase by five per cent by 2050 and by 10 per cent by 2080; the number of people at risk of hunger is expected to grow to 36 million (2020) and 50 million (2080). One reason for this growing risk of hunger will be increasing food prices.

As with other impacts there will be regional differences. Regions with little adaptive capacity, such as much of Africa, will be worse affected. The more extreme weather also predicted as a consequence of global climate change could well exacerbate the situation in these regions.

Sea-level rise

Predicting the effects of a general sea-level rise requires knowledge of the level of the rise, changes in coastal population and agricultural practice, and an assumption about the willingness and ability of peoples to improve flood defences. It would also be important

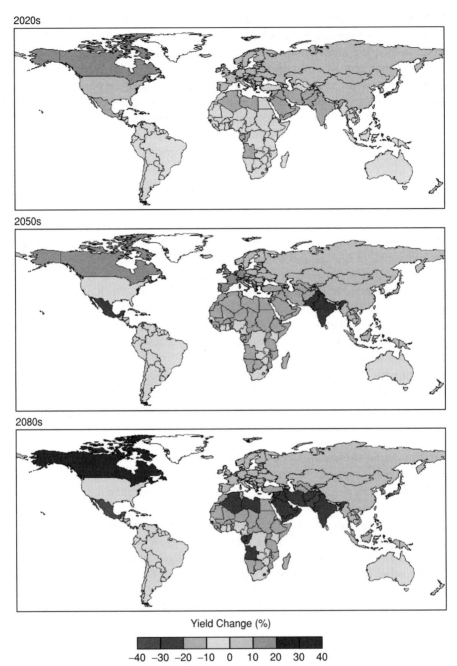

2020s

2050s

2080s

Yield Change (%)

−40 −30 −20 −10 0 10 20 30 40

Reprinted from UK Department of the Environment, Transport and the Regions, Climate Change and its Impacts – a Global Perspective, the Met. Office, 1997 © Crown copyright 2005, Published by the Met Office Hadley Centre

Figure 17.11 Percentage change in crop yield (wheat, maize and rice). © Crown copyright 2005, published by the Met Office Hadley Centre [MET97] (For a colour reproduction of the figure, please see the colour section located towards the centre of the book)

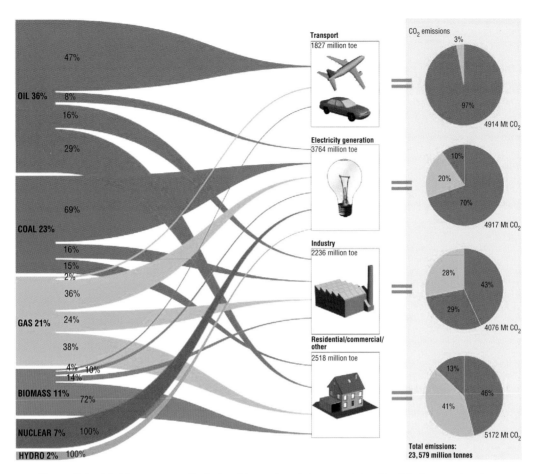

Figure II.2 World energy use (including biomass) and the sectorial split of fuel use and carbon emissions [NEW05a]

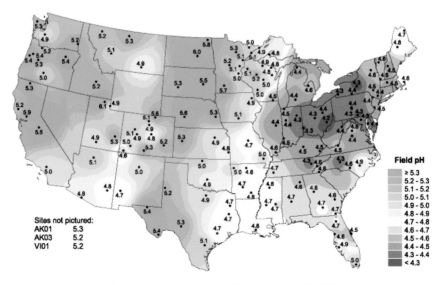

Figure 5.21 Surface pH within the USA [NAT99]

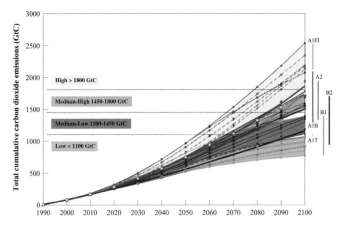

Figure 16.9 Cumulative emissions of SRES scenarios [IPCC00]. The figure shows individual alternative scenarios within the main family groups and the range of predictions for each family

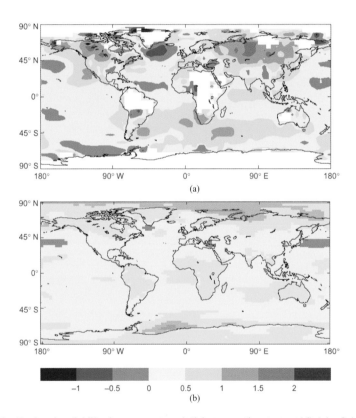

Figure 17.6 Regional variability in temperature (°C) between the start and finish of the twentieth century: (a) as observed and (b) as simulated by the Hadley model. Although hard to see without the aid of statistical analysis, there is broad general agreement about the form of the regional variability between the two images. © Crown copyright 2005, published by the Met Office Hadley Centre [MET97]

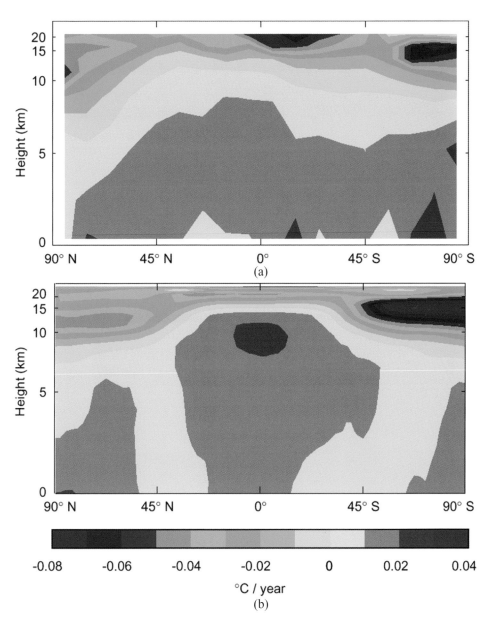

Figure 17.7 Vertical temperature trends 1961–1996: (a) as observed and (b) as modelled. The cooling of the stratosphere (i.e. above approximately 12 km) and warming of the troposphere (i.e. below approximately 12 km) can only be explained by human influence. © Crown copyright 2005, published by the Met Office Hadley Centre [MET97]

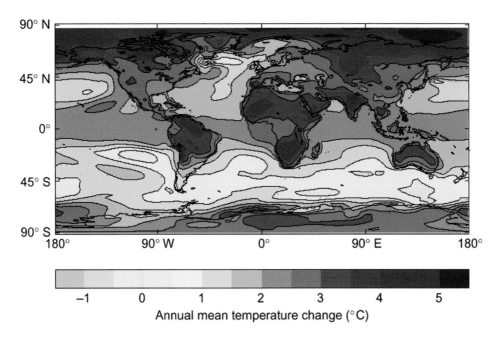

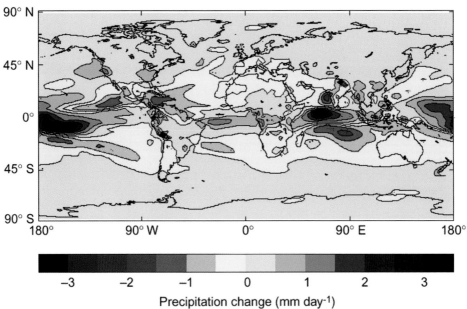

Figure 17.9 Predicted regional variability in temperature and precipitation for the 2050s as given by the Hadley model and an annual increase of one per cent in atmospheric carbon dioxide (i.e. less than its current rate). © Crown copyright 2005, published by the Met Office Hadley Centre [MET97]

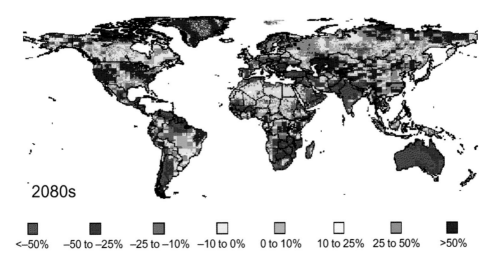

| <-50% | -50 to -25% | -25 to -10% | -10 to 0% | 0 to 10% | 10 to 25% | 25 to 50% | >50% |

Figure 17.10 Percentage change in runoff by the 2080s. © Crown copyright 2005, published by the Met Office Hadley Centre [MET97]

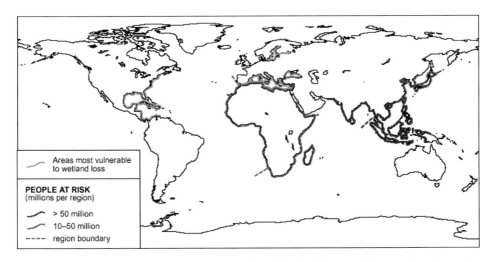

Areas most vulnerable to wetland loss

PEOPLE AT RISK
(millions per region)

> 50 million

10–50 million

region boundary

Figure 17.12 Number of people at risk from sea-level rise by the 2080s, by coastal region, under the assumption of constant (1990s) defences. Also shown are areas of greatest wetland loss. © Crown copyright 2005, published by the Met Office Hadley Centre [MET97] [Reprinted from UK Department of the Environment, Transport and the Regions, Climate Change and its Impacts – a Global Perspective, the Met. Office, 1997]

2020s

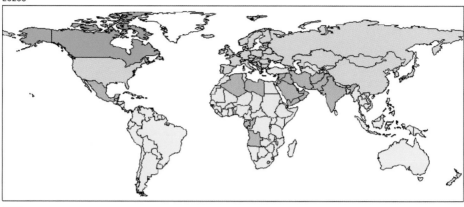

2050s

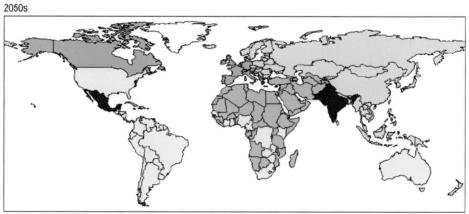

2080s

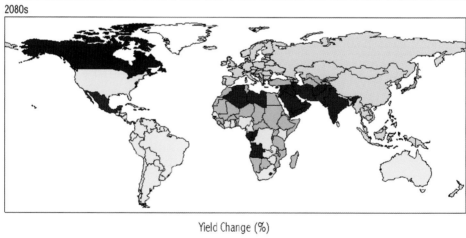

Yield Change (%)

-40 -30 -20 -10 0 10 20 30 40

Figure 17.11 Percentage change in crop yield (wheat, maize and rice). © Crown copyright 2005, published by the Met Office Hadley Centre [MET97]

Simulated present day

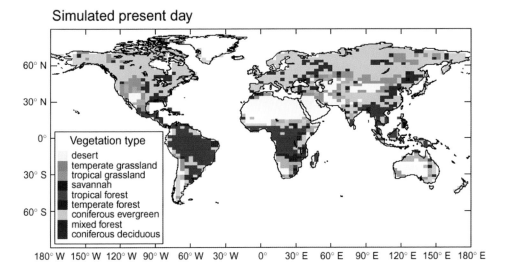

Predicted 2080s

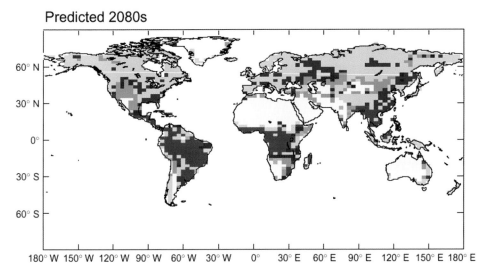

Figure 17.15 Global distribution of vegetation types predicted by an ecosystem model for the present day and the 2080s. Note the increase in mixed forest and the reduction in tropical grassland and tropical forest. © Crown copyright 2005, published by the Met Office Hadley Centre [MET97] [Reprinted from UK Department of the Environment, Transport and the Regions, Climate Change and its Impacts – a Global Perspective, the Met. Office, 1997]

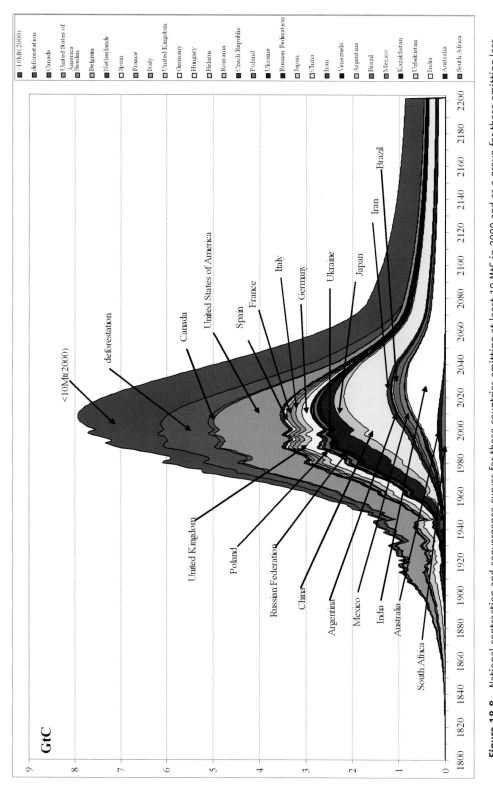

Figure 18.8 National contraction and convergence curves for those countries emitting at least 10 MtC in 2000 and as a group for those emitting less. (Contraction to 450 ppm$_v$ in 2200 and convergence by 2040) [Produced using software freely available from the Global Commons Institute's website]

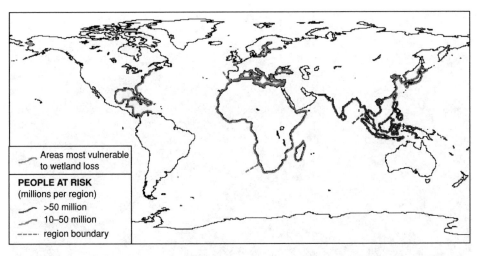

Figure 17.12 Number of people at risk from sea-level rise by the 2080s, by coastal region, under the assumption of constant (1990s) defences. Also shown are areas of greatest wetland loss. © Crown copyright 2005, published by the Met Office Hadley Centre [MET97] [Reprinted from UK Department of the Environment, Transport and the Regions, Climate Change and its Impacts – a Global Perspective, the Met. Office, 1997] (For a colour reproduction of the figure, please see the colour section located towards the centre of the book)

to be able to account for any impacts due to more extreme coastal weather, such as storms.

By 2080 sea-level rise due to thermal expansion and ice melt[4] is estimated to be 44 cm. Even without climate change more people would be at risk of flooding because of the high population growth witnessed in coastal areas – typically twice the national average. The evolution of the situation over the coming century can be described in terms of the total number of people at risk (defined as the average number of people flooded each year by storm surge) or as percentage changes. Currently 10 million people are at high risk of annual flooding; the reference scenario (i.e. no rise in sea level or increase in expenditure on defence) suggests that this will increase to 30 million by 2080. Allowing for sea-level rise increases this to 240 million by 2080: a 700 per cent increase over the reference scenario. However it is unlikely that an increasingly wealthy world will not increase expenditure on flood defences at least in line with rising GDP. Allowing for this reduces those at risk to 105 million [MET97].

Most of the people at risk will be concentrated into a few areas: the southern Mediterranean, Africa, and southern and southeast Asia. Although low in population, nations that comprise of low-lying coastal atolls such as those in the Indian and Pacific oceans will be greatly affected, as will countries with large populations living in and around river deltas. Figure 17.12 demonstrates the worldwide distribution of those at risk.

---

[4] This is from land-based ice. Sea ice melting has no direct effect on sea levels, as the volume of water generated equals the original volume of ice below the surface, so the displacement is the same.

Figure 17.13   Micronesia and other such states will be very sensitive to sea level rise and may all but disappear [USA05] [Reproduced from http://www.smdc.army.mil/SMDCPhoto_Gallery/Kwaj/KMR.jpg]

Figure 17.13 shows an example of just how sensitive to sea level rise some countries are. Most of Micronesia is only a few metres above sea level; therefore these islands may disappear almost completely.

People will not be the only casualties of rising sea levels. Flora and fauna within the coastal belt of salt marshes, mangroves and inter-tidal areas will experience changes to their environment. Northern and central America and Europe will be particularly affected due to the lower tidal range in some areas and limited potential for wetland migration. Overall, 40 to 50 per cent of wetlands are predicted to be lost by the 2080s, although the reference scenario suggests much of this would have occurred without rising sea levels because of the reclamation of land.

An interesting aspect of sea-level rise is its long timescale. Even if atmospheric carbon dioxide concentrations were to stabilize tomorrow, thermal expansion would continue as heat slowly penetrates to deeper ocean layers. As Figure 17.14 shows, this process would continue for centuries. Therefore greater loss of wetlands would seem to be inevitable beyond 2080 whatever our future energy policy, indicating that the continued existence of such areas will depend heavily upon conservation efforts. (Figure 5.16 shows how sea level and other parameters would be affected by a rapid reduction in carbon dioxide emissions.)

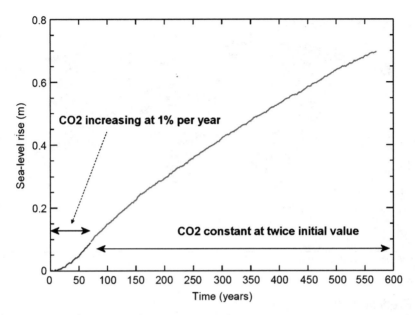

Figure 17.14   Rise in sea level due to thermal expansion alone (which accounts for approximately half of the total rise), following an increase in atmospheric concentration of carbon dioxide of one per cent per year for 70 years and then no increase for a further 500 years. © Crown copyright 2005, published by the Met Office Hadley Centre [MET97] [Reprinted from UK Department of the Environment, Transport and the Regions, Climate Change and its Impacts -- a Global Perspective, the Met. Office, 1997]

Natural vegetation

In theory, and as we will see in Chapter 25, it is possible to estimate the amount of likely plant growth from a physical and ecological model and a consideration of temperature, rainfall and amount of sunlight. Combining such a model with a model of climate change allows us to predict changes in the distribution of vegetation as the atmosphere becomes carbon dioxide rich and the planet warms.

Figure 17.15 shows the modelled current and future distribution of vegetation across the planet, a careful examination of which displays various changes particularly in the distribution of tropical grassland and desert. Figure 17.16 sums figures for these areas and reveals worldwide trends.

The predictions are quite dramatic. The majority of regions where tropical grasslands dominate will convert to areas of temperate grassland and desert by the 2080s, with approximately 90 per cent of tropical grasslands being lost. This process will begin in the 2050s when decreases in rainfall (in the relevant areas) are joined by a rise in average temperature of between five and eight degrees centigrade in northern South America and the Sahel (sub-Saharan Africa) and India. In general, there will be a major loss in global bio-diversity.

The story is very different for the world's temperate mixed and coniferous forests. Many of these will either expand or simply grow more rapidly (because of more

Simulated present day

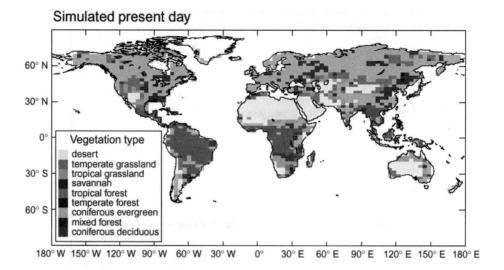

Predicted 2080s

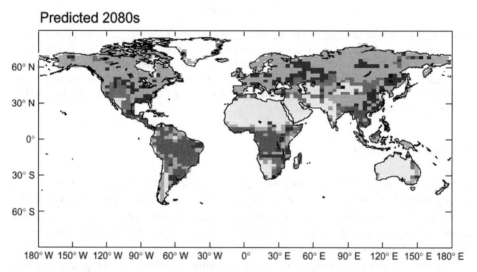

Figure 17.15 Global distribution of vegetation types predicted by an ecosystem model for the present day and the 2080s. Note the increase in mixed forest and the reduction in tropical grassland and tropical forest. © Crown copyright 2005, published by the Met Office Hadley Centre [MET97] [Reprinted from UK Department of the Environment, Transport and the Regions, Climate Change and its Impacts – a Global Perspective, the Met. Office, 1997] (For a colour reproduction of the figure, please see the colour section located towards the centre of the book)

favourable temperatures, more rainfall, carbon dioxide fertilization and more available nitrogen). Total land-based biomass may increase by 70 per cent between the present day and the 2080s at latitudes between 30 and 60 degrees north.

Such growth reflects a change in the rate at which carbon will be processed through the biosphere. Currently only one half of anthropogenic carbon dioxide emissions remain in the atmosphere and therefore able to function as a greenhouse gas. The rest is either

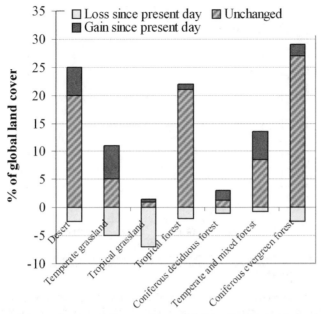

Figure 17.16 Simulated global vegetation for the present day and the 2080s expressed as a percentage of the global land area. Vegetation is shown divided into seven categories (by combining temperate and mixed forests, and assigning savannah to forest and grassland categories). The areas gained, lost or unchanged between the present day and the 2080s are distinguished (a negative value indicates a loss). © Crown copyright 2005, published by the Met Office Hadley Centre [MET97]

absorbed by the oceans or stored as organic matter. If the rate of photosynthesis exceeds the rate of decomposition the terrestrial sink will be positive and form a carbon store; when the opposite holds true, the sink will be negative. Currently the sink is believed to be positive to the degree of 1–2 GtC per annum. This puts a brake on the speed at which carbon dioxide is accumulating within the atmosphere and therefore reduces the level of radiative forcing. Figure 17.17 gives the predicted size of this sink until 2100. The results show the sign of the sink changing over the coming century. Arguably, this result is one of the most interesting insights into the details and complexities of climate change that such computer modelling has given us. Initially the sink will grow as biomass increases in tropical and northern temperate and boreal forests. Later the sink will shrink as climate change outstrips the ability of tropical forests to adapt, with a reduction in the impact of carbon dioxide and nitrogen fertilization on photosynthesis and growth.

Having estimated the size of the terrestrial carbon sink, we can compare it with estimated anthropogenic carbon emissions to see what fraction of our emissions is being offset (Figure 17.18). Although the sink is currently offsetting 30 per cent of our emissions and the size of the sink will continue to grow until 2050, the fraction of anthropogenic emissions it is storing is falling because of rising carbon dioxide emissions. By about 2080 the rate of decomposition will be greater than the rate of

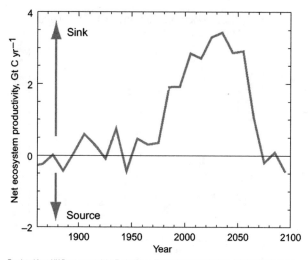

Figure 17.17   Rate of increase in the terrestrial carbon store, showing whether the terrestrial biomass is a sink or source of carbon. © Crown copyright 2005, published by the Met Office Hadley Centre [MET97]

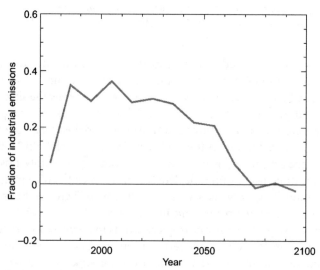

Figure 17.18   The faction of anthropogenic emissions absorbed by the terrestrial carbon sink. © Crown copyright 2005, published by the Met Office Hadley Centre [MET97]

photosynthesis and the sink will be reversed, making the terrestrial biomass a source rather than a store of carbon.

17.5 Costing the impact

With so many possible effects of future climate change there is clearly the need to estimate some final 'cost' to our emissions. This would allow us to compare on a rational basis the various emission scenarios discussed in the last chapter. We can do this by using the cost-benefit approach introduced in Chapter 6 and carrying out an *integrated assessment* of the costs of adaptation, A, (for example, building higher flood defences), mitigation, M, (for example, additional costs of sustainable energy technologies) and damage costs, D, (for example, storm damage and loss of life). The cost, Ψ, of a particular scenario with certain inherent levels of mitigation, m, and adaptation, a, is then given by:

$$\Psi(m, a) = M(m) + A(a) + D(m, a).$$

Ideally, we would like to select a scenario that minimizes Ψ.

When attached to models of emissions, climate change and impacts, such assessment tools are called *integrated assessment models* (IAMs). The use of IAMs might allow us to identify some flexibility over our future. We have the choice of selecting various pairs (m, a), i.e. combinations of mitigation or adaptation, that lead to the same overall cost, Ψ. However, others disagree with such economic approaches preferring to concentrate on reducing D: specifically loss of life. Current IAM estimates [PEA03] put the cost of climate change between 1.5 and 4 per cent of global GDP, with the cost being shared approximately evenly between the developing and the developed nations.

17.6 Student exercises

Computers in the future may weigh no more than 1.5 tons.

–Popular Mechanics (1949)

1. Describe the principles and assumptions behind a modern climate model.

2. Discuss why climate models help to prove the case for anthropogenic climate change.

3. Detail how temperature, precipitation, crop yield, water resources and sea level are likely to change by 2080 and how this will affect humanity. (This could either be in table or essay form.)

4. Describe the predicted size and role of the terrestrial carbon store over the period 2000–2080.

5. As a fraction of their wealth, who will pay the majority of the costs of climate change – the developed or the developing nations? (Hint: consider the results from IAMs.)

18

Politics in the greenhouse: contracting and converging

"Unless we stop dumping 70 million tons of global warming pollution into the atmosphere every 24 hours, which we are doing right now … the continued acceleration of this pollution would destroy the future of human civilization."

–Al Gore, September 2006

Having reviewed the estimates of future energy demand from the EIA, the predicted possible carbon futures given by the IPCC, and seen some of the worrying predictions from climate change models, we need to ask whether it might be possible to construct an alternative 'safe' future. Figure 18.1 shows the future atmospheric concentration of the major anthropogenic greenhouse gases for the various IPCC scenarios; Figure 18.2 the resultant radiative forcing; and Figure 18.3 the likely temperature rise caused by this additional forcing. (Positive feedbacks that might lead to greatly increased warming have been ignored.) Figure 18.4 shows a similar curve for sea-level rise. Finally, Figure 18.5 shows which sectors of the economy are currently responsible for these emissions in a typical developed nation (the UK) – from which we see that almost half the emissions are directly within the control of individuals rather than businesses.

The UK, our representative developed nation, produces around 178 MtC_{eq} per annum of anthropogenic greenhouse gases, 154 MtC [CCP05] of which is in the form of carbon dioxide. Carbon dioxide is responsible for 70 per cent of current anthropogenic greenhouse gas emissions, which implies that almost all of the emissions are energy related. Although it is also important not to forget warming due to methane (from landfill and agriculture) and that due to nitrous oxide (from agriculture and land-use change), tackling climate change means, to a large extent, tackling our energy related emissions. There are only two ways of doing this: we either use less energy, or we create it in some other, non-polluting way. Given that reductions of 60 per cent or more in

Energy and Climate Change   David A. Coley
© 2008 John Wiley & Sons, Ltd

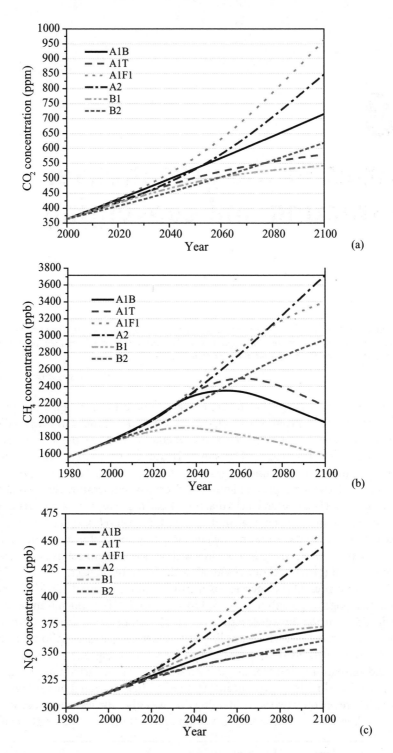

Figure 18.1a–c Historic and future atmospheric greenhouse gas concentrations (under SRES [data from IPCC01]). A1F1 approximately represents the business-as-usual growth senario of a 1% increases per annum in atmospheric CO_2 concentrations used in the modelling presented in Chapter 17. Business-as-usual is seen as the worst senario studied, it is therfore not extreme or unlikely and worse futures can be imagined.

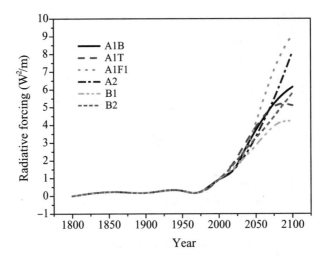

Figure 18.2 Historic and future radiative forcing (under SRES [data from IPCC01]). A1F1 represents the business-as-usual case. [Adapted from data in Houghton, et al, Climate Change 2001-Working Group I-The Scientific Basis, (Contribution of working group I to the Third assessment Report of the Intergovernmental Panel on Climate Change) ©IPCC, 2001]

our current levels of emissions might be needed to stabilize atmospheric carbon dioxide at a reasonable level, it is hard to see energy conservation delivering the whole of this reduction, especially given the rate of growth in demand from the developing world. It is therefore probably safe to conclude that alternative energy systems will need to form the backbone of the supply system in the second half of this century.

Such a switch will, at least in the short to medium time frame, cost money; fossil fuels are just so cheap. It will also require large-scale alterations to infrastructure, industry and commerce. Hence it is not surprising that governments, industry and individuals have resisted the challenge. In the following sections we will review various attempts to force the global community to meet this challenge and look in detail at one possible solution to the conflicting interests that have dogged the process. These attempts can be described as *preventive measures* (or *mitigation*), which attempt to reduce the impact of climate change by reducing the emissions that are responsible. Another class of policies are categorized as *adaptive measures*. These suggest following a route, largely dictated by economics, which includes the planting of crops that can more easily deal with a warmer world and increasing the number and size of flood defences, but to make very little effort to reduce emissions. A third approach is that of *technical fixes* (Chapter 29). These suggest using technological developments to redress the damage and return the climate to a pre-industrialized level, for example by fertilising the oceans to encourage photosynthesis, or reducing the amount of sunlight reaching the ground by releasing large quantities of dust into the atmosphere from high flying aircraft. Clearly, such fixes carry a risk because of our limited knowledge of any feedback processes they might trigger. For this reason the majority of scientists and politicians have ignored such propositions. It is likely that our response to climate change will be a mixed policy involving both preventive measures and adaptation.

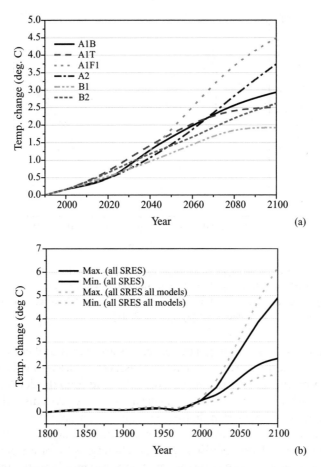

Figure 18.3   (a) Global mean temperature to 2100 under the SRES scenarios [data from IPCC01]. (b) The range of possible outcomes: shown are both the range for all SRES scenarios and all computer models used by IPCC, and the range for all SRES but using the average from all the computer models; both the scenario and the model used are seen to influence the predictions. (A1F1 represents the business-as-usual case.) [Adapted from data in Houghton, et al, Climate Change 2001-Working Group I-The Scientific Basis, (Contribution of working group I to the Third assessment Report of the Intergovernmental Panel on Climate Change) ©IPCC, 2001]

18.1 Climate negotiations

The first assessment report by the scientists within the IPCC was published in 1991. This proved to be a wake-up call to the world's governments. Amongst its many pages, two facts stood out: firstly, that we could not afford to emit more than 0.4 tonnes of carbon per person per annum if atmospheric carbon dioxide concentrations were to be stopped from rising above what was at the time thought be safe levels; secondly, that the corollary of this was that a cut in emissions of least 60 per cent was required, as the average emission per person was then one tonne per annum.

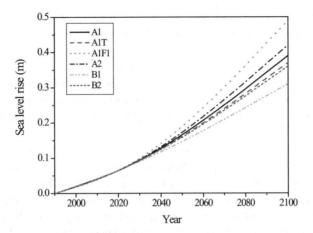

Figure 18.4 Global mean sea level rise to 2100 under the SRES scenarios [data from IPCC01]. The average for each scenario produced from several models is shown, together with the minimum and maximum from all SRES and all models. (A1F1 represents the business-as-usual case.) [Adapted from data in Houghton, et al, Climate Change 2001-Working Group I-The Scientific Basis, (Contribution of working group I to the Third assessment Report of the Intergovernmental Panel on Climate Change) ©IPCC, 2001]

In 1992 at the Earth Summit, 187 countries signed and subsequently ratified the United Nations Framework Convention on Climate Change (UNFCCC). The negotiations had taken two years, but the result was still only a *framework* in that it did not indicate who should make the reductions nor by how much. The reasons why a framework was produced, rather than a detailed convention, even after such a long period of negotiation, will become apparent as we study what happened after the Earth Summit and look at the positions taken by various groups and governments. The key clauses from the Convention are reprinted below (from [MEY00]).

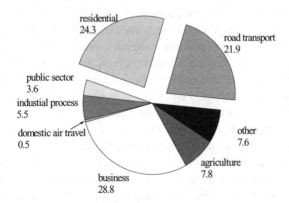

Figure 18.5 Percentage UK sectorial importance to radiative forcing of the major greenhouse gases. The exploded segments show emissions mostly directly connected to individuals [data from CCP05]

1. Parties to the UNFCCC 'acknowledge that change in the earth's climate and its adverse effects are a common concern of humankind'.

2. They are 'concerned that human activities have been substantially increasing the atmospheric concentrations of greenhouse gases, that these increases enhance the natural greenhouse effect, and that this will result on average in an additional warming of the Earth's surface and atmosphere and may adversely affect natural ecosystems and humankind'.

3. The convention 'is to achieve . . . stabilization of greenhouse gas concentrations in the atmosphere at a level that would prevent dangerous anthropogenic interference with the climate system'. (Today, this is generally taken as restricting the maximum temperature rise to less than 2 °C. Beyond this value "irriversible" climate change may occur, for example, the irriversible melting of the Greenland ice cap – and implies a maximum CO_2 concentration of 450 ppm)

4. The Parties 'should protect the climate system for the benefit of present and future generations of humankind, on the basis of equity'.

5. The Parties 'should take precautionary measures to anticipate, prevent or minimize the causes of climate change and mitigate its adverse effects. Where there are threats of serious or irreversible damage, lack of full scientific certainty should not be used as a reason for postponing such measures . . . ' (An example of the Precautionary Principle.)

Having produced a convention, the UN needed to ensure specific countries implemented specific policies to limit their emissions. Opposition was met with on several ideological fronts. Some people, led by groups such as the fossil fuel industry, tried to claim that climate change was not occurring and that we should wait until we have definite proof before implementing what might prove to be unpopular and costly policies. It was also argued that for some regions a warmer world might be of benefit. Others took a more moderate, but still largely negative, stance and suggested that an attempt should be made to reduce the build up of anthropogenic greenhouse gases in the atmosphere, but only if such actions made economic sense anyway. Examples being improved energy efficiency measures that had reasonable payback periods, or a switch to natural gas and away from oil (because of the higher efficiencies possible). Such a posture has the advantage that if climate change proved to be more myth than reality, one would have wasted very little money. These two positions can be seen as representing a business-as-usual ideology; however opposition also grew from the Green lobby.

Many environmentalists however felt that it was the responsibility of the developed world to solve the problem, as it was they who had grown rich by the use of fossil fuels and that massive cuts were needed immediately. Opposing this view were those in the developed world who asked the question, what is the point of us reducing our emissions when the developing world will not cut theirs? The rationale being that although the

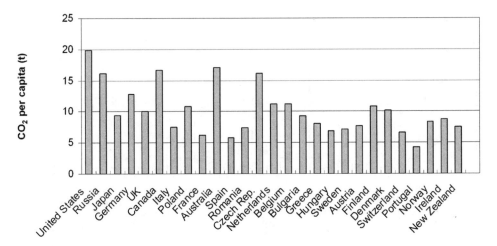

Figure 18.6 Examples of carbon emissions per capita. The USA is the heaviest polluter

developing world has much lower emissions per capita, their total emissions accounted for one quarter of the total carbon dioxide emission in 1990 and might equal those of the developed nations by 2030 (Figure 16.4).

On top of any ideological stances, individual countries fought against the setting of quotas for various reasons. Any simple system such as 'all developed countries should make reductions of x per cent' was felt to be unfair by countries like Japan who had made major investments in energy efficiency already and thereby greatly reduced its potential emissions by improving its energy intensity. Approaches that suggested fixing the maximum level of carbon dioxide that could be emitted per dollar of economic activity were rejected by countries such as Canada, which claimed that it needed a higher quota because of the lower temperatures it experienced (necessitating elevated levels of space heating). Unfortunately, the responsibility for action can be portrayed as lying with different groups depending on how the analysis is carried out. Using carbon emission per capita (Figure 18.6) indicates that the USA is the heaviest polluter; using emissions per unit GDP (Figure 18.7) produces a very different story.

Since 1991, much has changed in both the science and politics of climate change. Few now believe anthropogenic climate change is not real. In the words of the UK Government's 2003 White Paper on energy:

'Climate change is real. . . . over the 20th century, the earth has warmed up by about 0.6 °C largely due to increased greenhouse gas emissions from human activities' [DTI03a]

Emissions are growing so fast that simply making cuts deemed to be economic in the short term would have little impact, and although large cuts by the developed world would help, these would still not stabilize atmospheric concentrations. Even a long-term economic analysis of the problem proved impossible, as we cannot easily judge what

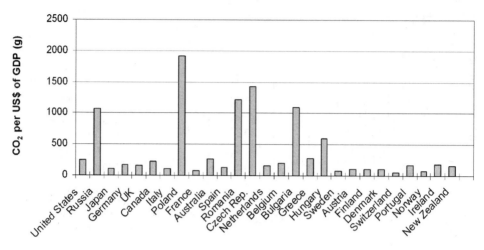

Figure 18.7 Carbon emissions per unit GDP. Poland is the worst polluter

the future costs might be. In addition, many felt this an unacceptable approach since some of the poorest people on the planet are predicted to suffer the greatest impacts. Such injustices tend not to be reflected in any economic analysis – and can be amplified. For example, the loss of one square kilometre of land in the developing world from rising sea levels carries a much lower economic value in dollar terms than the loss of an equivalent area in the developed world – because of much higher land values in the developed world – even though the impact on the local populace is likely to be far less in the developed world. Also, the use of dollars (or any other currency) hides the fact that although many nations have low economic output in terms of dollars, items such as food and housing are cheaper, therefore the value of land or any economic activity needs to be adjusted to take account of local purchasing power. When this is done, much of the developing world appears highly efficient in terms of economic output per unit of carbon emitted – the opposite of the common view.

Kyoto

By December 1997, the original Framework Convention had finally been turned into a protocol with specific reduction targets – the Kyoto Protocol[1]. These targets (Table 18.1) only cover the developed nations (termed Annex 1 nations) and for the less developed, *developed* nations (such as Spain) actually only amount to agreements to curb their *growth* in emissions.

The protocol requires each country to control their emissions to meet the values shown in Table 18.1 during the period 2008 to 2012 – reductions are relative to emissions

---

[1] The protocol seeks to reduce emissions of a basket of six greenhouse gases: carbon dioxide, methane, nitrous oxide, sulphur hexafluoride, hydrofluorocarbons (HFCs) and perfluorocarbons (PFCs). The requirement is to reach a pledged target averaged over the five years 2008 to 2012.

Table 18.1 The Kyoto protocol targets [data from MEY00, p78]

Country	CO$_2$ emissions in 1990 (Mt)	% of world emissions	Kyoto pledge	CO$_2$ per capita (t)	CO$_2$ per $ of GDP (grams)
United States	4957	23	7 % cut	19.83	246
Russia	2389	11	No rise	16.13	1071
Japan	1173	5.4	6 % cut	9.35	107
Germany	1012	4.7	21 % cut	12.79	169
UK	584	2.7	12.5 % cut	10.04	161
Canada	457	2.1	6 % cut	16.64	222
Italy	428	2.0	6.5 % cut	7.51	107
Poland	414	1.9	6 % cut	10.88	1919
France	367	1.7	No rise	6.23	84
Australia	289	1.3	8 % rise	17.12	268
Spain	261	1.2	15 % rise	5.79	126
Romania	171	0.8	6 % cut	7.37	1220
Czech Rep.	170	0.8	8 % cut	16.09	1431
Netherlands	168	0.8	6 % cut	11.21	161
Belgium	113	0.5	7.5 % cut	11.18	196
Bulgaria	83	0.4	8 % cut	9.24	1092
Greece	82	0.4	25 % rise	8.02	277
Hungary	72	0.3	6 % cut	6.93	591
Sweden	61	0.3	4 % rise	7.15	73
Austria	59	0.3	13 % cut	7.70	102
Finland	53	0.2	No rise	10.81	109
Denmark	52	0.2	21 % cut	10.12	110
Switzerland	44	0.2	8 % cut	6.60	54
Portugal	42	0.2	27 % rise	4.25	171
Norway	36	0.2	1 % rise	8.36	84
Ireland	31	0.1	13 % rise	8.76	187
New Zealand	26	0.1	No rise	7.59	160
Luxembourg	11	0.05	28 % cut	29.76	299

in 1990. As each country suggested its own targets, they are extremely modest. Adding up the second column in Table 18.1, we see that in 1990 emissions from the Annex 1 nations were 13 605 Mt of carbon dioxide. Applying the target reductions to each country and adding up the results once more we see that by 2012 Annex 1 emissions will stand at 12 844 Mt, i.e. the protocol will only deliver a 5.6 per cent cut. Considering emissions from the developing world are expected to grow by 5000 Mt during this time, the world is faced with an anticipated 30 per cent (4000 Mt) *rise* in emissions by 2012, despite the international community having deliberated for years to finally establish a protocol.

It could be argued that the protocol does deliver something worthwhile, in that without it Annex 1 emissions might have been expected to rise by around 29 per cent by 2012. Therefore the protocol, whilst not offering a genuine cut in world emissions, does offer a welcome short-term reduction in the rate of growth in emissions.

Several of the countries shown in Table 18.1 are not expected to need their full emissions quota – particularly those that are allowed to freeze or increase their emissions under the protocol – and will therefore have emissions to trade. The protocol allows this to happen, and it is expected that Russia for one will sell some of its quota. Such an *emissions trading scheme* is a *cap-and-trade* scheme similar to that introduced under *tradable permits* in Section 6.2 (i.e. it sets a *cap* on emissions then allows permits to be *traded* up to the value of this cap). It is also similar to the scheme presented in the next section (18.2), except that the Kyoto version does not allow emissions trading with the developing world[2] (as developing world countries are not signatories to the protocol), it is not based on keeping greenhouse gas concentrations lower than a predetermined safe level, and there is no requirement to reduce the size of the cap in later years.

The European Union (EU), which is responsible for around 22 per cent of world greenhouse gas emissions, has always been a strong supporter of the protocol. In late 2002, the EU designed an internal trading scheme to help encourage nations to meet the EU overall cut of eight per cent in the most economical way. Quotas were set for six major industries: energy, steel, cement, glass, brick, and paper/cardboard. Fines were also introduced for EU states that failed to meet their Kyoto obligations. It would appear that several EU nations deliberately overestimated their emissions from these key industries – thereby ensuring they would easily meet quotas in later years. It is also looking likely that some EU countries will fail to meet their Kyoto commitments.

The cost-benefit approach introduced in Section 6.2 has been used by economists to examine whether Kyoto places additional costs on the world economy, or will generate savings from reducing the need to adapt to climate change. Unfortunately, there has been no agreement on this issue, in part because we do not yet know the cost of this adaptation and the damage likely to be caused by climate change, and also because the result is affected by the choice of discount rate (Section 6.1).

Personal carbon credits

One thing not contemplated by Kyoto is the setting of limits on personal emissions, i.e. some form of cap-and-trade for individuals, as suggested by Mayer Hillman in his influential book *How We Can Save the Planet* [HIL04]. Such a scheme, as long as it only covered purchases of energy and travel, would, Hillman suggests, be surprisingly easy to introduce and police via the use of an electronic carbon credit card. This would be presented by the user when making energy or travel purchases. If an individual ran out of carbon credits, a top up could be purchased on the open market, again electronically, where those with unused credit would sell their excess quota. By slowly reducing the size of the annual personal quota, a growing level of reduction in emissions would be achieved.

---

[2] Although it is possible for signatories to fund projects that save emissions in the developing world and claim such carbon savings as their own.

18.2 Another approach

Several countries (most notably the USA, which is responsible for 23 per cent of the world's total carbon emissions despite having only 4.6 per cent of the world's population) did not ratify the Kyoto protocol[3]. America's position has been that unless the developing world is included in the solution very little can be achieved. Although their refusal to limit their own emissions and ratify the protocol might seem morally questionable, the above arithmetic (which showed Kyoto would not even achieve a reversal in the growth in emissions) demonstrates that they have a valid point. Clearly a new solution is needed.

A possible solution might include the following four steps:

Step 1 Reach an international agreement on an *acceptable* level of carbon dioxide in the atmosphere, i.e. a level that allows only an acceptable amount of harm, in that it does not jeopardize large numbers of individuals or exceptionally impact wildlife or economies.

Step 2 Decide a date by which this limit is to be reached.

Step 3 Given the concentration limit from Step 1, and knowledge of the fraction of anthropogenic carbon which is retained in the atmosphere, it is fairly straightforward to calculate a schedule of planet-wide emission reductions to reach the agreed limit by the agreed date. This process is called *contraction*.

These three steps are uncontroversial (with the possible exception of defining how much damage is acceptable and the problems of making predictions of the precise impact of greenhouse gases as discussed in Chapter 17). However, one final step is required; the apportioning of the reduction schedule amongst the world community.

The schedule of concentration limits needed for Step 3 defines the total amount of carbon humankind can emit in order to not exceed the maximum concentration. But how should we apportion this limit across all humankind? There is probably only one way of doing this that can claim an equitable basis:

Step 4 Share the right to emit carbon equally among the world's population, i.e. on a per-capita basis. When scheduled over a time frame this process is termed *convergence*: convergence to an equal per capita emissions quota.

---

[3] The USA together with South Korea, Australia, China, Japan and India have now formed a partnership – the *Asia-Pacific Partnership on Clean Development and Climate* – to investigate and share technologies to reduce greenhouse gas emissions. These nations use 45 per cent of the world's energy and account for more than half the world's carbon dioxide emissions, so this partnership might prove to be important. However at present (2007) it doesn't plan to set any targets on emissions and sees itself as a way of encouraging what it sees as clean technologies, particularly fission, carbon sequestration (Chapter 29) and advanced coal burning.

Together, these steps represent an approach called *contraction and convergence* (C+C). Although it does have an ethical basis, C+C is essentially a pragmatic approach. Given the need to create a world-wide solution, because of growing emissions from the developing world and the reluctance of the USA to contemplate an approach which is not world-wide, there is some logic in the suggestion by supporters of C+C that it is the only methodology that has a real chance of gaining acceptance by both the developed and the developing world. Although acceptance has by no means been universal, support for the approach has been growing.

It is important to realize that C+C does not suggest that all people should emit the same amount of carbon, only that they have the right to do so. Those countries that, due to lack of development or the adoption of non-carbon based technologies, emit less than the agreed per capita limits, would be allowed to sell their unused emission rights to other countries. This is the key to the possible acceptance of C+C by the developed world: they could to some degree buy their way out of the problem. Drastic cuts would still have to be made by the developed world, but the adjustment process would be gentler because of the possibility of trading carbon credits with the developing world. In turn, the approach has found much favour with the developing world because their emissions are currently below the likely per capita limit (assuming a final concentration of around 450 ppm$_v$ (about 0.4 tonnes of carbon per annum per person). Thus C+C only limits their future expansion of fossil fuel-derived energy technologies. Clearly, they will only accept such limits if payment is received in kind: which C+C provides for by the use of its tradable carbon credits.

In order for those selling credits to expand their use of energy as their economy grows, without exceeding the per capita limits (after subtraction of any sold credits) they will need to invest heavily in sustainable energy systems. C+C, via the use of tradable credits, gives such countries the financial resources to do this. Without financial help, they are unlikely to use alternative energy resources because the initial cost of doing so will probably be more expensive than using fossil fuels. As the number of credits would be reduced with time to exactly follow the contraction schedule, the value of the credits would increase, and therefore the incentive to introduce new carbon energy technologies within the developed world would increase. A similar approach could be applied to other greenhouse gases by working in terms of equivalent tonnes of carbon.

C+C, if adopted, would see large sums of money flowing from the developed to the developing world (and large numbers of carbon credits flowing in the other direction). Because of this flow of wealth to the less prosperous nations the approach has also found favour among many environmentalists and those on the political left, although the praise has not been universal amongst such groups. One criticism is that it takes no account of the history of emissions. Only future emissions are to be equalized on a per capita basis. Climate change is as much about historic anthropogenic emissions as it is about future ones, and the developed world has grown rich on these emissions. Some would argue that all countries should have the total same per capita rights over all time; it is just that the developed world chose to use theirs early.

Estimation of the reduction schedule itself depends on the chosen limiting atmospheric carbon concentration, and the date by which its achievement is desired. Figure 18.8 shows the historic emissions and future carbon emission rights on a

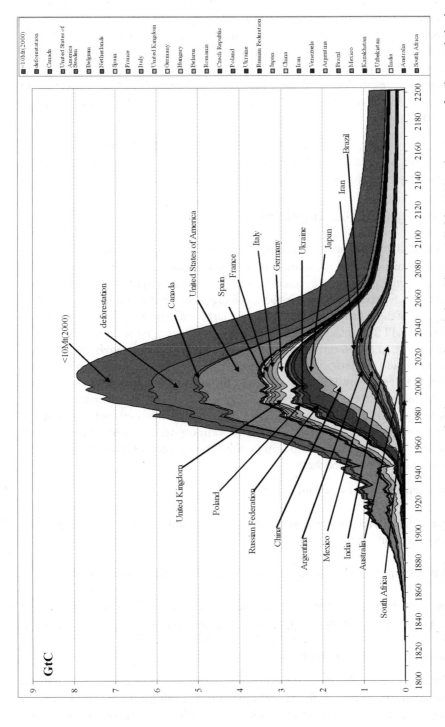

Figure 18.8 National contraction and convergence curves for those countries emitting at least 10 MtC in 2000 and as a group for those emitting less. (Contraction to 450 ppm$_v$ in 2200 and convergence by 2040) [Produced using software freely available from the Global Commons Institute's website] (For a colour reproduction of the figure, please see the colour section located towards the centre of the book)

country-by-country basis, assuming 450 ppm_v is to be realized and convergence is to be achieved by 2040 (the current concentration is over 360 ppm_v). Figure 18.9 shows the impact on per-capita emissions; Figure 18.10 the effect on atmospheric carbon dioxide concentration. Running the calculation for other final stable concentrations and other convergence dates is relatively easy using software freely available on the Internet (see the book's website for more details). You might have noticed the limiting CO_2 concentration of 450 ppm is less than any of IPCC senarios discussed in Chapter 16 and shown in Figure 18.1.

Problem 18.1   Visit the Global Commons Institute website (via the link given on this book's website) and use their software to plot similar graphs to Figures 18.8–18.10 for various convergence dates and limiting concentrations.

18.3 Bringing it all together

We have looked at four approaches to thinking about the world's future energy needs and carbon emissions and their impacts. The first, from the EIA (Chapter 16), just considered energy per se and predicted a growing energy requirement. The second, from the IPCC, worked in terms of carbon and suggested several possible futures. With the exception of those futures which predicted a large-scale move to non-carbon energy technologies (A1T), or a switch to service economies throughout the world (B1), all see substantial increases in carbon emissions in the long-term and thereby atmospheric concentrations and accelerating climate change. (All forecast such changes in the short- to mid-term.) We then saw (Chapter 17) that computer modelling indicates widespread impacts at this level of emissions. The final approach, C+C, began not from an energy or carbon viewpoint, and required no knowledge of future emissions, but from a mix of ethics and pragmatism. It defined the maximum atmospheric concentration and then asked us to create energy systems to achieve this limit by a certain date. Bringing these four analyses together we see that:

1. Energy demand is growing rapidly.

2. Climate change is real and will have an impact on humankind and the world's physical systems, flora and fauna.

3. If climate change is to be avoided a switch to non-carbon fuels is required.

4. As Figure 18.8 shows, the reductions that many countries will be required to make in order to stabilize emissions are large (possibly greater than 90 per cent). For example, look at how the emissions from the USA and the Russian Federation diminish in Figure 18.8 as 2100 is approached.

C+C gives us a timetable and a structure to manage our retreat from a carbon based economy to something new. As the majority of anthropogenic greenhouse gas emissions are energy related, this something new must be non-carbon based energy technologies.

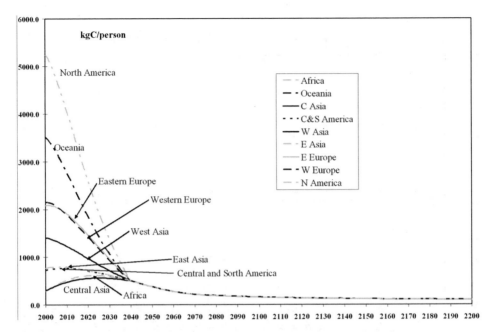

Figure 18.9 Regional contraction and convergence per-capita emissions. (Contraction to 450 ppm$_v$ in 2200 and convergence by 2040) [Produced using software freely available from the Global Commons Institute's website]

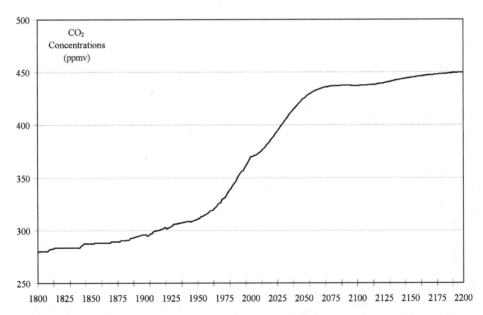

Figure 18.10 Carbon dioxide concentration under contraction and convergence. (Contraction to 450 ppm$_v$ in 2200 and convergence by 2040.) [Produced using software freely available from the Global Commons Institute's website]

Table 18.2 Required annual compound changes and total change required in carbon emissions per annum (from 2005 to 2040) assuming contraction by 2040 to 450 ppm$_v$

Country	Annual change %	Total change %	GW$_e$(2040)	Country	Annual change %	Total change %	GW$_e$(2040)
Afghanistan	4.5	380		Albania	2.7	160	0.112
Algeria	−0.63	−20	11.6	Antigua & Barbuda	−2.5	−59	
American Samoa	−0.97	−30	0.0488	Armenia	0.57	23	0.816
Angola	3.4	240		Aruba	−4.8	−83	
Argentina	−1.2	−36	28.4	Austria	−4	−77	25.6
Australia	−5.4	−86	154	Bahamas	−2.5	−60	0.583
Azerbaijan	−1.5	−42		Bahrain	−5.8	−88	8.87
Bangladesh	3.9	290	7.52	Barbados	−4.1	−78	0.923
Benin	4.0	310		Belarus	−3.5	−73	21.3
Bhutan	3.9	300		Belgium	−4.4	−80	41.5
Bolivia	1.3	61		Belize	0.17	6.3	
Botswana	−1.2	−34	0.768	Bermuda	−2.7	−63	0.155
Brazil	0.29	11		Bosnia-Herzegovina	0.47	18	
Burkina Faso	4.4	360		British Virgin Islands	−0.31	−11	0.00387
Burundi	4.5	380		Brunei	−3.9	−76	1.92
Cameroon	3.3	230		Bulgaria	−3.5	−72	14.6
Cape Verde	3.5	240		Cambodia	4.4	370	
Central African Republic	4.3	350		Canada	−4.8	−83	186
Chad	4.5	390		Cayman Islands	−1.8	−49	0.0758
Chile	−1.7	−45	17.2	China	−0.51	−17	283
Colombia	0.92	39		Croatia	−3	−67	7.63
Comoros	4.2	340		Cyprus	−3.6	−73	2.37
Congo	3.1	200		Czech Republic	−4.9	−84	46.5
Cook Islands	0.021	0.77		Denmark	−4.2	−79	19.7
Costa Rica	1.1	46		Dominica	0.92	39	
Cote d'Ivoire	2.3	120		Estonia	−6.3	−90	7.38
Cuba	−0.94	−29	4.29	Faroe Islands	−5	−84	0.279

Country			
Djibouti	2.9	180	
Dominican Republic	−0.70	−22	3.30
Ecuador	0.49	19	
Egypt	0.87	37	
El Salvador	1.8	89	
Equatorial Guinea	1.6	80	
Ethiopia	4.3	350	
Fiji	1.6	76	
French Polynesia	0.010	0.36	
Gabon	0.0074	0.27	
Gambia	3.9	300	
Ghana	3.5	250	1.80
Guam	−5.9	−89	
Guatemala	2.4	140	
Guinea	4.1	320	
Guinea Bissau	4.1	330	
India	1.5	72	
Jamaica	−1.6	−44	2.78
Kazakhstan	−3.5	−72	36.7
Kenya	3.3	220	
Kiribati	3.4	230	
Kyrgyzstan	1.5	69	
Liberia	4.2	340	
Libya	−2.5	−60	12.9
Madagascar	4.2	340	
Malawi	4.3	360	
Maldives	1.8	90	
Mali	4.5	380	
Mauritania	2.4	130	0.106
Mauritius	−0.20	−6.9	
Mexico	−1.4	−39	85.1

Country			
Finland	−4.8	−83	24.3
France	−3	−67	130
French Guiana	−1.3	−38	0.197
Georgia	−0.092	−3.2	0.128
Germany	−4.4	−80	326
Gibraltar	−1.3	−37	0.0147
Greece	−6.3	−91	40.2
Greenland	−4.4	−80	0.224
Grenada	−1.7	−46	0.0559
Guadeloupe	−1.8	−47	0.419
Guyana	−1.4	−40	0.387
Haiti	3.7	280	
Honduras	2.4	130	13.2
Hong Kong	−2.5	−60	21.5
Hungary	−3.6	−73	0.78
Iceland	−3.3	−71	
Indonesia	1.3	60	70.8
Iran	−1.5	−42	
Iraq	0.16	6	16.7
Ireland	−4.1	−78	23.3
Israel	−3.4	−71	169
Italy	−4.1	−78	473
Japan	−4.4	−81	77.9
Jordan	0.019	0.67	154
Korea (N)	−3.9	−76	20.6
Korea (S)	−3.9	−76	
Kuwait	−4.7	−82	
Laos	4.3	350	
Latvia	−2.1	−53	1.73
Lebanon	−2	−51	4.9
Lithuania	−2.7	−63	4.51

(Continued)

Table 18.2 (*Continued*)

Country	Annual change %	Total change %	GW$_e$(2040)	Country	Annual change %	Total change %	GW$_e$(2040)
Morocco	1.4	67		Luxembourg	−4.8	−83	3.24
Mozambique	4.2	350		Macau	−1.4	−39	0.365
Nauru	−3.0	−67	0.0476	Macedonia	−2.9	−66	4.14
Nepal	4.0	320		Malaysia	−1.8	−47	32.9
New Caledonia	−2.5	−60	0.538	Malta	−4.1	−78	1.42
New Zealand	−3.6	−74	12.3	Martinique	−2.7	−62	0.675
Niger	4.4	370		Moldova	−0.15	−5.3	0.211
Nigeria	3.4	240		Mongolia	−0.56	−18	0.795
Pakistan	2.8	170		Montserrat	−5	−84	0.0212
Palau	−3.5	−72	0.0906	Myanmar	3.7	260	
Panama	−0.73	−23	1.43	Netherlands	−4	−77	55.5
Papua New Guinea	3.2	210		Netherlands Antilles	−6.4	−91	2.47
Peru	1.5	69		Nicaragua	2.6	150	
Puerto Rico	−0.85	−26	1.37	Norway	−3.9	−76	15.2
Reunion	−1.0	−31	0.439	Oman	−1.8	−49	5.62
Rwanda	4.3	360		Other	−8.7	−96	636
Samoa	2.0	110		Paraguay	2.5	150	
Sao Tome & Principe	3.0	190		Philippines	1.8	88	55.5
Senegal	3.4	230		Poland	−4.1	−78	124
Seychelles	−0.79	−25	0.0311	Portugal	−3.3	−71	22.1
Sierra Leone	4.2	340		Qatar	−9.1	−97	24.8
Solomon Islands	3.5	250		Romania	−2	−52	20.9
Somalia	4.6	400		Russian Federation	−5	−84	594
South Africa	−4.0	−77	134	Saudi Arabia	−2.5	−60	72.8
Sri Lanka	2.5	140		Serbia & Montenegro	−2.6	−61	15.3
St. Helena	−2.6	−61	0.00771	Singapore	−5	−84	25.1
Sudan	4.2	350		Slovakia	−3.6	−74	14.3
Swaziland	2.7	160		Slovenia	−4.1	−78	5.95

Country				Country			
Tajikistan	2.1	110		Spain	−3.6	−73	104
Tanzania	4.3	350		St. Kitts-Nevis	−1.5	−42	0.0245
Togo	3.7	270		St. Lucia	−0.34	−12	0.0212
Tonga	1.0	43		St. Pierre & Miquelon	−3.8	−75	0.0179
Trinidad and Tobago	−6.2	−90	11.4	St. Vincent & Grenadines	0.43	17	
Tunisia	−0.032	−1.2	0.145	Suriname	−2.5	−59	0.677
Turkmenistan	−2.2	−54	8.40	Sweden	−2.9	−65	16.1
Uganda	4.4	380		Switzerland	−3.4	−72	15.6
Uruguay	−0.075	−2.7	0.109	Syria	−0.17	−6.1	2.01
US Virgin Islands	−10	−98	6.31	Taiwan	−4.3	−79	88.4
Uzbekistan	−1.7	−47	31.4	Thailand	−1.2	−35	39.8
Vanuatu	3.3	220		Turkey	−0.76	−24	29.4
Venezuela	−1.6	−45	31.3	Ukraine	−4.3	−80	138
Western Sahara	2.5	140		United Arab Emirates	−6.6	−91	42.4
Zaire	4.4	380		United Kingdom	−4	−77	211
Zambia	4.0	310		United States of America	−5.4	−86	2450
Zimbabwe	0.35	14		Vietnam	2.4	130	
				Yemen	3.1	200	

(Note some reductions are positive and some negative – negative implying a reduction is required). Because of the C + C methodology some of this reduction could be by emissions trading, particularly in the early years. Reductions would also occur from new installed capacity built by those who need to reduce emissions by 2040 and plan to do this by sustainable generating technologies alone (i.e. no changes in energy demand or efficiency). The required installed generating capacity to do this (GW_e) has been made on the assumption that sustainable energy systems generate at full capacity for half the hours in the year. Some of this requirement could be in the form of electricity or hydrogen imports from developing nations

Unfortunately, C+C does not tell us how to achieve this transformation, but it does tell us that the majority of our energy production infrastructure will have to be replaced.

Using the software mentioned earlier, it is possible to estimate, for a particular C+C scenario, the required contraction rate for any country. The results are shown in Table 18.2.

Table 18.2 shows that many developed countries will have to make reductions of around four per cent per annum and a total reduction of 80 per cent. Taking the USA as an example, Table 18.2 implies the need to remove seventy coal-burning power stations per annum. What will these be replaced by? (And as there are not in reality 2450 one-GW_e power stations[4] in the USA, other emitters such as vehicles will have to be replaced.)

18.4 Conclusion

The best available climate science tells us that we need to reduce our emissions of carbon dioxide so concentrations never exceed 450 ppm_v (as we are already seeing the impacts of climate change at 360 ppm_v, others suggest the figure might be as low as 280 ppm_v – the pre-industrial value). Reaching 450 ppm_v will require reducing worldwide emissions from 7.9 GtC per annum to 4.8 GtC, despite a large predicted increase in energy demand (Chapter 16) over the same period (2005–2040), and to even lower levels in later years. If this rise in demand is met with roughly the same mix of fuels as today, carbon emissions will grow and will equal approximately 12.1 GtC per annum. Therefore we can conclude that somehow or other, if carbon dioxide concentrations are to fall to a safe level, we need to save $12.1 - 4.8 = 7.3$ GtC per annum, i.e. roughly the same as our current carbon emission: implying that most of the world's energy will need to come from non-carbon sources. (Trying to achieve a 360 or 280 ppm_v target would mean that carbon emissions would have to almost stop.) Meeting this constraint will be a major industrial challenge. The rest of this book is about the energy systems that hopefully will expand to match this challenge.

18.5 Student exercises

"Global warming is too serious for the world any longer to ignore its danger or split into opposing factions on it."

–Tony Blair, speech, Sept. 27, 2005

1. Explain the logic behind the USA not ratifying the Kyoto protocol.

2. Outline why it is difficult to get agreement amongst nations to reduce emissions.

3. Outline the process of contraction and convergence.

4. Why might some consider C+C unfair to many developing nations?

---

[4] 35 years × 70 per year = 2450 power stations.

PART IV
Sustainable energy technologies

"*Preventing the transformation of the earth's atmosphere from greenhouse to unconstrained hothouse represents arguably the most imposing scientific and technical challenge that humanity has ever faced. It is local, national and international. It will affect all of us as well as all our children.*"

—British environment minister David Miliband, October 2006

In the previous chapters we have seen how our societies face a serious global environmental threat – climate change. We have also seen how this threat is entirely human-made and centres around our dependency on fossil fuels. In Part III we saw that predictions of future global energy use indicate climate change will accelerate unless we switch this dependency away from oil, coal and natural gas and towards alternative, non-carbon based technologies. This Part will present a series of these new technologies and attempt to estimate how relevant each is to supplying the world's future energy demand.

Although for an energy technology to be considered sustainable the resource must be unlimited (i.e. $Q_\infty = \infty$), this doesn't mean that the resource is infinite at any point in time. As the term suggests, renewable technologies rely on the resource being constantly renewed. This implies that there might be a maximum rate of use of the resource that would allow the technology to be classified as sustainable. However, as we will see most renewable resources are considerable and this theoretical limiting factor is unlikely to form a practicable limit for most technologies. Figure IV.1 shows the resources and fuels studied in Part IV.

The first alternative technology we will look at is energy conservation, followed by solar power – both as a way of warming our buildings and as a way of generating electricity. Our attention will then move on to water and wind-powered electricity generation, followed by an investigation into biomass technologies. (Although by definition biomass is carbon based, if it is produced in a sustainable way it can be seen as carbon neutral). A brief introduction will be given to geothermal power before discussing fast-fission, fusion and the sequestration of carbon dioxide – a technology that would

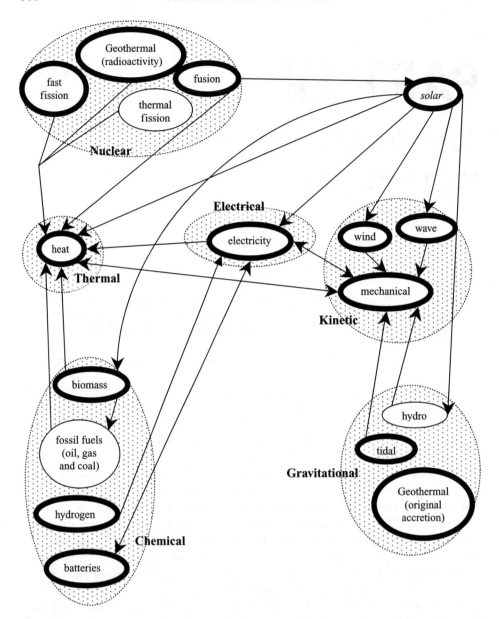

Figure IV.1 Energy flows, fuels and technologies studied in this book. Those discussed in Part IV are highlighted

allow us to continue our love affair with fossil fuels and still reduce our emissions of carbon dioxide. Finally, we will investigate how one government believes it can create a low-carbon, sustainable economy. For the reasons mentioned in Part II, transportation will be covered in a separate chapter.

Before we examine each technology in detail, we will quickly summarize the world's current use of sustainable energy.

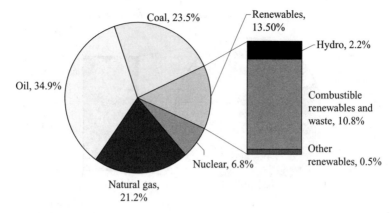

Figure IV.2 Fuel shares in world primary energy supply [IEA05]. The 'other renewables' are geothermal, wind, tide and solar

Unlike the non-sustainable energy sources we discussed earlier, most of the sources covered in this Part have not had consistent annual global resource estimates made of them. In part, this is because the true *resource* is all but infinite, but also because the size of the resource is difficult to judge given the infancy of the various technologies. So instead of giving true resource estimates we will generally present data on the current level of deployment and its rate of growth (some theoretical estimates will also be made). Most of these data are for the developed nations and is collected on an annual basis by the International Energy Agency (IEA) for the OECD. This is not to suggest that such technologies have not been deployed in developing nations – it is just that reliable annual data has not been collated. Indeed, much of the future resource for many sustainable technologies lies within developing nations – which is one of the reasons the subject is so interesting. Our apologies to readers whose countries have therefore been left out of these figures. Hopefully this lack of data will change as alternative technologies become more common. Because the uptake of many sustainable energy technologies is expanding rapidly it is suggested that interested readers visit the IEA website (www.iea.org) to check for the latest release of their 'Renewables Information'[1].

IV.1 Current world sustainable energy provision

Thirteen per cent of world primary energy is produced from renewable energy sources[2] including hydropower (hence the use of the term *renewable* rather than *sustainable*), compared to 35 per cent for oil, 24 per cent for coal, 21 per cent for natural gas and seven per cent for nuclear energy (see Figure IV.2). This suggests that renewable energy should be treated as seriously as other sources of energy.

---

[1] The data presented here is based on that given in Renewables Information 2004.
[2] See [IEA05] for how industrial waste and some fractions of domestic waste burning are classified.

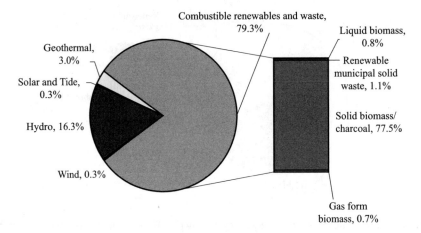

Figure IV.3 Shares in world renewable energy supply [IEA05] [Data from rom Renewable Information 2004 International Energy Agency]

Much (77.5 per cent) of the renewable fraction is the non-commercial use of biomass, which represents 10.4 per cent of world primary energy; the second largest is hydropower (16.3 per cent of the renewable fraction, or 2.2 per cent of primary energy). Geothermal represents 0.4 per cent of primary energy. Solar, wind and tide represent less than 0.1 per cent of the total world primary energy supply (Figures IV.2 and IV.3). Renewable

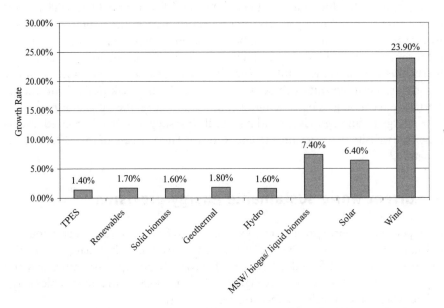

Figure IV.4 Annual growth rates of renewables, 1990–2002 [IEA05]. TPES = total primary energy supply; MSW = municipal solid waste [Data from rom Renewable Information 2004 International Energy Agency]

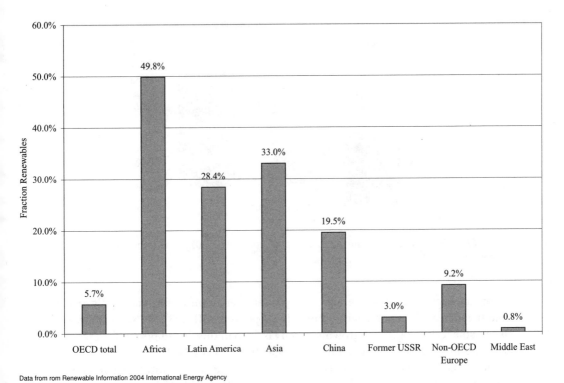

Data from rom Renewable Information 2004 International Energy Agency

Figure IV.5   Percentage of renewable energy used by world regions [IEA05]

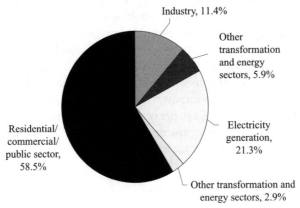

Data from rom Renewable Information 2004 International Energy Agency

Figure IV.6   World sectorial consumption of renewables [IEA05]. 'Other transformation and energy sectors' includes the transformation of one form of energy into another, for example the loss of energy from converting wood into charcoal

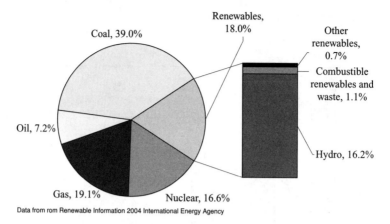

Data from rom Renewable Information 2004 International Energy Agency

Figure IV.7 Fuel shares in world electricity production in [IEA05]. The 'other renewables' are geothermal, wind, tide and solar

energy use is growing at an average of 1.7 per cent per annum, which is slightly greater than the increase in world energy demand (1.4 per cent) [IEA05]. However individual sustainable technologies are growing much faster than this. Wind power, for example, is increasing by 24 per cent per annum (1990–2002) (Figure IV.4).

Most solid biomass derived energy is produced and consumed within the developing world, where it is used for residential cooking and heating, with Africa producing 24 per cent of the world's solid biomass supply [IEA05]. Because of this high level of biomass use non-OECD regions account for 78 per cent of renewable energy supply, while the OECD (the most developed nations) produce only 22 per cent of the world's renewable energy, despite using 52 per cent of world primary energy. Clearly, the developing world is ahead of the developed as regards sustainable energy provision. In non-OECD countries, 22 per cent of energy use is on average from renewable sources, for OECD countries this falls to only 5.7 per cent (Figure IV.5). Some would consider this inexcusable.

Within the OECD countries, most renewable energy is used for generating electricity; worldwide, most is used for heating and cooking (Figure IV.6). Despite this, renewables are the third largest contributor to world electricity production [IEA05], accounting for 18 per cent (16 per cent for hydropower alone) of generation in 2002, against 40 per cent for coal, 19 per cent for gas and 16 per cent for nuclear (Figure IV.7). Unfortunately, renewable electricity generation is growing more slowly (1.8 per cent per annum) than overall generation (2.6 per cent per annum).

19
Energy efficiency

"Climate change is the most severe problem that we are facing today, more serious even than the threat of terrorism."

–David King, UK government chief scientific adviser, January 2004

Often termed 'the silent renewable', energy efficiency, or conservation, is the least glamorous of the sustainable energy technologies covered by this text. Just why it is seen as so unglamorous, even by some of the environmental movement, is hard to understand. Ecological campaigns, of the *Save the Whale/Rainforest/Tiger* type, often have conservation at their heart, yet energy efficiency is often left until the final chapter of a book on renewable energy technologies or climate change. However, in general it is the most promising sustainable energy strategy in the short to medium term, offering the promise of large carbon savings for relatively little expenditure and little environmental impact. For this reason, we will promote it by introducing it before we study other sustainable energy sources. In Chapter 12 we saw that one of the reasons thermal nuclear reactors could not be relied upon to solve climate change was their cost. Many sustainable energy technologies have similar costs (although usually much shorter construction times – reducing the cost of financing their construction somewhat); therefore there are similar economic questions over the feasibility of their deployment. In general energy efficiency is cheap, the technology is fully developed, there are unlikely to be public concerns over its use and construction/implementation times very short. The power of the approach when taken seriously can be gleaned from Figure 18.2. This shows how although Japan's wealth (in the form of its per-capita gross domestic product) is similar to the USA's, it is achieved by emitting less than half the mass of carbon dioxide per capita.

That energy efficiency measures have allowed Japan to greatly increase its wealth without a corresponding increase in its energy requirements should not surprise us, and reinforces what was said earlier, namely, that we are rarely interested in energy per se, only what it can do for us. We want to be warm, to have enough light, etc. We

Energy and Climate Change   David A. Coley
© 2008 John Wiley & Sons, Ltd

care little for how we are kept warm, or the technology of the light bulbs. Maybe this pragmatic or possibly almost complacent attitude is the reason energy efficiency fails to hold the attention of the public. A new large offshore wind farm development just seems so much more physical, so much more interesting. And let's face it, few prime ministers or presidents would be willing to spend much time at an opening ceremony for a new roof insulation campaign.

Problem 19.1   Japan uses 504.8 Mt_{oe} of primary energy per annum. If the introduction of a new energy efficiency technology could reduce this by five per cent, how many 500 kW_e wind turbines (turning half the hours in the year) would be required to make an equivalent saving? Which would be more likely to be the cause of public complaint?

In the following sections we will be studying cogeneration, energy recovery and energy conservation in buildings. There are however many other forms of energy conservation, particularly the adoption of new technologies within industry and the use of waste heat from one industrial process to drive another process, rather than simply losing the heat to the atmosphere. For an excellent introduction to such topics, readers are directed to reference [EAS90].

19.1 Cogeneration

Cogeneration, or combined heat and power (CHP) schemes, try to circumvent the poor efficiency seen in traditional electrical generation technologies. This is typically only 30 per cent with the remaining 70 per cent of the energy being lost as heat to the surroundings. As we saw in Chapter 7, much of this loss is an inevitable consequence of the Carnot efficiency of such technologies. However the low Carnot efficiency of a process does not mean we cannot use this waste heat as heat, it just means we cannot use it to provide work. There is nothing to stop us using the waste heat to warm our homes or offices. When the electrical generator is at some distance from the building to be heated such a scheme is referred to as *district heating* and has proved successful at various locations around the world, for example, Sweden. More popular have been CHP schemes based around a single facility, such as a hospital, with a large and reasonably constant need for both electrical power and heat.

Figure 19.1 shows the energy conversion diagram introduced in Part II and highlights the underlying poor efficiency of converting thermal energy into mechanical energy. This is the weak link. Not the conversion of chemical energy into thermal energy, or the conversion of mechanical energy into electrical energy. This efficiency is so low that the heat loss from UK power stations exceeds the total UK domestic heating requirement [ATK86].

The electrical/heat balance

A small cogeneration plant typically uses a diesel engine as the prime mover and then utilizes the heat from the exhaust gases to heat water for distribution to the heating

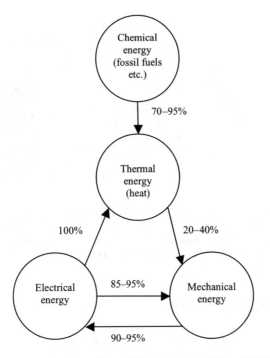

Figure 19.1 Energy conversion efficiencies [adapted from EAS90, p296]

system. Because there is no need to replace the heating system itself, such plants can be retro-fitted without the need for major refurbishment and often in the existing boiler room. Other cogeneration schemes make use of steam turbines or gas turbines (Figure 19.2). In future the use of fuel cells (Chapter 28) for cogeneration might become a reality. Because such units can be constructed on any scale this would open the possibility for a radical re-organization of the electrical generating industry, with the domestic boiler being replaced by a fuel cell producing both electricity and heat to warm a house. Excess electricity would be exported from the house via the grid. The national grid would then have similarities with the Internet and might link several million micro power stations.

Whether cogeneration makes economic or environmental sense has to be considered on a case-by-case basis. This is because although excess electricity can be exported from the site via the national grid, heat cannot be, so a site that uses much more electricity than heat will see little benefit from introducing cogeneration. A heat/power ratio, R_{hp}, of approximately 1 is the ideal situation if a diesel generator is used as the prime mover (i.e. the device producing the kinetic energy). Unfortunately it is unlikely that an end user has an R_{hp} ideally tuned to the cogeneration system, nor is R_{hp} likely to be constant, particularly if much of the heat is used for space heating, as there will be little need for such heat during the summer. Industrial or water heating applications (e.g. swimming pools) may offer a more constant R_{hp}.

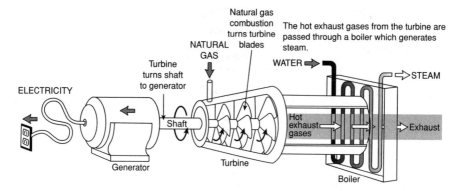

Figure 19.2 A cogeneration plant based on a gas turbine. Natural gas is burnt in the turbine, which turns an electric generator. The hot exhaust gases are then used to boil water to steam, which is then used (via a second heat exchanger) to provide space and water heating [Exxon Corporation in RIS99, p84] [Adapted from Ristinen, R.A. and Kraushaar, J.J., Energy and the Environment, John Wiley & Sons, Inc, 1999]

Example application: CHP

The Viborg CHP plant in Denmark was designed with the aim of achieving power generation with an efficiency of at least 85 per cent. The design also shows that such facilities can be of considerable scale and still have a distinctive and attractive look (Figure 19.3). This is important because if district heating schemes using CHP are to become popular they will need to be located within the communities they serve, in order to keep the heat distribution network to a reasonable size and minimize losses. It is therefore important that such *embedded* generation meets with minimum resistance.

The plant supplies heat to 11 substations via a 12 km heat transmission network. Four of these substations are fitted with natural gas boilers used to add capacity during periods of peak demand. The plant provides for nearly 80 per cent of the heat demand of the town. The design (Figure 19.4) centres on a combined cycle CHP containing a 42 MW$_e$ gas turbine and a 15 MW$_e$ steam turbine. Natural gas is preheated and then burnt at 1100 °C in the first of these and the flue gases (at 530 °C) used to raise steam for the latter. The flue gases are then cooled to 70–100 °C by the district heating system. Depending on demand, the hot water is either fed directly to homes or stored in a 19 000 m^3 heat accumulator.

Operational data for the system is given in Table 19.1. When compared to coal fired generation, the CHP plant is stopping the emission of about 30 000 t of carbon dioxide per annum and shows that the overall efficiency of electricity generation can be improved by more than a factor of two with an off-the-shelf solution.

19.2 Reducing energy losses

Thermal energy has a habit of escaping by conduction, radiation and convection. Chemical, potential and nuclear energy are far easier to constrain, and kinetic and electrical

Figure 19.3 The CHP plant in Viborg, Denmark [CAD430] [Reproduced courtesy of CADDETT, Technical Brochure, 430, IEA CADDET programme, http://www.caddet.org.]

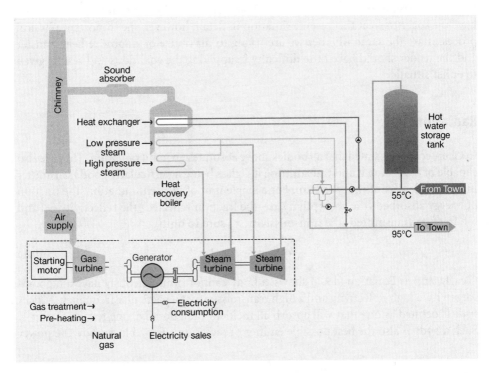

Figure 19.4 Schematic of the Viborg CHP plant [CAD430] [Reproduced courtesy of CADDETT, Technical Brochure, 430, IEA CADDET programme, http://www.caddet.org.]

Table 19.1 Annual statistics for the Viborg CHP plant [CAD430]

Statistic	Value
Gas consumption (Mm^3)	59
Heat production (MWh)	289 425
Electricity sales (MWh)	287 911
Overall efficiency (%)	87
Availability (%, i.e. the percentage of the time the plant is operational)	99
Heat sales (DKK million)	107
Electricity sales (DKK million)	119
Original cost of plant (not including heat transmission, DKK million)	388

energy are rarely generated much before their use. Because of this, the reduction of thermal losses is an important topic within energy conservation. Often the thermal energy needing to be conserved is contained within a waste product that needs disposing of, for example stale air within a building. In this case we will need to recover the heat first, and then pass it to another material – probably fresh incoming air.

Heat is lost from a body by radiation, convection and conduction. We will study each of these in turn. The assumption will be that we wish to reduce losses from these pathways. This is not always so. For a heat exchanger, or a radiative heating element, we usually wish to maximize the rate of heat transfer to the surroundings. We do not have the space to treat this reverse situation in detail, however, the important physical processes are the same whether we are trying to maximize or minimize heat transfer and the reader should have little difficulty in applying the equations and results given to either situation.

Radiation

In Chapter 3 it was stated that all bodies above absolute zero emit radiation. They are also capable of absorbing it and transmitting it – glass being a particularly good transmitter in the visible part of the spectrum. For a single unit of incident radiation, the fraction of energy absorbed (the absorptivity, α), the fraction reflected (the reflectivity, γ) and the fraction transmitted (the transmissivity, τ) sum to unity:

$$\alpha + \gamma + \tau = 1. \tag{19.1}$$

Clearly, and as Equation (19.1) shows, a body cannot simultaneously have a high absorptivity, a high reflectivity and a high transmissivity at a particular frequency. A *black body* is defined as one that will absorb all incident radiation, i.e. one for which $\alpha = 1$. Such a body is also the best possible emitter of radiation[1]. For a black body, the power

---

[1] If it were not it would absorb more and more radiation and its temperature would rise uncontrollably.

lost by it via radiation will be given by:

$$P = A\sigma(T_1^4 - T_2^4), \tag{19.2}$$

where: σ, the Stefan-Boltzmann constant, equals 5.67×10^{-8} W m^{-2} K^{-4}; T_1 is the temperature of the body (in kelvin); T_2 is the temperature of the surroundings (again in kelvin), and A the surface area of the body.

Not all bodies can be considered black. The emissivity, ε, of a body is defined as the ratio of energy emitted by the body to that emitted by a black body at the same temperature. ε is frequency dependent, but under many circumstances it can be considered constant over a wide range of frequencies and equal to α (a body for which this is true is termed *grey*). For a non-black body, Equation (19.2) can be re-written as:

$$P = \varepsilon A\sigma \left(T_1^4 - T_2^4\right). \tag{19.3}$$

Equations (19.2) and (19.3) assume that little of the radiation emitted is reflected back from the surroundings. This will usually be true if the surroundings are large compared with the object emitting the radiation.

Problem 19.2 How much energy would a brick ($\varepsilon = 0.93$) at 30 °C lose each day via radiation if its surroundings were at (a) 15 °C and (b) −10 °C? (Assume the brick is constantly heated by some unknown mechanism to make up this loss.)

Radiation losses are usually controlled by a combination of adopting low emissivity materials, or coatings, and placing insulation between a heated object and its surroundings. This reduces the surface temperature of the object and thereby its propensity to emit. If we have a body at 100 °C with an emissivity of 0.8 in surroundings at 10 °C, then reducing its emissivity to 0.6 will reduce radiative losses to:

$$\frac{0.6\,A\sigma\,(373^4 - 283^4)}{0.8\,A\sigma\,(373^4 - 283^4)} = 0.75 \text{ or } 75 \text{ per cent of their original value.}$$

Whereas, reducing the surface temperature of the object to 20 °C by applying an insulative layer will reduce the radiative losses to:

$$\frac{0.8\,A\sigma\,(293^4 - 283^4)}{0.8\,A\sigma\,(373^4 - 283^4)} = 0.074 \text{ or } 7.4 \text{ per cent of their original value.}$$

Therefore, in general, we are better off trying to reduce surface temperatures than tackling the radiative emissions of heated surfaces. The above calculation assumes that the area and the emissivity of the body change little with the application of the insulation. As we have seen, changes in ε are not that significant, the same is true under most circumstances for changes in A caused by the addition of insulation. Although it is just about possible for the application of a small amount of poor insulation to a thin pipe to actually increase its losses because of the relatively large increase in surface area.

Reducing ε can be useful when insulation would be difficult to apply without altering the functioning of an object, window glass being an example. A low emissivity coating is now standard for window glass in double glazed units in much of the world.

Conduction

Heat is transferred in solids by conduction as vibrational internal energy transported through the structure of the solid. The greater the difference in temperature between adjacent parts of the solid the greater the rate of heat (or internal energy) transfer, Q. Experimentally, we can show that the heat transfer is directly proportional to the temperature difference δt over the infinitesimal distance δx:

$$Q = -k\frac{\delta t}{\delta x}. \tag{19.4}$$

k, the constant of proportionality, is called the *thermal conductivity* of the solid and has the units $W\ m^{-1}\ K^{-1}$. For most solids, k is constant with temperature. For a wall of thickness l, integrating Equation (19.4) gives:

$$Q = -k\frac{(t_1 - t_2)}{l},$$

where t_1 and t_2 are the internal and external temperatures, respectively. For a real structure, such as a wall of known thickness, d, it is convenient to replace the thermal conductivity, k, by the thermal resistance R of the structure (per m^2 of area):

$$R = \frac{d}{k}(m^2\ K/W),$$

or the thermal transmittance, U:

$$U = \frac{1}{R}(Wm^{-2}K^{-1}).^2 \tag{19.5}$$

The thermal transmittance of a structure is often simply termed the *U-value*. The U-value gives the energy loss per m^2 of surface area of a structure and is often given with the engineering details of a material and is therefore easy to apply to real problems. For example, if a wall has a U-value of $0.1\ Wm^{-2}K^{-1}$, and if one side of the wall is at $20\ °C$

---

[2] At this stage we are ignoring surface resistances, see Equation (19.8).

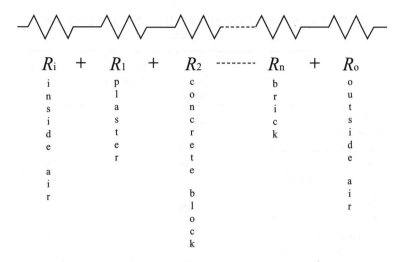

Figure 19.5 Combining thermal resistances in a composite construction

and the other at 5 °C then 4 m^2 of wall will conduct heat at a rate of:

$$0.1 \times (20 - 5) \times 4 = 6 \text{ W}.$$

For a composite construction, for example a wall, the overall U-value can be found by simply adding the individual thermal resistances ($R_1 \ldots R_n$) of the various materials:

$$U = \frac{1}{R_i + R_1 + R_2 + \cdots + R_n + R_o} \tag{19.6}$$

where R_i and R_o are the internal and external *surface resistances* of the structure and attempt to account for conduction and radiation losses. These are discussed in more detail in the next section. As Figure 19.5 illustrates, thermal resistances are analogous to electrical ones. Table 19.2 gives the thermal conductivity of various materials: from which we see that these cover two orders of magnitude – suggesting that the correct choice is very important.

Heat loss through floors in contact with the ground is slightly more complex. Most of the ground floor of a building will only lose heat to the underlying earth – the temperature of which will be modified by this heat loss and seasonal variations in soil temperature. The perimeter of the floor though will also lose heat directly to the air. If the exposed perimeter of the building is P_f and the area of the floor is A_f; d_w is the thickness of the surrounding wall (m), λ_g is the conductivity of the ground and R_f is the thermal resistance of the floor, then, using

$$L_f = \frac{A_f}{0.5 P_f} \quad \text{and} \quad d_{ef} = d_w + \lambda_g(R_i + R_f + R_o),$$

Table 19.2 Thermal conductivity of various materials [data from CIB86, p3.22]

Material	Density (kg/m^3)	Thermal conductivity (W/m K)	Specific heat capacity (J/kg K)
Cast dense concrete	2100	1.4	840
Brickwork	1700	0.84	800
Concrete block (medium weight)	1400	0.51	1000
Plaster board	950	0.16	840
Plaster (light weight)	600	0.16	1000
Roof tiles	1900	0.84	800
Felt on roof	1700	0.5	1000
Timber floor	650	0.14	1200
Concrete floor	2000	1.13	1000
Glassfibre quilt (insulation)	12	0.0404	840
Glassfibre wall-board (insulation)	25	0.035	1000
Glass	2500	1.5	670

the *effective U-value* will be:

$$U = \frac{2\lambda_g}{\pi L_f + d_{ef}} \log_e \left[\left(\frac{\pi L_f}{d_{ef}} \right) - 1 \right] \qquad \text{if } d_{ef} < L_f \qquad (19.7)$$

and

$$U = \frac{1}{0.457 L_f + d_{ef}} \qquad \text{if } d_{ef} > L_f \text{ (i.e. very small rooms)}$$

Likely values for ground conductivity, λ_g, are 1.5 Wm^{-1}K^{-1} for clay and silt (the most common possibility), 2.0 for gravel and 3.5 for solid rock.

Convection

A solid in contact with a fluid (for example, air or water) will lose heat into the fluid by conduction, but the warmed fluid may then have the opportunity to move away from the solid, taking this heat with it. Fresh colder fluid will then move into contact with the solid. As Equation (19.4) shows, the rate of this conduction increases with the temperature difference. The constant replacement of warm fluid by cooler means that the thermal bottle-neck that constrains the rate of heat loss by conduction within solids is overridden and heat loss by convection can be rapid. In general, the density of a fluid reduces with increasing temperature. If the warmed fluid moves of its own accord because of changes in density and the influence of gravity, then such movements

Table 19.3 Radiative heat transfer coefficient, h_r
[data from CIB86, p3.6]

Temperature of surface (°C)	$h_r(\mathrm{W/m^2\ K})$
−10	4.1
0	4.6
10	5.1
20	5.7

are termed *natural convection*. When the fluid is forced to move, for example by a fan or pump the term *forced convection* is used.

How much heat is lost by a structure from convection is a complex question and depends, amongst other things, on the density, thermal capacity, viscosity of the fluid, force applied and whether the flow is laminar or turbulent. For a detailed study of convection, the reader is directed to reference [EAS90]. As was suggested in the last section, for modest conditions, particularly those typically found at the surface fabric of buildings, we can summarize many of these complexities into a surface resistance, R_s:

$$R_s = \frac{1}{E h_r + h_c} \tag{19.8}$$

where E is the emissivity factor, h_r the radiative heat transfer coefficient and h_c the convective heat transfer coefficient. The emissivity factor of a surface depends on the shape of the radiating surface, its emissivity and the emissivity and shape of the surfaces towards which it radiates. Tables 19.3 and 19.4 give the typical values for h_r and h_c.

The values seen in Tables 19.3 and 19.4 can be combined via Equation (19.8) to give typical surface resistances. For the built environment it is convenient to do this separately for the internal and external surfaces primarily because the air speed inside the building will be close to zero, whereas external surfaces will experience much greater

Table 19.4 Convective heat transfer coefficient, h_c. Air speed at the surface is assumed to be not greater than 0.1 m/s [data from CIB86, p3.6]

Heat flow direction	$h_c(\mathrm{W/m^2\ K})$
Horizontal	3.0
Upward	4.3
Downward	1.5
Average	3.0

Table 19.5 Inside surface resistances, R_i. Air speed at the surface is assumed to be not greater than 0.1 m/s; surface temperature assumed to be 20 °C; high emissivity surfaces assumed [data from CIB86, p3.7]

Building element	Heat flow	Surface resistance (m^2K/W)
Walls	Horizontal	0.12
Ceilings and floors	Upward	0.10
Ceilings and floors	Downward	0.14

wind speeds. By convention, the internal air speed is assumed to be 0.1 m/s, and the external wind speed 3 m/s. Tables 19.5 and 19.6 give typical values.[3]

Given the inside and outside surface resistances, we can apply the resistive chain shown in Figure 19.5, i.e. Equation (19.5), and calculate the U-value of various constructions – Table 19.7.

Problem 19.3   Estimate the size (in watts) of a heating system required to keep an isolated room at 18 °C if the external temperature is zero degrees centigrade. Assume room dimensions are 3m by 4m and 3m high; the roof is tiled and includes 100 mm of glass fibre quilt; and the walls are of cavity brick with 75 mm of mineral wool in the cavity. Table 19.8 shows how to lay out such a calculation. (Ignore losses through the floor.)

As an example of just how useful insulation can be in reducing heat loss and why it should always be considered, one novel application uses hollow polypropylene balls as floats on top of heated liquids to reduce convective and evaporative losses in engineering processes. Such evaporation losses count as energy losses in that any lost heated liquid will have to be replaced by the heating of new cool liquid.

Table 19.6 Outside surface resistance, R_o. Air speed at the surface assumed to 3m/s, with high emissivity surfaces [data from CIB86, p3.7]

Building element	Emissivity of surface	Surface resistance (m^2K/W)
Wall	High	0.06
	Low	0.07
Roof	High	0.04
	Low	0.05

---

[3] See reference [CIB86 A3-7] for how to estimate the thermal resistance of air-filled cavities in detail. A reasonable estimate though is to assume $R = 0.18$ m^2K/W for horizontal or upward directions for a gap of more than 25 mm.

Table 19.7 U-values of common constructions. See the book's website, or search the internet, for the values of other materials and constructions

Material	U-value (Wm^{-2} K^{-1})
Walls	
105 mm brickwork, unplastered	3.3
(A) 105 mm brickwork, 25 mm air gap, 105 mm brickwork, 13 mm lightweight plaster	1.4
As (A) but with 25 mm of mineral wool in cavity	0.73
As (A) but with 75 mm of mineral wool in cavity	0.37
Roofs	
(B) 10 mm tile, loft space, 10 mm plasterboard ceiling	2.6
As (B) but with 25 mm glass fibre quilt above plasterboard	0.99
As (B) but with 100 mm glass fibre quilt above plasterboard	0.35
Floors	
Estimate using Equation (19.7)	
Windows (see Table 20.3 for additional values)	
Single glazed	5.9
Double glazed, argon filled, low emissivity	1.4

19.3 Energy recovery

As mentioned earlier, in many situations it is not possible to reduce heat losses from fluids by simply restricting the loss of fluid from the system. Often the fluid contains waste materials, such as stale air or other contaminates and we therefore only wish to retain the internal energy of the fluid and not the fluid itself. Usually, we wish to remove the heat from the waste fluid (liquid or gaseous) and use it to warm an incoming fluid. This is also what we are trying to achieve when, for example, we transfer heat between the possibly contaminated water that passes through the core of a nuclear reactor, and that which passes through the steam turbines, or between the turbine water and the final cooling water provided by a neighbouring river or ocean. A device that achieves this is called a *recuperative heat exchanger*. There are numerous possible designs, the choice depending on the state of the fluids (liquid or gaseous), the temperature and pressures involved and the chemical or corrosive nature of the substances. However the basic principles are simple: the two fluids need to be kept apart at all times (to avoid contamination) yet share a common interface with as large an area as possible so that as much heat can be transferred as possible. Figure 19.6 shows various types.

Such exchangers are used to pre-heat furnaces and boiler gases prior to combustion (Figure 19.6a), or to raise steam (Figure 19.6b). The plate-fin exchanger (Figure 19.6c) can be used for air-to air, air-to-liquid, and liquid-to-liquid heat recovery and is constructed from a series of corrugated plates sandwiched together.

If the two fluids cannot be brought close to one another, either because the design of the factory or process means that they are separated by a large distance, or because failure

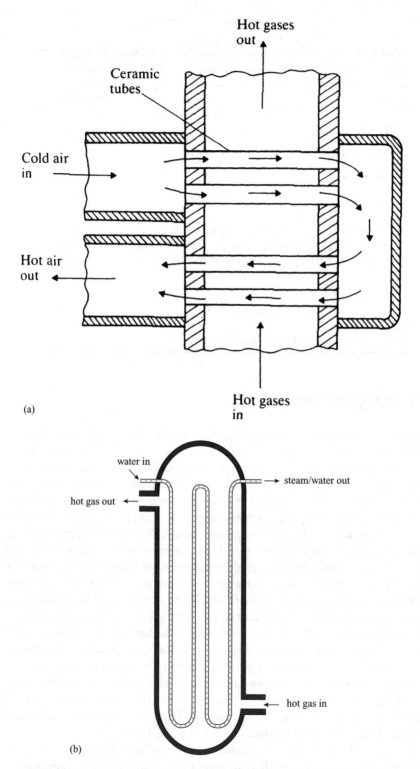

Figure 19.6 Various types of recuperative heat exchanger [EAS90, p166–167]

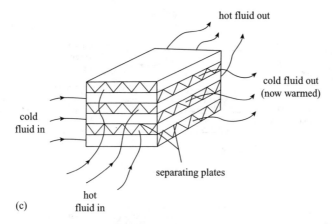

hot fluid out

cold fluid out
(now warmed)

cold
fluid in

separating plates

(c)   hot
fluid in

Figure 19.6   (*Continued*)

of the interface and the resulting contamination would have serious consequences, or even be explosive, then a 'run-around' system can be used (Figure 19.7). Here two exchanges are used together with a third, intermediate fluid. The first exchanger is heated by the waste fluid; it passes the recovered heat to the intermediate fluid, which is pumped to the second exchanger where it surrenders its heat to the incoming cold fluid.

Another form of heat recovery device is the *regenerative heat exchanger*. Such exchangers are only suitable for similar fluids where a small degree of contamination is not a serious problem, such as recovery of heat from stale air in buildings and the pre-heating of the incoming air.

Two forms of regenerative exchangers are common. Central to both is the idea of using the waste heat to warm a matrix of material, then the placing of the matrix in the flow of the cooler incoming fluid. This process is repeated cyclically. In stationary systems the matrix is stationary and the fluid path cycled (Figure 19.8), in rotary systems the matrix is mounted on a wheel, which is rotated slowly through the streams of hot and cold fluid (Figure 19.9).

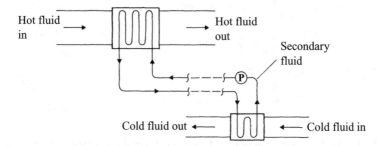

Hot fluid
in

Hot fluid
out

Secondary
fluid

P

Cold fluid out

Cold fluid in

Figure 19.7   Run-around system

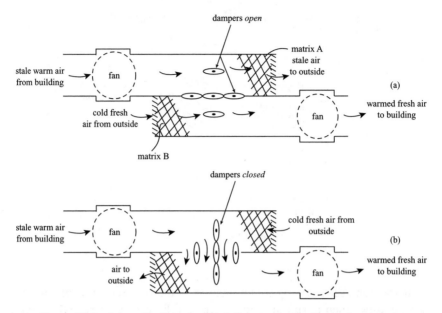

Figure 19.8 Schematic diagram of a stationary regenerative heat exchanger

19.4 Energy efficiency in buildings

Over one third of carbon emissions are associated with energy use in buildings and Equation (19.5) indicates that heat loss through the fabric of a building is proportional to the temperature difference between the internal and external environment. If we

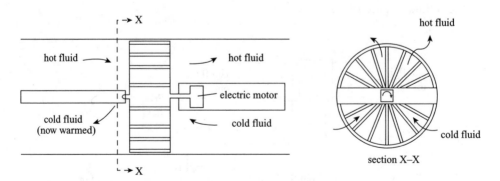

Figure 19.9 Rotary regenerative heat exchanger (or 'thermal wheel') for building ventilation

assume the external temperature is 5 °C on average during the heating season and the internal temperature is 20 °C then we are trying to maintain a temperature difference of $20 - 5 = 15$ °C. Therefore if we think a satisfactory internal environment can be provided at 19 °C then we will reduce the amount of energy required to heat the building by 1/15, or 6.7 per cent. The question now is how many people would notice this small change in their environment? I would suggest few.

The above example illustrates why energy conservation is potentially so important. If taken seriously by all of us it could offer reductions in carbon emissions at least as great as any individual alternative energy technology, mostly at very low cost and with little impact on other environmental systems. The near invisibility of energy conservation schemes is one of their greatest assets. Build a wind farm or a new hydropower scheme and the visual impact will cause those not wedded to the need to avert climate change to voice their concerns. Reduce energy use in a selection of buildings by a few per cent and it is unlikely anyone will even notice. Unfortunately the corollary of this is that it is very difficult to excite politicians, the press or public with the idea of energy conservation. This has had a negative effect on funding for energy efficiency projects when compared to the sums of money spent on other technologies such as nuclear power. Energy efficiency in buildings has, in addition, the advantage that we all live in buildings and therefore have the opportunity to reduce their carbon emissions immediately: it is harder for the reader to have an impact on national generating strategy.

Energy efficiency in use is not the only consideration. The energy used to manufacture the materials used in the building and to transport them to site also needs to be minimized in order to create a sustainable design. Traditionally only local materials were available and these were (and still are) transported no more than a few hundred metres (Figure 19.10). Most of today's designs no longer keep to this tradition with modern, high embodied energy materials such as aluminium, glass and plastic being extensively used. This need not be the case, with many architectural practices and enthusiasts constructing buildings from rammed earth, straw, cob, local stone and wood from sustainable forests.

Another reason heat loss from buildings is important is that various alternative energy generating technologies can be applied to buildings to greatly reduce the carbon emissions from their use, even to the point where they become carbon neutral, i.e. either needing no fossil fuels for energy input, or exporting the equivalent amount of non-carbon energy as they use. In either case this is much easier to achieve if the building base-line energy use has been minimized. In the next chapter we will see an example of this where solar radiation entering through windows can be used to keep buildings at comfortable temperatures during winter – but only if fabric and other losses are minimized.

Fabric losses

In section 19.2 we introduced U-values and saw that the lower the U-value of a construction, the less heat was lost through the fabric of the building. We can get a *very* rough

Figure 19.10 Traditional low embodied energy buildings that use local materials: (a) and (b) Pakistan, (c) Iceland (note the grass roof and the recycled wood walls). (Photos by author)

Table 19.8 U-values and heat loss calculation for two fictitious building standards. Floor U-values have been estimated using Equation (19.7), surface resistances from Table 19.3 and 19.4 ($R_i = 0.14$, $R_o = 0$), and conductivities from Table 19.2

Element	Area (m)	U-value ($Wm^{-2}K^{-1}$)		Area × U-value (W/K)	
		Building Standard 1	Building Standard 2	Building Standard 1	Building Standard 2
Walls	128	3.3	0.37	422	47
Roof	100	2.6	0.35	260	35
Floor	100	0.66	0.17	66	17
Windows	32	5.9	1.4	189	45
Sum (area × U-value), i.e. heat loss (W) per K.				937	144
Heat loss assuming 10 K temp. difference (W)				9372	1442
Seasonal heat loss (GJ)				148	23
Heat loss over whole building stock (PJ)				**2220**	**345**

idea of how important suppressing fabric losses might be to national energy policy, by estimating the heat loss from all the houses in a country with two fictitious building standards[4], then comparing this heat loss with the primary energy consumption of the country in question. We will use the UK as the example: population = 60 million, persons per household = 4 (an over-estimate), therefore number of households = 15 million (an under-estimate); assume all households are detached and 10 m by 10 m with walls 6 m high, assume 20 per cent of wall areas are glazed. Building Standard 1: single skin brick walls, single glazed windows, plasterboard roof and tiled with no insulation, 100 mm concrete slab floor with no insulation. Building Standard 2: brick cavity walls with 75 mm of insulation, double glazed low emissivity argon filled windows; tiled roof with 100 mm of insulation, 100 mm concrete floor with 75 mm of glassfibre wallboard.

Table 19.8 lists the U-values (taken from Table 19.7) of the various elements, together with the areas and the heat losses per degree centigrade of temperature difference between the inside of the building and the external environment. If we now assume the average internal-external temperature difference during the heating season (which lasts 6 months) is 10 °C, then we see that each building built to the first standard loses $10 \times 937 = 9370$ W, i.e. 148 GJ over six months, through the fabric alone. Over the whole building stock this equates to over 2 EJ, 85 per cent of which is seen to be saved if all buildings used the second standard. This would reduce national energy consumption and carbon emissions by around 20 per cent – without the introduction of a single new technology. In reality, housing is a lot more diverse in the UK, with many smaller

---

[4] In the case of the UK, The Building Regulations control various aspects of the design and performance of new buildings and cover subjects from insulation to foundation depth. For the current UK Building Regulations see reference [ODPM05]. Many other countries produce similar guidance.

properties; the mean internal-external temperature difference lower and most properties now have higher insulation levels than those implied by the first building standard. However, it is equally possibly to design buildings with higher levels of insulation than implied by the second building standard and the calculation clearly indicates the scale of the saving made possible by reducing energy losses from buildings. The only limit being the point where the embodied energy of the insulation equals the total energy saved during the lifetime of the insulation.

From this example we learn that most governments could greatly reduce national emissions of greenhouse gases by strongly encouraging the refurbishment of the current building stock to higher standards and the introduction of even higher standards for new buildings. Although this would cost money in the short term, much of the expenditure would be recouped from reduced energy costs. Such a policy would also have almost no environmental or visual impact – a truly silent renewable. This author finds it hard to understand why this approach has not been more aggressively adopted, and can only conclude that this is due to a deep reluctance within Government to believe climate change is a serious threat to the World. This position is in turn most probably a reflection of the public's, i.e. the voters', opinions.

Ventilation losses

Losses through ventilation are effectively convection losses and need to be minimized during the heating season. Some ingress of fresh air is necessary to ensure a healthy atmosphere; the amount will depend on the number of people in the room and whether any hazardous emissions are being produced in the space.

Apart from being a product of burning fossil fuels, carbon dioxide is produced by respiration at a rate of about 23 litres per person per hour. Guidance in [BB87] suggests that CO_2 levels in a classroom, for example, should be kept below 1000 ppm (parts per million). We can use these two pieces of data to get some idea of the necessary ventilation rate in a space and the heat loss to which this equates. For any room we have the following mass balance equation:

$$V\frac{dC_{(t)}}{dt} = G + QC_{ex} - QC_{(t)} \qquad (19.9)$$

where
$C_{(t)}$ = internal concentration of carbon dioxide at time t (ppm),
C_{ex} = external concentration of carbon dioxide (assumed constant at 360 ppm),
G = generation rate of carbon dioxide in the space (6.4 cm³/s per person),
Q = internal–external exchange rate (m³/s),
V = room volume (m³),
t = time (s).

At equilibrium the internal concentration will not be changing and therefore $dC_{(t)}/dt = 0$ and:

$$0 = G + QC_{ex} - QC_{(t)}, \text{ or } Q = \frac{G}{C_{(t)} - C_{ex}}$$

$$\Rightarrow Q = \frac{6.4}{1000 - 360} = 0.01 \text{ m}^3/\text{s} \tag{19.10}$$

If the CO_2 concentration is to be maintained below 1000 ppm, Equation (19.10) implies that we will need a ventilation of 36 000 litres per hour per person. This equates to a heat loss of $0.01\rho C_p$ per degree centigrade of temperature difference between the inside and outside of the building (where ρ is the density of air and C_p the specific heat capacity of air), i.e. $0.01 \times 1 \times 1000$ or 10 W/°C. However the metabolic (i.e. heat) output of a human is around 100 W. So we see that as long as we can keep ventilation rates low and control them for the number of people in a space, ventilation need not lead to substantial heat loss. Unfortunately, in many buildings ventilation rates are very much higher than this, often because of ill-fitting windows and doors. These losses are often so high, and the solution so easy and cheap to apply – draught excluders on all external doors and windows, and keeping internal doors closed between heated and non-heated spaces – that it should usually be the first act in trying to reduce energy losses from a building.

Problem 19.4   For the building and temperatures presented in Problem 19.3, estimate the heat loss from ventilation assuming ventilation rates of 2 air changes an hour and 0.1 air changes an hour. Compare these losses to the fabric loss in Problem 19.3. (The values of 2 and 0.1 are typical of a modern building during occupied and unoccupied periods respectively.)

Air conditioning

At this point it is worth giving special mention to air conditioning. Although it might seem obvious that there is a large market for air conditioning within equatorial regions and other locations where summertime temperatures are very high for sustained periods, there is also a demand for such systems in more temperate regions such as the UK. One of the main reasons for this is the large incidental gain from electrical equipment (including lighting) found in modern offices. The other is the desire for impressive, largely glazed facades, which promote excessive solar gain. In such temperate regions there is rarely much desire for air conditioning in domestic dwellings, where electrical and metabolic gains are small relative to the volume of the building. But in commercial properties, occupational and equipment densities tend to be much higher. Air conditioning should still be avoided wherever possible because it uses electrical energy created with a low efficiency and therefore high carbon emission to remove waste heat. This heat

Table 19.9 Efficiency of various lighting technologies [EAS90, p281]

Type	Illuminance (lm)	Power (W)	Efficacy (lm/W)	Life expectancy (h)
Tungsten GLS	1200	100	12	1 000
Tungsten GLS	20 000	1000	20	2 000
Tungsten halogen	50 000	2000	25	4 000
Compact fluorescent	1 200	28	43	8 000
fluorescent tube	4 500	70	64	10 000
High frequency tube	5 500	62	80	10 000
Low pressure sodiun	30 000	210	143	12 000
Metal halide	160 000	2000	80	10 000
High pressure sodium	25 000	280	90	12 000

is often from electrically powered items such as lights or computers, i.e. it also has a high carbon emission factor. If such heat can be prevented in the first place (by switching off equipment when not in use or by purchasing more energy-efficient equipment) carbon emissions will be reduced on two fronts.

Clearly, the solution is to make sure that no unnecessary energy consumption is taking place and that solar gains are controlled. Although this is true in any building, the above-mentioned double inefficiency increases the importance of this in buildings with air conditioning. Unfortunately, the installation of air conditioning can reduce the pressure on individuals to save energy as such acts will have no effect on internal air temperature. It also encourages poor building design.

Lighting

Traditional filament light bulbs are highly inefficient, consuming 30 times more electrical energy than the light they produce. Like most energy use we are only interested in what energy consumption can do for us, not in the amount consumed, therefore we have the option of moving to more efficient lighting technologies. The recommended illuminance for UK offices is 500 lux[5] (or 500 lumen per square metre).

Table 19.9 gives the illuminance and efficiency of various lighting technologies. Note how the efficiency depends not only on the technology but also on the rated power of the device. (Figure 4.12 puts these values into historical perspective).

---

[5] 1 lux is defined as an illuminance of 1 lumen per square metre, and 1 lumen as the luminous flux emitted within unit solid angle by a point source with a luminance of 1 candela. The candela is defined as the luminous intensity, in the perpendicular direction, of a surface of $1/600\,000$ m^2 of a black body at the temperature of solidifying platinum under a pressure of 101.325 N/m^2 [HAL88]. Such a definitional chain makes the application and use of British Thermal Units or horsepower seem almost trivial.

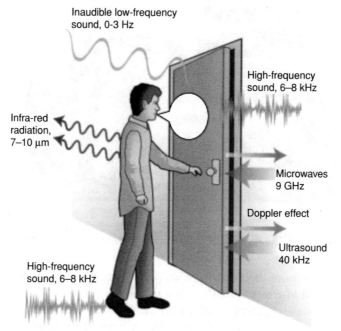

Inaudible low-frequency
sound, 0-3 Hz

High-frequency
sound, 6–8 kHz

Infra-red
radiation,
7–10 μm

Microwaves
9 GHz

Doppler effect

Ultrasound
40 kHz

High-frequency
sound, 6–8 kHz

Reproduced with permission from CADDETT, Technical Brochure, 431, IEA CADDET programme, http://www.caddet.org.

Figure 19.11 Possible lighting control signals [CAD431]

Problem 19.5 Estimate the annual CO_2 saving to be made by replacing half the domestic light bulbs in Singapore (population 4.4 million [CIA04]) with compact fluorescent ones.

Another way of saving energy from lighting systems is by turning the lights off! As you walk around your neighbourhood have a look at how often you see the lights on when they need not be. There have been many suggestions as to how to solve this problem, from education to timers which simply turn the lights off at preset times. Figure 19.11 shows some of the signals that could be used to turn lights on; the lack of which could therefore be seen as justification for turning them off.

Example application: green building design

This example application (taken from [CAD437]) illustrates that energy in use is not the only consideration when designing a sustainable building: we also need to minimize the energy used to manufacture and transport the construction materials, i.e. the *embodied energy*. The Mountain Equipment Co-op building in Ottawa, Canada was completed in June 2000 (Figure 19.12). The project's waste reduction goals included reusing at least 75 per cent of the existing structure found on site, using at least 50 per cent recycled and 10 per cent salvaged building materials, and creating a low energy-in-use design.

Reproduced courtesy of CADDETT, Technical Brochure, 437, IEA CADDET programme,
http://www.caddet.org.

Figure 19.12   The Mountain Equipment Co-op in Ottawa, Canada [CAD437]

These aims were more than met, with 86 per cent of the existing building being salvaged or recycled to create a 2484 m^2 structure with rainwater collection from the roof. The ground floor is timber framed (rather than concrete based[6]) using 30 × 30 cm reclaimed Douglas fir. The car park features light coloured gravel and trees to reduce solar gain. Vines provide south-facing façades with shade during the summer; these then die back during winter allowing daylight into the building.

The basic wall structure is shown in Figure 19.13. Air infiltration was kept low and 23 cm of cellulose insulation was laid between the vertical I-beams, with a similar depth of mineral fibre being used in the roof. Windows are either triple- or double-glazed. The building is heated using a boiler with a full load efficiency of 87.5 per cent. Annual energy use is 827 GJ of natural gas and 988 GJ of electricity. If these figures are compared with good modern design in the country[7], then energy use is only 44 per cent of that expected. In addition, the use of salvaged and recycled materials is believed to have saved a total of 2200 MWh. In terms of carbon dioxide, this represents an annual saving of 72 t from the low energy-in-use design and 617 t in total from using low embodied energy/reclaimed materials.

19.5 Student exercises

when the last tree has died and the last river has been poisoned and the last fish has been caught will we realize that we cannot eat money.

–Cree Indian saying

---

[6] The manufacture of concrete results in carbon dioxide emissions in excess of that from the energy used in its production. This is because one of the major ingredients of concrete is lime used to manufacture the cement. Lime, or calcium oxide (CaO), is created by heating calcium carbonate ($CaCO_3$), which is derived from limestone, chalk and other calcium-rich materials and releases CO_2 via: $CaCO_3 + Heat \rightarrow CaO + CO_2$.
[7] An MNECB/CBIP reference building of the same floor area.

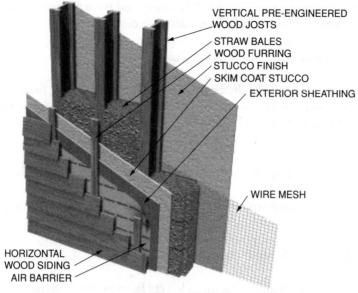

VERTICAL PRE-ENGINEERED
WOOD JOSTS
STRAW BALES
WOOD FURRING
STUCCO FINISH
SKIM COAT STUCCO
EXTERIOR SHEATHING

WIRE MESH

HORIZONTAL
WOOD SIDING
AIR BARRIER

Reproduced courtesy of CADDETT, Technical Brochure, 437, IEA CADDET programme, http://www.caddet.org.

Figure 19.13 Basic wall construction of the shop [CAD437]

1. Why is it particularly important to ensure that there is no unnecessary energy use in air-conditioned buildings?

2. What are the three forms of heat loss? Give examples of their reduction within buildings.

3. When is cogeneration likely to be successful?

4. Discuss the various types of heat exchanger and when they might be used.

5. Estimate the likely primary energy and carbon dioxide saving in your country if all the tungsten light bulbs found in domestic properties were replaced by compact fluorescents (remember to include the efficiency of electricity generation).

6. What is embodied energy?

7. A classroom of area 150 m^2 is illuminated by 20 standard incandescent lamps rated at 100 W with an illuminance of 1200 lumens and an efficacy of 12 lm/W. Each lamp costs 50 p and is switched on for 8 hours/day, 5 days/week, 40 weeks/year [adapted from SHE97].

   a. If electricity costs 7 p/kWh calculate the annual running costs.

b. If the life expectancy of a lamp is 1000 hours and the replacement labour cost is £5 per item calculate the annual replacement cost.

c. Calculate the total annual electricity costs.

8. An alternative plan to light the classroom in question 7 uses 10 fluorescent tubes, rated at 70 W with an illuminance of 4500 lumens and an efficacy of 64 lm/W. Each fluorescent tube costs £2 and has a life expectancy of 10 000 hours [adapted from SHE97].

a. Calculate the annual running costs if the electricity tariff is 7 p/kWh.

b. Calculate the annual replacement cost if the replacement labour charge remains at £5 per item.

c. Calculate the total annual electricity costs.

9. Compare the performance of the incandescent lamp system with the corresponding fluorescent lamp system [adapted from SHE97].

a. What would be the annual cost saving in moving to the fluorescent system?

b. The cost of the modified installation would be £660. Neglecting the effects of inflation and depreciation and ignoring any scrap value of the incandescent system, what would be the payback period of the modification?

c. Compare the illuminance of the classroom using the two systems.

20

Solar power

"Keep your eyes on the sun and you will not see the shadows."

–Aboriginal Australian proverb

Every 10 minutes the surface of the earth receives enough energy from the sun to provide the primary energy needs of humankind for a whole year. So why isn't solar power at the centre of current energy production? Part of the answer is that the 174 PW[1] we receive from the sun is spread over the surface of the planet – an area of 510 million km². So, although from the above numbers only 0.007 per cent of the surface area of the planet would be required to satiate our demand for energy, this is still equivalent to a land area of 38 000 km² or of 380 000 km² if we assume any technology might have an efficiency of only 10 per cent. This is slightly greater than the area of Malaysia, which illustrates the dilution of the resource.

Fossil fuels are a form of solar power. The original plants and animals that formed today's oil, coal and gas reserves grew because of sunlight. The secret of fossil fuels is that the carbon has been concentrated greatly over time. And this is why they are so useful. Trees are also forms of concentrated solar power (80 years' worth for a mature pine tree), as is all biomass, the wind, waves and rivers. Unfortunately even though such systems are concentraters, their original efficiency of production is extremely low – often less than one per cent. This means that vast areas of collection are required to provide for the levels of energy use to which we have grown accustomed. The carbon fuels now being expended took millions of years to be laid down and we are currently using them at a rate of at least 100 000 times the rate they were formed, and even faster compared to the rate they are now amassing in the much smaller areas where formation

---

[1] Roughly 30 per cent of which is directly reflected back into space by the atmosphere and surface.

Energy and Climate Change   David A. Coley
© 2008 John Wiley & Sons, Ltd

is still occurring[2]. So although ultimately fossil fuels are a renewable energy source, the problem is one of the difference in the rate they are being produced to the rate we are using them. Clearly we need to find systems where the rates of production of the energy resources are approximately equal to our required rate of use. Only then do we have true sustainability.

The first form of solar power we will consider is probably the simplest form of renewable energy – simply let the heat of the sun warm a large mass of material and then control the loss of heat as best we can. A conservatory on a sunny day is an obvious example. Such an approach is commonly termed *passive* solar heating; there is no working fluid except the air in the room, no pump (although a small circulating fan can be used) and the solar collector is part of the structural fabric of the building. Such systems are distinct from *active* solar technologies where a separate solar collector is used to raise a working fluid (often water) to much higher temperatures; an example being domestic solar water heating from roof mounted panels. The successful use of solar power in the built environment usually requires the minimization of unnecessary heat losses from the building. This topic was covered in detail in the previous chapter, but such is the importance of this point that we will restate a little of the material already covered for those readers visiting the text in a different order.

20.1 Passive solar heating

The basic features of a solar heated home include large south-facing (north facing in the Southern hemisphere) windows allowing sunlight to enter the room. This light is absorbed by the various surfaces in the room and then re-radiated as longer wavelength infrared radiation. Two, not so obvious, features are essential: the building must be constructed with a large amount of material that has a high thermal capacity, to store the heat; and the design must lose very little heat through the fabric. Low heat loss is essential because, compared to the amount of heat produced by a domestic boiler, very little energy will be received from the sun during the all-important winter months. High thermal mass is required because the sun will not shine every day or at night, and because we need the house to maintain an even and comfortable temperature. Without a high thermal mass, the space will be too hot on sunny days and too cold on overcast ones.

History

Space heating using solar gains has a long history. Pliny the Younger used windows built of thin sheets of mica to help warm a room in northern Italy around 100 AD, thus saving on firewood. In addition, many Roman bathhouses used large south-facing openings to access this free source of heat. The eighteenth century saw the first

---

[2] Such as the Great Dismal Swamp of North Carolina and Virginia, the Okefenokee Swamp in Georgia, and the Everglades in Florida [AME05].

Table 20.1 Specific heat capacity of common building materials. Note how the range is much less per m^3

Material	J kg^{-1} K^{-1}	Density, kg/m^3	kJ m^{-3} K^{-1}
Water	4200	1000	4200
Steel	450	7850	3533
Brick	800	1700	1360
Concrete	1000	2000	2000
Wood (hard)	1700 (approx.)	680	1156 (approx.)
Common stone	1000 (approx.)	2000	2000 (approx.)

widespread use of solar design with the introduction of conservatories for growing exotic fruits and as a way of creating a warm space before the general use of central heating. In the 1950s the architect Frank Bridgers designed the world's first commercial office building using solar space and water heating (the Bridgers–Paxton building in Albuquerque, USA).

The correct balance between window size (and orientation), heat loss and thermal mass must be met for the design to work. Although not overly complicated, the calculations required are more complex than those needed for a house heated only by a conventional system where the sole requirement is to balance boiler and radiator size with estimated heat loss. This is possibly one of the reasons architects have been slow at implementing solar design, although many good examples exist.

Basic design features

There is a wide choice of materials to form the heat storing high thermal mass elements of the design. Because the heat will be stored at very low temperatures (approximately the room temperature) large amounts of material are required. These can either be in the form of normal structural elements such as the floor and walls, or specific novel components especially intended to store heat, such as large drums of water.

Table 20.1 compares the specific heat capacity of various materials (i.e. the heat they store per degree centigrade they are raised above the surrounding temperature). Note the relatively high value for hardwoods including oak, and the even higher value for water.

The reduction of heat loss has two components – the minimization of fabric losses and the reduction of losses from unnecessary ventilation. The insulation needed can be made from various materials, including mineral wool and paper. Table 19.7 gives the heat loss per metre squared for a range of constructions, the amounts required will be highly dependent on the typical winter time external air temperature – the colder it is, the greater the depth of insulation required. Clearly there is an energy cost in the creation of any product, including insulation and therefore there is the need to ensure that more energy is saved by fitting the insulation than is used in the manufacture and transportation of the insulation itself. These production and transportation energy costs are termed the *embodied energy* of the material. Different materials have very different embodied energies, as indicated in Table 20.2. This means that, given a specified internal

Table 20.2 The embodied energy of building materials [data from BAI97]

Material	MJ/kg	Material	MJ/kg
Carpet	72.4	Linoleum	116
Cement	7.8	*Paint:*	90.4
Brick	2.5	solvent based	98.1
Tile	2.5	water based	88.5
Concrete:		Plaster, gypsum	4.5
block	0.94	Plaster board	6.1
pre-cast	2.0	Steel, recycled	10.1
Glass	15.9	Steel, virgin, general	32.0
Insulation:		Stone, local	0.79
cellulose	3.3	Straw, baled	0.24
fibreglass	30.3	Timber, softwood	1.6
polyester	53.7	Vinyl flooring	79.1
polystyrene	117	Aggregate, general	0.10
wool (recycled)	14.6	Aluminium, virgin	191
Asphalt (paving)	3.4	Aluminium, recycled	8.1
Bitumen	44.1		

and external temperature, it is possible to calculate the maximum depth of insulation it is sensible to apply. Beyond this depth we are not really improving the sustainability of the design.

As we saw in the last chapter, by calculating the area-corrected[3] average heat loss per degree celsius or *U-value* of a building and then multiplying the result by the average wintertime internal/external temperature difference, one has an estimate of the power needed to heat the building. Ventilation losses need to be added to this, although these can be kept small by using well-sealed windows and doors, thereby ensuring almost no ventilation occurs when the building is unoccupied. (Ventilation is important for health reasons during occupied periods.) In winter, ventilation might typically be one air change an hour (of all the air in the building) when occupied, if it is well controlled. Assuming the windows are well fitting, a value of one tenth of an air change per hour might be met if the windows are closed during unoccupied periods. Multiplying this by the thermal capacity of air (1 kJ/kg) we see that we will typically lose 1 kJ per metre-cubed of internal space per hour[4] during occupied periods. This is 0.3 W – or about 150 W for a typical house. For a poorly managed or maintained building the loss can be 10 times greater. For buildings such as offices and schools with more occupants per cubic metre of enclosed space, greater ventilation rates are needed than this to keep metabolic carbon dioxide levels and body odour at acceptable levels. In the last chapter we saw that such buildings might lose 100 W per person in order to achieve this.

---

[3] i.e. apportioning the correct area to each material.
[4] Because the density of air is 1 kg/m^3.

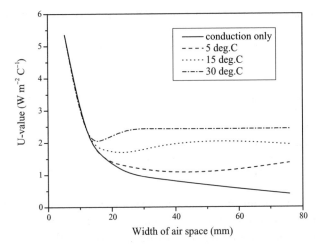

Figure 20.1 Heat loss via conduction + convection through a double glazed window as a function of width of air space between the two panes and the internal/external temperature difference [data from BOY96]

Problem 20.1 Sketch a three bedroom cavity brick house and estimate the energy required to heat it on a winter's day, firstly assuming it has single glazing and no insulation, and then with wall and loft insulation, and high performance double glazing. (Ignore losses through the floor and ventilation losses.) The details of the design are not important and we will adapt the design to make use of solar gains in a later problem.

Alongside being able to store the heat and stop it escaping, we need to ensure that a building gains enough free solar power to warm it. This relies on glass having several special properties. Most obvious amongst these is that it is transparent to visible light. However as previously mentioned glass is also opaque to long wave infrared radiation. Thus light can enter and exit through the window, but as any light that is absorbed by surfaces in the enclosed space will be re-radiated in the infrared, it will be trapped in the room. Even more importantly, but often forgotten, glass forms an airtight seal that is impervious to convection currents in the warm space and to wind from outside. This is the main reason greenhouses become so hot in summer, not just because they trap infrared radiation.

Normal glass allows over 80 per cent of the incident light to pass into the room and stops all but two per cent of the infrared radiation from escaping. However, heat can still escape by conduction through the glass, and once it is warmed by conduction the pane can lose heat from its outer surface by conduction, convection and radiation. Such convection losses can be greatly reduced by using *double glazing* (where a second pane of glass is installed in front of the original pane with a small air gap between). The effectiveness of such double glazing depends on the size of the air gap between the panes: if it is too large, convection currents will be established within the air space; if too small, then conduction losses will be large (see Figure 20.1). Losses can be further

Table 20.3 U-values of various vertical window constructions [data from CIBSE99, p3–28]

Type of glazing	U-value ($Wm^{-2}K^{-1}$)
Single	5.9
Double (6 mm air space)	3.4
Double (12 mm air space)	2.9
Triple (12 mm air space)	2.0
Low E double (12 mm air space), $\varepsilon = 0.1$	1.8
Low E double (12 mm air space), $\varepsilon = 0.1$, argon filled	1.4

reduced by replacing the air in the gap with argon (which due to its higher molecular mass demonstrates reduced convection currents compared with air).

Radiation losses can be minimized by coating the outside of the inner pane with a low emissivity coating. The emissivity of a material describes its propensity to radiate heat as a fraction of the theoretical maximum for any given temperature. Brick and glass have an emissivity of 0.9. Low emissivity glass (termed low-E class) can have an emissivity of only 0.1. Table 20.3 shows how, by reducing all three forms of losses – convection, conduction and radiation – the U-value of a window can be improved to the point where it has a lower (i.e. better) value than that of fibreglass loft insulation of the same thickness. This is a remarkable achievement. However, loft insulation is typically laid to a depth of 15 to 100 centimetres making it a far better insulator in practice. Because of this difference, windows, however good, lose a much higher proportion of the heat from a house then their area would suggest[5]. This means we have to balance carefully the solar heat gain from a window with the heat loss from the same window.

Getting this balance right is critical to passive solar design. Whether a window is a net provider of energy, or a net exporter depends on five considerations:

1. the amount of available sunshine at the location in question (this depends on latitude, average cloud cover and any shading from nearby buildings or trees, or caused by the local topography)

2. the internal and external temperatures

3. the transmittance of the window

4. the U-value of the window

5. the orientation of the window

---

[5] Problem 20.1 showed that the heat lost through the windows in a house using single glazing can be greater that that lost through the walls if they are well insulated.

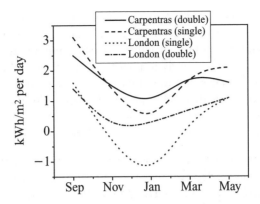

Figure 20.2   Energy benefit of a south-facing window at two locations in the northern hemisphere [data from Boyle, G., *Renewable Energy: Power for a Sustainable Future*, Oxford University Press, 1996, p65]

Figure 20.2 shows the energy balance for a south-facing window of typical glass located in London and the South of France and assuming an internal air temperature of 18 °C is maintained continuously. For the South of France a south-facing window will provide more energy than it loses year round (although care will be needed to stop overheating). In London a single glazed window will lose more heat than it gains from mid-October until March. This is the whole heating season. Energy gained outside of this period is of no use as it does not replace fossil fuel use and can cause overheating of the space. A double-glazed window fares much better and provides heat in all but the coldest months. However, a north-facing window, even if double glazed, will always have a negative energy balance in the northern hemisphere (east and west facing ones will be somewhere in between). However, windows can also be used to reduce the need for artificial lighting and therefore save energy.

Interestingly, the reason for the difference in performance in these two locations is not differences in the external temperatures – London and the South of France have similar average winter temperatures – it is the much larger amount of sunlight in France that is key. This observation also proves to be important to an analysis of solar water heating.

As we have just seen, it is not just how much energy a window provides that is important, but when. For a typical UK house, around 15 per cent of the heat required is provided by solar gains through the windows. Heat from lights, televisions, washing, and metabolic gains from people in the house (body heat) provide a further 25 per cent. The final 60 per cent comes from commercial, typically fossil fuel, sources.

By increasing the level of insulation in a house the annual energy demand can be greatly reduced, thus the percentage provided by free sources such as windows, people and appliances will be greatly increased. Which is more economic – reducing the heat loss of the building or increasing the solar gain with larger windows – will depend again on the location of the building and particularly on the amounts of sunlight. For many parts of the world insulation is likely to be the more cost-effective first step in reducing energy demand.

From the above considerations, it is clear that the use of passive solar energy is complex and not necessarily suitable for re-fitting to an existing building as it relies on the interaction of many parts of the structure, the occupants and the location. Needing to get all of this right simultaneously implies the introduction of a holistic approach to the design process that takes account of the synergy between environment, building and occupant. It also means that the impact that adoption of the approach can have over a short time frame is limited because the proportion of new buildings constructed each year is small compared with the existing building stock.

Problem 20.2   Estimate the amount and value of free energy provided by two adults living in a house 14 hours a day over the six months of the heating season. Assume the adults each have a diet of 2000 kcal a day and that the house is electrically heated at a cost of €0.11 (£0.08, US$0.14) per kilowatt-hour.

Problem 20.3   Using the house you sketched for Problem 20.1, make the largest windows double glazed, face south and double their size. Now estimate how much free heat is likely to be gained by these windows in November. Is this enough free energy to heat either the well insulated or the poorly insulated design? Hint: use the data in Figure 20.2.

Shading

One way to control the propensity of any large area of south-facing glass (or north-facing in the Southern hemisphere) to cause overheating in summer is to provide adequate shading. In winter the sun will be low in the sky and therefore unaffected by a correctly positioned shade. In summer the sun will be higher (almost overhead during the hottest part of the day) and a shade will greatly reduce the amount of direct sunlight striking the window (Figure 20.3). The size and position of the shade will depend on the height of the window and the position of the sun when shading is required.

For more control, the shade can be formed of rotating fins, or for an organic solution, a vine can be grown above the window: this will provide shade during the summer, but not during the winter when the leaves are lost.

Trombe walls

Before we leave the passive solar heating of buildings, one further use of architecture to harness solar power is worth mentioning – the Trombe wall. In a very sunny climate ensuring enough solar gain is not a problem and even during winter a large south-facing window can cause overheating. To make use of this energy what is needed is to store it before it gets into the majority of the building, causing overheating, and then release it slowly during the night and during any subsequent overcast days. By building a heavy-weight wall 10 to 20 centimetres behind a large glazed window (Figure 20.4) excess heat can be stopped from entering the room and stored in the thermal capacity of the wall. The wall can be made of any suitable high-capacity material: water in oil drums, concrete and, rocks held in cages have all been used (Figure 20.5). Heat then passes

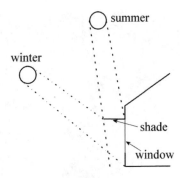

Figure 20.3 Schematic of a shaded window. The shade blocks the midday sun in summer but lets it enter the window during the winter

slowly into the rest of the building via conduction through the wall and by the use of top and bottom vents to circulate the very hot air in the cavity.

20.2 Heat pumps

One approach to solar energy collection is to extract heat from land, air or water that has been warmed by the sun. This warming need not be great and is carried out with the help of a heat pump (see Chapter 7). Figure 20.6 shows the general design of an electrically driven system, which extracts heat from relatively cold outside air and uses this heat to warm a room.

As was shown in Chapter 7, the maximum (i.e. Carnot) coefficient of performance (*COP*) of such a machine is given by:

$$COP_{Carnot} = T_{sink}/(T_{sink} - T_{source}) \tag{20.1}$$

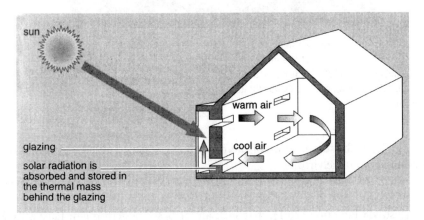

Figure 20.4 Schematic of a Trombe wall [BOY96, p57] [Reproduced courtesy of Boyle. G, Renewable Energy: Power for a Sustainable Future, 57 ©1996 The Open University]

(a)

(b)

Figure 20.5(a)–(c)   Various Trombe walls. Note the large glazed areas to allow sunlight in (a) Solar Energy Research, USA [NRE04a] [Reproduced from http://www.eere.energy.gov/buildings/database/ mtxview.cfm?CFID=4940022&CFTOKEN=13155306] (b) Zion Visitors Centre, USA [NRE04b] [Reproduced from http://www.eere.energy.gov/buildings/database/mtxview.cfm?CFID=4940022&CFTOKEN= 13155306]

Where T_{sink} is the temperature of the room being warmed and T_{source} is that of the outside air (both in kelvin).

The *COP* gives the power (in watts) extracted from the outside air and injected into the sink for every watt of power used by the machine. Realistic temperature values suggest, via Equation (20.1), that such an air-based system might have a *COP* of

(c)

Figure 20.5 (*Continued*) (c) National Renewable Energy laboratory Colorado, USA [NRE04c] [Reproduced from U.S. Department of Energy http://www.eere.energy.gov/buildings/database/mtxview.cfm?CFID=4940022&CFTOKEN=13155306]

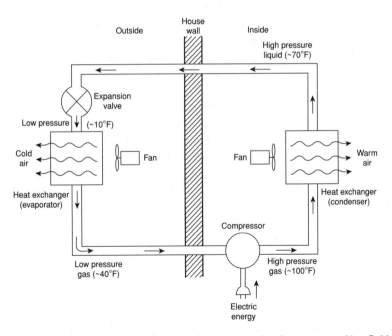

Figure 20.6 Schematic of an electrically driven heat pump using freon as a working fluid [RIS99, p82] [Reproduced courtesy of Ristinen, R.A & Kraushaar, J.J, Energy and the Environment, p82 ©1998 John Wiley and Sons Inc]

Figure 20.7 Photo of a domestic scale ground-source heat pump and the installation of the fluid-filled ground coil used to extract heat from the ground [HEAT05]

around 13. Unfortunately, engineering constraints limit realistic *COP*s more usually to between two and six. This implies that for every watt of electricity used by the system, six watts of heat might be transferred to the house. So, even if the electricity to run the machine is derived from fossil fuels (with a typical efficiency of only 25 per cent) such a device can often save carbon dioxide emissions compared to heating a building by using say a gas boiler with a much higher (typically 80 per cent) efficiency.

Similar devices are manufactured to extract heat from water or soil. These have proved very popular in Scandinavia and North America. Figure 20.7 shows a real device.

Heat pumps are unfortunately not a panacea for creating 'free' energy. Apart from the cost, and questions of reliability (compared with electrical heating) Equation (20.1) indicates that the *COP* will fall rapidly as T_{source} falls. This can mean that in winter, when heating demand is greatest, if they use air as the source, their performance can be poor. A ground-source heat pump that uses the more constant conditions found underground is less susceptible to these changes.

For all three cases – air, water, or ground sourced systems – it is important to realize that such systems are simply solar power extractors, but with one great advantage: the sun doesn't have to be shining.

Problem 20.4   Estimate the minimum COP required of an electrically driven heat pump for it to reduce carbon emissions when compared to gas-fired heating.

20.3 Solar water heating

History

Solar water heating has a history as long as that of solar space heating: the Romans built water channels lined with black slate to preheat water for bathhouses; in 1891 Clarence Kemp of Baltimore, USA, patented the first solar water heater and by 1897, 30 per cent of homes in Pasadena, USA, had Kemp's design fitted; in 1908 William J. Bailey invented a collector based on an insulated box containing copper coils – the basis of today's systems.

Unlike solar space heating, a large number of possible technologies exist, and the success of the approach is much less dependent on the architecture and construction of a particular building. This means the technology can easily be added to pre-existing buildings, or built in isolation; it also means that it is much easier to estimate how much fossil fuel-derived energy could be saved and carbon dioxide emissions reduced by the wide-scale adoption of the technology.

Although various individual technologies will be discussed in this section, there are two basic approaches: that used to heat water to modest temperatures (less than 100 °C) to provide hot water for use in swimming pools, the home and low temperature commercial processes; and that which superheats water to raise steam at high temperatures to drive turbines to generate electricity. We will discuss low temperature systems first, and then move on to power generation. The direct generation of electricity using photovoltaics will be discussed in the next chapter.

20.4 Low temperature solar water heating

As outlined in Chapter 3, the source of solar power is the energy released within the core of the sun as hydrogen is converted to helium. This releases energy in the form of photons, which struggle for many thousands of years, via innumerable collisions, to escape and finally take only a further eight minutes to reach earth. Although the core, or plasma region, which occupies one sixtieth of the solar volume, is at a temperature of 2×10^7 °C, the surface is at only 5800 °C. It is this surface temperature that defines the characteristics of the solar radiation reaching us.

The radiation reaching the earth's upper atmosphere covers a wide range of wavelengths but peaks strongly in or near the visible range 0.4 to 0.75 μm, with nearly half the available energy being in this range. At ground level the spectrum is complicated by absorption provided by various atmospheric gases including water vapour, oxygen and carbon dioxide (see Figure 5.4). At the top of the atmosphere the earth receives 1.36 kW/m^2 of *insolation*[6] on a plane perpendicular to the rays. The amount this is

---

[6] Insolation: the quantity of sunlight reaching the planet, usually expressed in units of W/m^2 or kW/m^2.

reduced by depends on the mass, m, of the atmosphere between any point on the earth's surface and the top of the atmosphere along a line of sight towards the sun. At the top of the atmosphere, $m = 0$; by convention if the sun is directly overhead $m = 1$; if it is at 15 degrees above the horizon (common on a winter's day) $m = 4$. For $m = 1$, the 1.36 kW/m^2 reaching the earth is reduced to only around 1 kW/m^2 at the surface. The angle, or inclination, of the sun not only reduces the surface heating because of atmospheric absorption, but also causes the radiation to be smeared over a greater area. It is important to distinguish these two processes because although we can do little about the mass of atmosphere above our heads, we can change the angle of the surface to be at a normal to the sun's rays by using a tilting plane. With suitable control gear, such a plane can be tilted and rotated to follow the sun, ensuring it always receives the maximum available light throughout the day.

For mid-latitude regions such as the UK the sun is overhead at midday in summer, implying $m = 1$ and therefore a peak insolation of 1 kW/m^2. Theoretically this equates to 24(hours) × 60(min) × 60(sec) × 1000(kW) = 864 MJ/m^2 per day. Unfortunately, the sun will not be overhead all day, nor out at night and some days it will be cloudy. A more realistic figure for the UK is 17 MJ/m^2 during June for a horizontal plane. About half of this will be in the form of direct radiation and half indirect, or scattered, light. At such a latitude, a south-facing plane at 45 degrees will receive almost the same amount of radiation (16 MJ/m^2). However in December such a plane will receive 3.6 MJ/m^2 per day, whereas a horizontal surface will only receive 1.9 MJ/m^2. This suggests that if the plane cannot be tilted as required, an angled surface is best. This is a fortunate observation, as many roofs are indeed angled and some point toward the south. For other locations the best angle will depend on the latitude. However, if the surface is tilted at an angle equal to the latitude at the location, it will be approximately perpendicular to the sun's rays during March and September and therefore form a good compromise solution year round: again implying that a roof will often form a sensible location for some form of collector. The requirement to face south is not too important and anywhere between southeast and southwest would be suitable (or northeast and northwest in the Southern hemisphere).

Problem 20.5   Estimate the roof area of all the houses in New Zealand (population four million) and the amount of free solar energy these roofs receive in summer (20 MJ/m^2 per day [LEA02]) and in winter (7 MJ/m^2 per day [LEA02]). Compare these figures with the average daily primary energy consumption of New Zealand.

Figure 3.3 shows the insolation for the whole planet. Surprisingly the highest values are not usually on the equator. This is because such areas often have frequent cloud cover. UK (London) receives on average 1000 kWh/m^{2}per annum – a surprisingly large amount of energy. Interestingly, the maximum is 8 kWh/m^2 per day in the UK – higher than the 7.5 kWh/m^2 maximum at the equator on a bright day. However even on a dull day the equator will receive 6.8 kWh/m^2, whereas the UK will only achieve 0.6 kWh/m^2 on a winter's day. Again this points to mid-latitude regions potentially having a problem with using solar power during the winter – just when demand is highest – unless they

Table 20.4 Absorbance and emittance of various coatings and materials [HOW79]

Surface	Absorbance	Emittance
Flat black paint	0.97	0.86
Grey paint	0.75	0.95
Red brick	0.55	0.92
Concrete	0.6	0.88
Galvanized steel	0.65	0.13
Aluminium foil	0.15	0.05
Zr on Ag	0.85	0.03
Copper oxide on copper	0.9	0.12

frequently have cloud free periods. Even so, for the USA or the EU, the average solar insolation is many times the total energy requirement.

In other regions the position is much more fortuitous. New Mexico for example has plenty of sunlight even during the winter, and importantly, the popularity of air-conditioning ensures that the demand for electricity is at its greatest in summer!

Problem 20.6 London receives on average 4.8 kWh/m^2 of solar radiation during the summer. If a 1 m^2 solar collector of some form collected all this energy with 100 per cent efficiency, how long would it take to heat water for a bath? In winter, the average is 0.6 kWh/m^2, how long would it take at this time of year?

Technologies for use

Several technologies exist for collecting solar energy and using it to heat water to modest temperatures (Figure 20.8). The simplest is just an unwanted domestic radiator painted black through which water flows and which is placed on a south-facing roof. This will give a temperature rise of up to 10 °C at mid-latitudes. By encasing the system in glass, heat loss can be greatly reduced and temperature rises of 50 °C can be obtained. (Typically a series of copper pipes bonded to a black backing plate is used.) By using an evacuated glass tube containing an evaporator pipe, temperature rises of 150 °C are possible. To reach higher temperatures, mirrors or other concentrating systems are required and are more typically found in systems which raise steam for electricity generation (Section 2.5).

Around one half of the incident radiation falling on a collector is re-radiated at a longer wavelength (4–15 μm) and can thus be trapped by a glass cover. Special coatings are available for increasing the absorbance off the collector and reducing its emittance, but as Table 20.4 shows there are many non-esoteric possibilities. Which is more important, the absorbance or the emittance, depends on whether a glass cover can be used.

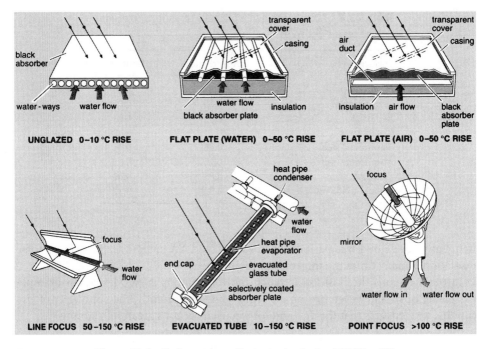

Figure 20.8 Various solar collector technologies [BOY96, p59]

Once the collector has reached an equilibrium temperature, the heat collected, E_{th} (W/m^2), will be the heat absorbed minus the heat re-emitted, i.e.

$$E_{th} = F\left(R\tau\alpha - L\left(T_c - T_a\right)\right) \tag{20.2}$$

where F is the collector heat removal efficiency factor (0.7 to 0.95), R is the incident radiation (W/m^2), T_a the temperature outside the collector (K), T_c the collector fluid temperature (K), L the heat loss temperature coefficient (1 to 7 W/m^{2}K) per degree of temperature difference ($T_c - T_a$), α the absorbance of the collector plate (0.8 to 0.97) and τ the transmittance of the enclosure cover (0.77 to 0.92).

From Equation (20.2) we can estimate the thermal efficiency, η_c, from the ratio of heat gained to the incident power from the sun, R:

$$\eta_c = \frac{E_{th}}{R} = F\left[\tau\alpha - \frac{L}{R}\left(T_c - T_a\right)\right], \tag{20.3}$$

i.e. the efficiency is reduced as $T_c - T_a$ increases. Implying that the plate should not be allowed to heat up too much. Empirically, it has been found that for a glazed collector

Table 20.5 Possible design alterations to improve the efficiency of a flat plate solar collector

Design parameter	Change	Effect
F increased	Improved collector design	Better heat transfer to working fluid
R increased	Orientation of system (possibly sun tracking)	Higher solar input
τ increased	Better glass or alternative transparent material ($\tau = 0.85$ for common glass, but 0.96 for some polymers)	Higher input into collector
α increased	Special absorber coating	Increased absorbance of the collector surface
L decreased	Double glazing	Reduced losses
T_c decreased	Reduced working temperature	Reduced losses

Equation (20.3) can be replaced by:

$$\eta_c = 0.8 - 8\frac{(T_c - T_a)}{R}, \tag{20.4}$$

for practical systems with a reasonable temperature difference ($T_c - T_a$). As an examination of Equation (20.3) shows, during cold periods, or at night such a system could lose more heat than in gains, so some way of stopping the movement of the collector fluid is required.

Equation (20.3) hints at ways we could improve the efficiency of the system. η_c will increase if: F, R, τ or α increase, or if ($T_c - T_a$), or T_c itself decrease. Table 20.5 suggests possible strategies.

All the changes in Table 20.5 carry an additional cost. For a practical system, such costs have to be balanced against the value of the extra energy produced. Improving the efficiency by reducing the temperature of the collector implies that we will end up with a greater mass of not quite so hot water. Clearly, there is a temperature below which such water will not be useful.

A practical system

Having optimized the collector in an economic way, there is the need to consider what to do with the heat produced. Usually the working fluid will be water and the required commodity hot water for domestic purposes, or to heat a swimming pool; although it is possible to use air or water as the working fluid and use this for space heating. Assuming we are heating domestic water, the main question to be addressed is that of storage. The more water we can store the better, as this will insure us against overcast days – as long as the solar collector area is large enough to heat this amount of water

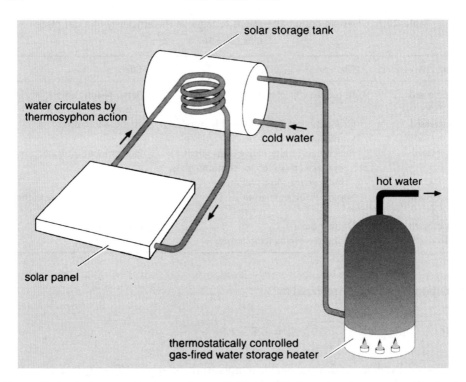

solar storage tank

water circulates by
thermosyphon action

cold water

hot water

solar panel

thermostatically controlled
gas-fired water storage heater

Figure 20.9 A thermosyphon water heater [BOY96, p43] [Reproduced courtesy of Boyle. G,
Renewable Energy: Power for a Sustainable Future, fig 2.3, pg 43 ©1996 The Open University]

to a useable temperature. However storing large volumes of water takes space and may
require structural changes to the building. Most systems rely on the modest storage
capacity of around 200 litres provided by a normal domestic hot water tank with 2–
4 m² of collector area. The two most common ways of connecting the collector to the
storage tank are either directly or via a circulating pump; both have certain advantages.
Figure 20.9 shows a directly connected system typical of that used in Israel and southern
Europe. This relies on the *thermosyphon effect* where water heated by the collector rises
by natural convection to the storage tank. The main advantages of the system are its
simplicity and that it does not use electricity to circulate the water. Its main disadvantage
is that the storage tank needs to be at a higher elevation than the collector, which can
be difficult if the collector is high on a roof. Where such systems are popular, flat roofs
are the norm and the tank simply sits on a frame in the open air. Clearly this would be
inappropriate in a colder climate, as would be the mounting of the tank in a cold attic
space. Although with suitable building design this problem can be solved.

Another approach is the pumped storage system (Figure 20.10). Here water is pumped
through the storage tank (which can be anywhere in the house). Advantages include
flexibility in siting the storage tank, the possibility of using a pre-existing hot water tank
and the need for only one tank (although multi-tank versions do exist). Disadvantages

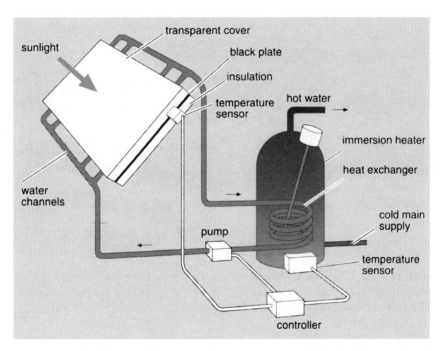

Figure 20.10   A pumped storage solar water heater [BOY96, p43] [Reproduced from Boyle. G, Renewable Energy: Power for a Sustainable Future, pg 43 ©1996 The Open University]

are the increased complexity of the arrangement and the need to use electricity to circulate the water.

In both the pumped and the thermosyphon solutions the water in the collector is separate from the water in the storage vessel. This allows the collector to be isolated at night and when the temperature in the collector is less than the temperature of the stored water. In the pumped system this isolation is achieved by a thermostatic control of the pump; in the thermosyphon it occurs naturally as the water in the collector will be denser than that in the storage tank under such conditions. Both systems also need a second method of heating the water, such as an immersion heater, for use when the sun cannot heat the water to a high enough temperature.

In many countries, including the UK, the costs of such an installation are so high that many would consider the approach uneconomic, even though it would be relatively easy to produce 50 per cent of domestic hot water for many households by this method at such latitudes. Given a combination of government subsidy and reduced manufacturing and installation costs due to increased sales, this position could well change. Some countries have adopted the approach far more enthusiastically, for example Israel and Greece, the reason being obvious – far greater levels of solar radiation during the winter months. There are many other states around the world where the technology would also be appropriate which have yet to seize the opportunity; the reason for this would appear to be the low cost of fossil fuels.

Figure 20.11 The solar field at Phoenix Federal Correction Institution. Source: IEA CADDET programme[7] [CAD01] [Reproduced courtesy of ETSU, CADDETT, Technical Brochure 140, 2001, IEA CADDET programme, http://www.caddet.org.]

Problem 20.7 Use Equation (20.4) to estimate how the efficiency of a flat plate collector changes as the water in the storage tank warms from 20 °C to 80 °C. Assume an external air temperature of 20 °C degrees and $R = 1$ kW/m^3. Plotting the results graphically with the aid of a spreadsheet would be ideal.

20.5 Example application: solar water heating, Phoenix Federal Correction Institution, USA

Due to its location (Arizona) the solar resource in the area is high – 5.2 kWh/m^2/day. As water heating at the institution was by electricity, rather than gas, costs were relatively high. Together these factors suggested that solar water heating might be commercially viable.

The prison houses 1100 inmates and has a staff of 350. Average daily consumption of hot water is 3408 kWh per day with a peak power demand of 228 kW.

The solar field (Figure 20.11) covers 0.6 hectares [CAD01] and consists of 120 parabolic trough collectors, each focusing the sun's light onto an oil-filled tube running along the focal line (Figure 20.12). Each collector has a 14 m^2 mirror made of acrylic-coated aluminium, giving a total reflective surface area of 1584 m^2. The oil reaches a maximum temperature of 280 °C, the heat from which is passed to the water, which is in turn stored in two water tanks with a combined capacity of 87 m^3.

---

[7] Web site: http://www.caddet.org.

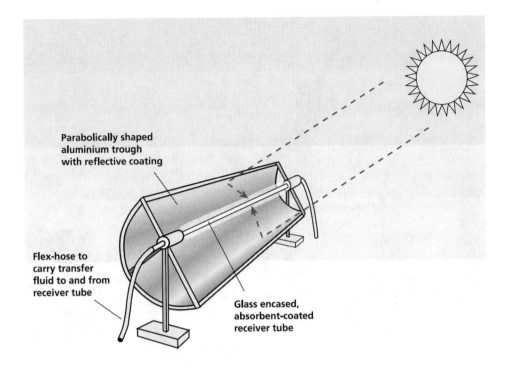

Parabolically shaped aluminium trough with reflective coating

Flex-hose to carry transfer fluid to and from receiver tube

Glass encased, absorbent-coated receiver tube

Figure 20.12 Parabolic trough concentrating solar hot water system with storage. Source: IEA CADDET programme [CAD01]

Because storage is included in the design, solar heated hot water is usually available even at night. The system is backed up by an electric water heater. Overall the system produces about 70 per cent of the hot water used by the prison [CAD01]. Installation cost was €502 000 (£346 000, US$649 000) and is saving around 650 000 kg of CO_2 from being emitted annually.

20.6 High temperature solar power

Warm water is a very low-grade form of energy and has few uses. However water or other fluids at much higher temperatures are far more useful and therefore more valuable, not least because they can be used to raise steam for electricity generation.

Higher temperatures are achieved by gathering sunlight over a wide area and concentrating it on to a much smaller area for absorption. The ratio of collecting to absorbing area is termed the concentration ratio (CR) and current systems have a value of between 10:1 and 10 000:1. Collection is via mirrors arranged in various arrays and with various

Figure 20.13   Abel Pifre's solar powered printing press [BOY96, p74]

forms, from troughs to parabola. All share the ability to be steered to follow the path of
the sun across the sky.

History

The Swiss scientist Horace de Saussure built the first solar collector in 1767, a design
that was later used by the explorer Sir John Herschel in the 1930s to cook food on
his travels in Africa. In the 1860s the mathematician Augustin Mouchot from Tours
designed a solar powered steam engine. During the 1870s and 1880s Mouchot and Abel
Pifre produced solar stills, cookers and even a printing press (Figure 20.13).

In 1913 Frank Shuman demonstrated a 41 kW (55 horsepower) steam engine in
Egypt that used five parabolic trough collectors, each 4 m wide and 80 m long, to focus
the light on a cast iron pipe. Shuman also suggested that by building 52 000 km^2 (20 000
square miles) of collector in the Sahara desert more energy could be harvested than
was being produced from coal at that time. 52 000 km^2 might seem a great area, but it
would only have required 1000 sites each with a side of length 7 km (4 miles) to achieve.
Unfortunately, since 1913 our use of fossil fuels has spiralled and the area would now
have to be very much greater.

Figure 20.14 Parabolic dish concentrators at White Cliffs with cooled PV cells at the focus of each mirror [WHI04] [Reproduced from [MET07] Metal Type: http://www.metaltype.co.uk/photos/images/92.jpg]

Problem 20.8 How much land would now be needed to realize Shuman's dream? Assume the efficiency of the technology has quadrupled and that the world coal demand was 300 Mt_{oe} in 1913[8].

Dish collectors

Mouchot and Pifre's design has now been advanced by placing a small steam or Stirling engine, a thermoelectric generator or a cooled PV cell at the focus of the parabolic dish (Figure 20.14). The advantage of using a Stirling engine (see below) is that it can operate at much greater temperatures (over 1000 °C), allowing a high concentration ratio. As each dish would only produce a small amount of power, a large field, or array, of dishes would be required to form the basis of a power station.

Parabolic mirrors are used because all rays coming from the direction in which the mirror is pointed are brought to a single focus. With either hemispherical or flat mirrors this will not occur. Unfortunately glass parabolic mirrors are very expensive to manufacture and there is a need to develop alternatives, possibly using aluminized plastic film moulded into the correct shape.

Stirling engines

Petrol (gasoline) and diesel engines and gas turbines are all examples of internal combustion engines where the fuel is burnt within the engine. Steam and Stirling engines are external combustion engines with the fuel being burnt outside of the engine and used to warm a working fluid (water/steam in the case of the steam engine). The Reverend

---

[8] This is an extremely rough estimate based on apportioning three quarters of the world primary energy consumption given in Chapter 4 and should not be used outside the context of this problem.

Figure 20.15 Barstow solar power plant in the Mojave Desert [NRE04d] [Reproduced from www.nrel.gov/data/pix/Jpegs/00036.jpg, US National Renewable Energy Laboratory]

Robert Stirling filed a patent for his engine in 1816. The main advantages the design offered over steam were safety (because of the avoidance of high pressure steam), the possibility of being scaled down for the domestic market without much loss of efficiency, and almost silent running.

A Stirling engine consists of one or more cylinders each containing two pistons. Air is shuffled from the hot end of the cylinder (which is continuously heated) where it expands to the cool end, where it contracts. This cycle of expansion and contraction converts heat into kinetic energy. The piston nearest the heat source fits loosely to the cylinder walls, the other tightly. Air between the bottom of the cylinder and the first piston is heated and expands, driving the first piston up, a mechanical linkage between the two pistons forces the second, sealed, piston upwards, creating a void between the pistons. Air from beneath the first piston is drawn into this void and cools at the same time as the linkage forces the first piston down to the bottom of the cylinder. The cycle is then repeated.

The design cannot easily respond to rapid changes in required power and is therefore best suited to continuous running at a single output. The design is only efficient at very high temperatures – above the failure temperature of traditional materials. Modern machines have tended to replace air with helium at up to 100 atmospheres.

Power towers

Rather than relying on a large number of separate collectors, power towers focus light from a very extensive area onto a single fixed receiver. This means that only the mirrors have to track the sun. Temperatures in excess of 3000 °C are possible from such systems (for comparison the sun's surface temperature is 5800 °C). The collector, or heliostat, area can be very large indeed. Figure 20.15 shows the heliostat field of the solar power station at Barstow, California. The central tower incorporates the receiver. The Barstow plant contains 1818 heliostats of approximately 40 m^2 each, focusing on a 91 m tower and producing 10 MW$_e$ of electricity from high-pressure steam at 560 °C. Because the focus (the tower) is so far away, each mirror need only be slightly curved. Excess heat is stored (in molten salt) to allow 7 MW$_e$ of electricity production to continue at night.

Theoretically the thermal power, P_{th}, available for power production from such a system will be:

$$P_{th} = RA\eta_{Carnot}$$

where R is the insolation (W/m^2), A is the ground area of the heliostat field, and η_{Carnot} the Carnot efficiency: $(T_{fluid} - T_{sink})/T_{sink}$.

The *actual* P_{th} generated will be less than this because the heliostats will not cover the whole land area. (For systems such as the Barstow one, they are only likely to cover a quarter of the site.) The true efficiency is likely to be half the Carnot efficiency and because of losses only half the incident energy will be converted to heat. Implying more realistically that:

$$P_{th} = \frac{RA\eta_{Carnot}}{16}. \tag{20.5}$$

Problem 20.9 Estimate the land area required by a power tower to produce 250 MW of thermal power for electricity generation given an insolation of 1 kW/m^2, a fluid temperature of 600 °C and a sink temperature of 100 °C.

The smaller 1 MW plant at Odeillo in France uses 63 heliostats covering a total area of 2835 m^2 focusing on a receiver of area 0.18 m^2. This gives a concentration ratio of 2835/0.18 $\approx$ 16 000. Assuming a standard insolation of 1 kW/m^2 implies a flux of 1000 W $\times$ 16000 = 160 MW/m^2.

Because of the very high temperatures created, the principal unwanted mode of heat loss from the absorber is a radiant one and can be estimated from (Chapter 19):

$$A\varepsilon\sigma \left(T_1^4 - T_2^4\right) \tag{20.6}$$

where A is the area of the absorber (m^2), ε is the emissivity ($\approx$ 0.05), T_1 and T_2 are the temperature (K) of the absorber and the surroundings respectively and σ the Stefan-Boltzmann constant (5.67×10^{-8} W/m^{2}K^4)

Equation (20.6) indicates that the temperature rather than the area of the absorber dominates the heat loss – because the temperatures are raised to the fourth power. The overall system design needs to be a compromise between achieving a high concentration ratio and not sustaining too great an absorber re-radiation loss.

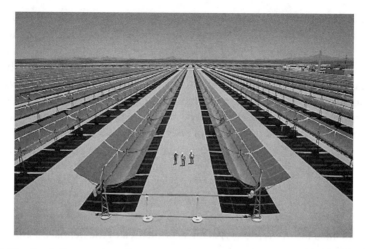

Figure 20.16 Parabolic trough collectors at the Luz power plants in California, USA; note the people for scale [OU07]. (Open University: http://openlearn.open.ac.uk/mod/resource/view.php?id=94988)

Trough collectors

Power towers focus all the collected solar energy on a central location. Trough collectors focus the light much closer to the mirror and along a pipe running the length of the mirror, which is in the form of a parabolic trough (Figure 20.16), and are very similar to that used by Shuman in 1913.

Systems reaching 80 MW_e have been built with up to 464 000 m^2 of collector area. Synthetic oil is pumped through the pipes and heated to around 390 °C to raise steam with a peak solar to electricity conversion efficiency of 22 per cent.

The concentration ratios are much smaller than those of power towers and lie in the range 10–100. If the axis of the trough lies east to west then there is no need to track the sun continuously as the rays will focus roughly along the pipe through much of the day if the tilt is adjusted every few days. If the axis lies north to south then continuous tracking will be required, but unlike a power tower, only in one plane and through exactly the same angle for all mirrors.

20.7 Low temperature water-based thermal energy conversion

One final approach to generating electricity from solar power is the use of small temperature gradients in large volumes of water. *Solar Ponds* use relatively shallow ponds containing a layer of high-density salty water lying beneath a layer of lighter fresh water. Sunlight is absorbed at the bottom of the pond and heats the salty water, which cannot rise due to its greater density and therefore carries on increasing in temperature (to around 90 °C). Hot water is extracted from and returned to the bottom of the ponds and cooler water is simultaneously extracted from and returned to the upper layer. The temperature

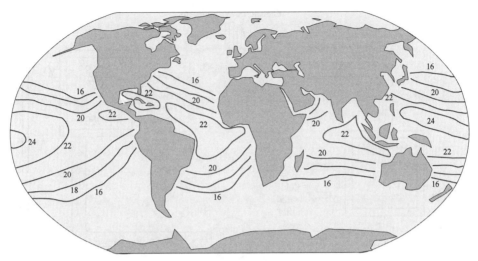

Figure 20.17 Global distribution of ocean thermal energy. The temperature difference shown is that between the surface and a depth of 1000 m [US National Renewable Energy Laboratory in RIS99]

difference between these two circuits can be used to drive a low-pressure vapour cycle engine – essentially a turbine driven by an organic vapour, rather than high-pressure steam. Demonstration plants of up to 50 MW$_e$ have been constructed. As might be expected, efficiencies have been very low (around two per cent), although the large thermal storage provided by the pond means that such systems can generate even at night.

On a far grander scale, and yet to be realized, ocean thermal energy conversion makes use of the modest temperature difference between the water near the ocean floor and that near the surface. This difference can be 20 °C over a depth of 1000 metres and although the efficiency of running a vapour cycle engine is likely to be very small (consider the Carnot efficiency at these temperatures) the amounts of water in the oceans, and therefore the resource, is colossal. Figure 20.17 shows the distribution of the resource. Because the colder deep ocean water must be pumped to the surface, the design sits balanced on a knife-edge where the energy created needs to be greater than the energy used in pumping. Pilot plants have been built and used to generate small amounts of electricity; Figure 20.18 shows a possible design.

Problem 20.10 Estimate the Carnot efficiency of an ocean thermal energy conversion plant assuming a deep ocean temperature of 5 °C and a surface ocean temperature of 25 °C.

20.8 OECD resource

OECD electricity production for thermal electric plants stands at 569 GWh per annum [IEA05]. The amount of passive solar energy provided by windows goes unestimated: perhaps the reader would like to try to estimate the value.

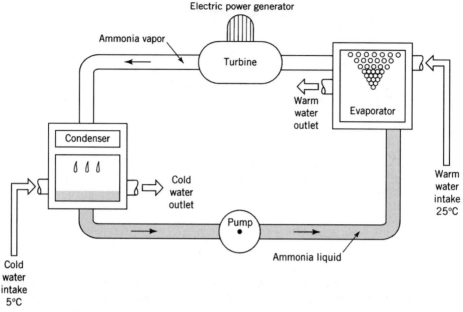

Adapted from R.A & Kraushaar, J.J, Energy and the Environment © 1998 John Wiley and Sons Inc

Figure 20.18 Operating principle of an ocean thermal energy converter [US National Renewable Energy Laboratory in RIS99]. The heat engine uses ammonia as a working fluid, with the turbine being driven by the ammonia vapour. The warm water used to vaporize the ammonia is from the ocean surface; the cold water used to condense the ammonia back to a liquid is from the deep ocean

20.9 Student exercises

The tree which moves some to tears of joy is in the eyes of others only a green thing that stands in the way. Some see Nature all ridicule and deformity, and some scarce see Nature at all. But to the eyes of the man of imagination, Nature is Imagination itself.

—William Blake, The Letters, 1799

1. Why does good passive solar design require a holistic approach to building design?

2. Why might a south-facing window be a provider of energy in winter in southern France but not in the UK?

3. The land area of Kenya is 570 000 km². Assuming the country receives solar energy at a rate of at least 250 W/m², estimate the total solar energy received and compare this value with total Kenyan energy demand (Appendix 1).

4. How could the performance of a solar water heater be improved? Does it always make sense to reduce the collector temperature to improve the efficiency?

5. Assuming an efficiency of 40 per cent, what area of collector might be needed to heat 200 litres of water in southern Australia by 30 °C in two hours?

6. Ignoring any inefficiencies or losses, how long would it take in theory for the Odeillo solar furnace to boil one kilogram of water originally at 20 °C?

7. Estimate the solar energy provided by all the world's windows. Don't forget to include heat losses from these windows. Compare your answer to world primary energy demand.

8. A solar collector is mounted at an inclination of 45° to the horizontal. If the sun rises to an inclination of 54° above the horizon what proportion of the radiation is then falling normally onto the collector? [From SHE97.]

9. A solar collector in northern England is mounted with its axis inclined at 54° to the horizontal. If the total annual radiation energy on a horizontal surface is 1000 kWh/m^2, divided equally between direct and diffuse components, what is the radiation received by the collector? [From SHE97.]

10. A flat-plate solar collector, mounted at the latitude angle and south facing, has an effective area of 2 m^2. Water is pumped in through this collector at the rate of 20 × 10^{-6}m^3/s and the mean temperature difference between the inflow and outflow is 18.4 °C. The collector is used to heat indirectly the water in a storage tank of capacity 50 gallons (227 litres) for 5 hours continuously. If the system operates at a typical efficiency, calculate the temperature rise in the storage tank. What is the power rating of the flat-plate collector in W/m^2? [From SHE97.]

11. A solar collector is to be mounted on the south-facing roof of a dwelling, feeding a storage tank with a capacity of 30 gallons (136 litres). The circulating pump is to operate at the rate of 20×10^{-6} m^3/s. On a warm, sunny day the difference in water temperature between the inflow and outflow at the collector is typically 15.5 °C and this difference exists for 6 hours. What is the operating efficiency of the system if the temperature of the water in the storage tank is increased by 20 °C? [From SHE97.]

12. It is proposed to use a roof-mounted solar water-heating system to supplement the energy input into certain industrial process. The south-facing solar collector is to be used to heat indirectly the water in a storage tank of capacity 5000 gallons (22 700 litres). Water can be pumped through the collector by a range of available water pumps. On a typical summer day, there are 5.6 hours of sunshine, which causes an average temperature difference of 16.5 °C between inflow and outflow of the collector [from SHE97].

   a. If the anticipated efficiency of the system is 42.6 per cent, what rate of water pump flow in m^3/s is needed to cause a temperature rise of 12 °C in the storage tank?

b. If this pump is used and the temperature of the storage tank becomes 14.5 °C, what is the efficiency of the collector system?

13. Give a broad specification for a solar water heating installation for a typical UK domestic dwelling (i.e. a three bedroom, semi-detached house) occupied by four persons. In particular, specify the necessary area of collector and the capacity of the supplementary water tank. The installed commercial cost is quoted at £3200. Estimate the payback period if the household is (a) a heavy user of hot water, (b) a light user of hot water [from SHE97].

14. A flat-plate collector of area 2 m² has a water inflow temperature of 15 °C and an outflow of 49 °C while the incident radiation is constant at 725 W/m². Calculate the approximate thermal efficiency [from SHE97].

15. The absorber of a solar concentrator system operates at 550 °C. The collector receives 200 W/m². If the concentrator ratio is 50 and the absorber emissivity is 0.05, what proportion of the input power is re-radiated? [From SHE97.]

16. A solar power tower plant receives an effective average radiation of 1000 W/m² from its concentrator collectors. The conversion efficiency of the collector heliostats into thermal energy is 53 per cent. If the plant fluid operates at 600 °C and the sink temperature is 100 °C, calculate the area of heliostats and the land area required to generate 100 MW of thermal power.

21
Photovoltaics

"Earth provides enough to satisfy every man's need, but not every man's greed."

—Mahatma Gandhi

The sun is the ultimate source of renewable energy and electricity the highest grade of energy. The Holy Grail of alternative energy is probably the ability to turn sunlight into electricity with high efficiency, at low cost, using common materials which have structural characteristics that allow them to replace other components (for example roofing tiles).

21.1 History

The search for this grail has been on for some time. In 1839, Edmond Becquerel noted that the voltage produced by a battery containing silver plates increased if the battery was exposed to sunlight. In 1833, Charles Edgar Fritts constructed the first true solar cell using slithers of selenium. His device had an efficiency of less than one per cent. By 1954, Darryl Chapin and others from Bell Labs had produced a silicon solar cell with an efficiency of six per cent. Improvements in design have now increased this to over 25 per cent.

21.2 Basic principles

The theoretical explanation of how photovoltaics (or PV) produce electricity was provided by Albert Einstein and earned him the Nobel Prize for Physics in 1921. Substances such as silicon and germanium are semiconductors – i.e. poor conductors of electricity, but not quite insulators. Both have four electrons in their outer shell, which are all used in bonding with four other silicon or germanium atoms to form a crystalline lattice.

Energy and Climate Change   David A. Coley
© 2008 John Wiley & Sons, Ltd

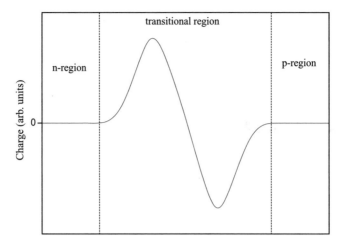

Figure 21.1 Schematic of the charge distribution across an n-p junction in a solar cell

Because all these four *valence* electrons are used in bonding, there are no free electrons left within the lattice to carry a current. However if heat is applied or light of short enough wavelength (i.e. high enough energy) strikes the crystal, some valence electrons can be released and the crystal then conducts. Being able to conduct electricity is not enough: however a mechanism is now needed to produce a potential difference across the crystal that will force the electrons to all move in the same direction. This is achieved using a technique called *doping*.

Doping allows two additional forms of the semiconductor to be made: *n-type* and *p-type*, n for negative, p for positive). To produce the n-type, a sprinkling (typically one in a million) of atoms (commonly phosphorus) with five electrons in their outer shell is introduced into the silicon. Because of the lattice structure of the crystal the phosphorus atoms are also forced to bond with exactly four other nearest neighbours. This leaves one electron spare. As these electrons are not used in bonding, it takes a lot less energy to free them and most end up roaming the lattice. To produce the p-type, atoms with only three outer electrons (e.g. boron) are introduced. This produces electron 'holes' in the lattice.

A solar cell uses both n and p-type semiconductors. A slice of p-type material is placed in contact with a slice of n-type creating an electric field. The free electrons in the n-type rush to fill the holes in the p-type. Before the two materials were placed into contact, both were electrically neutral: the sum total number of electrons exactly matched the number of protons in the nuclei of each material. Upon union this is no longer true. As the electrons move over the n-p junction they must leave behind them a net positive charge, and create a net negative charge in the p-type material. This negative charge creates an electric field and stops more electrons crossing the junction to fill all the holes in the p-type (Figure 21.1). In essence we have a device that stops electrons flowing in one direction, but is happy to let them flow in the other: a diode.

Any electrons freed by light striking the cell in the vicinity of the junction will be attracted toward the positive part of the field and will flow across the junction into

Table 21.1 Photon energy as a function of wavelength of solar radiation striking a solar cell

Range	Typical Wavelength (μm)	Energy per photon (J)
Ultraviolet	0.3	6.6×10^{-19}
Visible light	0.5	3.0×10^{-19}
Low infrared	1.0	2.0×10^{-19}
Mid infrared	1.15	1.7×10^{-19}
Deep infrared	2.0	1.0×10^{-19}
Very deep infrared	3.0	0.66×10^{-19}

the n-type material. By connecting together the sides of the n and p materials away from the junction (via an external load) these electrons will slowly return to the p-type semiconductor, a current will flow and useful work can be extracted.

In order to understand the detailed operation of a solar cell, and in particular the fundamental reasons why conversion efficiencies are less than 100 per cent we need to consider the nature of the incident radiation. Such considerations play much the same role as the Carnot efficiency does in heat engines. Key to the process is the quantum nature of light. Each photon will either free a single electron on none at all. (Occasionally more than one electron can be freed, but this is rare given the wavelengths found in sunlight.) The energy, E_{rad}, in joules of an individual photon is given by:

$$E_{rad} = hf \tag{21.1}$$

where f is the frequency in Hz and h is Planck's constant (6.626×10^{-34} Js). As the wavelength, λ, and the frequency of light are related to the speed, c, of light (3×10^8 m/s) by:

$$c = \lambda f,$$

Equation (21.1) can be re-written as:

$$\boxed{E_{rad} = hc/\lambda = 2 \times 10^{-25}/\lambda} \tag{21.2}$$

Table 21.1 applies this formula to typical wavelengths within the solar spectrum to estimate the energy per photon of the incident radiation.

The energy required to release an electron (the *band gap* energy) depends on the semiconductor in question. For silicon, 1.7×10^{-19} J is required, which equates to a wavelength of 1.15 μm. This means that almost one quarter of the solar spectrum (see Figure 5.4) cannot promote electrons into a free state if silicon is used. Such radiation will simply warm the cell. Because we are dealing with a quantized phenomenon, more energetic photons (i.e. with wavelengths less than 1.15 μm) can still only promote a single electron unless they have enough energy to release exactly two electrons. Thus

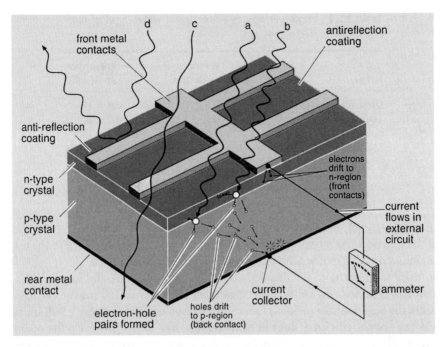

Figure 21.2 Schematic of a solar cell [BOY96, p98]. A silicon solar cell is a wafer of p-type silicon with a thin layer of n-type silicon on one side. When a photon of light with the appropriate amount of energy penetrates the cell near the junction of the two types of crystal and encounters a silicon atom (a), it dislodges one of the electrons, which leaves behind a hole. The electron thus promoted tends to migrate into the layer of n-type silicon, and the hole tends to migrate into the layer of p-type silicon. The electron then travels to a current collector on the front surface of the cell, generates an electric current in the external circuit and then reappears in the layer of p-type silicon, where it can recombine with waiting holes. If a photon with an amount of energy greater than the band gap strikes a silicon atom (b), it again gives rise to an electron-hole pair, and the excess energy is converted into heat. A photon with an amount of energy smaller than the band gap will pass right through the cell (c), so that it gives up virtually no energy along the way. Moreover, some photons are reflected from the front surface of the cell even when it has an antireflection coating (d). Still other photons are lost because they are blocked from reaching the crystal by the current collectors that cover part of the front surface [Reproduced courtesy of Boyle. G, Renewable Energy: Power for a Sustainable Future, p98 ©1996 The Open University]

their energy above 1.7×10^{-19} J is also wasted: again heating the cell. For sunlight, this equates to a loss of one third of the incident energy. Thus the maximum theoretical efficiency $\approx 100 - (25 + 33) = 42$ per cent. In reality the situation is worse. Because the semiconductors are indeed 'semi'-conductors, they have a relatively high resistance; therefore any current flowing through them will suffer losses. In addition there are losses at the cell junction and at the terminals. All in all, the maximum theoretical efficiency will be under 23 per cent, and in a working cell even less due to heating effects, reflections from the upper surface and blocking of sunlight caused by the terminals, which are in the form of a fine grid (Figure 21.2).

Improving the efficiency

Seemingly, the most obvious way to improve efficiency would be to reduce the band gap, i.e. choosing a material in which the electrons find it much easier to become free. This would mean sunlight with a wavelength less than 1.15 μm could also produce a current. Unfortunately there is a trade-off. The band gap also determines the voltage produced by the cell. Thus although a lower band gap will increase the current produced by the cell, the voltage produced will drop, and as power is the product of current and voltage, the power produced by the cell will fall. However, by using different semiconductors with differing band gaps – gallium arsenide, copper indium diselenide and cadmium telluride – a multi-junction cell can be constructed with the various semiconductors targeting different parts of the solar spectrum. More prosaic attempts to improve efficiencies include making the terminal grid transparent and adding an antireflective coating between the contact grid and the glass cover. Manufacturing costs can also be reduced by using *polycrystalline*, or *amorphous* rather than mono-crystalline silicon.

21.3 Technologies for use

Crystalline silicon is traditionally grown by very slowly drawing a seed crystal from molten silicon. As the crystal is drawn, an ingot, 10 cm in diameter and several metres long, is formed. The process is extremely difficult, expensive and prone to error: temperatures must be maintained at 1420 °C with an accuracy of ±0.1 °C. The p-type material is made by doping the molten silicon with boron, the n-type by using diffusion to form a thin surface layer of n-type semi-conductor less than a micron thick. A much thinner layer of n-type is required so that light can penetrate through to the n-p junction of the completed cell, which will have the n-type as the upper layer. The ingot is sawn using a diamond cutter. Unfortunately the cutter has approximately the same thickness as the slither of silicon cut (0.25 mm), so a lot of precious material is lost.

Complete cell efficiencies of 16 per cent are possible. However, because the silicon wafers are circular, an array of such cells can contain dead spots between the cells, which do not produce any electricity.

An alternative process is to draw a ribbon of silicon the required thickness out of the melt by pulling it through a die. The die is in the form of a nine-sided polygon, and thus the ribbon is really a nonagon tube, which is cut into its separate facets using a laser.

Polycrystalline silicon is easier to manufacture and large ingots can simply be cast and then cut up using fine wire saws. Unfortunately performance is reduced by about one third to give a complete cell efficiency of around 10 per cent. This reduction in efficiency arises because electrons and holes find it easy to recombine at the boundaries between the many separate crystals that make up the cell. A major advantage of the approach is that it can produce rectangular cells, thereby reducing inefficiencies due to dead spots.

The losses discussed above give rise to heat within the cell, and temperature rises of 25 °C are typical. If the ambient temperature is also high then the cell can become quite hot (>50 °C). These temperatures can cause a reduction in efficiency of about 0.4 per cent per °C, and if cell temperatures become very high, cooling the cell may be necessary. However if this strategy is adopted, then given enough capacity in the cooling system, it may be economic to focus light onto the cell from a larger area using a concentrating system, thereby increasing the power produced.

Gallium arsenide (GaAs) has a similar crystalline structure to silicon and a high light absorption coefficient; it also has the potential to offer a wider band gap. Cells made from GaAs are extremely efficient and are also able to operate at the higher temperatures found in solar concentrating systems, the downside being that they are more costly to produce than silicon-based cells.

One step beyond the crystalline disruption of polycrystalline silicon lies amorphous silicon. Because the atomic arrangement in such a material is much more distorted than within a crystal, many of the atoms do not bond with their four nearest neighbours, leaving unused (dangling) bonds that can absorb nearby free electrons, therefore negating the effect of an n-p junction. However if silicon with a small amount of boron dopent is mixed in a gaseous form with hydrogen and then used to deposit a very thin film on a stainless steel substrate, the hydrogen can bond with any dangling bonds. Figure 21.3 shows the structure of such a cell and shows that the p- and n-type materials are separated by a layer of undoped, or *intrinsic,* amorphous silicon. Such cells are considerably cheaper to produce, but their performance is much lower, often degrading to as little as four per cent after one year's use. This is of little consequence in items such as pocket calculators, where power use is minimal and cost of primary importance. Similar thin-film approaches have been used to produce cells from copper indium diselenide, gallium arsenide and cadmium telluride. Such materials have promising potential, but health concerns have been voiced over the use of various chemicals, including cadmium, during production.

As has been mentioned, two other ways of improving efficiency are by layering multiple n-p junctions to absorb photons from a greater percentage of the solar spectrum, or to use solar concentrators. Concentrating systems are essentially modifications of the designs given in Chapter 20 for solar thermal systems, although, as discussed, cooling of the cell may also be required. Figure 21.4 shows a schematic of a multiple junction PV cell. Alloying amorphous silicon with carbon allows the band gap to be moved to respond to the high-energy photons in blue light; alloying with germanium moves the band gap to respond more to the lower energy photons in red light. The carbon-alloyed junction then sits on top of one or more germanium-alloyed junctions, each with a band gap designed to absorb less and less energetic photons.

One final approach to increasing efficiency is the mounting of truly massive solar arrays in space in geo-stationary orbit. The electricity produced by such a system would be converted to microwave radiation and beamed to a receiving station back on earth. This would have several advantages. The solar array would be exposed to near constant maximal solar flux for around 23 hours a day, implying a gross annual efficiency much higher than that achievable on earth. Not surprisingly, there are serious questions over the safety and economic viability of such an approach.

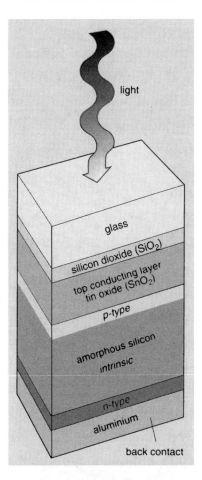

Figure 21.3 Structure of an amorphous silicon cell [BOY96, p107]. The top electrical contact is made of an electrically conducting, but transparent, layer of tin oxide deposited on the glass. Silicon dioxide forms a thin 'barrier layer' between the glass and the tin oxide. The bottom contact is made of aluminium. In between are layers of p-type, intrinsic and n-type amorphous silicon [Reproduced courtesy of Boyle. G, Renewable Energy: Power for a Sustainable Future, p107 ©1996 The Open University]

Problem 21.1 Estimate the area of a geo-stationary solar array needed to provide 1 GW_e of power back on earth. Assume a solar cell efficiency of 14 per cent and no losses in the microwave link. (Assume a solar insolation of 1370 W/m^2 above the atmosphere.)

Table 21.2 summarizes the performance of various current technologies.

21.4 Electrical characteristics

The performance characteristic, i.e. a plot of voltage, V, against current, I, produced by a typical solar cell is roughly rectangular (Figure 21.5). The short circuit current is

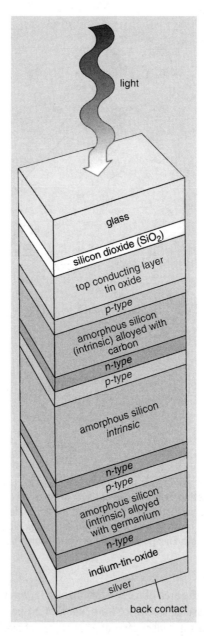

Figure 21.4   Schematic of a multiple junction amorphous silicon cell [BOY96, p110] [Reproduced courtesy of Boyle. G, Renewable Energy: Power for a Sustainable Future, fig 3.28 ©1996 The Open University]

proportional to the incident light intensity and the open circuit voltage varies logarithmically with this intensity. The power produced, the most useful measure for power generation, is the product of V and I. V and I will be related to the load resistance via Ohms law, and the maximum power will be obtained when the product IV is at a

Table 21.2 PV performance summary (percentage efficiency) [data from UNU05]

Cell type	Largest commercial standard module	Best prototype module	Best laboratory cell	Theoretical limit
CRYSTALLINE SEMICONDUCTOR				
Silicon				
Monocrystalline Si	15.3	20.8	24.0	30–33
Polycrystalline Si (p-Si)	11.1	17.0	17.2	22
EFG-band p-Si				50.00
GaAs				
Monocrystalline on other substrate				1.00
THIN FILMS				
Amorphous silicon (a-Si)	6.8	10.2	12.7	27–28
Polycrystalline silicon			15.7	
Polycrystalline GaAs			8.8	
CdTe	7.25	10.0	15.8	28
CIS		11.1	15.9	23.5
GaAs *(in concentrator system)*	22.0		0.50	

maximum:

$$P_{max} = I_{max\text{-}power} \times V_{max\text{-}power}. \qquad (21.3)$$

This will occur when the load resistance, R_l, is given by:

$$R_l = V_{max\text{-}power}/I_{max\text{-}power}.$$

In general, the power, P, produced by the cell will be given by:

$$P = I^2 R_l.$$

For the use of solar cells in specialized applications, such as pocket calculators, the power produced by the cell may not be as important as the maintenance of a minimum voltage, but for true power applications, we require the maximum power to be produced at all times. As Figure 21.5 shows, the maximum power point, i.e. the maximum of Equation (21.3) occurs at almost constant voltage and is therefore proportional to the current. This means that a cell connected to a resistive load will not be at its most efficient at all levels of solar radiation and that as illumination levels fall – with cloud cover, or at the beginning or end of the day – the resistive load required to ensure maximum efficiency ideally needs to increase.

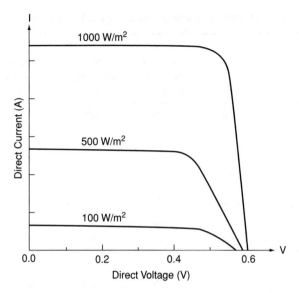

Figure 21.5 Performance characteristic of a typical solar cell exposed to various levels of sunlight[1] [SHE97, p335] [Reprinted from W. Shepherd and D.W. Shephard, *Energy Studies* , p150, ©1997 with permission from World Scientific Publishing Co. Pte.Ltd, Singapore]

21.5 Roof-top PV

Although up to now most PV technology has been used in relatively low-power applications, from pocket calculators to powering small pumps or telecommunications systems in remote locations, the two most interesting applications are in large-scale power generation and in building-mounted PV.

Problem 21.1 indicated that 5.44 km^2 of satellite-mounted PV would be required to produce 1 GW$_e$ of power. Solar intensity is lower at ground level due to absorption by the atmosphere, the angle of the surface to the sun and the fact that for half the hours in the year it is dark. This suggests[2] that approximately 30 km^2 would actually be required to produce as much energy per annum as a 1 GW$_e$ fossil fuel plant. This is a lot of land to cover in a human-made material and resistance from both the local population and environmentalists to such a facility might be expected. However, we are quite used to covering large areas of land with human-made materials when we build houses or other structures. If the roof of a building could be covered with PV cells, or even better, if the PV cells themselves could replace the roof structure, large amounts of electrical energy

---

[1] Reprinted from *Energy Studies*, Shepherd & Shepherd, 1998, with permission from World Scientific Publishing Co. Pte. Ltd, Singapore.

[2] Based on: average planetary insolation at ground level = 240 W/m^2; efficiency = 14 per cent; i.e. 240 × 0.14 = 33.6 W/m^2; therefore we need 1 × 10^9/33.6 = 30 km^2.

could be generated without having a big impact on the landscape (although the roof would look somewhat different from traditional designs).

Problem 21.2 Approximately how many houses would need to have their roofs replaced by PV cells in order to generate the equivalent to 1 GW$_e$ continuously?

The answer to Problem 21.2, around 600 000 houses, is a fascinating result. Ignoring electricity used for space and water heating (which could be replaced using gas or other means), the average UK home uses 3880 kWh of electricity per annum [UST05]. This is equivalent to 440 W continuously. So, our 600 000 houses would provide nearly four times more electricity than they need. It is worth pausing for thought in order to really digest this result: much of the world's 'necessary' electricity use (i.e. that not used for heating) could probably be replaced by an alternative that would have little impact on the environment without the building of any new large-scale structures – we just have to build our roofs from a different material. At present the cost of PV cells is too high for this to happen unless large subsidies are in place. There are also additional questions that need to be addressed, such as the production of the required area of semiconductor and the safe use of the chemical compounds used in the process.

As we have seen in Chapter 20 when we discussed passive solar design, for much of the world there is a seasonal discrepancy between the time of maximal resource and the time of maximal use for solar power. The resource is at its maximum around midday on sunny days in summer. By contrast, household electricity use is often greatest in winter, during cold dark days and nights. Thus any roof-based system capable of providing the majority of household electricity will over produce at times but under produce at others. Clearly, we either need to store the energy for a substantial period of time, or export the summer excess to other users. (For countries with a large demand for summertime air conditioning this seasonal discrepancy does not exist.) As the following problem illustrates, local storage is an option.

Problem 21.3 Ignoring that used for space and water heating, average UK household electricity use is 3880 kWh per annum. Estimate how many 12-volt car batteries (rating 80 amp hour[3]) would be needed to store one day's electrical energy for a household.

A more practicable option is to feed the excess into the national grid when the PV system is producing more electricity than the household can use. This is relatively simple and relies on a reversible electricity meter within the home, which either subtracts from or adds to the customer's bill depending on the direction in which the electricity flows. Other additional equipment required includes an inverter (to convert the direct current produced by the PV system to alternating current) and a transformer to increase the voltage to that of the mains supply.

---

[3] i.e. it can provide 80 amps for 1 hour at 12 volts, or 1 amp for 80 hours, or any such combination.

Figure 21.6 (a) Chulmleigh School, Devon, UK. This is a carbon neutral design, which combines PV and a ground source heat pump to provide both electricity and heat. (b) The PV modules are mounted on the south-facing sides of a pair of sloping turrets and on the roof. The northern side of the turrets are glazed to allow daylight into the building.

Another possibility is the cladding of commercial buildings in PV materials. As offices often have high daytime electricity use the advantages are obvious. There is also the possibility of combining a grid-connected PV system with a heat pump (Chapter 20) to effectively multiply the value of the electricity produced by the *COP* of the heat pump. However, it should be noted that this approach converts high-grade electrical energy into a large amount of low-grade heat energy – typically for space heating. Thus it may be environmentally better to export all the electricity possible from the building and use

Reproduced courtesy of ETSU, CADDETT, Technical Brochure 162, 2002, IEA CADDET programme, http://www.caddet.org.

Figure 21.7 Doxford Solar Office. Source: IEA CADDET programme[4] [CAD02]

an efficient boiler to heat the property if a convenient supply of gas or oil exists. However in areas not connected to a gas supply, a PV-powered heat pump might well be a sensible approach. It would also have the benefit of producing a carbon-free design, which might have a certain cachet, with some occupants. Figure 21.6 shows this approach applied to a UK school.

21.6 Example application: Doxford Solar Office, UK

The Doxford Solar Office Block in the UK (Figure 21.7) incorporates 532 m^2 of PV generating 73 kW (peak). The electricity is either used within the building, or exported

Table 21.3 Summary statistics for Doxford Solar Office [CAD02]

Incident solar radiation (horizontal plane)	950 kWh/m^2/year
Installation factor (accounts for incline, orientation and coatings)	1.04
Therefore solar radiation	988 kWh/m^2/year
PV area	532 m^2
Therefore total radiation	525.6 MWh/year
PV efficiency (at 25 °C)	0.14
Multiplier to account for operating temperature	0.9
Multiplier to account for losses from cables	0.98
Efficiency of inverter	0.85
Therefore estimate output	55.17 MWh/year
Therefore whole system efficiency	10.5%

---

[4] Web site: http://www.caddet.org.

to the grid. By implementing a passive ventilation philosophy, energy use within the building was also minimized, adding to the energy efficiency of the design.

The building is V-shaped with a 66 m long central south-facing inclined facade. Behind this facade is a three-storey atrium containing an internal 'street'. The central facade incorporates 352 polycrystalline PV modules within double glazed units. These are alternated in bands with conventional double glazing. In addition, the design includes the provision of cross-ventilation (this maximizes air flow for cooling through the building), high levels of natural light within the space, the use of stack effects for ventilation and cooling, and a well-sealed design to reduce heat loss during heated/unoccupied periods [CAD02].

In use the building is generating around 55.2 MWh/year compared with the predicted 56.5 MWh/year (see Table 21.3 for other statistics). Energy savings (including sales of exported electricity and reduced gas consumption) are estimated to be around €80 000 (£55 000, US$103 000) per annum. The cost of the PV installation, including glazing, invertors and cabling was around €1.4m (£950 000, US$1.78m). This suggests a simple payback period of 17 years. However, much of the energy saving is not from the PV system, suggesting the payback period of this element of the design is greater than this – possibly in excess of the lifetime of the PV modules, indicating that at present PV is not as economically viable as other sustainable energy technologies.

21.7 OECD resource

Many PV systems provide power to isolated equipment and are therefore off-grid. This makes it difficult to estimate total power production. The International Energy Agency [IEA05] states that PV produced 17 GWh in 1990 and that this increased to 361 GWh in 2002, thus achieving a 29 per cent annual growth. This makes it the fastest growing sustainable energy technology. Germany alone produces 188 GWh per annum from PV.

21.8 Student exercises

"In science there is only physics; all the rest is stamp collecting."

–Ernest Rutherford

1. Compare the efficiencies of a polycrystalline silicon photovoltaic cell at 20 °C (where its efficiency is 10 per cent) with that at 35 °C and 50 °C. Estimate the maximum energy that could be used by a cooling system in these two cases that would still be energetically worthwhile.

2. Describe how a photovoltaic cell works.

3. Describe and contrast crystalline, polycrystalline and amorphous silicon photovoltaic cells.

22
Wind power

"We simply must do everything we can in our power to slow down global warming before it is too late . . . The science is clear. The global warming debate is over."

–Arnold Schwarzenegger, bill signing ceremony for
California's strict anti-emissions law, September 26, 2006. [1]

In Chapter 3 we saw that over 10 TW is flowing in the world's winds – enough to meet the world's primary energy demand. This is all kinetic energy derived from sunlight and suggests that wind capture might be a sensible technology to develop in the search for a clean renewable source of energy.

In comparison to technologies such as photovoltaics, the core technology is simple, and has been used by humankind for several thousand years. The earliest machines were used to raise water for irrigation, grind corn and to propel boats. Many of these designs were found right across Asia, certainly from 250 BC and probably earlier. They were simple vertical axis machines with a central rotating shaft that incorporated a screen to ensure that only one side of the 'sails' felt the pressure of the wind. By the eleventh century AD, horizontal axis machines had been developed and introduced into Europe. The design was mainly used for grinding grain – hence the name *wind mill*. By the eighteenth century up to 100 000 such mills were operating in Europe [SMI94, DEZ78] and 10 000 in England alone by the nineteenth century (Figure 22.1 and 22.2). It is interesting to note that an equivalent number of modern wind turbines each with a rated output of 2 MW$_e$[1] would give an installed capacity of 20 GW$_e$: 30 per cent of current UK installed capacity. It might therefore be suggested that the landscape could again hold 10 000 wind mills without undue degradation (although they would be taller

---

[1] In this and subsequent chapters we will occasionally drop the use of the subscript 'e' when it is obvious we are referring to electrical output.

Energy and Climate Change   David A. Coley
© 2008 John Wiley & Sons, Ltd

Figure 22.1 Shipley horizontal axis windmill, England, which was built in 1879 [SHI05]

than their historic counterparts). This would also return us to large-scale embedded generation, where there is less geographic separation between generation and use. This *re-democratization* of supply would allow us all to experience both the benefits and the consequences of energy supply.

Horizontal axis machines have the advantage that no screen is required and hence all the sails – which traditionally might have been made of cloth or wooden slats – are exposed and able to extract energy through the whole cycle of rotation. They do however need to be able to point toward the oncoming wind. In traditional windmills this was achieved by rotating the tower manually or by a yaw mechanism that either rotated the whole structure, or just the top of the tower. Power output was controlled by feathering of the wooden slats or reducing the amount of cloth deployed. The output of such a mill was around 20 kW, today's machines develop between 200 W_e and 2 MW_e. However, large sailing vessels in the nineteenth century, which also represent a wind energy capturing technology, outclassed even today's machines.

Today, wind power is still used to pump water and grind corn in some parts of the world, however these are limited applications and the current resurgence of wind power has focused on the production of electricity. This is achieved by connecting the rotating axle to a gearbox and then a generator. The electricity produced can either be used locally, or more typically exported via a national grid. Almost all the machines (termed *wind turbines*) used today are horizontal axis machines (Figure 22.8) although some vertical axis machines have been developed (Figure 22.3).

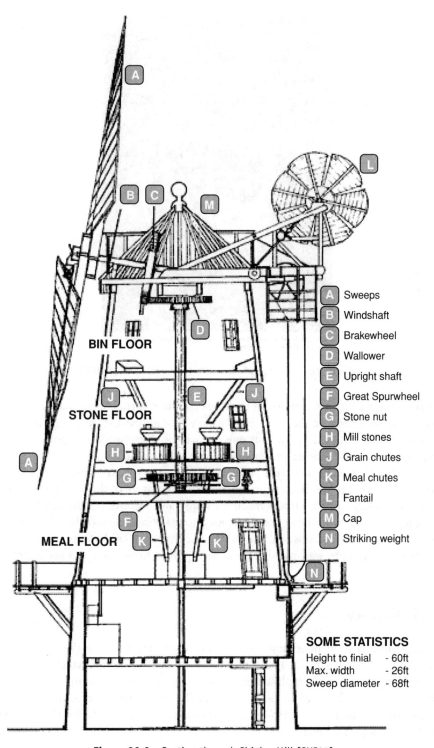

BIN FLOOR

STONE FLOOR

MEAL FLOOR

A Sweeps
B Windshaft
C Brakewheel
D Wallower
E Upright shaft
F Great Spurwheel
G Stone nut
H Mill stones
J Grain chutes
K Meal chutes
L Fantail
M Cap
N Striking weight

SOME STATISTICS
Height to finial - 60ft
Max. width   - 26ft
Sweep diameter - 68ft

Figure 22.2 Section through Shipley Mill [SHI05]

Figure 22.3 Darrieus-type vertical axis machines [MEM05]

Because wind turbines are one of the most common forms of alternative energy in place today and because their relatively widely distributed nature means that many readers will have the chance to visit one, we will be discussing wind power in more detail than some of the other sustainable energy technologies.

Ultimately there are limits to the percentage of a nation's electricity supply that can be generated by wind power because of the intermittent nature of the resource. However, if we move away from a carbon-based economy and toward a hydrogen-based one (Chapter 28), then wind power could be used to make hydrogen by the electrolysis of water. This would avoid this limiting factor. The hydrogen could then either generate electricity when there was little wind, or be exported to be used as a transportation fuel.

22.1 History

The first automatically operating wind turbine for electricity generation is believed to have been built by Charles F. Brush in 1888. He based his design, which was built of wood (Figure 22.4), on the high solidity devices normally used to pump water (i.e. with only a small space between blades). This kept the rotational speed of the 17 m rotor low and limited its efficiency. However it did produce up to 12 kW_e, which was used to charge batteries for 20 years.

Figure 22.4 The Brush wind turbine in Cleveland, Ohio, USA. Note the person to the right of the turbine for scale [DAN05]

The next breakthrough was the use of high-speed, low solidity machines, particularly by Poul la Cour who constructed a series of machines around 1897 at Askov Folk High School, Denmark (Figure 22.5). The electricity was used to produce hydrogen, which was used in turn to light the school. The idea proved popular and by 1918, 120 local

Figure 22.5 Turbines at Askov Folk High School, Denmark [DAN05]

Figure 22.6 The Smith-Putnam wind turbine, Vermont, USA (http://www.wind-works.org/photos/
Smith-PutnamPhotos.html)

utilities had wind turbines typically with a size of 20 to 35 kW$_e$, totalling 3 MW$_e$ of
installed capacity and supplying three per cent of Danish electricity consumption.

In the late 1930s a 1.25 MW$_e$ two-bladed, upwind turbine was constructed in Vermont, USA (Figure 22.6). This truly massive machine, with a rotor diameter of 53 m,
held the record of the largest turbine for 35 years and established the form of future
machines. When wind power returned toward the end of the century as a serious idea,
machines were smaller with the idea being that they would be collected together to
create *wind farms*.

During the early 1980s California underwent something of a wind-rush with thousands of machines being constructed in very large wind farms (Figure 22.7). This approach failed to find favour in the much more crowded European landscape where
250–300 kW$_e$ turbines were erected typically in groups of around 20.

Since the 1990s wind energy has developed in two directions: towards larger machines,
and offshore siting. Larger machines are more efficient and allow much smaller sites
to be used (one 1.5 MW$_e$ machine can replace five 300 kW$_e$ turbines, or thirty 50 kW$_e$
ones). They are also ideal for offshore generation where visual impact and noise is
less of a problem. Moving offshore also allows substantial numbers of large machines
to be grouped together in a windy environment. Figure 22.8 shows a NEG Micon
2 MW$_e$ machine with a rotor diameter of 72 m on top of a 68 m tower – a truly
massive machine, but still less than twice the capacity of the Smith–Putnam machine
of 1938. Figure 22.9 shows the largest wind farm in the world, which was completed
in 2002. It consists of eighty 2 MW$_e$ Vestas turbines located 14–20 km off the coast of
Denmark.

Figure 22.7 A 1000-turbine wind farm using Micon 55 kW$_e$ turbines, Palm Springs, USA [DAN05]

22.2 Technologies for use

The wind

As with any potential energy source, there is the need to understand both the form of the resource and the technology for extraction. Winds arise from differential heating of the earth's surface on a local, regional or worldwide scale. Near the equator the surface

Figure 22.8 NEG Micon 2 MW turbine, at Hagesholm, Denmark. Note vehicles for scale [DAN05]

(a)

(b)

Figure 22.9   (a) and (b) Horns Rev off-shore wind farm (Denmark) © Elsam A/S, 2006 [ELS05]

of the planet is perpendicular to the sun's rays; at the poles it is almost parallel. This leads to much greater heating near the equator with enormous quantities of heat being transferred to the air, which expands and rises. Conversely, air cools over the poles and subsides. This process results in a bulk movement of high-level air from the equator and its return in the lower atmosphere. This simple picture is complicated by the rotation of the planet and the non-uniform distribution of the continents and oceans.

Figure 22.10 shows the approximate distribution of these global winds. Three convection cells can be seen. The tropical cell extends from the equator to around 30°. This cell is powered by the strong heating effect in equatorial regions mentioned above. Warmed air rises to the top of the troposphere (16 km or 10 miles) and heads towards the poles. Near the equator the wind velocity is usually modest (as the air is generally rising) and unpredictable, hence the term *the doldrums* coined by mariners for the region. Such a region is unlikely to be ideal for wind farm development unless other, local winds can be relied upon. To the north (or south) the air returns to lower levels in a similarly slow and fickle manner, creating the *horse latitudes*, so named because sailors becalmed here for extended periods were forced to throw horses and other animals overboard to preserve supplies of drinking water. Again the wind pattern suggests this would also be a poor region to develop wind power unless good local winds existed.

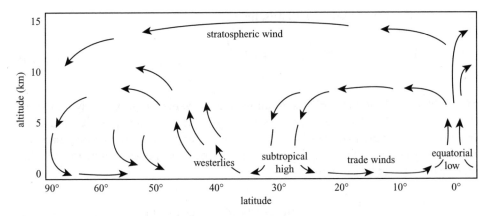

Figure 22.10   Section through the atmosphere showing large-scale winds [SMI99, p13]

The returning air moves outward both to the north and south creating much more constant surface winds: the trade winds. Between approximately 30° and 60° we find the temperate, or mixing cell where cold air from the poles mixes with warmer equatorial air, producing high-speed long-standing surface winds ideal for wind generation. This is also a highly populated region containing many of the richer, fossil-fuel hungry nations. Further north (south) the polar cell extends from around 60° to the poles. This cell also contains regions of consistent surface winds, but population density is typically low, suggesting that wind power in these areas might be mainly deployed for local generation and supply schemes for remote communities. Such a development is presented in detail in this chapter as an example application.

Global winds are not the only source of kinetic energy to drive wind turbines. Local winds that are also powered by differential heating but on a much smaller scale can be used. Sea breezes are formed when coastal land warms during the day more rapidly (due to its lower heat capacity) than the adjacent sea. Hot air rising above the land is replaced by cooler air over the sea, giving rise to an on-shore breeze. At night the situation is reversed, the land cools more quickly than the sea and the breeze is from the land. In areas like the Mediterranean where clear skies are common and heating substantial, such breezes arise daily, occurring with an almost clockwork regularity. One example is the island of Crete (population 600 000), which generates 10 per cent of its electricity from wind power. Localized differential heating can also create regular winds on mountainsides.

Energy in the wind

Estimating the energy flow provided by moving air is straightforward. Imagine a disk of diameter d, equivalent to the swept area of a typical wind turbine, of say 20 m in diameter. If the wind speed, v, is 10 m/s, then:

$$\pi \left(\frac{d}{2}\right)^2 v = \pi \left(\frac{20}{2}\right)^2 10 = 3140 \text{ m}^3/\text{s} \qquad (22.1)$$

of air will move through the blades. As air has a density, ρ, of approximately 1.2 kg/m³, this is $3140 \times 1.2 = 3768$ kg of air. This is a large mass, travelling quite fast and arises from the surprisingly high density of air, which we tend to consider almost mass-less in our everyday lives. If we could capture all the kinetic energy carried by this moving mass, m, we would be extracting:

$$\frac{1}{2}mv^2 = \frac{1}{2} \times 3140 \times 10^2 = 188\,400 \text{ J/s} = 188.4 \text{ kW}. \qquad (22.2)$$

Putting Equations (22.1) and (22.2) together we see that in general the energy per second, or power, $P_{kinetic}$, will be given by:

$$\frac{1}{2}\left(\pi \left(\frac{d}{2}\right)^2 \rho v\right) v^2$$

or

$$\boxed{P_{kinetic} = \frac{1}{2} A \rho v^3} \qquad (22.3)$$

where A is the swept area. If we use Equation (22.3) to calculate the power passing through one square metre for various wind speeds (Figure 22.11) we see that the wind is a naturally concentrated form of solar power for all but modest wind speeds; implying that wind power shares similarities with biomass, which is also a concentrated form of solar power.

Problem 22.1   Use Equation (22.3) to estimate the air speed in km/h required to give a 30 m diameter wind turbine an output of 70 kW. How common is this wind speed likely to be?

As we will demonstrate later, it is impossible to extract all of this energy; if we did the air would have been bought to a complete standstill, making it impossible for any more air to reach the turbine blades and impart its energy. However, what Equation (22.3) does show is that the power produced by a wind machine will be proportional to the swept area and proportional to the cube of the wind speed, i.e. doubling the swept area (by increasing the size of the machine, or adding more identical machines) doubles the power obtained, but doubling the wind speed increases the power eight fold. For this reason, small differences in wind speed between otherwise identical sites

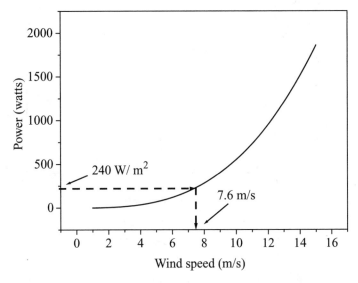

Figure 22.11 Comparison of the power contained in moving air with global average insolation levels (240 W/m^2). At all but very modest wind speeds, wind power is seen as a concentration of solar power, not a dilution

can have large implications for the relative economics of the sites. It also suggests that if wind power is only just economical, then developers will tend to choose only the windiest sites, many of which may well be in or close to mountainous or coastal areas where sensitivities about the landscape are greatest. In addition, a very accurate knowledge of local wind speed is clearly required to access the likely viability of a site. This will usually require long-term, on-site measurements. Even with such measurements, surprises can occur, as is illustrated by one of the example applications at the end of this chapter.

Figure 22.12a shows a typical frequency distribution of wind speeds and Figure 22.12b the power produced by a hypothetical, but typical, wind turbine as a function of wind speed. The wind speed at any location follows an approximately normal distribution. The power curve illustrates the cut-in speed, below which the turbine is stationary, a rise in output as the wind speed increases and a flattening out once the rated output (250 kW$_e$) has been reached, beyond which no additional power is obtained even if the wind speed increases. Clearly, despite Equation (22.3), the power extracted does not climb with the cube of the wind speed and at high wind speeds it is not even influenced by it. Therefore there must be some, wind speed-dependent, coefficient of performance for a real machine that limits its performance. Interestingly, it is possible to derive the maximal value of the coefficient from completely general arguments, and we shall do so later. The craft in matching a turbine to a particular site is in selecting a machine whose power curve matches that of the available wind. Any winds with greater speed than the rated wind speed (12 m/s in Figure 22.12b) represent energy lost; conversely, too high a rated power may mean that less energy is extracted at more common low wind speeds.

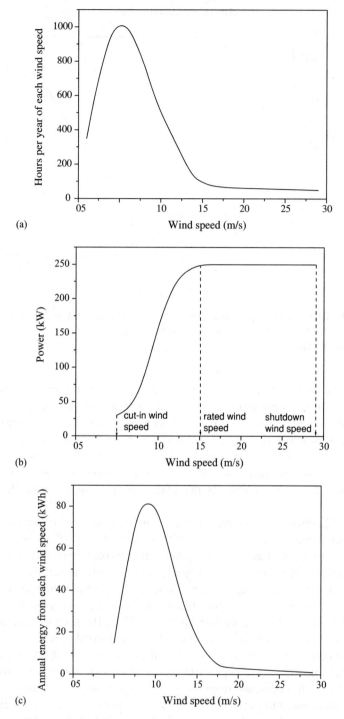

Figure 22.12(a)–(c) Wind speed distribution, power curve and energy distribution for a hypothetical turbine and location pairing [adapted from Boyle, G., *Renewable Energy: Power for a Sustainable Future*, Oxford University Press, 1996]. Note how the slope of the power curve does not follow the cube of the wind speed; this is because of engineering efficiencies and limits

By multiplying the power curve, point-by-point, by the wind speed distribution at each possible wind speed, we obtain the wind energy distribution for a particular mix of turbine and location (Figure 22.12c). Summing this distribution over all wind speeds gives the total annual energy production from the site.

All this might seem to suggest that ideally we should seek out the windiest locations possible and design turbines capable of harnessing such winds. However, the economic viability of a wind power site will also depend on other factors: the non-availability of the machines due to servicing, the cost of the machines and their installation, their life expectancy, and the distance and cost of connection to the electricity distribution grid. These variables are likely to be affected by the harshness of the environment, particularly if the wind farm is offshore.

Real wind generators

Even without precise knowledge of the details of a particular machine, we can get a reasonable idea of how much energy we might be able to obtain from a modern wind turbine and thereby what contribution it might be able to make to a local or national renewable energy strategy. From Equation (22.3) we know that:

$$P_{kinetic} = \frac{1}{2} A\rho v^3$$

and, as there are 7760 hours in a year and $\rho = 1.2$ kg/m^3, the annual energy, E_{annual}, of the wind through our turbine blades will be given by:

$$E_{annual,kinetic} \approx \frac{1}{2} A \times 1.2 \times v^3 \times 7760 \ Wh = 4.7 Av^3 \ kWh,$$

the result being only approximate as we do not know the coefficient of performance of the machine or the frequency distribution of wind at the location in question. As we will see later the coefficient of performance will always be less than 60 per cent, in addition, maintenance schedules mean that machines are not always available (typical availability is in the region of 95 per cent), energy is lost in converting the rotational energy of the system into electricity and, if a cluster of machines are installed to produce a wind farm, five per cent might be lost in wind shadows created between machines. In practice the constants in this equation can be replaced for typical wind turbines by the value 2.5, giving:

$$\boxed{E_{annual,electrical} \approx 2.5 Av^3 \ kWh}. \qquad (22.4)$$

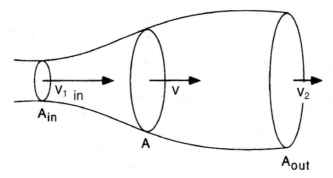

Figure 22.13 Wind-tube and speeds either side of a turbine [adapted from BOE99, p133] [Reproduced with permission from Boeker, E & Gondelle R., Environmental Physics, p133, ©1999 John Wiley & Sons, Ltd]

Betz' Law

We know that a wind machine cannot be 100 per cent efficient because this would mean bringing the air in front of the machine to a standstill, leaving no space into which more energy-carrying air could move, thereby stalling the blades. It is possible to calculate the maximum possible efficiency of a propeller-like wind turbine from theoretical considerations only. The result is 59 per cent, which agrees reasonably well with the 45 per cent measured efficiency found in real machines.

This maximum occurs when the wind is slowed to one third of its original speed. The proof of this, which applies to all wind converters with a disk-like rotor, was given by Albert Betz in 1919 [BET26] and is relatively straightforward. Figure 22.13 shows a tube of air intersected by a rotor, which is lowering the air speed and extracting energy from it. As the air on the right hand side of the rotor must be moving more slowly away from the rotor than that on the left (because it has given up some of its energy, and therefore momentum) it must be spread over a greater area. At the rotor the wind speed will be the average of the wind speed on either side (This was formally proved by Betz[2]), so the mass of air passing through the rotor per second will be:

$$m = \rho \left(\frac{V_1 + V_2}{2} \right) A \qquad (22.5)$$

where V_1 and V_2 are the wind speeds entering and exiting the turbine respectively.

The energy extracted per second by the rotor (i.e. the power, P) will be the difference between the energy carried per second by the in-coming and out-going wind:

$$P = P_1 - P_2 = \frac{1}{2} m V_1^2 - \frac{1}{2} m V_2^2.$$

[2] See the book's website for this element of the proof.

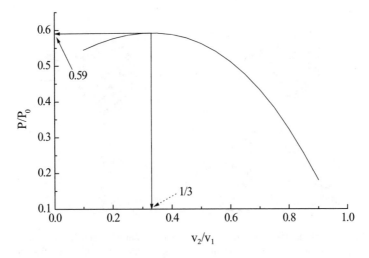

Figure 22.14 Plot of power extracted to power available for various combinations of V_2/V_1. The efficiency is found to peak at a value of 0.59

Substituting for m from Equation (22.5) then gives:

$$P = P_1 - P_2 = \frac{1}{2}\rho\left(\frac{V_1 + V_2}{2}\right) A\left(V_1^2 - V_2^2\right). \tag{22.6}$$

But from Equation (22.3) we know that the power, P_0, of an undisturbed wind (i.e. if the rotor was not there) of velocity, V_1, is given by:

$$P_0 = \frac{1}{2}A\rho\,V_1^3. \tag{22.7}$$

The ratio of the power extracted to the power of the undisturbed wind, from Equations (22.6) and (22.7) is therefore:

$$\eta_{Betz} = \frac{P}{P_0} = \frac{1}{2}\left(1 + \frac{V_2}{V_1} - \frac{V_2^2}{V_1^2} - \frac{V_2^3}{V_1^3}\right). \tag{22.8}$$

Plotting this ratio[3] as a function of V_2/V_1 (not forgetting that $V_2 < V_1$ and therefore V_2/V_1 is always <1), we see that the ratio of the power extracted to power available is 0.59 and is reached when $V_2/V_1 = 1/3$ (Figure 22.14). As mentioned earlier,

---

[3] Alternatively, we can differentiate Equation (22.8) with respect to $\overline{V} = V_2/V_1$ and set the result to zero to give the point where the ratio is at its maximum: $0 = \frac{d(P/P_0)}{d\overline{V}} = \frac{1}{2}(1 - 2\overline{V} - 3\overline{V}^2)$, or $0 = 1 - 2\overline{V} - 3\overline{V}^2$. Solving this quadratic for $\overline{V}$ gives the positive root $1/3$. Substituting this value into Equation (22.8) gives the maximum value of 0.59.

Figure 22.15 A small multi-blade high solidity device commonly used to pump water. By making the blades cover the majority of the swept area they interact with most of the air even at low rotational speeds allowing for greater torque to be generated at start up. This makes the approach suitable for designs where the rotor is directly connected to the pump. The low solidity devices used to generate electricity are 'spun-up' to an appropriate speed before generation commences. This is typically achieved by running the generator as a motor during this initial phase

this is a general result and applies to all disk-like rotor wind converters. It therefore plays a similar role in wind power systems as the Carnot efficiency does in heat engines.

Efficiency

It might seem strange that a device such as that shown in Figure 22.8 is capable of extracting very much energy at all from the wind. After all, won't almost all the wind just pass between the rotating blades? Surely wind turbines should be high solidity devices such as that shown in Figure 22.15. It is relatively simple to show that low solidity devices do interact with much of the wind passing through their swept area. Such machines rotate surprisingly fast, typically 60 revolutions per minute. Assuming the blades are a metre deep and the wind speed is 10 m/s, a parcel of air will take $1/10 = 0.1$ seconds to cross the depth of a blade. In this time the blades will sweep through 0.1 of a revolution, or 36 degrees. Given three blades, the machine will have interacted with $3 \times 36/360 = 0.3$, or 30 per cent of the air traversing it.

If the rotational speed is low the amount of interaction will be substantially less and the torque produced will also be much less, possibly not enough to rotate the machine. This suggests high solidity devices have an advantage at start-up, whereas low solidity devices might need to be spun-up using another source of power.

As of yet, we have not discussed the interaction of the wind machine with its environment. There will be drag forces acting on the blades and frictional losses from the blades passing through the air. Also, the rotation of the blades will disturb the general flow of air. These losses are subsumed into the dimensionless coefficient, COP, – the coefficient of performance (or power coefficient). There will also be other losses within the gearbox and the electrical generator, described by their coefficients, η_{gear} and η_{elec} respectively and allowing Equation (22.3) to be replaced by:

$$\text{electrical power,} \quad \boxed{P_e = COP\, \eta_{gear}\eta_{elec}\frac{1}{2}\rho A V_1^3}, \qquad (22.9)$$

where $COP = P/P_o$.

$$\text{That is, } COP = \frac{1}{2}\left(1 + \frac{V_2}{V_1}\right)\left[1 - \left(\frac{V_2}{V_1}\right)^2\right],$$

and as Figure 22.14 shows this will be at a maximum when $V_2/V_1 = 1/3$.

For real machines under real conditions COP never reaches 0.59, with a value of 0.4 being typical. We must not forget that this value will not be achieved at all wind speeds. Below the cut-in speed – typically 3.5 m/s – no generation occurs at all. To see how COP might change with the incoming wind speed we need to remove V_2 from the argument and replace it with the more natural rotational velocity of the turbine blades, ω. Some wind turbines use different rotational velocities at different wind speeds, but many modern designs use a single value. Therefore ω can often be seen as a constant. The tip speed ratio (TSR) is defined as the ratio of tangential velocity, v, of the tip of the blades to the incoming wind speed:

$$TSR = \frac{v}{V_1} = \frac{r\omega}{V_1},$$

where r is the radius of the swept area. If the rotational velocity is given as n revolutions per minute, then

$$TSR = \frac{2\pi nr}{60 V_1} \approx \frac{nr}{10 V_1}$$

Figure 22.16 shows COP plotted as a function of TSR, for various designs. It is clear that different designs obtain their maximum values of COP at different TSRs, with modern high speed twin bladed designs peaking at around 5.5 (i.e. the speed of the blade tip is 5.5 times the wind spped.); whereas multi-bladed designs with a high solidity factor, such as that pictured in Figure 22.15, peak at $TSR < 1$.

Figure 22.16 gives another clue as to when different machines might be appropriate. The solidity of a design is defined as the total blade area divided by the swept area normal to the wind. High solidity machines give lower power coefficients at all but the lowest wind speeds – when high-speed turbines would not rotate. We would therefore expect high solidity machines to be used when wind speeds are often light and efficiency is not

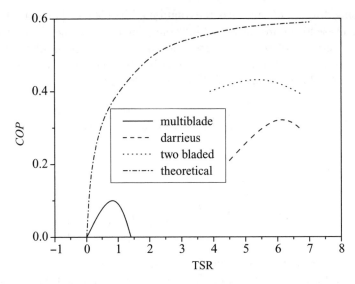

Figure 22.16 Power coefficient, *COP* as a function of tip speed ratio. TSR, for real and theoretical machines [data from Shepherd, W. and Shepherd, D.W., *Energy Studies*, World Scientific, 1997]

at a premium. An ideal application would be the pumping of water to a relatively small storage tank.

For commercial machines designed to generate large quantities of electricity, the efficiency of the whole system is paramount. We have already seen how the amount of power that can be extracted from the wind is dominated by the wind speed, and that the wind turbine chosen for a particular site must be selected so as to match this speed in order to minimize losses. Other losses arise in the generator and the gearbox, with the electrical power being given by Equation (22.9).

Table 22.1 gives typical values for the parameters in Equation (22.9). In practice the values suggest that large machines are capable of greater efficiency.

Problem 22.2 Using the values in Table 22.1, estimate the best and worst likely overall efficiency of a commercial wind turbine.

Table 22.1 Typical efficiencies of the various stages in an electrical generating wind turbine

	High (%)	Low (%)
COP	50	20
η_{gear}	95	70
η_{elec}	95	60

Table 22.2 Percentage costs of the various components of a 60 m diameter 1.5 MW$_e$ wind turbine [P. Fuglsang, and K. Thomsen in BUR01]. Reproduced courtesy of Burton, T., Wind Energy Handbook ©2001 John Wiley and Sons

Component	Cost as a percentage of total	Component	Cost as a percentage of total
Blades	18.3%	Controller	4.2%
Hub	2.5%	Tower	17.5%
Main shaft	4.2%	Brake system	1.7%
Gearbox	12.5%	Foundation	4.2%
Generator	7.5%	Assembly	2.1%
Nacelle	10.8%	Transport	2.0%
Yaw system	4.2%	Grid connection	8.3%
		TOTAL	100%

22.3 The modern horizontal axis wind turbine

Modern large grid-connected machines are produced in a variety of electrical capacities from 250 kW$_e$ to 2 MW$_e$. In this section we look at each of the major components of such machines and how they might be optimized for a particular range of conditions.

Table 22.2 shows the percentage costs of the various components of a 1.5 MW$_e$ machine. Some of these costs grow rapidly with machine size, such as blade cost; others increase more slowly, for example connecting the machine to the grid.

Rotor diameter

As we have seen, the energy capture of a wind turbine is proportional to the swept area. Wind speed also increases with height above the surface. Therefore a smaller number of larger machines would seem to be the sensible choice for any wind farm. However, blade mass, and therefore cost, increases roughly with the cube of the rotor diameter [BUR01]. This is because longer blades must also be made thicker and stronger to withstand the greater forces. The optimum balance between these factors is thought to be at a diameter of 44–52m [BUR01], although this could change with new fabrication technologies.

Rooftop generation in the urban environment from very small turbines is likely to prove unpopular because of the need to use reasonable rotor diameters. Even assuming high efficiencies were achieved (which is unlikely for a small machine), Equation (22.9) suggests that a turbine with a 1 m diameter would produce (at a wind speed of 8 m/s):

$$P_e = \frac{1}{2}\rho AV^3 COP\, \eta_{gear}\eta_{elec} = \frac{1}{2} \times 1.2\pi \left(\frac{1}{2}\right)^2 \times 8^3 \times 0.5 \times 0.95 \times 0.95 = 308\,\text{W}.$$

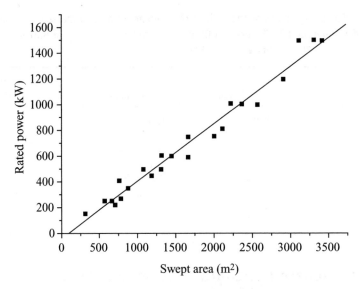

Figure 22.17 Rated power versus swept area [adapted from BUR01]

If we use Equation (22.4) to estimate the annual electrical output allowing for a typical wind distribution, we find $E_{annual} = 1005$ kWh, which dividing by the number of hours in the year gives an average power production of 115 W. Unless this small amount of power could be sold back to the grid, via a reversible electricity meter, its installation would be of little use.

Figure 22.17 shows the rated power of a large number of real machines as a function of swept area. As indicated by Equation (22.9) the relationship is a linear one, proving that the theory works in practice.

Problem 22.3 (a) Use Figure 22.17 to estimate the mean specific power (the rated power divided by the swept area) for real machines. (b) Use the value you derived in (a) to estimate how big a wind turbine would have to be to provide 1.5 kW$_e$ of power to a house. (c) By how much is this reduced if we allow for the non-constant nature of the wind? Hint: compare with the answer given by Equation (22.4) for a mean wind speed of 8 m/s. (d) By calculating the diameter of the turbine blades, state what the result suggests about the possibility of small roof-mounted wind turbines providing electricity within the urban environment.

Machine rating

The machine rating – the amount of energy produced when operated at the rated wind speed – needs to be large enough so that the opportunity to extract energy from substantial winds is not missed. However if the rating is set too high the machine will rarely generate at this rate and the extra cost of the larger machine will be unjustifiable.

Table 22.3 Dependency of weight/cost of components on rated wind speed [data from BUR01]

Object	Dependency of cost on rated wind speed
Foundations	Independent
Assembly	Independent
Transport	Independent
Blades	Directly proportional
Tower	Directly proportional
Gearbox	Varies as the rated wind speed squared
Generator	Varies as the rated wind speed cubed

The costs of some of the components are dependent on the machine rating; some grow in direct proportion to the rating, others grow more rapidly (Table 22.3). It turns out however that an error in the rated wind speed has a relatively small effect on the cost of electricity finally produced, with about a one per cent increase in cost for an error of ± 200 kW$_e$.

Rotational speed and number of blades

The speed of rotation does not define the rated power. It is possible to produce the same output with lower speeds by changing the design and number of blades, but this will increase the torque, requiring a stronger and therefore more expensive drive chain. However blade weight and cost increases with rotational speed. So, once more we find ourselves needing to select a balance.

There is also the need to consider the amount of noise pollution any wind farm might cause. The aerodynamic noise produced is approximately proportional to the fifth power of the tip speed. When the wind speed is low the background noise from the wind passing through trees and vegetation will also be low and blades rotating at a high speed will be clearly audible over a large distance. In practical terms this limits the tip speed to around 65 m/s. It has also been suggested [BUR01] that turbines are more visually intrusive the faster they rotate.

Problem 22.4 Given a tip speed limit of 65m/s for acoustic reasons, what is the maximum permissible number of rotations per minute of a 60 m diameter turbine?

Almost all modern machines operate with either two or three blades. Although costs can be reduced by opting for only two blades, the rotational speed is likely to need to be increased in order to extract the same amount of power. This may mean that the tip speed limit of 65 m/s will be exceeded. However, for designs operating in remote locations this is unlikely to cause any problems. It is also believed [BUR01] that three bladed machines are less visually intrusive.

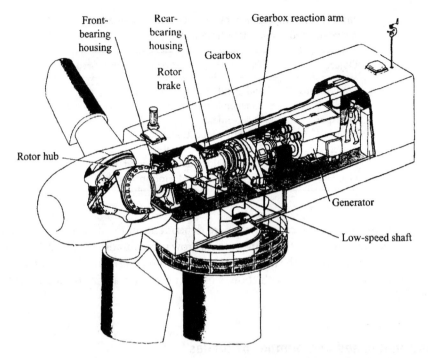

Figure 22.18   View of the Nacelle (note the person for scale) [BUR01] [Reproduced with permission from Burton, T., Wind Energy Handbook, p367 ⓒ2001 John Wiley and Sons, Ltd]

The generator

One of the most expensive components of a wind turbine is the generator and gearbox combination. These are housed in the *Nacelle* at the top of the tower and account for around 20 per cent of the overall cost. Figure 22.18 shows the basic arrangement. (Note the person for scale). All this has to be suspended 50 or more metres above the ground and capable of surviving storm force winds. Figure 22.19 shows a similar design with the lid removed and clearly reveals the combination of heavyweight mechanical engineering and more delicate electrical engineering that is at the heart of most power generating technology.

22.4 Environmental impacts

Wind power has one of the smallest environmental footprints of any generating technology. No rivers need to be dammed, no radioactive waste is stockpiled and no oil spills can arise. However, there are other concerns about how the look and sounds of the countryside can be altered by the construction of even a single wind turbine.

Figure 22.19 Turbine assembly of a 1.5 MW machine (note the person for scale) [BUR01] [Reproduced courtesy of Burton, T., Wind Energy Handbook, p 369 ©2001 John Wiley and Sons, Ltd]

Visual impact

Modern machines can be 70 m or more from ground to blade tip and need to be sited in locations with high wind speeds often near the top of slopes, which means they will be visible for a considerable distance. Because wind has a low energy density, wind farms need to be large and the requirement to space machines at least three to five rotor diameters apart to avoid problems with turbulence, means that they are even more extensive.

Problem 22.5 Estimate the land area required by a 200 MW wind farm (a very large installation).

If a wind turbine is viewed with the sun behind it, a stroboscopic effect will occur. In general the effect will only cause a shadow flicker on walls or the ground, although some may find this annoying. If the frequency is between 2.5 and 20 Hz more serious problems can arise particularly for sufferers of epilepsy. As the following problem shows, modern machines are designed to operate at lower frequencies than this.

Problem 22.6 Given the 65 m/s tip speed limit, find the maximum flicker frequency caused by a three bladed 60 m diameter machine.

Electromagnetic interference

Through a combination of their size, the materials they are made from and their location (often on the tops of ridges and hills) wind turbines have the potential to interfere

Table 22.4 Estimated number of birds killed in Solano County, USA from collisions with wind turbines [data from P. Gipe, in BUR01]

	Low estimate	High estimate
Birds/year	17	44
Turbines	600	600
Birds/turbine year	0.029	0.074
MW	60	60
Birds/MW year	0.29	0.74

with radio, television and radar signals and with microwave communication links. The amount of interference depends on various factors including [ETSU97]: rotor diameter, hub height and materials used. Good design can reduce the potential for interference, although it can be cheaper for wind farm developers simply to cover the cost of a cable television connection to those affected.

Ecological impact

In our crowded world, sites ideal for producing wind power are often areas of ecological importance. Direct loss of habitat due to a wind farm is small (approximately three per cent of the site area [GIP95]). Although additional damage will occur during construction, which will be short lived. During the early days of wind power development it was thought that large numbers of birds might be killed in collisions. Experiments have proved the opposite and collisions are rare. Table 22.4 gives the number of birds killed in Solano County, USA during 1995 by turbines. At other sites where the clusters of turbines are smaller collisions may well be lower as the site will be easier to fly around [BUR01]. At other sites with greater bird populations, for example coastal regions, collisions might be more frequent.

At Blyth Harbour, UK, 31 collisions were reported over three years from the nine 300 kW$_e$ turbines. As a comparison, more than one million birds are killed by road traffic in Denmark each year (human population 5.4 million) [ETSU96]. Of greater concern is the possibility that poorly sited offshore wind farms might discourage birds from using traditional feeding areas [NEW05].

Problem 22.7 India's fossil fuel-based generation is 457×10^9 kWh per annum (2002). Estimate the number of birds that might be killed per annum in India if 20 per cent of its fossil fuel-generated electricity was obtained from wind power.

Noise

Leaving aside subjective, aesthetic concerns over the visual pollution caused by a wind farm development, concerns over noise emissions are the greatest restraint on the spread of onshore wind power. We will therefore consider noise pollution in some detail.

Wind turbines produce mechanical noise mainly from the gearbox and the generator. By careful design and the use of anti-vibration couplings such noise can be greatly reduced. Aerodynamic noise is much more fundamental and in part arises from the changes in wind speed experienced by the blades as they pass the tower; however the frequency of this sound is low and the human ear is not particularly sensitive to it. The interaction of the blades with eddies in the air stream causes a swishing noise with a frequency of up to 1000 Hz. Aerofoil noise is generated by the blade as it slices through the air and is broadband in nature (i.e. composed of a wide range of frequencies) mainly between 750–2000 Hz (where the ear is highly sensitive) and is produced most notably by the trailing edge of the aerofoil. For various reasons, the aerodynamic noise is modulated, i.e. fluctuates in a regular way, and it would appear that this makes it potentially more annoying [PED03].

The amount of aerodynamic noise can be reduced by lowering the rotational speed or by reducing the angle of attack of the blades – both of which would reduce the power extracted by the machine. Variable or two-speed machines can do this when the wind speed is low and natural masking or background noise is at a minimum – an advantage of such machines.

The sound power of a device refers to the amount of energy in the form of sound radiated by the device each second. This energy is radiated into the air around the device and as we move away from the source the loudness of the sound will decrease as the energy spreads itself out over an ever-increasing surface area. For a point source either high above the ground or above a sound-absorbing surface, the sound will be spreading itself over a sphere or hemisphere whose area will be increasing four fold for every doubling of distance and we will perceive the source becoming rapidly quieter as we walk away from it. Therefore wind turbine noise is usually only a problem relatively close to the turbines, although the rural environment in which they are often placed can be extremely quiet at night and disturbance is a real possibility. For line sources such as roads, noise drops off much less quickly with distance because the energy is spread over an expanding cylinder, not a sphere.

Sound power is measured in watts, but because the human ear is sensitive to sound energy over a range of 13 orders of magnitude, a logarithmic scale relative to the quietest sound we can hear is used. The sound power level, L_w, of a device is given by:

$$L_w = 10 \log_{10} \left(\frac{W}{W_0} \right),$$

where W is the sound power and W_o the reference level of 10^{-12} watts.

The ear (or a sound level meter) perceives pressure, P, not energy and the sound pressure level L_p is given by:

$$L_p = 10 \log_{10} \left(\frac{P^2}{P_0^2} \right),$$

where P_0 is the pressure reference level of 2×10^{-5} Pa (which equates to the quietest sound we can hear).

Manufacturers quote the sound *power* of a source because this number is unaffected by the distance from the source, the surrounding terrain or the number of sources. The units of L_w and L_p are both decibels (dB).

If we have more than one source (as we do in a wind farm) then the sound pressure level of the collection of N sources will be:

$$L_p = 10 \log_{10} \sum_{i=1}^{N} 10^{L_p(i)/10}. \qquad (22.10)$$

For each source, i, the sound pressure level at a distance, r, will (if spherical propagation is assumed) be given by:

$$L_p(i) = 10 \log_{10} \left(\frac{W_i}{4\pi r^2 10^{-12}} \right) = L_w - 10 \log_{10} \left(4\pi r^2 \right) \qquad (22.11)$$

where W_i is the sound power of the ith source. The sound power level of a 600 kW wind turbine can be 100 dB(A) [BUR01] (the (A) refers to A-weighting which simply accounts for the differing sensitivities of the human ear to differing frequencies). Therefore Equation (22.11) suggests that at 100 m this source might be perceived as a sound pressure level of:

$$L_p = 100 - 10 \log_{10} \left(4\pi \, 100^2 \right) = 49 \text{ dB(A)}.$$

For comparison, the sound pressure level of an office is around 50 dB(A) and the rural nighttime background level only about 25 dB(A). Therefore the turbine will be clearly audible and a potential nuisance.

The key to avoiding this problem is not simply good engineering but distance. The further the wind farm is from people the less chance there is for concern. As the following problem shows, this distance need not be all that great.

Problem 22.8 Estimate the sound pressure level 300 m from a single turbine with a sound power level of 100 dB(A). Estimate the sound pressure level 600 m from 10 turbines each with a sound power level of 100 dB(A).

Figure 22.20   A small Ropatec WRE.005 500 W vertical axis wind turbine (Nuuk, Greenland)[ROP05] [Reproduced from Ropatec Ltd, http://www.ropatec.com/en/reference_gallery]

At 600 m the answer to Problem 22.8 is probably less than the background noise level in most locations (except at night at very low wind speeds in open countryside) and therefore unlikely to be a cause for concern. (Although it might be perceivable during other conditions as its tonal components will differ from those of the background noise.)

The figure of $L_w = 100$ dB(A) mentioned above was measured at a modest wind speed of 8 m/s. The sound power of a turbine will typically increase by 0.5 to 1 dB(A) per m/s increase in wind speed. However, the background noise (from the wind passing through undergrowth and trees) will usually rise faster than this with increasing wind speed and it is therefore low wind speed conditions that are likely to cause complaint.

One advantage of vertical axis machines is that they can avoid the compression waves caused by the passing of the blades in front of the tower, and thus remove this source of noise. Machines such as that shown in Figure 22.20 might therefore be more suitable for highly noise-sensitive urban locations.

22.5 OECD resource

In 2002 wind produced 3.3 per cent of the renewable energy in the OECD (assuming all hydropower is classified as renewable). Between 1990 and 2002 wind power increased from 3.8 TWh to 47.6 TWh, an average annual growth of 23.3 per cent [IEA05]. Growth in absolute (rather than percentage terms) was greatest in Germany, the USA and Spain.

Table 22.5 Important parameters of
the Harøy turbines [data from CAD03]

Parameter	Value
Nominal output	750 kW each
Power regulation	Stall
Nominal wind speed	16 m/s
Cut-in wind speed	4 m/s
Cut-out wind speed	25 m/s
Rotor diameter	48.2 m/s
Rotor speed	15 or 22 rpm
Drive chain ratio	1:67.5
Nominal voltage	690 V
Hub/tower height	50 metres

22.6 Example application: Harøy Island Wind Farm, Sandøy, Norway

Sandøy consists of a series of islands on the west coast of Norway. Given an increasing demand for electricity towards the end of the 1990s, the local electricity company faced the decision to either strengthen the grid and import more electricity, or generate locally. Although other options were investigated, local generation using wind power proved the most economical. The example of Harøy shows not only that wind power can be economical even in harsh conditions, but also that local generation has additional benefits.

Five NEC Micon 750 kW turbines were installed, starting in August 1999, and the system was connected to the grid in December 1999. Table 22.5 lists the main parameters of the turbines used. The total generating capacity is: $5 \times 750 = 3750$ kW.

Before installation, the wind farm was expected to generate 11 GWh/year, unfortunately lower than expected winds in recent years have meant that less than 9 GWh/year has been realized. However, reducing the need to bring as much electricity long distances has reduced transmission losses by 1.1 GWh/year – this is more than 10 per cent of the realized energy production of the wind farm.

The wind farm cost €3.8m (£2.7m, US$4.9m), but attracted a 25 per cent subsidy from the Norwegian Water Resources and Energy Directorate. A further subsidy of €0.074/kWh (£0.053, US$0.096) generated was also granted. Against this, around €0.039/kWh (£0.028, US$0.50) is needed to cover operational and capital costs.

The potential for noise disturbance and the impact on birds was studied before the project was started and found to be acceptable, but is continuing to be monitored. So far bird life has been unaffected. Assuming a generating mix typical of OECD countries, the wind farm is offsetting about 4400 tonnes of carbon dioxide per year.

22.7 Student exercises

"Ignoring climate change will be the most costly of all possible choices, for us and our children."
−Peter Ewins, British Meteorological Office

1. Summarize the effect of various parameters on the cost of a modern wind turbine.

2. Explain quantitatively why a wind turbine can never have an efficiency of greater than 50 per cent?

3. At what rate does aerodynamic noise from a wind turbine increase with tip speed? What limit does this place on tip speeds and under what wind conditions is this most important? How does this limit reduce the possibility that such a machine can trigger an epileptic event?

4. Outline the major concerns over wind power (one paragraph on each).

5. Outline your personal feeling about wind power and its impact on the countryside.

6. What is meant by the re-democratization of the energy supply?

7. How many birds might be killed in your country per annum if 10 per cent of electricity were produced from wind power? (Hint: find your nation's electricity demand by searching the web.)

8. (a) Given an average wind speed of 10 m/s and a wind turbine with a swept diameter of 20 m, and assuming an efficiency of extracting energy from the wind of 40 per cent, how much power might such a turbine produce? (b) If the wind speed were to increase to 15 m/s, how much might now be produced? (c) Assuming the turbine were to operate 70 per cent of the time and the electricity produced be sold at £0.05 per kWh, how much extra money per annum might be generated by this relatively modest increase in wind speed?

9. Estimate the sound pressure level at a house 400 m from the edge of a 49-turbine wind farm laid out on a square grid. Assume the machines have a swept radius of 40 m and produce a sound power level of 100 dB(A). Might this level be likely to cause complaint?

10. Estimate the diameter of a wind turbine that would generate 10 MW of electrical power in a 15 m/s wind [adapted from SHE97].

11. A large propeller-type wind turbine has a diameter of 70 m. If the speed of rotation at full load is regulated to 32 rpm when the speed is 48 km/h, what then is the value of the tip speed ratio? [Adapted from SHE97.]

12. A two-blade propeller is used as a wind turbine directly on the shaft of a small electric generator. Assign typical efficiencies to the wind turbine and the generator and calculate the blade diameter required to generate 500 W_e in a wind of average speed 25 km/h. [Adapted from SHE97.]

23

Wave power

"America has not led but fled on the issue of global warming."

–John Kerry

Like wind power, and unlike tidal power, wave power is an indirect form of solar power. Because water has a higher density than air, much higher energy densities are realized: 70 kW per metre of wave front or greater at some sites. Figure 23.1 shows this energy density for various locations around the planet; the UK has particularly high densities and is therefore potentially a good location for generating electricity from wave power.

Many of the wave power devices prototyped so far would be placed a considerable distance from the shore and therefore it is probably the alternative energy technology with the least environmental impact. Even designs which need to be on the seashore are likely to have scant impact on the marine environment or on nesting and feeding bird populations, as they have little effect on the general flow of water around them. The central issue for the development of wave power is neither the size of the resource, nor possible environmental impacts, but the problem of harnessing it economically within the harsh marine environment. Tidal power faces a similar challenge, but has the advantage that the technology is likely to be deployed within sheltered estuaries and bays and hence in a less hostile setting. For wave power to work we need sites with regular waves. This creates problems for designers and also for construction teams attempting to deploy the technology on site.

A plethora of designs have been proposed with no one design dominating at either a theoretical or practical level, and it would be fair to say that the whole field is far less commercially developed than wind power. There are so many possibilities and prototypes that a general text like this cannot hope to do justice to all these designs and we will review just five devices. Two are onshore designs – the TAPCHAN and the Oscillating Water Column – and three are offshore technologies – the Sea Clam, the

Energy and Climate Change   David A. Coley
© 2008 John Wiley & Sons, Ltd

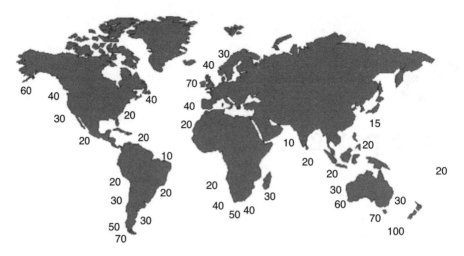

Figure 23.1 Average global energy densities of wave fronts (kW/km of wave crest) [EESD02]

Salter Duck and the Sea Snake. To appreciate the differences between these approaches necessitates studying the energy content of waves in deep and shallow waters.

23.1 Wave characteristics

If the height of a deep-water wave (defined as having $d \gg h$ in Figure 23.2) is h, the density is ρ and the wavelength is λ, then the total energy content, E, of the wave will be

$$E_{pot} + E_{kinetic} = E = \frac{\rho g h^2 \lambda}{8} \text{ J/m}$$

per metre of wave front, where g is the gravitational constant. In terms of the time period, T, of the wave (i.e. the time taken for consecutive peaks to pass a pole or other fixed object, such as a wave power device) we have:

$$E = \frac{\rho g^2 h^2 T^2}{16\pi} \text{ J/m.}$$

Dividing by T, then gives the power, P, per unit length of wave crest:

$$P = \frac{\rho g^2 h^2 T}{16\pi} \text{ W/m.} \tag{23.1}$$

Using the values, $g = 9.8$, $\rho = 1000 \text{ kg/m}^3$ and $\pi = 3.14$ allows us to write Equation (23.1) as

$$P = 2h^2 T \text{ kW/m.} \tag{23.2}$$

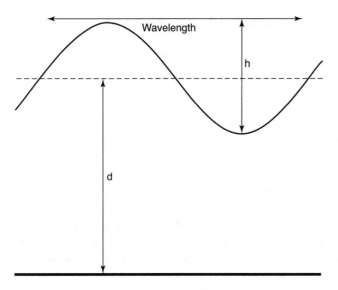

Figure 23.2 Ideal representation of a wave

In reality, the state of the sea will not be a continuous wave of wavelength λ and period T, but a complex amalgamation of waves of varying λ and directions of travel. In order to get an accurate idea of how much power there is over a reasonable length of time, measurement becomes the only option. A typical result [SHE97, TWI86] is that:

$$P_{measured} = 0.55h^2T \text{ kW/m}. \tag{23.3}$$

The power, P_{elec}, which can be extracted and converted to electricity by any wave machine, is likely to be even less, possibly only a third so:

$$P_{elec} = 0.55h^2T/3,$$

i.e.

$$\boxed{P_{elec} = 0.18h^2T \text{ kWh/m}}. \tag{23.4}$$

Dividing Equation (23.4) by Equation (23.2):

$$\frac{P_{elec}}{P} = \frac{0.18h^2T}{2h^2T} \cong 0.1.$$

Which demonstrates that we will only be able to extract 10 per cent of the total power indicated by Equation (23.2).

In deep water the velocity, v, of the wave will be proportional to the time period:

$$v = \frac{gT}{2\pi},\qquad(23.5)$$

and because $v = \lambda / T$, $\lambda = \frac{gT^2}{2\pi}$.

Interestingly, Equation (23.5) implies that the larger ocean waves travel faster than the smaller ones.

Near the shore in shallower waters, $d < \lambda/4$, and:

$$v = \sqrt{gd},\text{ i.e. the velocity is independent of the wavelength or period.}$$

Problem 23.1   Assuming a wave height of 7 m (typical of waves found in the Atlantic) and $T = 10$ s, use Equation (23.4) to estimate the length of wave power device needed to supply the non-heating electrical needs of 100 000 UK homes. (Assume 440 W is required per house – Chapter 21.)

23.2 Technologies for use

As mentioned in the introduction, wave power devices can be categorized as offshore or onshore. We will look at onshore devices first, partly because these have actually been built and used to deliver electricity to a national grid.

Onshore devices

In shallow water (defined as $d < \lambda/4$) the wave velocity, v, is independent of the wave period and is given by:

$$v = \sqrt{gd} \cong 3\sqrt{d}.$$

Thus, as the water becomes increasingly shallow toward the shoreline, the velocity of the wave will reduce and the energy content of the wave decrease due to frictional coupling between the wave and the sea floor. This suggests that only locations where deeper water is maintained all the way to the shoreline will be suitable for the siting of wave power devices. Such locations are typified by cliffs or rocky landscapes, a point exemplified by the following two examples.

Figure 23.3a shows the basis of a tapered channel wave energy converter, and Figure 23.3b a real system near Bergen, Norway. The device is a potential energy extractor containing many of the features of the tidal barrage systems studied in Chapter 24. The design consists of a tapered collector channel, a small raised reservoir and a turbine. Waves entering the 40 m wide collector increase in height as they run the length

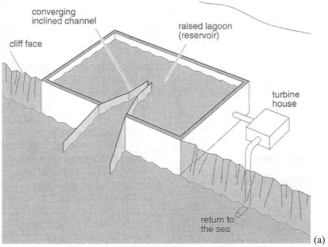

converging
inclined channel

raised lagoon
(reservoir)

cliff face

turbine
house

return to
the sea

(a)

(b)

Figure 23.3   (a) Schematic of a tapered channel wave energy converter (or TAPCHAN) [BOY96].
(b) Tapered channel wave energy converter near Bergen [SIN05]

of the narrowing channel eventually spilling over the sides and filling the reservoir,
thereby creating a store of potential energy. The reservoir empties through a low speed
Kaplan turbine and powers a 350 kW$_e$ induction generator connected to the national
grid.

The most obvious restriction to the widespread adoption of the technology is the
need for a very limited tidal range.

Several proposed wave energy converters make use of the incompressible nature of
water, which allows the sea to be used as a piston to force air through a Wells turbine.
Figure 23.4 shows the operating principles and Figure 23.5 the oscillating water column

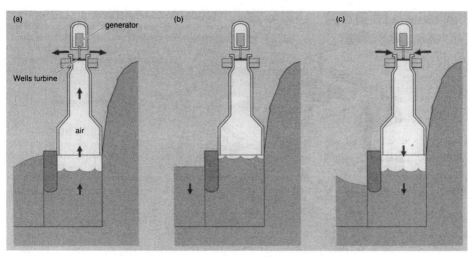

Figure 23.4 Schematic of an oscillating water column showing the three basic stages of operation: (a) compression and expulsion of the air in the column as the wave strikes; (b) equilibrium; (c) rarefaction as the wave returns out to sea and air is drawn in through the turbine [BOY96]

Figure 23.5 Limpet oscillating water column device on the island of Islay, Scotland. The water depth at the entrance to the OWC is typically seven metres and design has been optimized for annual average wave intensities of between 15 and 25 kW/m. The water column feeds a pair of counter-rotating turbines, each of which drives a 250 kW$_e$ generator, giving a nameplate rating of 500 kW$_e$ [WAV05]

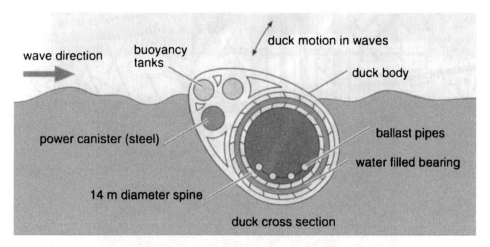

Figure 23.6   Salter Duck [BOY96]. The device rocks back and forth as waves pass [Reproduced with permission from Boyle. G, Renewable Energy: Power for a Sustainable Future ©1996 The Open University]

(OWC) device on the island of Islay, Scotland. The general idea of using an oscillatory water column and an air turbine has been adopted by various designers and applied to other wave energy converters both on and off shore.

Offshore devices

Offshore wave energy converters should be able to harvest more energy than their onshore cousins, as the size and reliability of the waves around them will be greater. However installation and maintenance costs are likely to be greater, reducing the chances of such devices being economic. Amongst the possibilities is the Sea Clam – a 60 m diameter toroid (think of a ring doughnut) consisting of 12 interconnected rubber cells. Waves passing the toroid compress the cells, forcing air through Wells turbines into adjacent cells. An alternative version of the Clam is linear, with the air sacks following each other in a long line – 275 m for a 10 MW$_e$ device. This might seem a great distance, but is only half the distance that is commonly left between 2 MW$_e$ offshore wind turbines. However, it would be a truly massive structure with a displacement of 44 000 tonnes. Another much studied design is the Salter, or Edinburgh, Duck (Figure 23.6). This consists of one or more cam-shaped devices designed to rotate under the force of incident waves. This rotation needs to match in frequency that of the waves. A 15 m diameter duck would absorb the majority of the energy from waves with a period of 8 s.

The Cockerell Raft (or Sea Snake) has been taken beyond the theoretical, and large-scale prototypes are under construction for testing. A series of snake-like floating artic-ulate structures would be deployed in a grid-like formation (see example application).

Figure 23.7 A Pelamis P-750 wave energy converter undergoing testing [OCE05] [Reproduced from http://www.oceanpd.com/docs/OPD%20Brochure%202005.pdf]

23.3 Example application: the Pelamis P-750 wave energy converter

The machine is a semi-submerged, articulated structure composed of cylindrical sections linked by hinged joints, and was developed by Ocean Power Delivery Ltd. The current design (Figure 22.7) is 120 metres long, 3.5 metres wide and 700 tonnes in weight [DTI05c].

The passing waves force the joints to pump high-pressure oil through hydraulic motors. These motors then drive the electrical generators and produce 750 kW$_e$. Power from the device is fed down a cable to the seabed, then to shore. The device is held in position by a combination of floats and weights, which allow the machine to point itself perpendicular to the waves [OCE05].

The Pelamis is designed to be moored in waters 50–70 m deep, typically 5–10 km offshore. A 'wave farm' covering a square kilometre of ocean could provide a power output of 30 MW [DTI05c].

23.4 Student exercises

> *"The fact that some geniuses were laughed at does not imply that all who are laughed at are geniuses. They laughed at Columbus, they laughed at Fulton, they laughed at the Wright brothers. But they also laughed at Bozo the Clown."*
>
> –Carl Sagan

1. Estimate the minimum length of wave power device capable of replacing all of Norway's electricity generation (107 TWh).

2. Describe the major wave power technologies.

3. Compare, using Equation (23.4), the power extractable from the following typical Atlantic wave conditions (period, s; height, m): storm (14, 28), average (9, 7), calm (5.5, 1). What does this say about the robustness of any design?

4. What is the power extractable from a deep-sea wave system of wavelength 140 m and height 3 m? [From SHE97.]

5. The western coast of Scotland is struck by Atlantic waves of theoretical maximum power 70 kW per metre of wave width. If a typical wave height is 2 m what are the corresponding frequency and periodic time? Estimate the realistic power available onshore [from SHE97].

6. Estimate the necessary length of a proposed wave-power system that collects 10 MW of usable power if the maximum theoretical power is 70 kW/m [from SHE97].

24

Tidal and small-scale hydropower

"At its core, global climate change is not about economic theory or political platforms, nor about partisan advantage or interest group pressures. It is about the future of God's creation and the one human family. It is about protecting both 'the human environment' and the natural environment."

–United States Conference of Catholic Bishops statement [2]

The regular rise and fall of the tides represents a shifting store of potential energy powered by the drag of the moon and the sun. This drag is gradually slowing the rotation of the earth (by 0.016 seconds per thousand years [BRO78]), but by any reasonable definition the resource is a renewable one. In order to make use of this energy source we either need to somehow collect the water at high tide, then wait until the receding tide has created a sufficient head and drain the basin through a water turbine to generate electricity, or use the kinetic energy of tidal currents to spin turbines directly in much the same way as the kinetic energy of the wind is harvested by wind turbines. In the following we will examine both. Tidal generation shares many features with low-head hydro, particularly the size of the head and the rotational speed of the turbines; so small-scale hydro schemes are examined in this chapter too.

In Chapter 3 we analysed tidal energy from the viewpoint of how an object (or the ocean) might gain and lose potential energy on a regular cycle as the earth and moon interact. We can carry out an alternative analysis by considering how tides affect the rotation of the earth. The mass of the earth is 6×10^{24} kg and is rotating at one revolution per day, or $1/(24 \times 60 \times 60)$ per second. The kinetic energy of a homogenous rotating sphere (of mass m and radius r) is given by:

$$E_{kinetic} = \frac{1}{5}mr^2\omega^2$$

where ω is the angular velocity ($= 2\pi \times$ revolutions per second).

Energy and Climate Change   David A. Coley
© 2008 John Wiley & Sons, Ltd

Therefore the loss of 0.016 s of rotational speed per thousand years equates to a loss of kinetic energy of 95 EJ per annum. This is equal to a quarter of current world primary energy use simply being converted into heat by friction on the seabed. Much larger quantities could be extracted by slowing the tidal flow and slowing the rotation further – but still by an incredibly small amount.

24.1 Tides

Tides arise from the interaction of the gravitational pull of the earth, moon and sun, varying over time. The gravitational force, F, between any two bodies of masses, m_1, and m_2 is given by:

$$F = G \frac{m_1 m_2}{d^2}$$

Where d is the distance between the centres of the masses and G the universal gravitational constant. Therefore the force between a 1 kg body on the earth and the sun is (see Appendix 4 for the values of the various constants):

$$F_{E-S} = 6.67 \times 10^{-11} \frac{1 \times 1.99 \times 10^{30}}{(1.5 \times 10^{11})^2} = 6 \times 10^{-3} \text{ N},$$

and between the same mass and the moon:

$$F_{E-M} = 6.67 \times 10^{-11} \frac{1 \times 7.36 \times 10^{22}}{(3.8 \times 10^8)^2} = 3.3 \times 10^{-5} \text{ N}.$$

We therefore might expect that tides are caused more by the pull of the sun than that of the moon. However any mariner will tell you that the opposite is true. The reason is that the gravitational pull on either side of the earth from the sun is almost the same, whereas because the moon is much closer than the sun, the pull from the moon is substantially less on the side of the earth not facing the moon.

We can see this by calculating the difference (ΔF) in the pull of the moon (mass m_M), or sun (mass m_S), on a 1 kg weight on the side of the planet facing (or facing away) from the moon or sun. For the sun and a 1 kg mass on the earth (see Figure 24.1):

$$\Delta F = G \frac{1 \times m_S}{d_{E-S}^2} - G \frac{1 \times m_S}{(d_{E-S} - 2r_E)^2} = -5.01 \times 10^{-7} \text{ N}, \tag{24.1}$$

where Figure 24.1 identifies the various parameters. Whereas for the moon and a 1 kg mass on the earth:

$$\Delta F = G \frac{1 \times m_M}{d_{E-M}^2} - G \frac{1 \times m_M}{(d_{E-M} - 2r_E)^2} = -0.2290 \text{ N}. \tag{24.2}$$

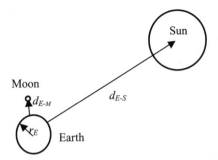

Figure 24.1 Meaning of the various parameters in Equations (24.1) and (24.2). (Not to scale!)

As ΔF for the moon is much greater than ΔF for the sun, we would expect the moon to play the major role in tides around the planet, as indeed it does. We can also see that a weight placed on the side facing, or facing away from, the moon will be attracted more, or less, toward the moon. This is equally true for a body of water.

Although we tend to think that the moon rotates around the earth as though the earth were held stationary in its orbit around the sun, the moon and earth actually orbit each other, much like a spinning pair of dumbbells. Such a system rotates around the combined centre of gravity. If each half of the dumbbell weighs the same, the centre of gravity will be in the middle of the bar. As the earth weighs much more than the moon the centre of gravity will be pushed toward the earth. In fact, it is just inside the earth 1719 km (1068 miles) below the surface [NOAA05].

Water will tend to be attracted towards the moon because of its gravitational pull, so we would expect to find a bulge of water facing the moon, under which the earth rotates, giving one tide a day. However, because the system is rotating at a point just inside the earth, points on the surface facing away from the moon will feel a greater centrifugal force and the oceans will form a second bulge on that side. The earth therefore

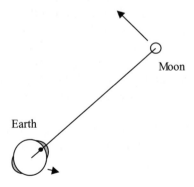

Figure 24.2 Rotation of the earth-moon system around its centre of gravity (the black dot) and the pair of tidal bulges formed (not to scale)

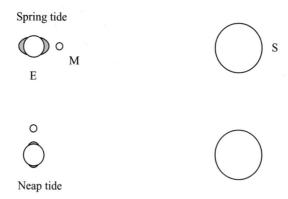

Figure 24.3 Spring and neap tides (E = earth, M = moon, S = sun; not to scale)

experiences two tides a day allowing tidal power dams to be filled approximately every 12 hours.

The size of the tidal range (the increase in water height as the bulge passes) is influenced by the local coastal geography, the gravitational pull of the sun, the weather (low pressure increases water height), the wind and resonances that can stretch across whole oceans. All of these will affect the amount of power that can be derived from a particular tidal power scheme.

The smaller pull of the sun will amplify the tidal bulges if the sun, earth and moon are in line (spring tides) and suppress it when a right angle is formed (neap tides) (Figure 24.3).

Resonances will occur if the width of the ocean is such that a standing wave is created by the 12-hour frequency of the tidal driving force. The Atlantic Ocean happens to be the correct width for this and the tidal range is enhanced by around 2.5 m because of it. In the Pacific the more irregular shape of the continents makes it difficult for resonance to form, and as a consequence tides are much smaller.

Because all the driving forces, with the exception of the weather (which only plays a minor role), are either static or can be predicted, tidal power could provide a regular source of energy and if some form of pumped storage facility were included in a scheme, continuous power would be attainable. This regularity of supply is one of the major advantages of tidal power compared with wind. Another is the density of the working fluid, water, as vast amounts of power could therefore be extracted through a small number of turbines.

Despite this, tidal power, unlike wind, is not a common form of power generation. Reasons for this are cost (because of the need to build large dams) and environmental concerns about altering tidal habitats.

Problem 24.1   Calculate the wind speed required to make the kinetic energy of a stream of air equivalent to a flow of water of 1 m/s.

Technologies for use

Tidal basins

Leaving aside the possibility of using tidal streams (currents), which we will return to later, harnessing tidal power requires holding a large amount of water behind a dam at high tide and letting it out through a water turbine at low tide. This observation allows us to predict the maximum possible energy output from a tidal barrage (dam). If the basin behind the barrage has constant surface area, A, and constant depth, d, then the stored potential energy, E_{pot}, at low tide will be given by:

$$E_{pot} = \textit{mass of water} \times \textit{mean height} \times \textit{acceleration due to gravity}$$

$$= (\rho\, Ad)\left(\frac{d}{2}\right)(g),$$

or

$$\boxed{E_{pot} = \frac{\rho\, Ad^2 g}{2}}. \tag{24.3}$$

So, if $A = 2\ \text{km}^2$ and $d = 5\ \text{m}$, we have:

$$\frac{1000 \times 2\,000\,000 \times 25 \times 9.8}{2} = 1.2 \times 10^{11}\ \text{J}$$

every 12 hours. Or a mean power output of

$$\frac{1.2 \times 10^{11}}{60 \times 60 \times 12} = 2.8\ \text{MW}.$$

However, much larger basins are possible. Damming the Severn estuary (in England), as has been suggested several times, would create a $480\ \text{km}^2$ basin with an annual output of 61 PJ. The depth of water is very important, with Equation (24.3) indicating that the stored energy is not only proportional to the enclosed area, but also proportional to the square of the water depth.

Power could also be extracted during the filling of the basin if the incoming water was allowed to flow through the turbines.

The head of water behind a tidal barrage will be much smaller than the head provided by a normal hydroelectric dam. This implies that the rotational speed of the turbines will be a lot slower (50–100 rpm). Another difference is that the whole basin needs to be emptied if possible in a single tidal cycle (approximately 12 hours), therefore a larger number of turbines is required – 216 in the case of the proposed Severn barrage.

Figure 24.4 Artist's impression of a row of tidal current turbines, with one of them raised for maintenance. © Marine Current Turbines TM Ltd [MCT03] [Reproduced courtesy of www.marineturbines.com/technical.html ©Marine Current Turbines TM ltd]

Importantly, tidal barrages could also operate as pump-storage facilities, with the turbines being used as pumps to increase the water level in the basin for later release. This suggests a natural symbiosis between wind and tidal power. Another possibility is to use multiple interconnected smaller basins. Water could then be pumped between basins and released sequentially in order to even out energy production.

Tidal streams

Rather than trying to store large quantities of potential energy behind a barrage, it is possible to generate electricity from the kinetic energy of tidal currents near landmasses. The general tidal flow is greatly altered by local coastal geography and this leads to higher than average water speeds past headlands and through constrictions. If these tidal streams have a speed greater than about 2.5 m/s (9 km/h) (mean springs peak flow) and are in water with a depth of about 20 to 30 metres, then it is possible to use a device very similar to a wind turbine to produce power (Figure 24.4).

The design shown in Figure 24.4 consists of twin axial flow rotors of 15 m to 20 m in diameter, each driving a generator via a gearbox. The twin power units of each system are mounted on wing-like extensions either side of a tubular steel monopile some 3 m in diameter, which is set into a hole drilled into the seabed from a jack-up barge. Such machines should be able to generate between 500 and 1000 kW$_e$ each with a rotor speed of between 10 and 20 revolutions per minute. Because tidal streams do not

Figure 24.5 300 kW$_e$ tidal steam turbine, Lynmouth. © Marine Current Turbines TM Ltd [MCT03] [Reproduced courtesy of www.marineturbines.com/technical.html ©Marine Current Turbines TM ltd]

change direction (except through 180°) the machines do not need to be able to rotate, however they do need to be able to extract energy from both incoming and outgoing tides.

Although the tops of such machines would be visible to shipping, the amount of the structure above the surface would be small and therefore probably less visually obtrusive than an offshore wind farm close to the coast.

2003 saw the installation of the first large monopile-mounted experimental 300 kW$_e$ single 11 m-diameter rotor system off Lynmouth, Devon, UK (Figure 24.5). This is an experimental machine, which uses a dump load in lieu of a grid connection and will only generally operate with a single tidal direction.

Figure 24.6 shows locations where both the water depth and the tidal speed are deemed suitable for such devices around the UK and northern France. Assuming 34 UK sites, each with twenty 1 MW$_e$ machines, the total installed capacity would be 680 MW$_e$ – around one per cent of total current UK generating capacity (although they would not generate continuously at that level).

Example application: La Rance tidal barrage, France

Constructed between 1961 and 1967 the 740 m long La Rance dam contains 24 two-way pump turbines each with a capacity of 10 MW$_e$ [BOY03]. The dam doubles as a

Figure 24.6 Suitable locations (circles) for tidal stream generation in and around the UK. © Marine Current Turbines TM Ltd [adapted from MCT03] [Reproduced courtesy of www.marineturbines. com/technical.html ©Marine Current Turbines TM ltd]

road crossing and contains a separate lock for ships to pass through (Figure 24.7). The system was designed to allow generation during both incoming and outgoing tides, or pumping into the basin to supplement generation on the ebb during neap tides.

The tidal range is approximately 12 metres and the typical head around five metres. This allows the generation of an average of 480 GWh per year (1.73 PJ).

Problem 24.2   Estimate the true installed electrical capacity of La Rance tidal barrage. (Hint: divide the average annual generation by the energy it would generate if it worked at full capacity continuously.) Does this suggest the system is efficient?

Example application: the Severn Barrage, UK

The proposed Severn Barrage consists of a 16 km (10 mile) tidal barrage and a major road between Lavernock Point, near Cardiff and Brean Down, near Weston-super Mare, and would contain a basin of 480 km^2 (185 square miles) (Figure 24.8). The scheme uses ebb generation and reverse pumping at high tide to further increase the head and volume of trapped water.

Figure 24.7 La Rance tidal barrage [ACT05]

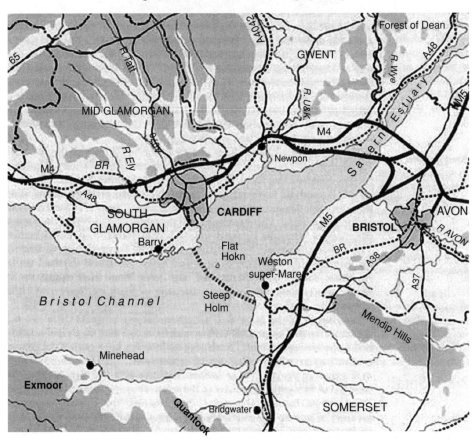

Figure 24.8 Location of the proposed Severn Barrage (dotted line). The barrage is 16 km long [TAY02]

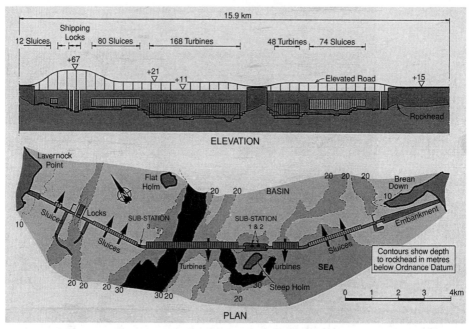

Figure 24.9   Elevation and plan of the barrage [TAY02]

Turbines would be housed in caissons built off-site and towed into position. The design includes 216 turbines rated at 40 MW$_e$ each, providing a total of 8640 MW$_e$ with an annual average output of 17 TWh, and 166 sluices (Figures 24.9 to 24.11).

Construction is estimated to take seven years, at a cost of £8283m with an additional £1230m for strengthening the electricity grid.

Following construction, tidal currents might be half those at present landward of the barrage. This would cause a major change in sediment transportation and hence an increase in primary biological production and food available for birds and fish. Taylor [TAY02] suggests the need for further work into the damage to small life forms caused by the turbines and the design of fish deterrents and passes.

Employment during the project is estimated to amount to 200 000 employee-years amongst 35 000 jobs. Following construction, 10 000 to 40 000 permanent jobs are predicted.

24.2 Small-scale hydropower

In Chapter 13 we saw that large hydropower schemes had been criticized for a number of reasons:

- the need to resettle large numbers of people

- the loss of important archaeological remains

(a)

(b)

Figure 24.10 Installed turbine (a) [Reproduced from http://www3.dti.gov.uk/renewables/publications/pdfs/t0900212ra.pdf, Department of Trade and Industry, crown copyright] and sluice caisson (b) for the Severn Barrage [TAY02]

- loss of habitat

- loss of rare species

- major impacts on the river wildlife and people on the downstream side of the dam

- methane production from rotting vegetation in the flooded area

- loss of human life from dam failures, and

- amplification of interstate tensions from diverting water resources

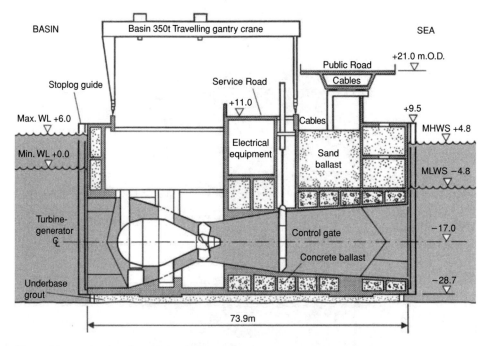

Figure 24.11 Section through Severn Barrage turbine caisson [TAY02]. MHWS = mean high water springs; MLWS = mean low water springs; WL = water level. All vertical distances are relative to a datum mid way between MHWS and MLWS [Reproduced from http://www3.dti.gov.uk/renewables/publications/pdfs/t0900212ra.pdf, Department of Trade and Industry, crown copyright]

Many small schemes avoid these impacts and have the additional advantage that they can be applied in numerous locations where the local geography means that a major scheme would not be economic.

Although there is no defining boundary between large and small-scale hydropower, the classification scheme shown in Table 24.1 is commonly applied. Figure 24.12 shows a typical location that might be suitable for a micro-scale scheme.

Another advantage of small-scale hydro is its low capital cost. Due to their high initial cost and long construction times, large hydro schemes are amongst the most expensive forms of energy infrastructure. This means that hydropower is often deemed uneconomic due to the interest payments that must paid on such large sums. We can get

Table 24.1 Classification of hydropower schemes [adapted from OPE94]

Scale	Rated power
Large	>25 MW$_e$
Small	3–25 MW$_e$
Mini	100 kW$_e$–3 MW$_e$
Micro	<100 kW$_e$

some idea of the extent of the problem by carrying out a present value calculation of the type given in Chapter 6. From such calculations it is often seen that large-scale hydro plants cannot repay their costs despite the fact that they have access to an almost endless source of fuel. Small-scale hydro can also be uneconomic in the current business climate. However, costs are reduced for schemes that do not require the creation of a dam and other large works. If costs are low enough, individuals, communities or companies might be open to not making a strictly economic decision as the capital risked is small, and they may install a small hydro plant regardless. This latter point also applies to energy efficiency measures, solar heating and small single wind turbine installations where individuals or companies might rate the issue of sustainability high enough to make the necessary investment. The reliability of small-scale hydro also makes it attractive for remote communities, particularly parts of the developing world where fuel importation would be difficult (Nepal being an example).

A typical fossil fuelled, nuclear or large-scale hydro plant may be rated at 1 GW_e, a typical small-scale scheme might only produce one thousandth of the power. As was the case with wind power, this implies that if the technology is to play a sizeable role within a national energy strategy and reduce carbon emissions, a very large number of machines would be needed and a similar number of sites developed. This in itself would have an impact upon the environment, even if it was mainly a visual one. It would however re-democratize the energy supply, in that many more of us might own or live near a power plant. As has been mentioned before, and will be returned to when we discuss the possibility of a hydrogen economy, a national grid connected to so many small generators might look very different, and be financed differently, from today's system. In many respects, it may start to take on some of the characteristics of

Figure 24.12 Filleigh Sawmill weir, North Devon, UK. This picture clearly shows the modest fall and river size needed. Courtesy of Hydro Generation Ltd [HYD04] [Reproduced courtesy of www. hydrogeneration.co.uk, Hydro Generation Ltd]

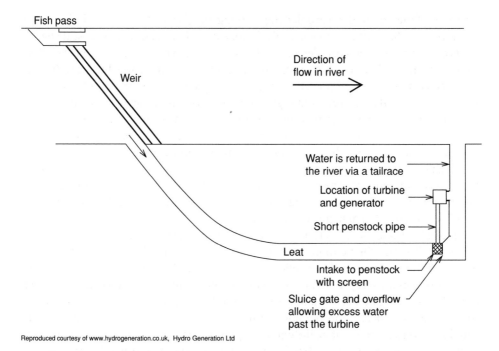

Fish pass

Weir

Direction of flow in river

Water is returned to the river via a tailrace

Location of turbine and generator

Short penstock pipe

Leat

Intake to penstock with screen

Sluice gate and overflow allowing excess water past the turbine

Reproduced courtesy of www.hydrogeneration.co.uk, Hydro Generation Ltd

Figure 24.13 Layout of a typical run-of-river micro-hydro scheme. Note the figure is drawn in plan, i.e. from above. Courtesy of Hydro Generation Ltd [HYD04]

distributed computer systems such as the Internet and generate a highly robust system not dependent on a single industry, workforce or fuel.

Technologies for use

Figure 24.13 outlines the basic form of a micro-hydro scheme. The figure is drawn in plan, i.e. from above, and we see that the plant sits to one side of the river, rather than at the base of a large dam. To generate a reasonable head without engineering a long leat, the scheme is centred on a small weir. A small waterfall could also be used, or neither, if the fall of the river is substantial (as is common in mountainous regions such as the Himalayas). A fish filter is placed at the end of the leat and the water supplied to the turbine via a short penstock. Alternatively a small holding tank and a much longer penstock (possibly several hundred metres in length) could be used to provide the required head. Water then returns to the river. Because only a small amount of water is held back and diverted, the basic dynamics of the river are unaffected. A device mounted directly on the bottom of the riverbed and relying on the kinetic energy of the flowing water could also be used. Although this is likely to produce less power, very little ground works would be required and the plant would be almost invisible.

    The head and the volume of water that can be diverted without disrupting the river need to be ascertained in order to size the turbine and to prove the viability of any project. If the head is h metres and the flow rate r litres per second, then the electrical

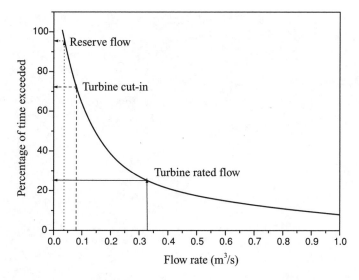

Figure 24.14   Flow rate curve for Cutters Mill, UK [adapted from data in HYD04]

power, P_e (in watts), generated will be given by:

$$P_e = \text{mass per second in kilograms} \times g \times h \times \eta = r \times g \times h \times \eta$$

where η is the efficiency of the system and g the acceleration due to gravity. η is typically 0.5 for such schemes [HYD04]. The amount of water that must be left flowing in the river without disruption between the point of extraction and the return point is termed the *reserve flow*. This is typically chosen as that flow which exists for at least 95 per cent of the year. Figure 24.14 shows the flow rate curve for a small river. Maximum power will be produced at the rated flow of the turbine. However a higher rated power will usually mean a higher cut-in flow, below which the turbine will fail to generate. Typically the cut-in flow will be half to one quarter of the rated flow [HYD04]. This implies that for the curve shown in Figure 24.14 the turbine will run at rated power for around 25 per cent of the time and fail to generate for around 30 per cent of the time. Rivers with flatter flow curves will give greater values and provide a more guaranteed source of power. This is important for remote locations wholly dependent on a single scheme, but less so for locations using the scheme as only one of their sources of power.

Example application: Linne low-head hydro plant, River Maas, Netherlands

The 11 MW_e low head hydroelectricity generation station (Figure 24.15) at Linne on the River Maas was constructed between 1987 and 1989 adjacent to a pre-existing weir with an average level difference of 4 m. The plant utilises four 2.87 MW_e (nominal, 3.5 MW_e maximum) Kaplan turbines (Figure 24.16) inside a 24 000 m^3 caisson sunk

Figure 24.15 Linne low-head hydro plant [CAD95] [Reproduced from ETSU,CADDETT, Technical Brochure, 19, IEA CADDET programme, http://www.caddet.org]

to a depth of 21 m, with most of the structure below water level. The turbines run at 88 rpm [CAD95] and are connected to the generator by a gearbox that increases this to 750 rpm.

Production has frequently been less than the 52 GWh/year expected, often only reaching 30 GWh/year. This was because of the river experiencing extended periods of low flow (implying a very low head) or high flow (which does not give rise to a corresponding increase in head, as the weir height is limited).

The cost of the project was €32m (£22.2m, US$41.7m). Annual electricity production is valued at €2.4m (£1.7m, US$3.1m) (1995 prices), implying a simple payback period of 13 years. Other summary statistics are given in Table 24.2.

24.3 OECD resource

With the exception of France which produces 536 GWh per annum and Canada which produces 311 GWh per annum [IEA05], tidal power has yet to make an impact, and international data on small-scale hydropower are difficult to obtain – although countries such as Nepal, India and Canada have made extensive use of the approach. However we must not forget the contribution from traditional hydropower. Although we classified such large-scale installations as non-sustainable, many sources, such as the OECD and the IEA see them as renewable.

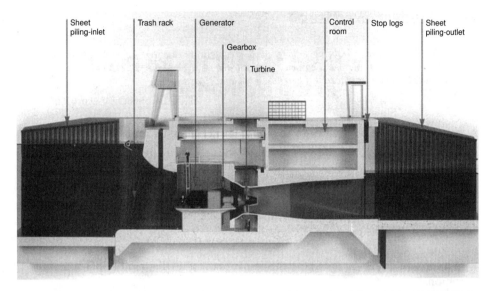

| Sheet piling-inlet | Trash rack | Generator | Control room | Stop logs | Sheet piling-outlet |

Gearbox

Turbine

Figure 24.16 Section through a Kaplan turbine [CAD95] [Reproduced from ETSU,CADDETT, Technical Brochure, 19, IEA CADDET programme, http://www.caddet.org]

Large-scale hydropower has all but reached its potential limit in most OECD countries. Capacity increased between 1990 and 2002 from 1169.7 TWh to 1230.5 TWh, i.e. an annual average increase of only 0.4 per cent. Despite this increase, the fraction of renewable electricity met by hydro decreased from 89 to 84 per cent over this period due to the rapid expansion of other sustainable sources, such as wind [IEA05]. The largest OECD hydropower producers are Canada, USA, Norway, Japan, Sweden and France (see Appendix 1 or [BP05] for national data).

Table 24.2 Statistics of the Linne low-head hydro project [data from CAD95]

Statistic/object	Value
Maximum head	4 m
Average river flow	250 m^3/s
Flow through turbines	0.125 m^3/s
Percentage of flow captured	0.05
Turbines	4 at 2.87 MW
Design life	40 to 50 years
Runner diameter	4 m
Generator rating	4175 kVA
Voltage	10.5 kV
Speed	750 rpm

24.4 Student exercises

"The European records, being so long, make a convincing case that we're already seeing changes ... This is not like 'Centuries from now the ice sheets will melt.' This is 'In a few decades it will be dramatically different.' To me, that's alarming."

–Drew Shindell, NASA physicist, August 2006

1. Why are the tides dominated by the moon rather than the sun?

2. Why is the tidal range greater in the Atlantic than the Pacific?

3. What advantages and disadvantages does tidal power have over wind power? (Don't forget to consider the densities of the media.) Explain the natural symbiosis that could exist between the two.

4. Why do tidal dams require a greater number of turbines than traditional hydropower stations?

5. In a pumped storage scheme combined with a tidal power project, water is pumped by wind generated electricity from the high-tide level to an additional height of 1 m. The tidal range is 6 m. Calculate the proportion of extra energy gained [from SHE97]:

   a. neglecting losses

   b. if the pump motor is 80 per cent efficient

6. The mean tidal range of the Rance scheme is 8.45 m and its basin covers 22 km^2. If the mean output is 75 MW, what proportion of the theoretical power capacity does this represent? [From SHE97.]

7. In North America the Bay of Fundy-Passamaquoddy area near the border of New Brunswick and Nova Scotia in Canada, and Maine in the USA, is considered to be a prime tidal power site. It has a basin area of 700 km^2 with an average tidal range of 10.8 m. Calculate the theoretical maximum power capability and the estimated realistic power available [from SHE97].

25

Biomass

He who plants a tree plants a hope.

–Lucy Larcom

Biomass can be considered as solar energy locked into the global carbon cycle. We can remove carbon temporarily from this cycle by growing biomass, then burn it to produce carbon dioxide and return it to the cycle. We have simply short-circuited a natural process. Left to itself, such carbon resources, in the form of plant matter, would decay and decompose to carbon dioxide and low-grade heat, or be consumed in forest fires. The situation is analogous to that of hydropower, where water which would naturally drop through a given vertical height over a long distance and length of time on its journey from mountain to sea, producing low-grade heat along the way, is captured and forced to surrender some of its potential energy over a much shorter period – resulting in a higher grade of energy.

25.1 History

Of all our energy feedstocks, biomass has the longest history. It is thought that humankind had the ability to create fire at least 250 000 years ago [GOU92, JAM89, PAT95]. The use of animals to pull loads and help with agriculture was common before 3200 BC when wheeled carts were introduced. Until the rise of coal during the industrial revolution, biomass, together with a small amount of waterpower, provided the majority of the world's energy. This situation is still true in some developing countries. Nepal and Ethiopia both meet the majority of their energy needs from biomass and to others such as India and Brazil it is a more important source than oil. Given this long history, and current high levels of use, energy derived from biomass can not really be

Energy and Climate Change   David A. Coley
© 2008 John Wiley & Sons, Ltd

classified as an alternative energy; however it is nearly carbon neutral and its use has the potential to be greatly expanded, therefore fulfilling the definition of a sustainable energy technology given in Chapter 5. As was mentioned in Part III, it is difficult to obtain reliable estimates of biomass use in much of the world: a great deal of it is produced locally, often by the user, or is in the form of work done by draft animals, and therefore there is no paper trail of trades from which to gather data. In the developed world, where the majority of biomass use is in the form of wastes or crops grown specifically as a feedstock and traded, the data are more reliable. Approximately one third of energy use derives from biomass in the developing world but this falls to only three per cent in the industrialized nations. Because of the much greater per-capita energy use in the developed world, this difference is reduced if one looks at the total energy derived or tonnage of biomass used. Developing world consumption is believed to be around 0.75 t per person per annum, and that of the developed world 0.3 t per person per annum – a much smaller difference. The UK and similar countries have lower levels of use than the developed world average; heavily forested nations such as the USA, Sweden, Austria and Switzerland have a greater than average use.

25.2 Basic principles

The key to this part of the carbon cycle is the ability of plants to photosynthesize. Within plants, chlorophyll (a green pigment) absorbs sunlight to split water (H_2O) into oxygen (O_2), hydrogen ions (H^+) and electrons (e^-):

$$2H_2O + light \rightarrow O_2 + 4H^+ + 4e^-. \qquad (25.1)$$

The oxygen is released to the atmosphere; the hydrogen ions and electrons are involved in the chemistry of various molecules including the creation of ATP, which is used in the construction of carbohydrate from the hydrogen ions and atmospheric carbon dioxide:

$$4H^+ + 2CO_2 \rightarrow 2[CH_2O]. \qquad (25.2)$$

This is a generic reaction for the formation of carbohydrates, CH_2O is just the simplest; other common varieties include glucose ($C_6H_{12}O_6$), and sucrose ($C_{12}H_{22}O_{11}$). All are connected through the formula, $C_x(H_2O)_y$. Combining (25.1) and (25.2) we have the overall reaction:

$$CO_2 + 2H_2O + light \rightarrow CH_2O + H_2O + O_2.$$

The plant provides energy for itself during respiration by using its stores of carbohydrates and thereby returns some of the carbon it stored back to the atmosphere. This return also occurs when the plant decays, or if it is burnt. Over about 300 years, all atmospheric carbon dioxide is cycled through this process, i.e. turned into plants which then decay; the far greater amount of oxygen in the atmosphere takes longer to complete the cycle – approximately 2000 years. This is recycling on a grand scale in both space and time. By using biomass as an energy feedstock, we are simply shortening the length of this cycle.

Photosynthesis is highly inefficient. We can make an estimate of the efficiency from the following bottom-up calculation. The majority of the photons driving photosynthesis are from the red end of the spectrum with a wavelength of around 0.7 μm. As we saw in Chapter 2,

$$E_{rad} = \frac{hc}{\lambda}$$

where h is Planck's constant and c the speed of light. So, this corresponds to an energy of 2.72×10^{-19} J (1.7 eV) per photon. Around eight photons of this energy are needed to mediate the creation of a single unit of carbohydrate and as a carbohydrate unit (CH_2O) stores approximately 5 eV (8×10^{-19} J) of energy, the storage efficiency is:

$$100\frac{5 \text{ eV}}{8 \times 1.7 \text{ eV}} \approx 35\%.$$

Problem 25.1 Calculate the amount of energy in the form of sunlight needed to create 1 kg of CH_2O. What mass of oil stores the same quantity of energy?

An efficiency of 35 per cent sounds reasonable, but ignores various losses. Which losses we choose to include is a bit arbitrary, and depends on where we wish to set the boundary for the calculation. Approximately 50 per cent of solar energy reaches the ground, of which only a quarter is of the right wavelength to initiate photosynthesis – this suggests a very approximate efficiency of $0.35 \times 0.25 = 0.086$ or 8.6 per cent – less than that of photovoltaic technologies.

Approaching the calculation from a slightly less fundamental perspective highlights other losses: only one third of the ground-incident solar radiation will occur during the growing season (assuming a temperate zone), say 20 per cent strikes the leaves of plants, with only a quarter being of the correct wavelength; a storage efficiency of 35 per cent (as estimated above) and allowing for the 40 per cent or so of the final energy that the plant will need to keep itself alive, suggests an efficiency of $0.33 \times 0.2 \times 0.25 \times 0.35 \times (1 - 0.4) = 0.0035$ or 0.35 per cent, i.e. much less. Given land with an annual solar radiation of $1000 \text{ kWhm}^{-2}\text{y}^{-1}$, only $0.0035 \times 1000 = 3.5 \text{ kWhm}^{-2}\text{y}^{-1}$ will be stored and available for harvesting. This equates to 126 GJ per hectare (10 000 m^2). Estimates based on real energy crops suggest harvests of between one and 30 tonnes of dry matter per hectare per annum, which equates to between 15 and 500 GJ. Table 25.1 lists the calorific values of various biomass feedstocks. For comparison coal has a calorific value of approximately 29 GJ/t and natural gas a value of 55 GJ/t.

Table 25.1 Calorific value of various biomass feedstocks [data from BOY96, p143]

Fuel	GJ/t	GJ/m^3
Grass (fresh-cut)	4	3
Domestic waste (as collected)	9	1.5
Straw (baled)	14	1.4
Sugar cane (air-dried stalks)	14	10
Wood (air-dried, 20% moisture)	15	10
Paper (stacked newspaper)	17	9
Commercial wastes (UK average)	16	Highly varied
Dung (dried)	16	4

25.3 Technologies for use

As Table 25.1 suggests, biomass feedstocks come in a variety of forms and it is the form of the feedstock and the use to which the energy will be placed that dictates the degree of processing required and the form of combustion. Given reasonably dry biomass, one can simply burn it to provide domestic heating, or on a larger scale to raise steam for electricity generation. However there are other possibilities that produce liquid or gaseous fuels for later combustion; Figure 25.1 outlines these. We will look at some of this potential for fuel processing and also study some of the available feedstocks for these processes in this chapter. Chapter 28 reintroduces some of the processes where

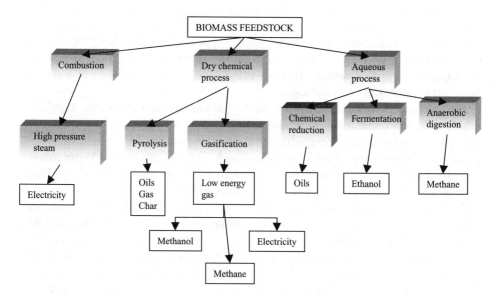

Figure 25.1 Biomass conversion possibilities

Figure 25.2 Biomass-powered restaurant, western China (photo by author)

the technology is seen as a possible way of cutting the dependancy of transportation on fossil fuels.

Combustion

Combustion is the simplest biomass conversion process, and is easily scaled to the size of the resource. It is still the major energy source in much of the world, particularly in rural communities (Figure 25.2). Specially grown feedstocks such as willow trees can be used, as can industrial, agricultural or domestic waste. The heat produced can be used within the industry itself, or exported for space heating, or if the scale of the operation is large enough, steam can be raised for electricity generation; the latter being a popular option for municipal waste incinerators. The consumption of industrial and municipal waste can lead to the release of hazardous by-products such as heavy metals; the emission of which can cause concern to local communities and needs to be controlled by the design of the incineration, the operational temperatures and cleaning of the exhaust gases.

One advantage of combustion is that no intermediate steps are required between biomass and heat. However the low energy content (either by unit mass or volume) of wastes and other biomass feedstocks means that large quantities of material may need transporting long distances (probably using fossil fuels) to facilities that can burn

Figure 25.3   *Miscanthus* being harvested [DEFRA01]

the material on an economic scale. This is particularly so if electricity generation is envisaged. Therefore the approach is often used at locations where the waste material arises, or where it would have to be moved anyway if an alternative waste-disposal approach such as landfill were to be used. For example, Shepherd [SHE97] suggests 15 km^2 (5.8 square miles) of land might be needed to produce 1 MW of power as heat. One of the reasons for intermediate processing of biomass to produce a high quality, energy-dense resource is to remove this transportation problem. Another possibility is to produce the fuel locally and then only move it a relatively short distance to local consumers. Short rotation coppicing of willow can generate large quantities of wood over a period of five years. *Miscanthus* can produce an annual crop (Figure 25.3). Local natural drying and processing of the harvested timber can be used to derive wood pellets suitable for burning in small commercial or domestic boilers (Figure 25.4). Such an approach places energy production directly within the local energy-using community, both in terms of the resource (wood pellets) but also in terms of the surrounding landscape. This can create an atmosphere of ownership often missing from technologies such as wind power.

Municipal solid waste (MSW), as collected, makes a poor fuel. It typically contains 20 per cent water and much non-combustible material such as metals and glass, resulting in a volumetric energy density of 1.5 MJ/m^3 (three per cent that of coal) (Table 25.2 and Figure 25.5). Although it can be burnt directly in specially designed systems, another possibility is to treat the raw material using a combination of sorting, drying, shredding and compression to give refuse-derived fuel (RDF). This process can produce a fuel with a volumetric energy density of 15 MJ/m^3 (one third that of coal), which is suitable for burning in conventional plants. There are several hundred waste-to-energy plants now

Table 25.2 Composition of UK municipal solid waste (Mt) [1994 data from BRO99]

6.87	Paper and card	1.10	Film plastics
4.18	Putrescible	0.43	Textiles
1.93	Glass	0.37	Misc. non-combustible
1.68	Misc. combustible	0.33	Non-ferrous metal and Al cans
1.38	Non-inert fines	0.03	Inert fines
1.22	Dense plastics	0.00	Composted putrescible
1.18	Ferrous metal	20.71	Total

operating worldwide, with countries such as Japan treating the majority of their waste in this way: Figure 25.6 shows a typical plant's layout. Industrial wastes, such as chicken litter, are often less contaminated and therefore need less processing before combustion. (The power station at Eye, Suffolk, UK being an example, this processes 180 000 t of chicken litter per annum to produce 12.5 MW_e which approximately equates to 1 W per chicken.) Forestry and timber processing wastes are amongst the purest wastes and obviously suitable for combustion. The USA, central Europe and Scandinavia are now making extensive use of this waste stream for both heat and power generation. *Dry*

Reproduced courtesy of CADDETT, Technical Brochure, 109, IEA CADDET programme, http://www.caddet.org.

Figure 25.4 Wood pellets and pellet-burning boiler. These are used to heat the Energy Centre at the Agricultural University of Norway. An annual demand of one million litres of oil has been cut by 90 per cent by using 2200 t of wood pellets consumed in a 3.6 MW_{th} plant. Source: IEA CADDET programme[1] [CAD109]

---

[1] Web site: http://www.caddet.org.

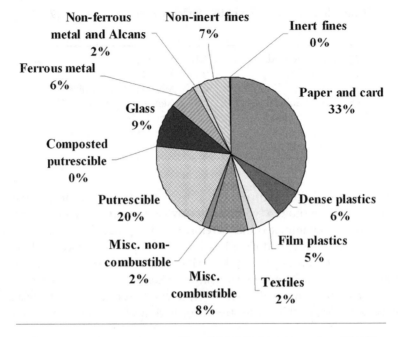

Figure 25.5 Fractional composition of UK MSW [1994 data, from BR099]

agricultural wastes, such as straw can be combusted, and systems have been commercialized in several countries including the UK. *Wet* agricultural wastes, such as vegetables and fruit, may be better processed by anaerobic digestion or fermentation.

Pyrolysis

This is the process used to produce charcoal from wood, but can be used with other feedstocks. By heating the biomass in the presence of insufficient oxygen, or a non-oxygen atmosphere such as nitrogen, the moisture and volatile compounds are driven off to leave a mix of carbon and inert materials collectively known as *char*. The volatile and liquid products are also energy sources and can either be captured or used to provide heat for the process itself. Charcoal is formed at around 250 °C; at 500–900 °C methanol, carbon monoxide and hydrogen can be produced leaving only 10 per cent of the material as solid char. Without volatile/liquid component capture an efficiency of 50 per cent is possible; with capture this can rise to 80 per cent.

At 33 MJ/kg, charcoal has twice the energy density of dry wood by mass (and much greater than this of freshly cut wood) and can therefore be economically transported much greater distances. It also burns at a higher temperature and the rising number of urban dwellers in the developing world makes charcoal a highly prized fuel.

(a)

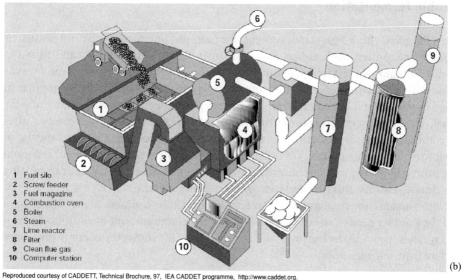

1　Fuel silo
2　Screw feeder
3　Fuel magazine
4　Combustion oven
5　Boiler
6　Steam
7　Lime reactor
8　Filter
9　Clean flue gas
10　Computer station

(b)

Figure 25.6　Photograph and schematic of the energy from waste plant at Trondheim, Norway. This burns waste from the paper industry to produce heat and has a thermal output of 6 MW and the ability to handle over 20 000 t of waste annually [CAD97]

Gasification

Like pyrolysis, gasification requires heating the biomass to high temperatures, but this time with a mix of steam and air or oxygen to produce a gaseous fuel. This was the process used to produce town gas before the wide scale introduction of natural gas. Typically temperatures higher than those used in pyrolysis are applied and pressures of up to 30 atmospheres may be required. Very pure fuels can be produced by such methods and if the system is designed to produce a mix of carbon monoxide and hydrogen called *synthesis gas* almost any hydrocarbon can be synthesized from the mix. This allows for the production of high value liquid fuels. Lower value *producer gas* (a mix of nitrogen, carbon dioxide and hydrogen) can also be produced, which although relative energy poor with a calorific value only one sixth of natural gas, can be used locally to produce heat.

Liquid fuels offer the advantages of high energy densities, ease of transportation and the ability to be used in internal combustion engines. As has been described above, gaseous and liquid fuels can be produced using pyrolysis or gasification. These are both 'dry' processes; the following two techniques represent 'aqueous' processes.

Aerobic fermentation

Given biomass with a high starch content, for example, potatoes, grains and cereals, or a high sugar content, such as beet and cane sugar plants, fermentation can be used to produce ethanol (C_2H_5OH), without many of the complications introduced by synthesis. This makes an excellent substitute for petrol (gasoline) and it highly popular in Brazil in particular. Oilseed rape can be converted to rape methylester, a substitute for diesel. The growing of crops for fermentation, rather than the localized use of wastes for combustion, raises concerns over monocultures, transport distances and the use of fossil fuels at various points. Yet because a liquid transport fuel is created, it should be possible to greatly reduce concerns about transport distances – as the transporting vehicles can also use this fuel and thereby be carbon neutral.

Unfortunately, the resultant mix after fermentation only contains 20 per cent ethanol and the product needs to be distilled. This requires a large amount of additional heat – which can be provided by burning waste crops. Table 25.3 shows the energy derived per hectare from various crops.

Anaerobic digestion

A sludge of organic matter and suitable bacteria held in an airtight container (hence anaerobic) at a modest temperature of around 50 °C will decompose to produce large quantities of *biogas*, a mix of 50–70 per cent methane and carbon dioxide. This can be used onsite or burnt in internal combustion engines to produce electricity. A similar process will also occur in the semi-anaerobic conditions found in many landfill sites.

Table 25.3 Ethanol yields [data from BOY96]

Crop	Litres per hectare per annum
Sugar cane (harvested stalks)	400–12 000
Corn (maze, grain)	250–2000
Cassava (roots)	500–4000
Sweet potatoes (roots)	1000–4500

The bacteria can either be those naturally existing in the slurry, as is the case with animal manure, or artificially introduced. The remains left after digestion can be returned to the land and include fertilising compounds rich in nitrogen. This is a clear example of temporarily borrowing from the carbon cycle: natural waste that would have decomposed (aerobically) largely to carbon dioxide and leaving important non-carbon remains locally to sustain other growth, is replaced by anaerobic digestion to methane which is then burnt to produce carbon dioxide with the non-carbon elements again returned. Biogas production can operate over a range of scales: from single farm/community digesters (1–10 m^3) such as those found in the developing world (China had over seven million during the 1970s [ICEEE]), to commercial electricity-producing digesters (2000 m^3), or even from the mining of gas from landfill sites.

The efficiency of methane production and the quantity of gas produced will depend on the scale of the operation, the feedstock and the temperature at which the digester operates: the colder the climate the more of the resultant methane is needed just to keep the digester warm. Outputs of around 300 m^3 per dry ton of feedstock are possible. Figure 25.8 shows an example. Efficiency in generation is not the only important factor in deciding whether digestion is economically feasible. The processing of the waste into a form more easily disposed of is often the driving force behind the construction of a facility – processing human waste being an example.

Within a landfill site water content is much lower than within a slurry (which might be 95 per cent water) and waste is not kept artificially warm, so decomposition takes place much more slowly, typically over several decades. The rate of production of methane varies from site to site and depends on the content and the form of the site. The actual rate is important, not just for estimating the size of the resource in any site and therefore the economics of extraction, but because methane is a powerful greenhouse gas (23 times more powerful than carbon dioxide by mass) and therefore its emission from landfill needs to be accounted for in any country's inventory of greenhouse gases. Being such a strong greenhouse gas gives an extra justification for extracting the gas and burning it to produce less harmful carbon dioxide. Figure 25.7 shows the proportions of oxygen, carbon dioxide and methane within a hypothetical landfill site: as the air is used up methane production takes off and goes on until it exceeds carbon dioxide production. Eventually production drops as the density of organic matter left falls.

Before extraction the site is covered with a layer of clay to keep air out and to keep the gas from escaping. The landfill gas (typically 55 per cent methane) is extracted via

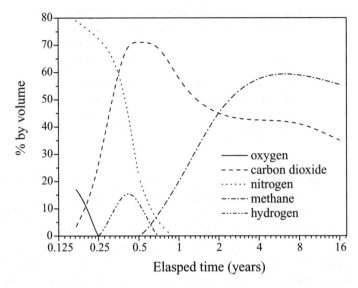

Figure 25.7 Concentration of various gases within a typical landfill site

a grid of interconnected pipes buried within the waste. Methane production rates of 3500 m^3 per hour are possible from a large well-developed site, with around 3 GJ of methane being produced per ton of waste.

Problem 25.2   Given a methane output of 3500 m^3 per hour, how much electricity could be generated from such a site? Compare this to the likely output of a 2 MW$_e$ wind turbine.

25.4 Example application: anaerobic digester, Walford College Farm, UK

The farm on which the digester (Figure 25.8) is sited covers 260 hectares with 130 dairy cows and 160 pigs. These produce around 3000 tonnes of manure per annum. The digester, which was built in 1994, was primarily designed to deal with this waste, but is also used to generate heat and electricity via a CHP plant (Chapter 19).

Slurry is fed from the animal units to a reception pit and then to a chopper/pump, which transports the slurry to the 335 m^3 digester (see Figure 25.9). Digestion takes 16–20 days and produces 450 m^3 of biogas per day. This is fed to the CHP unit with heat recovery from the unit's exhaust and coolant systems (part of which is used to maintain the digester at a suitable temperature). After digestion, the slurry is sieved and any fibrous product removed to a composting shed. The liquor is held in a 950 000 litre storage tank before spreading back on the pastures.

The CHP unit is rated at 35 kW$_e$ and 58 kW$_{th}$, however actual output is around 18.2 kW$_e$ for 19.5 hours a day. Of the output, 30 kW$_{th}$ is required to warm the digester

Figure 25.8 Walford College Farm's digester. Source: IEA CADDET programme[2] [CAD60] [Reproduced courtesy of ETSU,CADDETT, Technical Brochure, 60, IEA CADDET programme, http://www.caddet.org.]

to between 35 and 37 °C. In addition, the unit produces 15 m³ per day of liquor and three tons per day of fibrous material, the former being used on the farm as a fertilizer and the latter being sold to garden centres.

Table 25.4 lays out the economic case, which suggests a simple payback period of six years.

25.5 Global resource

Global average plant production is believed to be around 320 g/m² per annum, with half of this biomass production taking place in the oceans. The radius of the earth is 6400 km; so annual land-based biomass production must be of the order of:

$$4\pi (6.4 \times 10^6)^2 \times \frac{0.32}{2} = 8.25 \times 10^{13} \text{ kg}.$$

Assuming a calorific value of 4 GJ/t then gives, from a totally theoretical perspective, a natural biomass potential resource of 330 EJ per annum.

Problem 25.3   Indonesia has a land area of 1.8 million km² and uses 105 Mt_{oe} of fossil fuel energy per annum. Could it (in theory) replace all of this by natural biomass-derived energy? Could the UK (area 242 000 km², annual fossil fuel use 202 Mt_{oe}) do the same?

---

[2] Web site: http://www.caddet.org.

Table 25.4 Economic analysis of the Walford plant

	Euro	£	US$
Costs			
Digester	124 900	89 349	161 600
CHP unit	48 500	34 700	62 750
Composting unit	13 420	9 600	17 360
Total	**186 800**	**133 649**	**241 700**
Income/savings (per annum)			
Electricity	23 880	17 082	30 890
Hot water	3 673	2 628	4 752
Fertilizer	2 796	2 000	3 617
Reduction in slurry spreading	3 495	2 500	4 521
Total	**33 840**	**24 210**	**43 780**
Annual running and maintenance costs	2 935	2 100	3 797

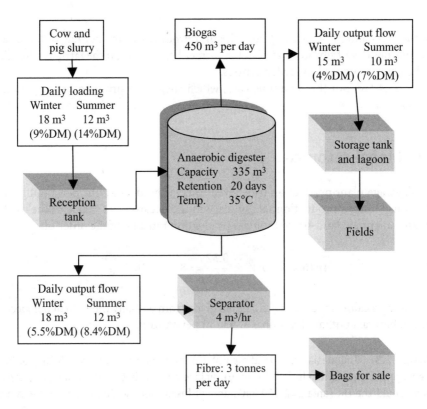

Figure 25.9 Schematic of Walford College Farm's digester [based on CAD60]. DM = dry mass

Table 25.5 Potential world biomass-derived energy production, 2050 [data from BOY96]

Resource	Energy production (EJ)
Energy crops	128
Dung	25
Forestry residues	14
Cereal residues	13
Sugar cane residues	12
Existing forests	10
MSW	3

Clearly not all 8.25×10^{13} kg of natural biomass is available for use and as was mentioned above, reliable data in this area are notoriously difficult to assemble. The true resource will depend on a combination of economics, willingness to take up new technologies and overcoming various practical difficulties, for example the transportation and use of biomass-derived fuels. The figure of 320 g/m^2 quoted earlier is a global average for natural biomass. Some areas produce little biomass, deserts and glaciers being obvious examples. Cultivated land can produce much greater amounts than most natural locations, so we would expect correspondingly higher yields. It has been suggested that by 2050 over 200 EJ per annum could be provided by cultivated biomass (Table 25.5). This equates to around half the world's primary energy consumption. In order to achieve this 400 million hectares of land would be needed (2.5 per cent of the world's land area) and around half of all agricultural and forestry residues would need to be processed by energy recovery technologies. Much of this energy would come from North and South America, Africa and parts of Asia, suggesting a great change in the distribution of energy production, which could in turn have geo-political consequences.

25.6 OECD resource

We will distinguish between solid, gaseous and MSW sources of biomass.

Solid

Electricity generation from solid biomass grew from 77.2 TWh to 92.3 TWh between 1990 and 2002 [IEA05], i.e. an annual average growth of 1.5 per cent, which made it the largest source of sustainable energy (second largest if hydropower is considered sustainable) [IEA05]. The USA is the greatest advocate of this approach in terms of installed capacity, followed by Japan, Finland and Canada.

Gaseous

Electricity production from gaseous biomass rose from an estimated 7 TWh in 1990 to 22.9 TWh in 2002 [IEA05]. The largest individual producer is Japan (6.8 TWh). The highest growth is found in the European Union with Germany producing 3.2 TWh (with a 26.3 per cent annual growth rate) and the UK producing 3.0 TWh (with a 15.8 per cent annual growth rate).

MSW

In 2002, 32.7 TWh of electricity was produced from renewable MSW, with the USA being the largest producer (16.5 TWh). The second largest producer is Japan (3.3 TWh), then France (2.9 TWh) [IEA05]. An additional 5.1 TWh of electricity was produced in 2002 from non-renewable municipal waste (i.e. non bio-degradable waste) and a further 16.3 TWh from industrial waste.

Problem 25.4   Repeat Problem 25.3, but assume the biomass is cultivated biomass and instead estimate the percentage of each country's land area required to eliminate the need for fossil fuels.

25.7 Student exercises

Forests precede civilizations and deserts follow them.

–Chateaubriand

1. Outline how photosynthesis works.

2. Show how the basic efficiency of photosynthesis can be estimated from first principles, then continue the estimation to include other less fundamental factors.

3. From a biomass perspective, describe combustion, pyrolysis, gasification, fermentation and digestion.

4. Compare biomass use within the industrialized nations on a per-capita basis and a percentage-of-national-energy-use basis.

5. How much of your country's land area would be required to replace fossil fuel use by biomass?

6. In a country where there is about 28 million tonnes of domestic waste each year and the energy efficiency of collection is 50 per cent, how much energy in kWh is potentially available? [Adapted from SHE97].

26

Geothermal

"The prosperity we have known up to the present is the consequence of rapidly spending the planet's irreplaceable capital."

–Aldous Huxley

With the exceptions of nuclear and tidal power, all the sustainable energy sources in this book are solar in origin, as are fossil fuels. Geothermal energy however is mainly derived from radioactive decay of isotopes (principally Thorium-232, Uranium-238 and Potassium-40) deep within the earth. The remaining one third is from heat left over from the original coalescence of matter that formed the earth, where the kinetic and potential (gravitational) energy of the accreting particles was converted into heat. This build up of heat melted the original material and thereby allowed the various elements and compounds to separate and settle according to their density, with the very lightest forming the original atmosphere (which has subsequently been substantially altered by biological processes). Figure 26.1 shows the current internal structure of the planet.

26.1 Background

As Figure 3.13 shows, 32 TW_{th} wells up from the depths to the surface. Comparing this to our current energy requirement of 13 TW_{th} suggests that geothermal energy might be a good candidate for a renewable source of heat. Unlike tidal, solar, wind, biomass or hydropower it is not dependent upon a time-varying or seasonal source. However, although the flow of energy is staggering, the following exercise shows the power density is disappointing.

Energy and Climate Change   David A. Coley
© 2008 John Wiley & Sons, Ltd

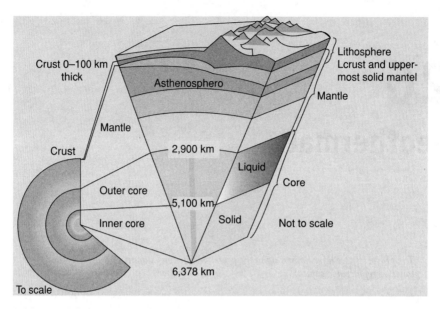

Figure 26.1 Structure of the earth. Layers include (a) the dense *inner core* composed largely of solid iron and subordinate nickel, with a radius of about 1250 km, (b) the molten *outer core* composed largely of liquid iron, with subordinate sulphur, with a width of about 2200 km, (c) the *mantle*, composed of relatively dense rocky materials, with a width of about 2800 km, and (d) the *crust* which comprises the thin, relatively light outer skin of the earth, divisible into two types: the *oceanic crust* (approximately 7 km thick) and the *continental crust* (approximately 35 km thick). Whereas the oceanic crust is composed of basaltic rock, the less dense continental crust is made of a great variety of rock types, but has an overall average composition akin to granite [USGS05] [Reproduced from http://www.washington.edu/burkemuseum/geo_history_wa/The%20Restless%20Earth%20v.2.0.htm]

Problem 26.1   The surface area of the earth is 510×10^{12} m². Estimate the geothermal power density in W/m² at the earth's surface.

The answer to Problem 26.1 is only one five-thousandth of the 240 W/m² typically provided by the sun. Such a low figure also indicates that ground source heat pumps (Chapter 20) do not utilize geothermal energy, but simply access the solar store provided by the ground.

All is not lost, however; there are two ways that this immense store of heat can be accessed. We can either drill down closer to the radioactive sources that produce the heat (so-called *hot dry rock technologies*), or use locations (such as that shown in Figure 26.3) where the natural output of geothermal power is much higher than the average because of the presence of water (i.e. use *wet rock technologies*). This latter route is used in Iceland to satisfy much of its energy requirement. Iceland uses its geology to provide both hot water for domestic heating and to raise steam for electricity production and is hoping to produce enough hydrogen from the electrolysis of water to replace its use of oil as a

Figure 26.2 (a) Natural Pool of boiling water, Iceland. (b) Geyser in Iceland. This erupts due to trapped steam every few minutes ejecting boiling water (photos by author)

transportation fuel. Such systems use natural up-wellings of magma to heat either water in the water table or pumped water, to high temperatures and pressures. So cheap is this energy that it is not uncommon for heating and hot water costs to be unmeasured in domestic situations, but simply paid for by a fixed charge.

Such a happy coincidence of magma, and people to use the heat, is a rare thing. Generally either the energy produced must be turned into something more

Figure 26.3 Geothermal power station, Iceland, which shares its energy source with a semi-natural outdoor swimming pool (photo by author)

portable – electricity and then possibly hydrogen – or we must access deep hot rocks, usually granite, and pump water down to them.

The earth has an average geothermal gradient of 30 °C/km with areas of 40 °C/km not being uncommon. Experimental sites have been established in several countries that have seen drilling to depths of several kilometres and used explosions or extreme pressures to fracture the granite to provide a large heat exchange area. Figure 26.4 shows a schematic of the general concept. Unfortunately, to date the technology has yet to reach commercialization.

The degree to which geothermal energy is renewable depends on the balance between how fast energy is removed compared to how fast it is replenished by the earth. It is conceivable that many geothermal stations will, if built, mine heat at a far greater rate than it is replaced by the surrounding geologic resource, requiring new boreholes to be built on a regular basis (possibly every 10 years). It is therefore questionable whether it is truly a renewable process.

However, this has not stopped 58 countries worldwide from exploiting the resource [NEW03]: two examples being the Southampton and the Paris geothermal district heating schemes. In Southampton, UK, a borehole was drilled to a depth of 1.8 km and brine at 70 °C encountered within a permeable sandstone layer. Because this aquifer is naturally pressurized the brine rises unaided almost to the surface with pumping only required to complete the last 100 m. A heat exchanger is then used to extract 1 MW_{th} of heat, which is increased to 2 MW_{th} by the use of heat pumps (Chapter 20). The resultant heat is used within several city centre buildings for space and water heating.

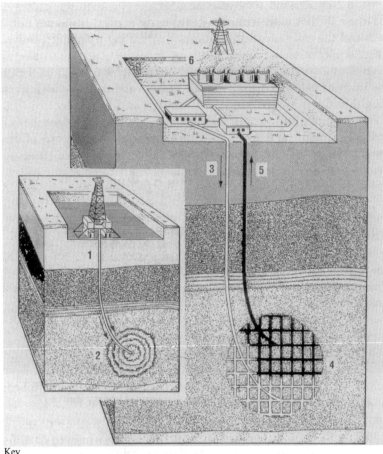

Key
1	Drilling two holes for injected water
2	Explosive fracturing
3, 4, 5	Injected cold water returns hot
6	Geothermal power plant

Figure 26.4 Schematic of a hot dry rock geothermal power station[1] [SHE97, p150] [Reprinted from W. Shepherd and D.W. Shephard, Energy Studies , p150, © 1997 courtesy of World Scientific Publishing Co. Pte.Ltd, Singapore]

26.2 History

Like many energy resources, humankind's use of geothermal energy has a longer history than one might suppose. Roman spa towns were an early example as were Polynesian

---

[1] Reprinted from Energy Studies, Shepherd & Shepherd, 1998, with permission from World Scientific Publishing Co. Pte. Ltd, Singapore.

settlements in New Zealand 1000 years ago. Such exploitation was only possible in locations where the hot water naturally welled to the surface. It was not until the late nineteenth and the early twentieth centuries that drilling techniques evolved to the point where deeper resources could be tapped, and in 1913 a 250 kW_e electric power plant was built at Larderello, Italy (since expanded to over 400 MW_e). In the 1950s and 1960s the Wairakei field was opened in New Zealand and the massive Geysers field in Southern California, USA, which has an installed capacity of 2.8 GW_e.

Since these examples, many nations around the Pacific have developed the technology for electrical generation, as has Iceland. Other countries, such as parts of the former USSR, produce large amounts of heat from their geothermal resource. However it is fair to say that on the whole the technology has not been applied on anything like the scale suggested by the size of the resource.

26.3 Resource and technology

The form of the resource and the technology for extraction depend on whether we are tapping into a natural aquifer, or attempting to access radioactive heating within dry rocks directly.

Wet rocks

Four geologic elements are required for a wet rock geothermal source to be useful: a heat source (for example crystallising granite), a covering of impermeable rocks, topped by an aquifer (typically recharged by rainfall) and finally a cap of impermeable rock (Figure 26.5).

The two impermeable layers constrain the water within the aquifer allowing it to be heated and pressurized. The rock within the aquifer itself must be permeable, allowing water to flow freely – examples being volcanic ashes and some sandstones and limestones. The amount of heat that can be extracted will depend on the permeability, which can be described by the *hydraulic conductivity*, K_h of the rock. Darcy's law states that the speed of flow (v) of water in a porous medium is given by:

$$v = K_h \frac{h_p}{d},$$

where h_p is the effective head of pressure (measured in metres of water) over distance d. The mass, M, of water flowing through an area of A m^2 is then

$$M = \rho_w A v = \rho_w A K_h \frac{h_p}{d},$$

where ρ_w is the density of the water (or brine). K_h can take values of less than 0.01 m/day for clays to over 10 000 m/day in gravel. Table 26.1 gives example values and shows that

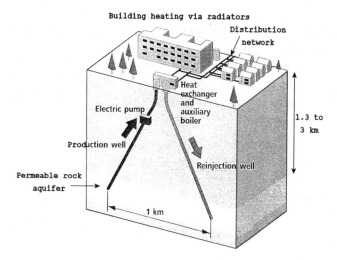

Figure 26.5 Schematic of a wet rock geothermal site[2] [SHE97, p149] [Reprinted from W. Shepherd and D.W. Shephard, Energy Studies, 1860940420, p150, 1997 courtesy of World Scientific Publishing Co. Pte.Ltd, Singapore]

for many rock types the possible range is substantial, but that clays and mud rocks are ideal capping rocks.

For the aquifer to be useful, heat needs to be conducted into it and then retained. This suggests that ideally we need rocks of high thermal conductivity, K_q, below the aquifer and low conductivity above it. The heat flow, P_{th}, (in W/m^2) into or out of the aquifer will be given by:

$$P_{th} = K_q \frac{\delta T}{d},$$

where δT is the temperature change over a height of d (metres). K_q takes values of between 2.5 and 3.5 Wm^{-1}°C^{-1} for sandstone, limestone and most crystalline rocks, and 1–2 Wm^{-1}°C^{-1} for clays and shales – again suggesting these make good capping rocks as they will retain the heat and lead to higher temperatures within the aquifer.

Hot dry rocks

Making use of the geothermal resource within hot dry rocks, rather than within the wet sedimentary strata used to form aquifers, requires the creation of an artificial heat

---

[2] Reprinted from Energy Studies, Shepherd & Shepherd, 1998, with permission from World Scientific Publishing Co. Pte. Ltd, Singapore.

Table 26.1 typical hydraulic conductivities [BOY96, p362]. Data from Boyle. G, Renewable Energy: Power for a Sustainable Future, 1996, Oxofrd University Press

Material	Porosity (%)	Hydraulic conductivity (m day^{-1})
Unconsolidated sediments		
Clay	45–60	$<10^{-2}$
Silt	40–50	10^{-2}–1
Sand, volcanic ash	30–40	1–500
Gravel	25–35	500–10 000
Consolidated sedimentary rocks		
Mudrock	5–15	10^{-8}–10^6
Sandstone	5–30	10^{-4}–10
Limestone	0.1–30	10^{-5}–10
Crystalline rocks		
Solidified lava	0.001–1	0.0003–3
Granite	0.0001–1	0.003–0.03
Slate	0.001–1	10^{-8}–10^{-5}

exchange surface and the injection of water by pumping. In theory drilling could be carried out over much of the surface of the earth, making the resource considerably more extensive than relying on wet rocks. However drilling to great depths is expensive and only those areas with temperature gradients in excess of 25 °C/km and within 6 km of the surface are likely to even be considered economic.

Within such rocks we have two sources of heat, the upwelling of heat from the earth's mantle and the heat produced from radioactivity within the drilled layer. We can get an idea of the size of this additional heat flow from assumptions about the composition of the granite and its thickness. There are five heat-producing radioactive decays of interest:

$$^{40}\text{K} \rightarrow {}^{40}\text{Ar} + \text{heat or } {}^{40}\text{Ca} + e^- + \text{heat (half-life} = 1.25 \text{ billion years)}$$

$$^{238}\text{U} \rightarrow {}^{206}\text{Pb} + 8\,{}^4\text{He} + 6e^- + \text{heat (half-life} = 4.5 \text{ billion years)}$$

$$^{232}\text{Th} \rightarrow {}^{208}\text{Pb} + 6\,{}^4\text{He} + 4e^- + \text{heat (half-life} = 14.0 \text{ billion years)}$$

$$^{235}\text{U} \rightarrow {}^{207}\text{Pb} + \text{heat (half-life} = 1.25 \text{ billion years)}$$

These decays double the flow of heat within granite. This means that the temperature gradient and therefore the actual temperature is likely to be greater at the depth in question, suggesting that hot dry rocks would be ideal for power generation – not just hot water production.

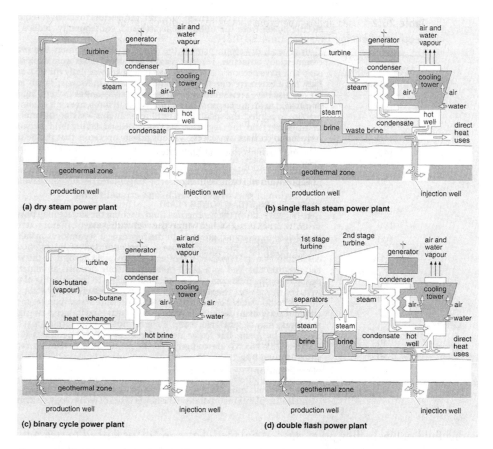

Figure 26.6 Types of geothermal power plant [BOY96, p373] [Reproduced courtesy of Boyle. G, Renewable Energy: Power for a Sustainable Future, p373 ©1996 The Open University]

26.4 Technologies for use

Wet rock technology

The details of the technology involved will depend on the rock being drilled through and the temperature and pressure of the underlying fluid to be tapped. Typically a 0.5 m diameter hole is drilled which tapers to around 0.15 m at the production depth. The well is lined with steel and concrete and a valve fitted to control the flow of fluid to the surface. A re-injection well is also drilled. If the well produces hot dry steam (at a typical temperature of 180 °C and eight atmospheres) then this can be passed directly through a turbine, condensed and re-injected into the ground (Figure 26.6a). If brine, or a mix of brine and steam (at a typical temperature of 155 °C and six atmospheres) is produced then either this must be passed through a heat exchanger, normally filled

Table 26.2 Uses and temperature ranges for the direct use of geothermal energy [adapted from BOY96]

Approximate Temperature °C	Use
180	Digestion in paper pulp
160	Drying of fish meal
150	Alumina via Bayer's process
140	Drying farm products at high rates
	Canning of food
130	Evaporation in sugar refining
	Extraction of salts by evaporation and crystallization
120	Fresh water distillation
110	Drying and curing of light aggregate cement slabs
100	Drying of organic materials
	Washing and drying of wool
90	Drying of fish
80	Space heating
60	Animal husbandry
	Greenhouse heating
50	Mushroom growing
40	Soil warming
30	Swimming pools
	Fermentation
20	Hatching of fish

with a fluid of low boiling point, such as butane (Figure 26.6c), or flashed to steam in an expansion vessel (Figures 26.6b and c). The amount of hot fluid required depends on the technology and efficiency of the system. Dry steam plants may require only 2 kg/s of steam, whereas a system using a heat exchanger might require 10 times this. This energy can be used for a variety of purposes depending on the supply temperature (Table 26.2).

Hot dry rock technology

Although wet rock technology is well established, with numerous plants operating worldwide, hot dry rock (HDR) technology has yet to be applied commercially. This lack of progress is for several reasons, the first being the very high drilling cost of boring to 3–6 km, in part through hard crystalline rocks, rather than the more modest depths and softer rocks encountered by wet rock technologies. Unlike most wet rock systems, an HDR plant cannot rely on pre-existing water in an aquifer as the working fluid. Water must be injected into the HDR layer then allowed to percolate through fractures to a second bore hole and brought to the surface. Although some natural fractures will exist in the granite, the effective area of the heat exchanger needs to be greatly increased if large quantities of water are to be heated to reasonable temperatures. Additional fracturing is achieved by forcing water through the system at such high pressures that further faults are created (or by using explosives). Figure 26.4 illustrates the basic concept.

There have been some recent interesting developments in HDR technology with the advancement of single borehole techniques. These use a single well for both injection and return by inserting a two-way pipe down the well, with the cold injection water flowing down the outer pipe and the hot return water travelling up the inner pipe. Return water temperatures are expected to be lower (due to the reduced heat-exchanger area) and therefore such systems are only likely to be suitable for space heating, but cost should be dramatically lower and the technology suitable for use in most locations, with the bore hole being within the grounds of the building.

Several problems with the approach have been identified at experimental HDR plants. The fracture networks have offered much higher resistance to the working fluid than expected, meaning that higher pumping pressures have been required to circulate water through the system. Short-circuits have developed that allowed some of the circulating water to travel through more major fractures and thus interact with less of the hot granite and therefore be returned to the surface at a lower than expected temperature. Finally, considerable water losses (>10 per cent) have occurred during circulation. However, despite these problems circulating loops have been constructed in granite with hot water being returned to the surface at temperatures of up to 140 °C and small amounts of electricity generated.

26.5 Environmental problems

Geothermal power would seem to be a largely environmentally benign approach to energy production. The only concerns are the possibility of unfortunate odours being produced due to the release of hydrogen sulphide, the chances of land subsidence and the risk of increased seismicity. By re-injecting the up-welling water, pollution of surface waters can largely be avoided.

26.6 World resource

One way of estimating the theoretical resource is to imagine how removing heat from the interior of the planet might cool the earth. Earth has a mass of 5.98×10^{24} kg, and if we assume it has an average heat capacity of 750 J/kg/K (the same as granite), then cooling the planet by 0.01 °C would release 4.5×10^7 EJ, one hundred thousand times current annual primary energy demand. Of course such an extraction would be impossible, but it does indicate that the size of the resource is unlikely to limit the technology.

Figure 26.7 Hacchobaru geothermal power station. Source: IEA CADDET programme[3] [CAD99] [Reproduced courtesy of ETSU, CADDETT, Technical Brochure 109, IEA CADDET programme, http://www.caddet.org.]

26.7 OECD resource

Within the OECD geothermal electricity production grew from 28.7 TWh to 32.9 TWh between 1990 and 2002, i.e. an annual average increase of only 1.1 per cent – a very slow rate of growth. However geothermal is still the second largest source of sustainable energy (or the third if all hydropower is considered sustainable). The largest producer is the USA followed by Mexico, then Italy. Japan and New Zealand also have major facilities [IEA05].

26.8 Example application: Hacchobaru geothermal power station, Kokonoe-machi, Japan

Kokonoe-machi has three geothermal power stations with a combined output of 147.5 MW$_e$. Hacchobaru is the largest with a capacity of 110 MW$_e$ (Figure 26.7). The facility not only produces electricity but also hot water for horticulture, space heating and domestic hot water.

Despite being in the southern part of Japan, Kokonoe-machi experiences a harsh climate due to the mountainous terrain in which it sits. This climate has put a premium on hot water for space heating and horticultural use.

Hacchobaru power station consists of two units; both rated at 55 MW$_e$ and each using a double-flashover technique. In this system steam and hot water from the geothermal wells pass through two-phase fluid transfer pipes and into steam separators, which

---

[3] Web site: http://www.caddet.org.

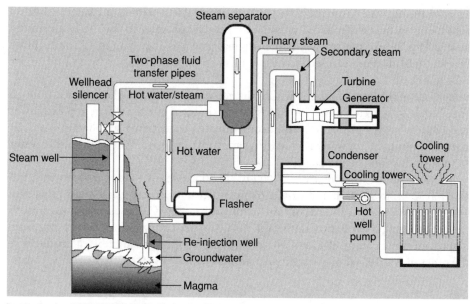

Figure 26.8 Schematic of Hacchobaru geothermal power station. Source: IEA CADDET programme [CAD99]

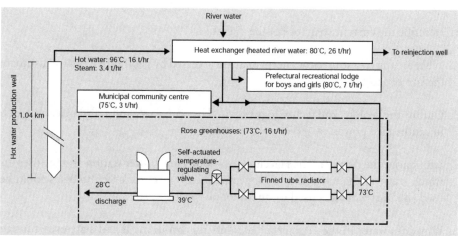

Figure 26.9 Heat supply system for the First Sensui Rose Park (assuming an outdoor temperature of 8 °C). Source: IEA CADDET programme[4] [CAD99]

---

[4] Web site: http://www.caddet.org.

separate the hot water from the high-pressure steam. The separated hot water then flows into flashers where the reduced pressure causes the hot water to flash to low-pressure steam. This steam, together with the original steam is then passed to the turbines (Figure 26.8). The use of a second flasher increases electrical output by around 20 per cent [CAD99].

Because of the level of arsenic within the geothermal wells, the hot water cannot be supplied directly to consumers but needs to be re-injected into the well. A heat exchanger is used to solve this problem, with uncontaminated water being supplied by a nearby river. The resultant hot water is used for a variety of purposes, from domestic hot water to rose production (Figure 26.9).

Two rose parks (totalling over 25 000 m^2 in 30 glasshouses) are provided with hot water at 73 °C to 90 °C at a rate of 22.6 tonnes per hour. These glasshouses grow over two million roses annually at a constant temperature of 18 °C. The use of hot water from the geothermal power station saves over 30 000 GJ per annum which would otherwise be provided by burning 884 kl of oil. In addition, the power station itself produces in the order of 3 × 10^6 GJ of electricity per annum

26.9 Student exercises

"Global warming is not a conqueror to kneel before – but a challenge to rise to. A challenge we must rise to."

–Joe Lieberman

1. Describe the current structure of the planet.

2. What are the two sources of energy that are accessed by geothermal power? Estimate the size of each.

3. Outline the geology and technology required by both a wet and a dry rocks approach to geothermal power.

4. Summarize the design of the Hacchobaru geothermal power station and the uses for which water in the temperature range provided by wet rock technologies could be used (not just what it is used for in Kokonoe-machi).

5. Why does water in the geothermal aquifers remain in the liquid state even though its temperature may be much higher than 100 °C? [From SHE97.]

6. A geothermal district-heating scheme issues a flow rate of 22.5 litres/sec with a wellhead temperature of 70 °C. It rejects water at 40 °C and runs for a period of 162 days/year. If the overall combustion efficiency of an oil burner is 73 per cent, how much oil is saved per year?

27

Fast breeders and fusion

"It is not too much to expect that our children will enjoy in their homes [nuclear generated] electrical energy too cheap to meter."

–Lewis Strauss, Chairman, US Atomic Energy Commission, 1954

Electricity generation using thermal nuclear reactors was discussed in Part II, i.e. under conventional, rather than sustainable energy sources. This was because it failed numbers two and three of our three requirements for a sustainable energy source, which were:

1. it contributes little to anthropogenic climate change

2. it is capable of providing sustainable power for many generations without significant reduction in the size of the resource, and

3. it does not leave a burden to future generations

We also saw that at the current rate of use the easily accessible uranium deposits would only last for 70 years. There are two other forms of nuclear reactor which by-pass this problem: the *fast breeder reactor* and the *fusion reactor*. Either could form the backbone of twenty-first century electricity generation in much of the world. However, the former has yet to be truly exploited for commercial generation, although several working reactors have been built, and the latter has yet to be proved even feasible.

27.1 Fast breeder reactors

As we discussed in Chapter 12, a thermal reactor requires the neutrons produced by fission to be slowed down by a moderator before they can fission more uranium. We

Energy and Climate Change   David A. Coley
© 2008 John Wiley & Sons, Ltd

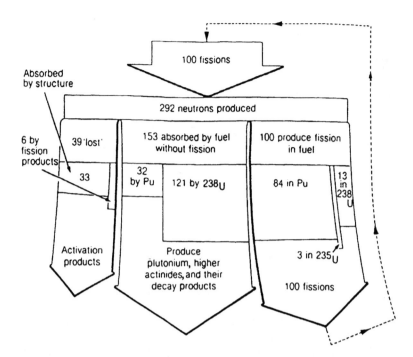

Figure 27.1 Creation, reaction, and absorption of neutrons during fission in a fast reactor [HEW00] [Reproduced from Hewitt, G.F. and Collier, J.G., Introduction to Nuclear Power (2nd edition) ©2000 Taylor and Francis]

also saw that it was only uranium-235 that underwent fission, and that this forms only 0.7 per cent of natural uranium – the rest being uranium-238. Uranium-238 does have a useful role within a thermal reactor: it absorbs neutrons to produce plutonium, via the route shown in Figure 12.5, the subsequent fission of which contributes about a third of the heat generated within the core. However, most of the plutonium remains unused, as does the majority of the uranium-238.

A reactor can be built that relies on plutonium for its chain reaction. Such a design no longer requires a moderator, as fast rather than thermal neutrons drive the process. There are two advantages to this: the power density is much higher, and more efficient use of the fuel can be made. A mix of around 20 to 30 per cent plutonium (from the reprocessing of waste fuel rods from conventional thermal reactors) and 80 to 70 per cent uranium is used as the fuel within a core of only one to four cubic metres – much smaller than a car. As the plutonium undergoes fission the resulting neutrons are absorbed by a surrounding blanket of uranium-238 to create yet more plutonium (Figure 27.1). Thus the reactor actually breeds more plutonium than it was originally stocked with, hence the name *breeder*. The utilization of uranium is about 60 times greater than that of a thermal reactor. This gain is such that concerns over the global reserve of uranium would be removed by the widespread adoption of the technology.

Figure 27.2 Prototype fast Breeder Reactor at Monju, Japan [KU07] (Research Group of Plasma Physics and Engineering Department of Nuclear Engineering Graduate School of Engineering Kyoto University, Kyoto 606-8501, Japan http://p-grp.nucleng.kyoto-u.ac.jp/~honda/viewer.php?name=monju&sel=j)

Any country attempting this route to nuclear power would need access to three technologies: thermal reactors to produce the seed plutonium, fuel reprocessing facilities and fast breeder reactors.

The higher power density requires the rapid removal of substantial quantities of heat from the core. Because of its ability to be used at much higher temperatures (at which it is a liquid), its excellent heat transfer, its relative abundance and its low neutron absorption, sodium has been used as the coolant in a series of fast breeder reactors. As anyone who has mixed sodium with water will know, the result of such an act can be explosive. As the coolant needs at some point to be used to raise steam to drive turbines this is a serious concern. Another concern is that large amounts of plutonium are required. Plutonium is extremely toxic to humans and could be used by terrorists or others to pollute water supplies or possibly used within a nuclear device.

The first fast reactor was built at Dounreay, UK, in 1959. Since then such reactors have been built in France, Japan and the USA. India is also interested in developing the technology. Figure 27.2 shows the breeder reactor at monju, Japan

World resource

Natural uranium only contains 0.7 per cent fissionable ^{235}U. A fast breeder would allow us to harness the other 99.3 per cent. This would effectively expand the world resource to at least:

$$\frac{99.3}{0.7} \times 70 \text{ years} = 9930 \text{ years at the current rate of use;}$$

a truly astronomical figure equivalent to 70 000 EJ. This should be compared to our reserves of oil and gas and current world primary energy demand (409 EJ).

27.2 Fusion

As we have stressed several times, most sustainable energy technologies rely on the energy produced by the sun. Wind power for example, relies on the sun to heat the surface of the planet, the ground and oceans to warm the atmosphere and create motion in the form of winds and then wind turbines to convert this kinetic energy into electricity. This realization begs the question, why not try and short circuit this chain and recreate the sun here on earth? The generally low concentration of renewable energy resources is one of the main stumbling blocks to sustainability. Fusion would give us a concentrated source of energy.

Figure 12.1 indicates that the binding energy of low-mass nuclei could be released if two such nuclei were fused, and indeed, the sun fuses hydrogen nuclei (protons) to form helium by the following chain of events:

$$^1_1H + {}^1_1H \rightarrow {}^2_1H + \beta^+ + \upsilon + \text{energy}$$

$$^1_1H + {}^2_1H \rightarrow {}^3_2He + \text{energy}$$

$$^3_2He + {}^3_2He \rightarrow {}^4_2He + 2{}^1_1H$$

so overall, we have:

$$4{}^1_1H \rightarrow {}^4_2He + 2\beta^+ + 2\upsilon + \text{energy}$$

That is, hydrogen nuclei are transformed into helium nuclei, positrons (β, positively charged electrons) and neutrinos (υ).

These reactions take place at enormous temperatures and pressures within the core of the sun. Here on earth we face the problem of re-creating such temperatures and then confining the hot nuclei. As a temperature of many million degrees centigrade would be required, the use of any physical container would be impossible since contact with a surface would lead to rapid cooling. In the sun this confinement problem is solved by gravity. On earth we would be dealing with much smaller amounts of material and gravitational forces would be unable to offer confinement. At such high temperatures the electrons that normally surround nuclei will have separated themselves from the atoms and we will have been left with a sea of charged particles called a *plasma*. Because they are charged, these particles can be manipulated by magnetic fields and thus, by surrounding the confinement vessel with electro-magnets, the hot plasma can be held away from the confining walls. The favoured shape for the containment vessel, or reactor, is a torus, or ring doughnut. Such a machine is called a tokamak (see Figure 27.3 and the JET example application).

One further difference between the sun and any useful fusion reactor is that of power density. As we saw in Problem 2.10, the power density of the sun is very unimpressive. Given a solar radius of 6.96×10^8 m and a solar output of 3.846×10^{26} W we have a volumetric energy density of $3.846 \times 10^{26} / (4/3)\pi (6.96 \times 10^8)^3 = 0.27 \text{ W/m}^3$. Thus a reactor with an equivalent power density and a volume of 10 m³ would only

Figure 27.3   The experimental JET fusion torus, UK [JET04] [Reproduced from Joint European Torus http://www.jet.efda.org/documents/brochures/jeteuropeansucess.pdf]

produce 2.7 W. Clearly we would need much higher power densities, introducing further engineering challenges.

In theory, the successful and economic application of nuclear fusion to generate heat, raise steam and thereby generate electricity could solve all the problems associated with our reliance on fossil fuels. This is because the basic fuel, hydrogen, could be provided in almost limitless quantities from seawater and nuclear fusion could provide four times as much energy per unit of fuel as nuclear fission. However, amongst physicists it has been a continuing joke for the last 40 years that fusion power will only take *another* 40 years to realization. And it is true that even today we have yet to build a reactor that can sustain fusion for a lengthy period or generate more energy than it takes in confinement and initial heating. However the potential rewards are so great that many billions of dollars have been spent on fusion projects. Unfortunately, even if we make the assumption that fusion may be possible in 40 years, the technology will not be a solution to our current problems with anthropogenic climate change, which are immediate. Most scientists would agree that we will have done widespread and irreparable damage to the planet before then if we continue our accelerating use of fossil fuels. In addition, there is the question of how quickly, and at what cost, a worldwide fusion energy infrastructure could be rolled out and how it could be used to reduce transport emissions. It is increasingly looking as though fusion may be the energy source of the future, but that future is some way off, and even then not guaranteed.

Technologies for use

The most promising reaction for sustaining nuclear fusion on earth is the combination (at around 100 million celsius) of deuterium (an isotope of hydrogen containing a neutron and a proton in its nucleus) and tritium (another isotope of hydrogen containing two neutrons alongside the proton), to form helium:

$$^2_1H + ^3_1H \rightarrow ^4_2He + n + 2.8 \times 10^{-12}\,J \qquad (27.1)$$

The amount of energy released by this reaction can be confirmed by consideration of the atomic masses involved. Deuterons, tritons, alpha particles (^{4}He) and neutrons have atomic masses of 2.0136 u, 3.0160 u, 4.0015 u and 1.0087 u respectively (One atomic mass unit (u) is approximately the mass of a proton or 1.660×10^{-27} kg). Balancing the masses on either side of Equation (27.1) gives:

$$2.0136 + 3.0160 = 4.0015 + 1.0087 + \text{missing mass},$$

or a missing mass of 0.0194 u.

Applying $E = mc^2$ we see that this missing mass equates to:

$$E = 0.0194 \times 1.66 \times 10^{-27} \times (3 \times 10^8)^2 = 2.8 \times 10^{-12}\ J.$$

Most of the energy is carried away by the neutron in the form of kinetic energy. A blanket containing liquid lithium would surround the plasma and slow these neutrons, become warm and thereby provide heat to raise steam for conventional steam generation. It may seem surprising that this futuristic, highly technological generating device in the end comes down to a machine for heating water and therefore suffers the Carnot limitations of any heat engine.

The lithium blanket would do more than provide the heat removal path. Tritium does not occur naturally (it is radioactive with a half-life of 12 years) and has to be manufactured by the following process:

$$n + ^6_3Li \rightarrow \alpha + ^3_1H + 7.68 \times 10^{-13}\,J.$$

So overall in the reactor we have:

$$^2_1H + ^6_3Li \rightarrow 2^4_2He + 3.58 \times 10^{-12}\,J.$$

Initially the plasma is heated by passing currents of several millions of ampere through it, then further heating it with high-energy neutron beams. The deuterium required by Equation (27.1) is easily obtained directly from seawater, in which it exists at a concentration of one in seven thousand atoms of hydrogen. The confinement process requires a large energy input and the present record for the ratio of energy in the plasma to the magnetic field energy is only 0.4 [NEW02a], i.e. the confinement currently uses more energy than the plasma creates.

Problem 27.1   Estimate the length of time that one per cent of the world's supply of deuterium could furnish all the world's primary energy needs (assume current use of primary energy, an average ocean depth of 3.18 km, two thirds of the planet is covered in water and the radius of the earth is 6370 km).

Figure 27.4 The Joint European Torus tokamak machine during construction; the iron transformer cores are clearly visible [JET04]

World resource

Fortunately, lithium-6 is also relatively common, and the world's extractable land-based reserves are thought to be in excess of 10^{33} atoms [SCI71]. The oceans are known to hold an even greater reserve, although this could be costly to extract. Like fission reactors, a fusion reactor would also produce radioactive wastes. Tritium itself is radioactive and the reactor would become radioactive from neutron bombardment and other reactions. However, most of the products would have short half-lives and only exist in low concentrations.

Another approach to fusion reactor design and confinement is that of *inertial confinement*. This uses an intense blast of radiation from lasers to compress and heat a fuel pellet containing tritium and deuterium. The approach is currently less advanced than the tokamak design.

27.3 Example application: JET Torus, Culham, UK

The Joint European Torus (Figure 27.4) uses 32 large D-shaped copper coils equally spaced around the torus-shaped vacuum vessel (Figure 27.5) to give the plasma its toroidal cross-section. In the centre of the machine sit the primary windings (inner

Figure 27.5 Inside the JET's vacuum vessel (note the person for scale) [JET04]

poloidal field coils) of the transformer, which are used to induce the plasma current that produces the poloidal component of the field. The plasma itself acts as the single-turn secondary of a transformer (Figure 27.6). Just as in a normal transformer, the fields within the primary and secondary coils must be coupled, and, again as in normal

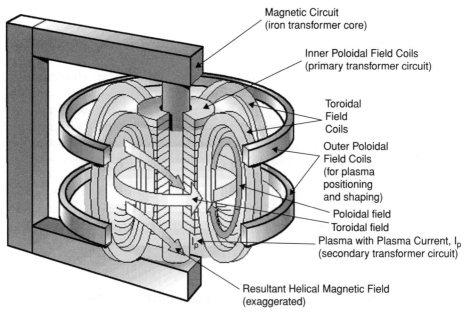

Figure 27.6 Schematic showing the coils and magnets used to constrain the plasma with the vacuum vessel [JET04]

Table 27.1 JET's mechanical and electrical parameters [data from JET04]

Parameter	Value
Plasma major radius	2.96 m
Plasma minor radius	2.10 m vertical, 1.25 horizontal
Weight of iron core	2800 t
Toroidal field coil power	280 MW
Toroidal magnetic field	3.45 T
Plasma current	3.2–4.8 MA
Additional heating power	25 MW

transformers, this is achieved by iron bridges – in this case eight massive, equally spaced limbs. In addition, around the outside of the machine are six more field coils (the outer poloidal field coils) used for positioning, shaping and stabilising the plasma [JET04].

The machine operates in a pulsed mode at a maximum rate of one pulse every 20 minutes, with each pulse lasting a maximum of one minute. The maximum mass of gas introduced into the vessel is only one tenth of a gram per pulse. Table 27.1 lists some of the other specifications.

27.4 Student exercises

"Nothing in this world is to be feared ... only understood."

–Marie Curie

1. Why are fast breeder reactors so called?

2. What advantages do fast breeder reactors have over thermal fission reactors?

3. Outline the reactions behind fusion in the sun and on earth.

4. Discuss how heat might be extracted from a fusion reactor and the scale of the world resource.

5. Why is fusion unlikely to be a 'solution' to climate change?

6. What are the two main roles lithium plays within a fusion reactor?

28

Alternative transport futures and the hydrogen economy

"There will never be a bigger plane built."

–A Boeing engineer, after the first flight of the 247, a twin
engine plane that holds ten people

In Chapters 14 and 15 we looked at the amount of energy used by a typical developed nation to transport its citizens and goods – about a third of the national energy budget. Because almost all of this energy is in the form of oil burnt within internal combustion engines, often within urban confines, we saw that this has resulted in environmental concerns being voiced. This chapter will investigate alternative transport technologies and fuels including hydrogen. Hydrogen will be found to have a role beyond transportation and could form a major component of twenty-first century energy supply.

The current environmental footprint of the transportation sector could be reduced in any of four ways:

1. we travel less and transport fewer goods less far

2. we switch to fossil fuels that cause less air pollution

3. we switch to non-fossil fuel based technologies

4. we design more energy efficient vehicles

Implementation of the first of these will require deep social change. How this can be achieved is not clear, but until less polluting and ultimately non-fossil fuel based technologies can be developed it will be necessary. The developing world has much lower levels of transportation use and consequently lower emissions. If the majority of

Energy and Climate Change   David A. Coley
© 2008 John Wiley & Sons, Ltd

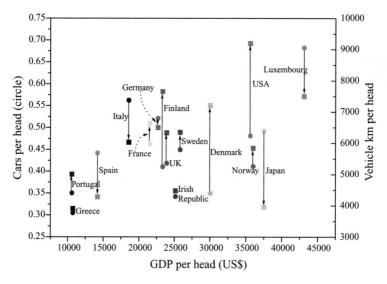

Figure 28.1   Comparison of cars per head of population (circles), annual distance driven (squares) and GDP [adapted from UK Department for Transport, *Transport statistics Great Britain 2004*, HMSO, London, 2005.]

the world's six billion citizens had equal access to developed world-like transportation systems, global carbon dioxide production would increase greatly at a time when we need to be making reductions.

Problem 28.1   Estimate the growth in carbon dioxide emissions that might arise from the world's populace having the developed world's level of transport use.

Emissions are not the only concern. In the UK approximately 3500 people die in traffic accidents each year. An equivalent rate throughout the world would imply 370 000 deaths per annum; a number exceeding that from almost any single natural disaster and a state of affairs that seems to pass without comment. Whether this lack of concern will always exist we cannot say. Another problem is one of land area. As more roads, runways and rail lines are built, the more such infrastructure comes into conflict with areas of natural beauty, or simply locations where people live. Such developments not only affect the area they occupy, but also have an impact on wildlife and human wellbeing through both pollution and noise.

Problem 28.2   The UK (population 60 million) has 392 000 km of roads [DFT05]. Given that the UK has a land area of 242 000 km², estimate the percentage of the country occupied by roads. Compare this with an estimate of the area taken by housing in the country.

A detailed investigation into how we might be weaned off our love of transport is beyond the scope of this book, so we will concentrate on looking at how feasible any of the other three options for change might be. However we do know that there are substantial differences in car use between the industrialized nations. Figure 28.1 shows that there is little correlation between the wealth of a nation, or its size, and the number of vehicles owned or the amount they are used. This indicates that there are alternative ways society can be structured to accommodate the need to travel.

28.1 Improving energy efficiency

The most obvious way to improve efficiency does not even require any technological advances – simply move more of our transportation demands to whichever current form is the most energy efficient. Table 28.1 shows the relative efficiency of various common forms of transportation. Clearly human power wins, and both buses and trains (when most seats are occupied) are more efficient than cars.

Other improvements in energy efficiency are possible by reducing various losses. A car needs a source of energy to overcome the forces acting on the vehicle. The first force is that needed to accelerate the vehicle from a stationary state:

$$F_{acceleration} = ma, \tag{28.1}$$

where m is the mass of the vehicle and a the acceleration. If the car is accelerated more aggressively, more force will be required, but once a constant speed is reached $a = 0$ and therefore $F_{acceleration} = 0$. Conversely, if the vehicle is decelerated (by applying the brakes) then a will be negative as well as $F_{acceleration}$. If such a negative force can be captured and stored as energy, it can be redeployed later. $F_{acceleration}$ is proportional to m; therefore if the mass of the vehicle can be reduced, losses due to $F_{acceleration}$ will also be reduced.

Table 28.1 Energy efficiency of various transportation modes [adapted from RIS99]

Passenger Transportation	Passenger km/GJ
Bicycle	5400
Walking	3210
Bus	1010
Train	641
Car	321
Aircraft	236

Freight Transportation	Tonnage km/GJ
Pipelines	3710
Waterways	3210
Railroads	2530
Trucks	675
Aircraft	45.6

Any hill that needs to be climbed will require more energy to work against gravity. The force required, $F_{gravity}$ is given by

$$F_{gravity} = mg\frac{h}{d},$$
(28.2)

where h is the height gained, d the horizontal distance travelled and g the acceleration due to gravity (≈ 10 ms^{-2}). If the vehicle is travelling down hill then the slope will be negative, making $F_{gravity}$ negative and again making it possible to store energy for later use.

At all but the most modest speeds aerodynamic drag will be important. At speeds in excess of 60 km/h (40 mph) drag becomes the most important inefficiency outside of the inherent inefficiencies of the engine. The Force, F_{drag}, needed to overcome this resistance depends on the shape of the vehicle and its size. In practice it has been found that, approximately,

$$F_{drag} = \frac{C_{drag}\,\rho\,Av^2}{2},$$
(28.3)

where C_{drag} is the aerodynamic drag coefficient, which depends on how streamlined the vehicle is (typical values being 2.1 for a smooth brick, 0.9 for a bicycle with cyclist, 0.32 for a modern car and 0.4–0.45 for a large four wheel drive vehicle or SUV), ρ is the density of the fluid (in this case air) and A is the frontal projected area of the vehicle including the windscreen and any projections. Clearly, smaller more streamlined vehicles travelling at lower speeds reduce the losses due to F_{drag}.

Other, smaller losses due to rolling resistance in bearings and of the tyres on the road can be subsumed into F_{roll}, where approximately:

$$F_{roll} = C_{roll}\,mg,$$
(28.4)

where C_{roll} is a dimensionless coefficient that describes the performance of the tyres and bearings used. C_{roll} is typically around 0.015. Again, reducing m reduces this loss. The total force required at any instant will be

$$F_{acceleration} + F_{gravity} + F_{drag} + F_{roll}.$$
(28.5)

This analysis suggests it would be desirable to avoid rapid acceleration, keep the mass of the vehicle low, make it small and aerodynamic and drive it at low speeds[1]. All of which points away from the fashion for aggressive driving of large, square fronted, four wheel drive-type vehicles at high speeds on today's motorways, highways and autobahns.

---

[1] However, at speeds below about 80 km/h (50 mph) efficiency can sometimes deteriorate because of inefficiencies within the engine at lower speeds.

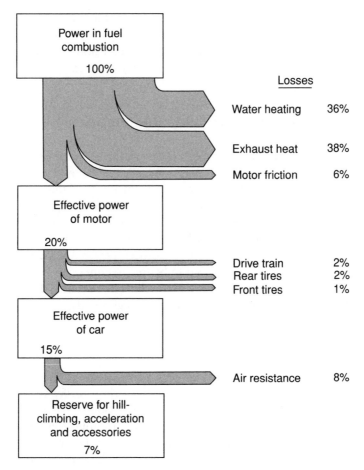

Figure 28.2   Relative contribution of losses to the inefficiency of a motorcar being operated under typical conditions [RIS99, p251]. The large losses to water heating and exhaust gases are unavoidable (and are in part due to the Carnot efficiency-limited processes involved) [Adapted from Ristinen, R.A & Kraushaar, J.J, Energy and the Environment, p82, 1998 ©John Wiley & Sons, Ltd John Wiley and Sons, Inc]

Problem 28.3   Estimate the fuel efficiency of a car by applying Equation (28.5), assuming $C_{drag} = 0.35$ and $C_{roll} = 0.015$ (typical values) and that the car is travelling on flat ground at a constant speed of 120 km/h. (Note: the power required to move the vehicle is the product of the force required and the speed of the vehicle.)

The answer to the above problem, 59 km/l, is much greater than today's cars commonly achieve: values near 14 km/l (40 miles per gallon) are typical. So there must be some other source accounting for the majority of the inefficiency. The reason for this discrepancy lies in the inefficiency of the internal combustion engine, which is subject to a Carnot limiting efficiency. Figure 28.2 shows the relative contributions of various losses to the inefficiency of a motorcar. Clearly, although the losses we have studied so far are important, even greater losses are experienced by the engine itself. How these losses might be overcome by replacing the internal combustion engine, or powering it with some other type of fuel, will occupy the rest of this chapter.

28.2 Alternative transport fuels and engines

We saw in previous chapters that transportation typically accounts for approximately 20 per cent of anthropogenic greenhouse gas emissions (and up to 30 per cent of carbon emissions) from industrialized nations, and that the majority of this arises from road transport. We will therefore limit our analysis of the future of transport to the future of road vehicles. Central to our aim will be to cut carbon emissions whilst maintaining the efficacy of the motorcar or lorry (truck). Three approaches will be discussed in detail: electric vehicles, non-fossil carbon fuels and hydrogen powered vehicles. To fulfil the efficacy requirement, these solutions need to be free roaming, i.e. carry their energy source on board, be cost-effective, capable of travelling a substantial distance between refuelling, refuel relatively quickly and be capable of high speeds.

Battery powered vehicles

One of the central problems in trying to compete with carbon-fuelled vehicles is matching the high energy density of such liquid fuels. Petrol (gasoline) has an energy density by mass (i.e. the calorific value) of 43 MJ/kg (net) and by volume of 36 MJ/l. This means that 1 kg (1.2 litre) of petrol will allow a typical car to travel 17 km (11 miles). A cube of only just over 10 centimetres (four inches) on each side is needed to hold 1 kg of petrol and such fuels also only need lightweight, thin-walled storage vessels or tanks. By comparison, the lead-acid batteries typically used in prototype electric vehicles have an energy density of around 90 kJ/kg – one 500th of that of petrol. On a volume basis, lead-acid batteries provide about 180 MJ/m^3, around one 250th of the energy density of petrol.

Problem 28.4   Estimate the weight of batteries needed to store as much energy as a 50 litre tank of petrol. Compare this to the typical weight of a car (around 1000 kg).

As three of the four sources of energy loss in a vehicle discussed in Section 28.1 include a mass term, a battery-powered vehicle is at a severe disadvantage. Clearly, either better battery technologies are needed, or the market in electric vehicles will be restricted to short range vehicles requiring fewer batteries. Although batteries are being developed with higher energy densities, they will probably only achieve a few times the current energy densities, not the several hundred fold improvement required by the above analysis.

Figure 28.3 Congestion and air quality problems are not just a problem in the developed world (Pakistan, photo by the author)

Another serious problem faced by electric vehicles is the question of where the electricity to charge such vehicles is going to come from if the concept were ever to gain common acceptance. Electric vehicles have a natural advantage in that they can achieve efficiencies of around 40 per cent compared with the 15 per cent of those using internal combustion engines. However if a fossil fuel electric plant is generating the electricity, the overall efficiency will be the product of the two efficiencies. So if the power plant is operating with an efficiency of 35 per cent, then the overall efficiency will be

$$\frac{40}{100} \times \frac{35}{100} = 0.14, \text{ or 14 per cent,}$$

comparable to that of a normal car. This shows that such vehicles might be of use in reducing emissions of air pollutants in congested urban environments (which occur in both the developed and developing world – Figure 28.3), because the power plant is likely to be in a less populated area. But they will have no impact on reducing emissions of greenhouse gases unless a sustainable energy source is used to provide the required electricity. As the following analysis shows, this will be difficult to achieve.

The current UK demand for primary energy to drive the transport sector is 2.4 EJ per annum; that used by the electricity sector 1.4 EJ per annum. Thus if 50 per cent of current vehicles were replaced by electric ones, the UK would need to almost double its generation capacity. Is it likely that this much renewable plant can be commissioned in a reasonable time? The reality is that renewable generation is growing at such a slow rate that its main use will be in displacing the supply of electricity used for more traditional purposes, such as lighting, and that switching to electric vehicles will be of no benefit to the global environment. One small caveat is needed here. If the use of nuclear energy were to substantially increase, electric vehicles would be able to make use of the electricity such power stations produce at night, when demand for electricity is traditionally low, to recharge their batteries. This would indeed reduce carbon emissions.

In general charging at night would seem a sensible approach. If the vehicle is to have a range of one tenth of that of a petrol equivalent we will need to provide

$$\frac{15}{40} \times \frac{1}{10} \times 640 \text{ (km)}/17 \text{ (km/kg)} \times 42.8 \text{(MJ/kg)} = 60 \times 4 \text{ MJ} = 60\,400 \text{ kJ}.$$

(The 15/40 term arising from the relative efficiencies of electric and petrol vehicles.) A domestic supply socket can only provide 13A, or $13 \times 240V = 3.12$ kJ/s. Therefore refuelling times would be of the order of $(60\,400)/3.12$ seconds or 5.4 hours. (As the charging process is not 100 per cent efficient, longer would be needed in reality.) Increasing the charging current will reduce the lifetime of the battery. This indicates that electric vehicles are unlikely to be able to match the refuelling rates of fossil-fuelled vehicles. The following problem illustrates this well.

Problem 28.5   Estimate, in units of kW, the rate of energy transfer between a petrol pump and a car's fuel tank. If an electric vehicle could be charged at this rate, how much current would be drawn at 240 volts? (It is worth noting that the maximum current that can be provided by a domestic supply is typically less than 100 A, thus such a vehicle would have to be recharged at a dedicated facility.)

Flywheel-powered systems

Some of the disadvantages of electric vehicles can be overcome by storing the energy to drive the vehicle not as chemical energy within batteries, but as kinetic energy within a rapidly rotating mass, or flywheel. However, the fundamental problem of providing the electricity needed to set the flywheel spinning from an environmentally benign source remains. Flywheel-powered buses have been developed, but several technological challenges need to be solved before the concept becomes commonplace.

The kinetic energy of an object is given by:

$$\frac{1}{2}mv^2, \tag{28.6}$$

where m is the mass of the object and v the velocity. For a rotating object, those parts furthest from the axis of rotation will be travelling faster, so they will make a higher contribution to the overall kinetic energy of the object. Thus flywheels are commonly flat, disc-like objects with as much mass concentrated near the periphery as structural constraints will allow.

Problem 28.6   Estimate the kinetic energy stored by a 100 kg flywheel with a diameter of 1 m, rotating at 500 revolutions per second. (Assume all the mass is concentrated on the circumference.) How far might this power a car before it needs recharging?

The material used to form the flywheel needs to be very strong because of the high speeds. As Equation (28.6) shows, the speed rather than the mass is the critical factor as the energy stored is proportional to the square of the velocity, but only directly proportional to the mass. This suggests that materials with very high strengths are likely to perform better than weaker, denser materials such as lead. If the wheel rotates on magnetic bearings within a vacuum, it will continue to rotate for a year or more, implying that such a vehicle would have much of the get-up-and-go convenience of a normal car. Re-charge times are also likely to be substantially faster than batteries – if the required current can be delivered at a dedicated facility. Problems that might be caused by gyroscopic forces within the flywheel changing the manoeuvrability of such a vehicle can be eliminated by using pairs of flywheels rotating in opposite directions.

Hybrid vehicles and alternative fossil fuels

At present, battery powered vehicles do not, in the public's eye, seem to satisfy the efficacy requirement mentioned above. Their range is too short: except possibly as a second vehicle. There are also concerns over charging times and the possibility of being stranded without power. One way around these problems is to fit the vehicle with a small liquid fuelled engine that would run at a single speed and have the dual role of topping up the batteries during travel and of providing additional power when needed. Because the engine runs at a constant ideal speed its efficiency will be at its greatest and its emissions at their lowest, and as the batteries are frequently recharged, a much smaller number are needed.

Example application: hybrid car – Toyota Prius

Figure 28.4 shows a Toyota Prius; this design and others from the same manufacturer have sold over 130 000 units. The vehicle has an estimated fuel consumption of 21 km/l (60 mpg) for city driving, a 1.5 litre 16-valve VVT-i 4-cylinder petrol engine with an output of 76 hp @ 5000 rpm (57 kW), together with an AC synchronous electric motor with a power output of 67 hp @ 1200–1540 rpm (50 kW) running at a maximum voltage of 500V, and a 201.6V Nickel-Metal Hydride battery with a power output of 28 hp (21 kW). The design includes regenerative braking (the energy from which is stored in the battery) and electronically controlled continuously variable transmission.

(a)

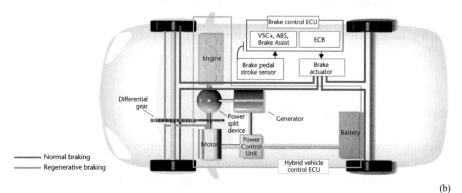

(b)

(c)

Figure 28.4   (a) A Toyota Prius hybrid vehicle. (b) Component layout. (c) The engine/drive system. (d) The flow of energy during various driving phases, note how the engine switches off when not required. (All with permission from Toyota [TOY05])

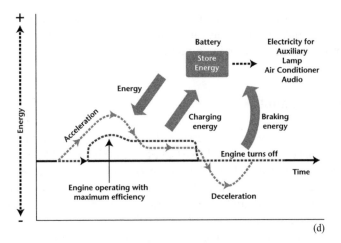

(d)

Figure 28.4 (*Continued*)

Another possibility is to keep the internal combustion engine but switch to a less polluting alternative than petrol. Liquid petroleum gas (LPG[2]) is a possibility, as is natural gas. Table 28.2 shows the carbon emissions from these fuels together with the efficiency of various other 'fuels' discussed in this chapter. Although by no means carbon-neutral, they tend to have lower carbon and other emissions. Conversion of petrol vehicles to LPG is relatively simple and it is possible to run vehicles on both fuels, so

Table 28.2 Energy densities and direct carbon emissions for alternative transport fuels and technologies (charging, engine and other losses are not included; electricity, methanol and ethanol are assumed to be from a sustainable source – if not, they will give rise to substantial carbon emissions)

Medium	Energy Density, MJ/kg	Carbon Dioxide Emission, kg/MJ
Hydrogen	137	0
Petrol/gasoline	46	0.074
LPG	49.4	0.058
Methane (CH_4)	48	0.056
Methanol (CH_4O)	24	0
Ethanol (C_2H_6O)	31	0
Fused silica flywheel	870	0
Lead-acid battery	25	0

---

[2] LPG is a mixture of propane and butane and arises in the extraction of natural gas and as a by-product of refining. Because of the simple chemical compounds it contains, it produces a more homogeneous mixture with air than petrol. This allows for more complete combustion and therefore reduced emissions.

refuelling is not a problem – although the need for a second fuel tank usually means some space in the boot (trunk) of the car is lost. Because of tax advantages making these fuels economic, they are growing in popularity.

Ethanol and methanol

Fossil fuels are not the only source of carbon. By converting biomass to a liquid form a carbon-neutral fuel can be produced (see Chapter 25). The most likely possibilities are ethanol (C_2H_6O) or methanol (CH_4O). Both have already proved themselves as suitable, with large numbers of vehicles using alcohol fuels (although not always derived from non-fossil sources) in North and South America. Because methanol can be produced from coal the approach suggests a way of putting the world's large coal reserves to good use and reducing the rate of oil reserve depletion. Once alcohol has become a commonly accepted fuel, moves could be made to produce an ever-increasing fraction from non-fossil sources. This might avoid the natural resistance to changing technologies that electric vehicles or hydrogen power may meet with. However, the scale of biomass production required to replace a substantial proportion of the world's use of oil should not be underestimated.

28.3 Hydrogen powered vehicles and the hydrogen economy

A commonplace school science experiment is to extract hydrogen from water using electrolysis and then ignite the hydrogen with a match. The hydrogen burns in the presence of atmospheric oxygen to form water once more:

$$2H_2 + O_2 \rightarrow 2H_2O + \text{heat.}$$

This is true recycling. Water is converted to hydrogen and oxygen, which burn to produce water! The process produces no pollutants during combustion except possibly nitrogen compounds formed at high temperatures from the abundance of nitrogen in air. The electrolysis could be carried out at central facilities and the hydrogen transported as a liquid at very low temperatures for final distribution to customers' vehicles at filling stations. As a simple consideration of the law of conservation of energy shows, the original electrolysis must use as much energy as is released, and in practice such electrolysis is only around 67 per cent efficient. For the process to make environmental sense, the original energy must be from a carbon-neutral source, for example wind power. In many ways the creation of hydrogen by electrolysis can be seen to perform a similar role as pumped storage. An intermittent renewable source – wind, wave, tidal or solar – can be converted to a time-invariant one by using hydrogen as a temporary store. The possibility is particularly exciting because unlike most other renewable sources, hydrogen is (at low temperatures and high pressures) a liquid – thus making it ideal as a transportation fuel. In the industrialized world, transport has the fastest growing emissions of any economic sector (for many other sectors carbon emissions are falling), and has proved itself stubbornly resistant to major developments in energy efficiency. For many environmentalists, to be able to produce a liquid non-carbon fuel from an environmentally sustainable process is almost a dream come true. Some have suggested that the implications of this process go way beyond just the transport sector, and that hydrogen has the potential to be *the* fuel of the second half of the twenty-first century. This, it is suggested, would allow our modern carbon-based economy to be replaced by an equally modern hydrogen economy, without a reduction in economic growth, lifestyle choices or wealth.

An alternative source of hydrogen would be from fossil fuels, with the carbon fraction being burnt to generate electricity and form carbon dioxide. The carbon dioxide would then be sequestrated underground or in the oceans (Chapter 29).

At three times that of petrol (gasoline), the energy density of hydrogen also makes it attractive. Apart from the problem of needing to increase the renewables base to provide the electricity to carry out electrolysis, the main problem that hydrogen faces is storage, as it is only liquid below $-253\,°C$. The cryogenic storage vessels required are bulky and would probably reduce the true energy density of the fuel plus storage system to below that of petrol. One possible way around this storage problem might be to store the hydrogen within a metallic hydride. This works by using a refillable metallic 'sponge'. Gaseous hydrogen is pumped under pressure into a metal matrix, which absorbs the hydrogen and fixes it as solid hydride. Because the hydrogen is stored as a solid, many of the problems and concerns over using liquid or gaseous hydrogen are removed. Heating the hydride (possibly by using waste heat from a vehicle's exhaust) releases the hydrogen once more. Currently storage densities of 0.03 kg/l have been achieved in vehicles. Another possibility is just to pressurize the hydrogen, although this limits the storage density and therefore the range of any vehicle. However, because the technology is easy to apply it has been used to power several demonstration vehicles.

Another advantage of hydrogen is that it makes an ideal feedstock for fuel cells. These devices, which combine oxygen and hydrogen in the presence of a catalyst, produce electricity directly and are highly efficient (>85 per cent). Burning hydrogen in an internal combustion engine would result in an efficiency of about that of burning petrol (<20 per cent); combining a hydrogen-oxygen fuel cell with an electric motor results in an efficiency of around $0.85 \times 0.80 = 0.68$, or 68 per cent.

Example application: liquid hydrogen bus, Germany

The liquid hydrogen SL 202 demonstration bus (Figure 28.5) accommodates 92 passengers and has a maximum speed of 85 km/h [HYW05]. It uses a six-cylinder 4-stroke internal combustion engine with spark ignition and a displacement of 12 litres. A three-way catalytic converter is fitted to reduce NO_x emissions. The engine is suited for gaseous

Figure 28.5   A cryogenic hydrogen powered bus (http://www.hyweb.de/Wissen/NHF97.htm)

hydrogen and unleaded gasoline and provides a power output of approximately 140 kW at 2200 rpm when operated with hydrogen.

The hydrogen system consists of three vacuum-superinsulated cryogenic-tanks each with a volume of 200 litres connected by superinsulated hoses. A heat exchanger operated by the cooling water from the engine heats the gaseous hydrogen to ambient temperature before it is sent to the engine. The tanks and heating equipment are mounted under the floor between front and rear axles offering some protection in case of accidents; in addition, a reinforced frame offers additional protection.

Refuelling takes 15–20 minutes. A conventional gasoline tank and fuelling system is included to allow for optional dual fuel operation.

Demonstration of the bus started in the Bavarian city of Erlangen in April 1996 and continued until February 1997. The bus was then transferred to Munich together with the liquid hydrogen refuelling station. The accumulated driving experience with liquid hydrogen fuel in public operation during this time amounted to around 42 000 km.

28.4 Fuel cells

Fuel cells side step the need to convert chemical energy to thermal energy and then to rotational kinetic energy in a generating set in order to simply produce electricity. This is achieved by tapping directly into the electron exchange that occurs during a chemical reaction. By forcing the electrons through an external circuit part way through this exchange, useful work can be extracted from the flow of electricity. Central to the technology, and at the core of a fuel cell, is a membrane that will allow positive charges to flow through it, but not negative ones (electrons), which are then forced to take the long route via the external circuit – in this case an electric motor.

William Grove invented the fuel cell in 1839 when he noted that not only could electricity be used to produce hydrogen and oxygen from water, but that these two elements could be combined to produce electricity. So, the technology has a history almost as long as many of our fossil fuel-based technologies. Sporadic interest was shown in the idea throughout the late nineteenth and early twentieth centuries. In 1959 Francis Bacon demonstrated a 5 kW alkaline fuel cell, which formed the basis of the fuel cells used by NASA for its Apollo missions.

Proton exchange membrane fuel cells (PEMFC)

The most promising current technology is the proton exchange membrane fuel cell. The PEMFC forces the reaction:

$$2H_2 + O_2 \rightarrow 2H_2O \tag{28.7}$$

to occur in two parts. On one side of the membrane (the anode):

$$2H_2 \rightarrow 4H^+ + 4e^-, \tag{28.8a}$$

whilst on the other (the cathode):

$$O_2 + 4H^+ + 4e^- \rightarrow 2H_2O. \tag{28.8b}$$

Figure 28.6 is an illustration of the core of a fuel cell. The anode and cathode sit either side of the proton exchange membrane; as Equation (28.8) suggests, this is permeable to protons, but not to electrons. The anode and the cathode have a series of etched channels,

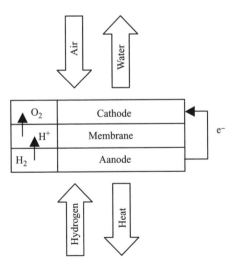

Figure 28.6 Basic design of a proton exchange membrane fuel cell. Note: the electrons are returned via the external circuit

which allow the pressurized hydrogen and oxygen gases to be equally distributed over the whole catalytic surface of each. Air is used as the source of oxygen. The proton exchange membrane is an electrolyte that looks similar to ordinary kitchen plastic wrap. The catalyst is made of carbon paper or cloth very thinly coated with platinum powder. The rough, porous nature of the catalysts ensures maximum contact area between them and the hydrogen or oxygen. As the cell only produces around 0.7 volts, many fuel cells are usually combined into a *stack*. The reaction is highly efficient, and because the whole process runs at about 80 °C the warm-up time is short, making the approach suitable for use in vehicles.

Ideally, and without a load attached, a hydrogen-oxygen fuel cell should produce 1.2 V, however several losses combine to make the true voltage less than this, create heat and make the efficiency less than 100 per cent. Which loss dominates depends on the size of the external load. *Activation losses* dominate if the cell is operated at a high voltage and low current and occur because around 200 mV of potential is lost in dissociating the hydrogen and driving the chemical reactions. *Resistive losses*, due to the internal resistance of the cell, dominate in the centre of the current range. As would be expected, the size of this resistance increases according to Ohm's law ($R = V/I$, where R includes the external load resistance). At very high currents, as might be experienced if a high demand for power suddenly occurred, *mass transport losses* dominate. These occur when the required fuel input and water output exceed the physical capabilities of the cell. A final source of inefficiencies is *fuel crossover losses*. These arise from leakage, or diffusion, between the anode and cathode (through the electrolyte) forming water and heat directly via Equation (28.7), and failing to contribute to electricity production.

Any heat produced need not be wasted. If very high temperatures are produced steam can be raised for electricity generation by steam turbines where a large stationary fuel cell is being used as part of a power station. More modest temperatures could be used for space heating. One of the most exciting potential applications of fuel cells is in home electricity generation. A fuel cell running off natural gas could provide both heat and light for a home, it could also export electricity to the grid if demand from the home was low. Taken to its natural limit, this might mean a centralized generating plant could become almost a thing of the past, with electricity being generated by millions of micro-stations interconnected by the grid in much the same way as computers are interconnected via the Internet. As both the electricity and the waste heat from the device would be utilized, overall efficiencies might be much higher and carbon emissions much lower than with conventional generation.

PEMFCs producing 50 kW$_e$ have proved relatively easy to manufacture and 250 kW$_e$ units are under development. The main disadvantages with PEMFCs are the cost of the membrane and catalysts and the need to provide almost pure hydrogen. This need for a pure feedstock is related to the low operating temperature, which requires the use of highly sensitive catalysts susceptible to poisoning by carbon monoxide and other trace impurities. Unfortunately, if the hydrogen is produced not by electrolysis of water, but by reforming carbon fuels, the resultant hydrogen is likely to contain such impurities.

There are several other types of fuel cell being developed, mostly designed as stationary sources generating much larger quantities of electricity, for distribution via the national grid. The various types are classified according to the electrolyte that separates the electrodes.

Alkaline fuel cells

Used extensively by NASA, this design is similar to a PEMFC but with an aqueous solution or stabilized matrix of potassium hydroxide as the electrolyte. Hydroxyl ions (OH^-), rather than protons, migrate from cathode to anode through the electrolyte:

At the anode: $2H_2 + 4OH^- \rightarrow 4H_2O + 4e^-$.

At the cathode: $O_2 + 2H_2O + 4e^- \rightarrow 4OH^-$.

Alkaline fuel cells operate at about the same temperatures as PEMFCs but have a power density of only one tenth; therefore they are unlikely to become a commercial reality for vehicles, where space is always at a premium. They also need very pure hydrogen. However they are the cheapest form of fuel cell to manufacture.

Phosphoric acid fuel cells

These use liquid phosphoric acid as the electrolyte, typically constrained in a silicone carbide matrix, and operate at between 150 and 200 °C. Because they still use platinum catalysts they are susceptible to poisoning from impure feedstocks but slightly less so than PEMFCs and can tolerate one to two per cent carbon monoxide and a few parts per million of sulphur. The chemical reaction is the same as that used by PEMFCs.

Commercial units have been built with outputs of up to 20 MW$_e$ and are typically used to provide power for buildings.

Molten carbonate fuel cells (MCFCs)

These use what might be best described as an active electrolyte. The electrolyte is made of lithium potassium or lithium sodium salts, which, when heated to around 650 °C, melt and produce carbonate ions that flow from the cathode to the anode. Once at the anode they combine with hydrogen to generate carbon dioxide, water and electrons (which return via the external load to the cathode):

At the anode: $H_2 + CO_3^{2-} \rightarrow H_2O + CO_2 + 2e^-$.

At the cathode: $CO_2 + \frac{1}{2}CO_2 + 2e^- \rightarrow CO_3^{2-}$.

Because these fuel cells operate at such high temperatures they can reform hydrocarbons (see the section on fuel processors below), including natural gas and petroleum, to generate hydrogen within the fuel cell itself. At these temperatures there is no problem with poisoning from carbon dioxide (although sulphur is still potentially a problem) and the expensive catalysts used in the PEMFC can be replaced with cheaper nickel species. However, the corollary of a high temperature cell is a longer warm-up period, making the approach unsuitable for transport systems. This is not a consideration for centralized power generation, which would also allow the high temperature waste heat to be collected and used to raise steam for conventional electricity generation. With steam collection, MCFCs could prove to be 80 per cent efficient. Demonstration stacks have been built with a capacity of 2 MW_e.

Solid oxide fuel cells (SOFCs)

SOFCs use a solid ceramic electrolyte such as a mix of zirconium and yttrium oxides and operate at very high temperatures of between 800 and 1000 °C. Oxygen anions migrate from the cathode to the anode to oxidize the fuel stock (typically a mix of hydrogen and carbon monoxide) and release electrons. These travel via the external load back to the cathode where they once more reduce the oxygen feedstock:

$$\text{At the anode:} \quad H_2 + O^{2-} \rightarrow H_2O + 2e^-,$$
$$CO + O^{2-} \rightarrow CO_2 + 2e^-.$$
$$\text{At the cathode:} \quad O_2 + 4e^- \rightarrow 2O^{2-}.$$

Because of the very high temperatures, poisoning by carbon monoxide is not a problem (as it is simply oxidized to carbon dioxide) and such cells are relatively resistant to sulphur damage. Being able to use hydrocarbon fuel directly is a bonus; however, as the above reaction shows, they emit carbon dioxide and are therefore of limited interest in reducing carbon emissions (although they might well be a more efficient way of using fossil fuels). Their most likely application will be stationary electricity and heat generation.

Direct methanol fuel cells (DMFCs)

These are variants of the PEMFC and operate at the slightly higher temperature of 120 °C. Their main advantage is that they use methanol directly without the need for reforming the fuel first:

$$\text{At the anode:} \quad CH_3OH + H_2O \rightarrow CO_2 + 6H^+ + 6e^-.$$
$$\text{At the cathode:} \quad {}^3\!/_2 O_2 + 6H^+ + 6e^- \rightarrow 3H_2O.$$
$$\text{Overall reaction:} \quad CH_3OH + {}^3\!/_2 O_2 \rightarrow CO_2 + 2H_2O.$$

Because of the use of methanol, carbon dioxide is again emitted, meaning the approach cannot be seen as sustainable. The most interesting application of the technology is likely to be in very small fuel cells used to power laptops or mobile phones.

Fuel processors

If fuel cells are to gain acceptance within the transportation sector, the vehicles using them will need to be refuelled easily and the fuel will need to occupy the minimum possible space; it would therefore seem sensible to provide the fuel in a liquid form that does not need cryogenic storage. For home-based electricity generation hydrogen would have to be supplied to dwellings; providing natural gas via the pre-existing gas pipelines is an obvious solution. Both these approaches need the feedstock to be processed, or reformed, so as to provide the required hydrogen. Unless this is done by the fuel cell itself (as in the case of MCFCs, SOFCs and DMFCs) an external fuel processor will be required.

Hydrogen can be extracted from both methanol and natural gas (methane) by using a steam reformer. The aim of the reformer is to extract as much of the hydrogen as possible whilst minimising the quantity of carbon dioxide and other pollutants. Waste heat from the process is used to vaporize water, which is mixed with the feedstock (either vaporized methanol, or methane) and passed over a heated catalyst. This encourages the methanol to split into hydrogen and carbon monoxide and the water vapour to split into hydrogen and oxygen with the oxygen combining with the carbon monoxide to produce carbon dioxide:

$$CH_3OH \rightarrow CO + 2H_2$$

and

$$H_2O + CO \rightarrow CO_2 + H_2.$$

Or, if natural gas is used:

$$CH_4 + H_2O \rightarrow CO + 3H_2$$

and

$$H_2O + CO \rightarrow CO_2 + H_2.$$

Unfortunately, the process is not perfect and some feedstock and carbon monoxide pass through the reformer without reacting. These are burnt in the presence of a catalyst and additional oxygen to give carbon dioxide and water.

28.5 Example application: the greening of natural gas

The Netherlands has the highest level of natural gas market penetration in the world [IEA00] and has suggested a novel approach to reducing carbon emissions: the direct injection of small amounts of hydrogen made from sustainable sources into the national gas network. This would provide a quick way into the hydrogen economy and be equivalent to converting a very large number of individual systems to hydrogen, without the associated problems of storing and transporting hydrogen within a dedicated infrastructure.

Gaseous hydrogen has a lower volumetric calorific value than natural gas and a 10 per cent substitution with hydrogen would only translate into a three per cent energy substitution, meaning more gas would be consumed overall. Or the amount of nitrogen currently added to Dutch natural gas to give the current calorific value could be reduced [ENE02].

For a 10 per cent substitution (the maximum amount unlikely to cause problems to boilers and other systems), 5×10^9 m^3 of hydrogen would be needed per annum. However, one of the attractions of the approach is that it is completely scaleable. The project could be started with a single site producing a very modest amount of hydrogen and injecting it locally into the network. Which method of hydrogen production was chosen would depend on cost and resource availability. Again, the approach would seem attractive. A variety of technologies could be applied in varying locations, including wind power, providing a way of integrating unpredictable time-varying sustainable energy generation technologies into the national energy supply. Table 28.3 illustrates the cost per GJ of producing the hydrogen using various technologies. This shows the most sustainable technologies to be the most expensive, probably prohibitively so. However the production of hydrogen from methane or coal could itself be 'greened', by re-injecting the carbon dioxide produced, back into the natural gas well (see Chapter 29).

The financial cost would not be the only limitation. Electrolysis would require around 5 kWh/m^3 [BEL02] per annum and thus if we assume photovoltaic panels might deliver 80 kWh/m^2, the production of 5×10^9 m^3 of hydrogen would require an area of 313 km^2. Biomass gasification might require around 4×10^9 kg of biomass, or a land area of 2600 km^2.

Table 28.3 Costs of hydrogen production per GJ using various approaches [data from PAD99]

Method	Cost ($€$)	Method	Cost ($€$)
Steam methane reforming	5–7	Electrolysis	21–25
Coal gasification	10–12	Electrolysis (wind)	11–20
Non-catalytic partial oxidation	7–10	Electrolysis (solar)	25–42
Biomass gasification	9–16	Electrolysis (concentrated solar)	34–65

28.6 Student exercises

"Physics is like sex: Sure, it may have practical results, but that is not the reason we do it."

–Richard Feynman

1. Estimate the charging time required to give an electric car a range of 300 km, if charged through a domestic 13 amp socket.

2. Why does a switch to fossil-fuel derived methanol as a road transportation fuel make environmental sense and pave the way to a sustainable alternative?

3. Detail the four main types of losses from a hydrogen-oxygen fuel cell.

4. Produce a table that contrasts the reactions, tolerance to impurities, temperatures, electrolytes, catalysts and other pertinent data for the various fuel cells discussed in the chapter.

5. Explain what is meant by 'the greening of natural gas'. What is its main attraction as a way of reducing carbon emissions compared with other sustainable energy technologies?

29

Carbon sequestration and climate engineering

"Today, we can see with our own eyes what global warming is doing. In that context it becomes truly irresponsible, if not immoral, for us not to do something."

–Joe Lieberman

All the sustainable energy technologies we have investigated have been just that – energy technologies. In this chapter we will be studying a very different approach to the problem of carbon emissions related to our reliance on fossil fuels. Although the production of carbon dioxide is a necessary result of burning such fuels, there are alternatives to simply venting any carbon dioxide to the atmosphere. And even if we do release it into the atmosphere there exists the possibility of capturing carbon dioxide and storing it. To do this two problems need to be solved: firstly, how to capture the carbon dioxide and secondly how to store it for a considerable length of time (possibly thousands of years). Several ways have been proposed to solve both problems and we will investigate some of the most well developed. Two methods, the direct injection of carbon dioxide into geologic features and the planting of trees, have already been deployed. Others are at various stages of development and at present it is unclear which ones are likely to be realized on a substantial basis. However, what is clear is that, at least in theory, the storage capacity of the biosphere and the deep earth is vast.

In addition to sequestration, another proposed approach to engineering the climate is to reflect incoming sunlight back into space before it reaches the planet, thereby reducing the global temperature.

Energy and Climate Change   David A. Coley
© 2008 John Wiley & Sons, Ltd

29.1 Capture technologies

The atmospheric concentration of carbon dioxide is currently around 370 ppm$_v$ (parts per million by volume). This low concentration suggests that capture might be better targeted at the much more concentrated carbon dioxide exhaust streams of power stations. Several possibilities have been put forward.

Flue gas separation

Currently carbon dioxide is separated and captured from flue gases at a number of plants around the world – not with the aim of tackling climate change, but in order to produce carbon dioxide for later use in various industrial processes. One approach uses chemical absorption (i.e. absorbing the gas in a liquid solvent forming new chemically bonded compounds). The gas is first bubbled through the solvent within an absorber column. The carbon-loaded solvent is then sent to a regenerator unit where the absorbed carbon dioxide is removed from the solvent by using steam at 100–120 °C. This can achieve a concentrated carbon dioxide stream with a purity of 99 per cent. The solvent is then cooled and returned to the absorber column. The most commonly used absorbent is monoethanolamine ($C_2H_4OHNH_2$).

Oxy-fuel combustion

If the amount of excess air required for combustion can be minimized and the mix of nitrogen and oxygen that makes up natural air replaced by pure oxygen then the exhaust stream will only contain carbon dioxide and water – the latter being easy to condense – leaving almost pure carbon dioxide for storage and later disposal. Unfortunately the air separation stage requires a considerable energy input – around 15 per cent of the power plant's final electrical output [HER04].

Pre-combustion capture

Another possibility is to combine carbon dioxide capture with a coal gasification combined cycle plant. Such plants gasify the coal to produce a synthesis gas of carbon monoxide and hydrogen. The carbon monoxide could then be reacted with water to produce carbon dioxide for storage and the hydrogen burnt within a gas turbine. This approach has the advantage that the hydrogen would also make a suitable transportation fuel. An example [HER04] is the Great Plains Synfuels Plant, North Dakota, USA, which gasifies 16 000 t of lignite coal per day, captures 2.7 million m^3 of carbon dioxide and pipes this 325 km to Weyburn, Saskatchewan where it is injected into the ground as part of an enhanced oil recovery process.

Table 29.1 Planetary carbon dioxide storage potential [data from HER04]

Option	Approximate worldwide capacity (GtC)
Ocean	1000–10 000+
Deep saline aquifers	100–10 000
Depleted oil and gas reservoirs	100–1000
Coal seams	10–1000
Terrestrial (reforestation)	10–100
Utilization	Currently <0.1 per annum

29.2 Storage technologies

Three storage technologies have been suggested: geological storage deep underground, storage within the oceans at depths of over 1 km, and biological storage within forests or marine organisms. Table 29.1 lists the estimated storage potential of these various options.

Problem 29.1 Using the data in Table 29.1, estimate the number of years of capacity each sequestration option could provide for our current rate of anthropogenic carbon emission.

Geologic storage

Carbon dioxide has been injected into depleted oil and gas fields for some time, the main purpose being to dispose of 'acid gas' – a mixture of carbon dioxide, H_2S and other residues from the oil and gas industries. To be successful such schemes require a reasonably nearby reservoir of suitable porosity isolated from production reservoirs. Carbon dioxide is also used for enhanced oil recovery. Unfortunately much of the carbon dioxide is later released back to atmosphere when the oil reservoir is de-pressurized during decommissioning – although this part of decommissioning could be avoided with only a small reduction in the oil recovered.

Carbon dioxide diffuses through coal and is adsorbed by it. (Adsorbtion describes a process whereby a gas is taken up by the surface of solid, either by molecular bonding, or by van de Waals forces.) This suggests disused coal seams could be used to store carbon dioxide. Because coal more readily adsorbs carbon dioxide than methane at a ratio of 2:1, CO_2 injection can be, and is, used to enhance methane recovery from coal beds.

Saline aquifers were introduced in Chapter 26 because of their potential to provide hot water for geothermal power production. Another possible use for such formations is for carbon dioxide storage. As Table 29.1 shows, such aquifers offer the greatest geologic storage potential. However the approach will only work if the injected carbon dioxide is stopped from rising to the surface. Aquifers capped by impermeable cap-rocks or geologic traps would be suitable, but these are relatively rare. Although, if the return path to the surface is sufficiently long and torturous, much of the carbon dioxide will react with minerals and fluids to form other stable compounds thereby becoming permanently stored.

To date there is only one commercial geologic storage project designed specifically to store carbon dioxide: the Statoil aquifer project in the Sleipner West gas field in the North Sea, 250 km off the coast of Norway. This project is examined in more detail as this chapter's example application.

There are some concerns over the safety of geologic storage, particularly the possibility of large quantities of carbon dioxide returning to the surface rapidly, and the potential for induced seismicity. There has been at least one large-scale natural release of carbon dioxide in recent history – that at Lake Nyos in Cameroon. This caused 1800 human fatalities in 1986 when an upwelling of carbon dioxide displaced the surrounding air [BBC02].

Oceanic storage

Oceanic storage at depths in excess of 1 km offers the greatest storage facility for anthropogenic carbon dioxide. Two storage mechanisms have been proposed, firstly dissolving the carbon dioxide at mid-depths (1.5–3 km) and secondly injecting the carbon dioxide at depths in excess of 3 km where it would form lakes of liquid carbon dioxide.

Below the surface layer, the oceans are unsaturated with respect to carbon dioxide, suggesting such depths could be used to store large quantities of carbon dioxide. It has been estimated that if all the anthropogenic carbon dioxide required to double the current atmospheric concentration were injected into the deep ocean, it would change the ocean carbon concentration by less than two per cent and lower its pH by less that 0.15 units [HER04]. However it is unclear how this might change the ocean's ecology.

It is worth noting that because of the global carbon cycle (Chapter 5), 80 per cent of today's anthropogenic carbon dioxide emissions will be naturally sequestered in the oceans over a period of several thousand years. Therefore we can see oceanic storage as simply accelerating a natural process in order to smooth out a spike in atmospheric carbon dioxide concentration caused by our current use of fossil fuels.

Because of the considerable distance between suitable oceanic sites and fossil fuel power stations, carbon dioxide is likely to be transported as a liquid, or power stations moved to coastal areas. If injected above 500 m, i.e. at a pressure of 50 atmospheres or less, this liquid will boil and bubble back to the surface. If released at depths of between 500 and 3000 m the carbon dioxide would remain a liquid but still be less dense than the surrounding water and gradually ascend. However if the carbon dioxide were injected through a diffuser so that it only formed droplets of 1 cm or less in diameter, it would fully dissolve before ascending even 100 m [HER04]. Below 3 km liquid carbon dioxide would be denser than seawater and would sink to the ocean floor. If released at such depths as small droplets, a solid hydrate would be formed where one or more carbon dioxide molecules would be surrounded by a 'cage' of water molecules (cf. methane hydrate, Chapter 11).

Both mid- and deep-injection have benefits and drawbacks. Dissolution at mid-depths would be relatively easy to achieve and could not lead to a catastrophic release of concentrated carbon dioxide once injected. However the storage facility offered by dissolution is not permanent as there is a slow exchange of surface and deep waters within the oceans. Storage below 3 km would effectively be permanent, but would need new technologies to be developed and is likely to be costlier. It would also leave a waste deposit that would require monitoring by future generations, and therefore clashes with clause 3 of our definition of sustainability.

As none of these technologies have been fully implemented it is difficult to estimate their likely costs and therefore how competitive they might be compared to the use of alternative energy sources. However [HER04] suggests capital costs might be in the range of €0.77 to €3 (£0.5 to £2, US$1 to US$4) per tC with an associated energy penalty of 14–28 per cent. Transportation by pipeline from a 1.5 GW$_e$ power plant might cost around €39 (£27, US$50) per tonne per 100 km.

The cost of injection could be positive or negative. If done as part of an enhanced oil recovery scheme, the value of the recovered oil is likely to exceed the cost of injection by possibly €9 (£6, US$12) per tonne of carbon dioxide. Injection into aquifers might have a cost of around €2.3 (£1.6, US $3) per tonne of carbon dioxide and injection directly into the ocean a cost of €4.3 (£3, US$5.5) per tonne of carbon dioxide. Ignoring the case of carbon dioxide injection for enhanced oil recovery, the above figures suggest that carbon capture and sequestration will not be economic unless the 'value' of carbon is set above €77 (£53, US$100) per tC. This is a considerable amount and would greatly affect the cost to consumers of fossil fuels, in turn making alternative energy technologies appear more economic.

Problem 29.2   Assuming a cost of electricity of €0.1 per kWh (£0.07, US$0.13), estimate the percentage increase in the price of electricity of implementing a carbon capture and sequestration policy based on a carbon price of €77 (£53, US$100) per tC.

Biological storage

Using the biosphere for carbon storage offers several advantages: firstly there would be no need to capture the carbon dioxide at the point of production, secondly if reforestation were used, many would see the approach as of net environmental benefit. This is unlikely to be the case for geologic or oceanic storage.

On land

Any form of growing biomass is a method of carbon capture, however, the carbon will be returned to atmosphere when the biomass decomposes. Thus the store will be of limited duration unless it is constantly replenished. Reforestation, followed by careful management would naturally provide such replenishment. Another possibility is to grow biomass for burning as a replacement for fossil fuels. Forests could be used for this, as could crops or microalgae grown in artificial ponds (see Chapter 25). It has been estimated [BOY03] that it would require the reforestation of an area the size of Europe to absorb the carbon dioxide likely to be emitted from fossil fuel use during the period

2000–2050. Worldwide there are large areas suitable for reforestation, and although the approach could not be applied on a scale capable of solving the problem of climate change, it may form part of a national or international strategy to reduce the concentration of carbon dioxide in the atmosphere in the short term. The IPCC have calculated that a 50 year programme to reduce deforestation, encourage regeneration of tropical forest and reforestation could sequester 60–87 billion tonnes of atmospheric carbon, equivalent to 12 to 15 per cent of projected carbon dioxide emissions during this period.

It is worth noting that although the replacement of fossil fuel use by the burning of biomass can help to reduce the degree of climate change, the combustion of biomass is not without problems. Approximately 11 per cent of world primary energy use is provided by the consumption of traditional biomass, often simply the burning of wood on open fires [BOY03]. This can release methane and other greenhouse gases with global warming potentials (Chapter 5) much greater than carbon dioxide. Such incomplete combustion of biomass can therefore have a greater overall global warming effect than the more easily controlled combustion of fossil fuels.

In the oceans

Phytoplankton within the oceans absorb carbon dioxide from the atmosphere via photosynthesis in the cold dense waters found at high latitudes. Predation by zooplankton and then fish and marine animals stores this carbon until the animals die and a proportion (estimated to be around 30 per cent) of their carbon descends to the ocean floor; where over thousands of years it is converted back to carbon dioxide by bacteria and returns to the surface. This is the biological pump within the global carbon cycle referred to in Chapter 5.

As we discussed in Chapter 5, there is a constant transfer of carbon dioxide from the atmosphere to the oceans. Unfortunately this exchange is hindered by the poor connection between the oceans and the atmosphere – a two-dimensional surface – and the general lack of mixing between the layers in the ocean. These mechanisms are known to be slow, in part because if they acted more quickly the rapid growth in atmospheric carbon dioxide concentration we have witnessed would not have occurred – the oceans would have simply absorbed it. This has encouraged scientists to consider if this transfer process could be made more rapid.

We know that in general, carbon dioxide is held very effectively within the oceans by conversion to carbonate and bicarbonate ions:

$$\text{carbon dioxide} + H_2O \leftrightarrow HCO_3^- + H^+ \leftrightarrow CO_3^{2-} + 2H^+.$$

The equilibrium is such that for every molecule of carbon dioxide held as a gas within the upper ocean, 99 are held in carbonate or bicarbonate forms. As the dissolved gas will be in equilibrium with the atmospheric concentration, water at atmospheric pressure can contain much larger quantities per volume of carbon dioxide than the air.

So, what biological processes might be encouraged to increase the rate of transfer from the atmosphere to the oceans and then to depth? We saw in Chapter 4 that one of the limiting factors for early agricultural systems was the lack of nutrients in the soil. This was solved by adding nutrients – in the form of dung – to the fields. Applying this logic to the oceans suggests we consider 'fertilising' the water with whatever we think might be missing. Nitrate and phosphate fertilization has been considered, however the most developed idea is the addition of iron. This makes sense since the amount of iron in the oceans is small, as it tends to precipitate out and is only slowly replaced by atmospheric dust. In the first large-scale experiment, in 1993, iron was added to an 80 km^2 patch of the Pacific Ocean west of the Galapagos Islands. The work was repeated in 1995 with a growth in algae and a fall in carbon dioxide concentration within the water observed [WAT94, COO96]. The algae also produced quantities of dimethyl sulphide [TUR96]. This might alter cloud processes over the oceans and provide additional cooling.

Another possibility would be to dissolve lime (CaO) in the oceans. This would increase the alkalinity of the oceans and mop-up dissolved carbon dioxide, which would in turn be replaced by atmospheric carbon dioxide. Unfortunately lime is obtained by heating limestone ($CaCO_3$): a process that releases almost as much carbon dioxide as the lime would absorb, although this carbon dioxide might be captured and stored by direct ocean pumping to depth as described above.

An alternative would be to try and use fertilization techniques to grow seaweed on a massive scale, thereby mopping up carbon dioxide directly. Or to genetically modify sea-life so it had a greater propensity to absorb carbon dioxide.

Many are unhappy with these ideas largely because their impacts are little understood; the most obvious impact being that on ecological systems around the fertilization areas, which would need to be very large indeed. Others question the amount of lime or minerals needed. The effect of mass fertilization on the ecology of the oceans is also unknown, as is the percentage of carbon that would finally be naturally exported to the deep ocean by such fertilization. Yet it has been suggested the cost of implementing a storage programme based on oceanic fertilization could be extremely modest compared to other capture and sequestration technologies, at €0.77–7.77 (£0.53–5.3, US$1–10) per tC.

29.3 The reflection of solar radiation

Rather than trying to absorb excess carbon dioxide, we might attempt to reflect some of the sunlight striking the planet back into space. This would have an obvious cooling effect. It has been suggested that the amount of reflection needed may be modest, possibly only one per cent [MAT96]. It is believed that this process is already happening, in that our emissions of sulphate aerosols from power generation are reflecting sunlight thereby reducing the impact of climate change at industrialized latitudes. Such aerosols are precipitated out in rain, so to be long lasting the reflectors would have to be in the stratosphere, or higher.

Unfortunately, such reflectors are unlikely to completely 'solve' global warming because they would achieve a general cooling of the planet but not a pattern that matches the expected regional variation presented in Chapter 17: because anthropogenic climate change is not due to an overly bright sun, or a reduction in natural reflectors.

It was first suggested [BUD74] that rockets could be fired into the stratosphere to carry the necessary aerosols aloft. Later it was postulated this might be better achieved by a modification to commercial jet fuel. Amongst the benefits would be the reversible nature of the effect; the corollary being that injection would need to be done on a continual basis. Objections arise again from concerns over possible unknown effects, and the clash with principles of sustainability, as it leaves future generations with the problem of continuing the process because it does nothing to reduce the underlying problem – the high concentration of carbon dioxide and other greenhouse gases.

At an even higher altitude, giant reflectors could be assembled in space. They would cast a shadow on the planet that would transverse the surface, again reflecting one per cent of the overall incident light. The cost, or even feasibility, of doing this is unknown. One possibility for the far future might be to capture some of the radiation intercepted by such a system and use photovoltaics to convert it to electrical energy for beaming back to earth in the form of microwaves, for further conversion to electricity for distribution.

29.4 Example application: Statoil, Sleipner West gas field, North Sea

The natural gas extracted from the Sleipner field is contaminated with nine per cent carbon dioxide. This needs to be reduced to a maximum of 2.5 per cent for the gas to meet export restrictions [TOR98]. This equates to a need to extract and dispose of one million tonnes of carbon dioxide per annum. If simply released to atmosphere – as is standard industry practice – it would increase Norway's carbon dioxide emissions by nearly three per cent.

To avoid this release it was decided to re-inject the separated carbon dioxide into a saline aquifer (Figure 29.1). Re-injection started in October 1996. The carbon dioxide is initially absorbed in an amine contact tower pressurized to around 100 bar. The amine is then stripped of carbon dioxide in another tower. The offshore module to do this measures $50 \times 20 \times 35$ m and weighs over 8000 tonnes [TOR98], making it the heaviest module ever lifted offshore. The injection aquifer lies some 1000 m under the sea bed and is not connected to the gas reservoir (which lies at 3.5 km).

Although not precisely known, the capacity to store carbon dioxide underground in Europe alone might be in excess of 800 billion tonnes, much being under the North Sea [TOR98].

Problem 29.3   Europe[1] uses 976 Mt_{oe} of natural gas per annum [BP04]. Estimate how many years worth of carbon dioxide from this source Europe could store underground.

The answer to Problem 29.3 (over 300 years) indicates why carbon capture and sequestration has captured the imagination of politicians. This has translated into several countries (including the UK) suggesting that it will be at the centre of their policies to manage carbon emissions, particularly from the otherwise difficult to tackle

---

[1] As defined as Europe and Eurasia in Appendix 1 and [BP04].

Figure 29.1 Sleipner field, North Sea. (Photo: Kim Laland – Bitmap, Statoil [BEI05])

transportation sector (by producing hydrogen directly from fossil fuels and sequestering the carbon dioxide so produced), and we will investigate this policy further in the next chapter. The USA together with South Korea, Australia, China, Japan and India have formed a partnership – the *Asia-Pacific Partnership on Clean Development and Climate* – to investigate and share technologies to reduce greenhouse gas emissions with carbon sequestration and advanced coal burning being the principle technologies suggested.

29.5 Student exercises

"The general root of superstition is that men observe when things hit, and not when they miss, and commit to memory the one, and pass over the other."

–Sir Francis Bacon

1. Compare and contrast the various sequestration options available.

2. Is it possible to solve carbon dioxide-related climate change by sequestration? (Hint: compare anthropogenic emissions of carbon dioxide with the world's sequestration potential for each storage approach.)

3. If sequestration were fully implemented, would the world still have a potential climate problem from the emission of other greenhouse gases?

4. Why would the reflection of sunlight before it reaches the earth not truly solve climate change?

30

A sustainable, low carbon future?

"The danger is that global warming may become self-sustaining, if it has not done so already. The melting of the Arctic and Antarctic ice caps reduces the fraction of solar energy reflected back into space, and so increases the temperature further. Climate change may kill off the Amazon and other rain forests, and so eliminate once one of the main ways in which carbon dioxide is removed from the atmosphere. The rise in sea temperature may trigger the release of large quantities of carbon dioxide, trapped as hydrides on the ocean floor. Both these phenomena would increase the greenhouse effect, and so global warming further. We have to reverse global warming urgently, if we still can."

–Professor Stephen Hawking – ABC News interview, August 16, 2006

In the last 11 chapters we have looked at a variety of ways in which some of our future energy demands might be met by sustainable means. For many of the technologies studied, the size of the resource is such that it would, in theory, be possible to meet all our requirements using a single resource. However, is this possible in practice? In its Energy White Paper [DTI03], and its supporting documents, the UK Government sets out a plan to move the country towards a low carbon economy. It presents the outline of a programme to reduce UK carbon emissions by 60 per cent or more by 2050. Although many have criticized the White Paper for lacking detail, it is based on modelling of future energy demands and the possible costs of future technologies; this allows us to catch a glimpse of what the future might hold.

The UK has a comparatively high population for its size, is highly developed, has a mid-latitude location and has no obvious advantage as far as sustainable energy technologies are concerned (except possibly wind). This makes it a useful test bed for carrying out this type of analysis: if a switch to sustainability is possible for such a country, it should be possible for other better energy- and land-resourced countries.

Energy and Climate Change   David A. Coley
© 2008 John Wiley & Sons, Ltd

Possibly the most important result from this analysis is that it will be more cost effective to make reductions early, even though the costs of sustainable technologies will probably be at their greatest due to their infancy and low production volumes [AEA03, p44].

The following results are based on the studies presented in [AA04] and [AEA03]. These analysed a series of low-carbon futures with cuts in emissions of up to 70 per cent (not including international aviation). We will discuss the assumptions and form of the analysis, see which energy sectors (demand or efficiency) are assumed to be most important, study the final energy mix (gas, electricity, etc.), revisit energy efficiency, and see which sub-sectors (domestic, industry, etc.) will be responsible for most savings. We will then investigate how the problem of growing emissions from road transport could be practically solved in the period before a hydrogen economy is introduced. Finally we will examine probably the most interesting part: how our need for electricity could be met by a range of non-carbon alternatives. We will do this twice: once assuming nuclear power is expanded and once assuming it is all but removed.

Having seen that it is possible for a developed nation to greatly reduce its carbon emissions, we need to ask whether it is possible for the world as a whole to make similar reductions. Each country is different, so such an analysis can only be very broad-brush in nature, but we will see that it would appear possible for the world to make cuts in its emissions such that excessive climate change can be avoided. Of course, this does not mean that the political will to do so exists.

30.1 Methodology and assumptions

Trying to predict the future 50 years hence is no easy task. The prediction of several factors is key: economic growth over the whole period, as this will determine wealth and thereby demand for energy; the spectrum of industries operating, particularly the relative sizes of more energy-intensive manufacturing industries and less energy-intensive service industries; the population density and transport and planning polices followed; and the range and cost of possible energy technologies at any point in time. (In the results presented here the costs of energy technologies were assumed to be amenable to the benefits of mass production – i.e. a lowering of cost.)

30.2 Results

The overall energy and carbon dioxide savings arose in the studies from a combination of energy efficiency, reduction in the size of energy-intensive industries and changes in transportation systems.

Energy efficiency

Despite a predicted annual growth in GDP of between 2.25 and 3 per cent, it was assumed for these studies that energy use will not grow. This is despite an expected increase in demand of 47 per cent by 2050. Supply is predicted not to grow in line with

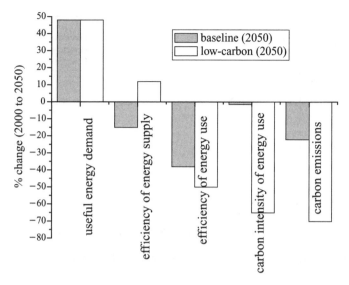

Figure 30.1 Contributions to the change in carbon emissions over the period 2000 to 2050 [data from AEA03]. (Note: a negative movement in efficiency indicates an improvement.) By useful energy we mean energy that is genuinely useful in a particular situation: i.e. the fraction that heats a house or moves a car, rather than that lost due to inefficiencies in the house or car. *Useful energy demand = efficiency factor × final energy demand* and is therefore amenable to changes in technology. Efficiency factors are published for various sectors, and make use of 'ideal' houses, cars, appliances, lighting levels, etc. against which real devices and behaviours are compared

this because it is assumed that efficiency of supply (the production and generation of energy) will improve by 16 per cent and the efficiency of final energy use is expected to improve by 26 per cent. Together, these imply that carbon dioxide emissions should fall without intervention by 22 per cent. (This is unlikely to occur in developing nations – it might not even occur in the UK and other developed nations.)

We will concentrate on what the energy provision will look like if we choose policies aimed at reducing carbon emissions by 70 per cent. (This is in contrast to the IPCC scenarios studied in Chapter 16 where no such policy was implied.) Figure 30.1 shows the assumed, natural (i.e. from expected changes in industrial practices, etc.), reductions, together with those required if policies aimed at achieving a 70 per cent reduction in carbon dioxide emissions by 2050 are implemented. (We shall term these two scenarios the 'baseline' and the 'low-carbon' scenarios respectively[1].) Interestingly this shows that the efficiency of energy supply reduces! This is because technologies such as hydrogen production and carbon capture are energy hungry and therefore reduce supply- side efficiencies – a fascinating result.

Because the baseline case has moderately improving supply and demand-side efficiencies, whereas the low-carbon scenario has reductions in supply-side efficiency, but

---

[1] Reference AEA03 used a different terminology, which distinguishes between various desires for sustainability. Our baseline scenario equates to their BL0 and our low-carbon to their BL70.

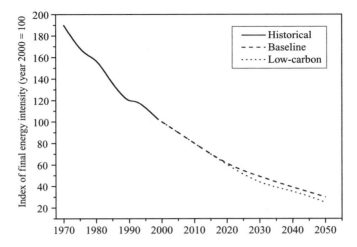

Figure 30.2   Historical (actual) and projected final energy intensities [data from AEA03]

greater demand-side improvements, the energy intensity (the energy required to produce one monetary unit of economic output) falls by similar amounts (Figure 30.2)
and would seem to follow a simple extrapolation of the historical situation.

Fuel mix

Clearly if carbon dioxide emissions are to fall by 70 per cent the mix of fuels used will
have to change radically. Figure 30.3 shows the relative proportions of fuels used in

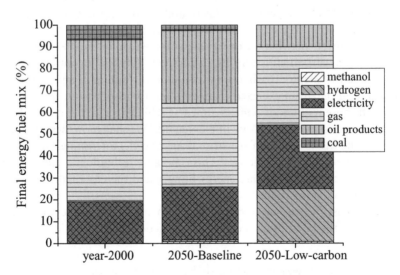

Figure 30.3   Final energy mix [data from AEA03]. ('Methanol' includes ethanol and biodiesel)

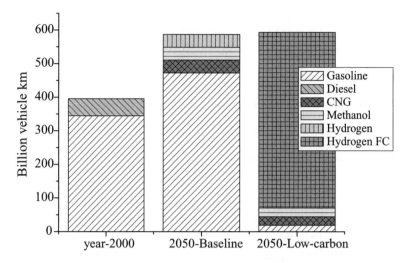

Figure 30.4 Car usage and engine technology [data from AEA03]. Note: for this plot, 'low-carbon' refers to a 60 per cent reduction, rather than 70 per cent; CNG = compressed natural gas; all fuels are assumed still to be consumed in internal combustion engines except that identified as 'hydrogen fuel cells (FC)'

2000 compared to that in 2050 for both the baseline and low-carbon scenarios. In the low-carbon case, relative oil use is reduced and replaced by hydrogen.

Major changes are required in the fuel-mix used by the transportation sector to achieve such a large cut in carbon emissions (Figure 30.4), with gasoline-powered internal combustion engines almost disappearing (to be replaced by hydrogen systems). As was discussed in Chapters 14 and 28, transportation may prove the most difficult sector to engineer reductions in emissions, in part because, as Figure 30.4 shows, demand is expected to continue to increase. (Note: No attempt is made to reduce emissions from international aviation in the White Paper – a clear failing.)

Electricity

The mix of future electricity generating technologies developed will be highly sensitive to their relative costs. The studies in [AEA03] investigate this sensitivity and in particular the effect of nuclear power being viewed as cheap (€0.035, £0.025, US$0.045 per kWh) or expensive (€0.049, £0.035, US$0.063 per kWh). Most analysts see nuclear power falling from favour, with the capacity lost, when old nuclear stations stop generating, being replaced by highly efficient combined cycled gas turbines (possibly with carbon capture). However, this is not the only possibility, and as Figure 30.5 shows, if it proves economic (and acceptable to the public) then nuclear power could play an important role within a low-carbon economy unless preferential treatment is given in the form of subsidies to new sustainable energy technologies. Although it would seem from the modelling (presented in reference [AEA03]) that we can achieve a 70 per cent reduction in carbon emissions without nuclear power.

Table 30.1 Increases in energy costs due to a carbon reduction of 60 per cent by 2050 [data from AEA03]

Sector	Increase in cost by 2050 (%)
Domestic – change in average cost per household	20
Services – change in total annual cost	23
Transport – change in average cost per km	54
Industry – change in total annual cost	22

Costs

Such an abatement in emissions will have costs for consumers. Table 30.1 shows the estimated increase in costs for a 60 per cent carbon dioxide reduction. These values are not that large (less than 0.5 per cent per annum) and the effect on the economy is found to be even less.

In much abatement modelling the climate change implications of following different paths to a low-carbon future are not considered. (By 'path' we mean the exact time series of reductions.) And indeed, the science suggests that over the time frame we are considering, the total, integrated amount of carbon emitted by any future date is the most important factor, not in which years various amounts were emitted. Economic models

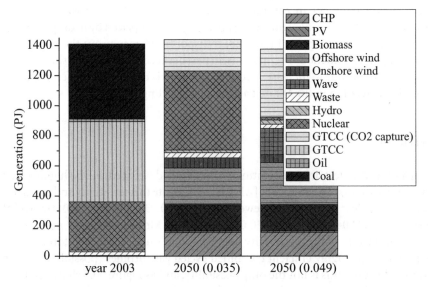

Figure 30.5 The impact of alterative nuclear power generation costs (either €0.035 or €0.049 per kWh) on the mix of fuels used for electricity generation [data from DUKES04 and AEA03]. (GTCC = combined-cycle gas turbine.) Data for 2003 have a slightly different recording basis and should only be considered illustrative, for example all renewables not specifically mentioned have been ascribed to 'waste'

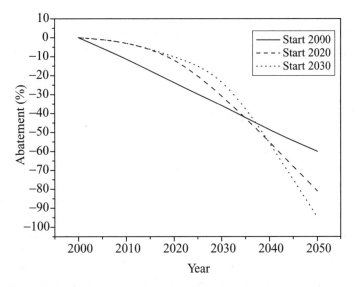

Figure 30.6 Time series of possible abatements that give the same cumulative carbon emission to 2050 [data from AEA03]. (These are the bldef1, bldff5 and bldff6 abatement paths given in reference [AEA03].) Note the near impossibility of delaying action to 2030, as this would imply reducing emissions by more than 90 per cent

cannot make this simplification, as differing paths may have differing costs. In turn, carrying out a path-dependent analysis allows us to ask the question, should we act now to reduce emissions, or leave it until later when sustainable energy technologies may be cheaper – even if this means making greater cuts to ensure the same total cumulative emission?

It would appear [AEA03, p44] that it is far more cost effective to make reductions early, even though the costs will probably be at their greatest due to the sustainable technologies not being fully developed and production volumes as yet being low. The main reason for this is that if we delay the reductions then to have achieved the same total carbon emissions over the period 2000 to 2050 we will have to make much larger cuts in the final years. This point is worth emphasising. It is the total cumulative amount of carbon (and other greenhouse gases) between now and any future date that is key. Therefore the later we leave it the greater the required reductions will be, and beyond a certain date we will face the reality of either making cuts of over 100 per cent, which is clearly impossible[2], or face a great degree of warming.

If a linear reduction path is followed then the total cost of a 70 per cent reduction in emissions to the UK economy is expected to be around €81 billion (£58 billion, US$104 billion). This grows to €172 billion (£123 billion, US$222 billion) if action is

---

[2] Unless carbon storage technologies, such as increasing land or ocean biomass, were taken to extremes and could absorb each year more carbon that we emit.

delayed to 2020[3]. If we delay to 2030 then it is not clear if it is possible to make the required reductions with known technologies, as this implies a reduction of more than 90 per cent (Figure 30.6). The message is clear: from both an economic and a practical viewpoint we need to start reducing our emissions as soon as possible.

30.3 Worldwide reductions

Having seen that it is possible for a single developed nation to greatly cut its carbon emissions, we need to see if such an analysis can be extended to the rest of the world. Pacala and Socolow [PAC04] have gone some way to doing this and they have shown that it could in part be achieved without the need for new, unproven, technologies.

Pacala and Socolow identify 15 technologies or adaptations each capable of saving 1 GtC per annum by 2054. By deploying seven of the fifteen, 7 GtC per annum would be saved. This is equal to the predicted growth in carbon emissions by the business-as-usual path presented in Part III (i.e. a growth in world carbon emissions of around 1.4 per cent per year). So, such a saving would leave carbon emissions at the level they now are, which in turn would mean carbon dioxide levels would increase to around 500 ppm$_v$ by the middle of the twenty-first century (i.e. a doubling from pre-industrial levels). To stop levels increasing beyond this toward the end of the century further reductions would be necessary. Because the approach is only trying to match the expected growth in emissions the rate of deployment only has to equal this growth. This makes the suggestions plausible as we can expand the introduction of each technology/adaptation slowly.

In the following each of the suggestions is outlined. Remember, only seven of the fifteen need to be adopted to stabilize carbon dioxide concentrations to 500 ppm$_v$. However not all changes are independent. For example, if car use were to drop dramatically, then changes to vehicle efficiency would have less impact.

1. Improved car fuel efficiency

Assume car ownership grows to two billion (roughly four times today's number) and average annual distance travelled stays as 16 000 km (10 000 miles), but fuel efficiency doubles from 13 km/l to 26 km/l (30 mpg to 60 mpg).

2. Reduced reliance on cars

Assume car ownership grows as in Option 1, but they are used only half as much, i.e. annual average travelled distance falls to 8000 km (5000 miles).

---

[3] This is true irrespective of discount rate.

3. Increased energy efficiency in buildings

Assume energy use in buildings drops by 25 per cent.

4. Improved power plant efficiency

Assume coal-based generation efficiency were to grow from 40 to 60 per cent, whilst the amount of electricity generated by coal doubles.

5. Replace coal-based generation by gas-based generation

Replace 1400 GW of coal-based power stations by natural gas-based ones (even though coal-based generation efficiency is assumed to increase to 50 per cent). The savings arise from the lower emissions of carbon per joule of energy from gas-based generation.

6. Carbon storage from power plants

Assume storage and capture at 800 GW of base-load[4] coal power stations (or 1600 GW of gas power stations) and that this prevents 90 per cent of carbon dioxide being released from these plants. Pacala and Socolow consider that the most likely technology would be pre-combustion capture of carbon dioxide with hydrogen being burnt to generate electricity.

7. Carbon capture from hydrogen generation

Assume 250 Mt of hydrogen per annum is generated from coal and storage is used for the resultant carbon dioxide, however the hydrogen is not burnt to generate electricity, but to replace other fossil fuel use – for example, that within cars.

8. Storage of carbon captured in synfuels plants

Assume oil-like fuels are produced from coal on a large scale (30 million barrels per day) with half the carbon entering the plant being sequestered.

9. Fission

Assume thermal nuclear power capacity is tripled to 700 GW.

---

[4] i.e. ones that run almost continuously.

10. Wind electricity

Assume 2000 GW of wind power capacity replaces coal generation, i.e. the installation of one million 2 MW turbines. This allows for each GW of wind power still requiring 0.33 GW of base-load generation from other sources. This is fifty times today's level of deployment.

11. PV

Assume PV deployment increases 700-fold to 2000 GW. This would require two million hectares of land (or 2–3 m^2 per person). Roofs would be ideal for this level of adoption.

12. Green hydrogen

Assume, for example, 4000 GW of wind power were used to produce hydrogen for vehicle use (with the hydrogen being used in fuel cells). This is twice the number of turbines required by Option 10 and suggests that we are better off using hydrogen for electricity generation than as a vehicle fuel.

13. Biofuels

Assume the production of 34 million barrels a day of biofuels replacing an equivalent amount of gasoline use. This is 50 times larger than current world production. This would require 250 million hectares of land – an area equal to one sixth of the world's cropped land and assumes no fossil fuel inputs are required.

14. Forest management

If tropical deforestation were reduced to zero (rather than halved) by 2054 then 0.5 GtC per annum would be saved. An equal amount of carbon could be saved by reforesting or afforesting 250 million hectares in the tropics, or 400 million hectares in the temperate zone. Current areas of tropical and temperate forest cover are 1500 and 700 million hectares respectively.

15. Agricultural soils management

The tilling of previously natural land releases up to half the stored carbon in the soil, primarily because such tilling increases the rate of decomposition of organic matter in the soil. *Conservation tillage*, i.e. drilling seeds into the soil rather than ploughing, and erosion control can reverse this loss of carbon. By 1995, 110 million hectares of the world's 1600 million hectares of cropland were being planted in this way. 1 GtC per annum would be saved if these practices were expanded to all cropland.

30.4 Conclusion

From both these analyses, we can see that the move to a low-carbon future is unlikely to be achieved by the adoption of a single new technology, but by a multitude of approaches. This is nothing new. Today we make use of coal, oil, gas, hydro and nuclear as well as sustainable technologies. In future the diversity is likely to be just as wide. The real question is whether the world will take the issue of climate change seriously enough, soon enough.

30.5 What can I do?

This is a question often asked by students and the public. Firstly, you can lead a more sustainable life by reducing your use of energy both at home and at work. Secondly, you can encourage others you meet to follow suit. Thirdly, you can campaign for the political establishment, industry and commerce to take climate change more seriously. Finally, you can choose a career dedicated to reducing humankind's impact on the environment. The choice is wide and covers almost all of the sciences – from physics and chemistry to psychology – and in organizations as disparate as universities, oil companies and charities.

As a first step, you should estimate your current level of carbon emissions by adding up all your emissions from fuel use (for air travel multiply the answer by 2.8 to account for the radiative forcing index of aviation). Then set a target – say, a 60 per cent reduction by some future date – and use this to establish a required annual reduction in order to meet your personal target. The level of detail you will be able to go into will depend on whether you use one of the many carbon calculators available on the web, or (better) write your own spreadsheet. Each year repeat the calculation and see if you are on track.

Problem 30.1 (a) Produce a spreadsheet that allows you to estimate your carbon emissions. Include heating/cooling your home, all use of electricity in the home, your transport to and from work (or place of study) and all other travel. (b) Plot a series of pie charts that allow you to analyse where most of your emissions arise (home, commuting, air travel, etc.) and which fuel is most responsible (gas, electricity or petroleum): you might be surprised by some of the answers. (c) Examine, truthfully, where you could most effectively make savings and put together a strategy to reduce your emissions by 60 per cent by 2020. (Hint: use the various emission factors given in the book and a radiative forcing index of 2.8 for air travel.)

30.6 Student exercises

"I believe that a scientist looking at nonscientific problems is just as dumb as the next guy."

— Richard Feynman

1. Why does the above analysis assume that energy use in industrialized countries such as the UK will not increase greatly in future years? Present percentage change figures for GDP, energy supply and energy use to support your case.

2. Produce a table that compares the UK fuel mix in 2000 with that expected in 2050 under the low-carbon scenario.

3. Why is the cost of nuclear power critical to a discussion of the future energy mix? Present, qualitatively, possible future fuel mixes given nuclear power costs of €0.035 and €0.049 per kWh.

4. Why is it important for us to make reductions early? Quantitatively, what is the economic benefit to the UK of doing so?

5. Why might supply-side efficiency decline under a low-carbon scenario?

6. For each option given by Pacala and Socolow, list any issues or problems that you see.

References

AA04   Climate Change, Energy Technology Support Unit, Harwell, UK, (http://www.future-energy-solutions.com, accessed August 2006).

ACT05   Action Renewables, http://www.actionrenewables.org/RenewableImages/Tidal/ti04. jpg (accessed June 2006).

AEA03   Marsh, G. *et al.*, *Options for a Low Carbon Future – Phase 2. A report produced for the Department of Trade and Industry*, AEA Technology Report, ED01806, Harwell, UK, 2003.

AME05   American Coal Foundation, http://www.teachcoal.org/aboutcoal/articles/coaljourney. html (accessed June 2006).

AMI75   Ford, K.W. (ed.), *Efficient Use Of Energy: A Physics Perspective (AIP Conference Proceedings)*, AIP Press, pp49–50, 1975.

ANT02   British Antarctic Survey, http://www.antarctica.ac.uk/News_and_Information/Press_Releases/2002/20020319.html (accessed June 2006).

AST05   Astronomy Notes, http://www.astronomynotes.com (accessed June 2006).

ATK86   Atkins, G., *the Advantages of CHP Systems*, Proceedings of I Mech E Symposium on CHP, Sheffield, UK, 1986.

BAI97   Baird, G. *et al.*, the Embodied Energy in Building Materials – Updated New Zealand Coefficients and their Significance, *IPENZ Transactions*, vol. 24, no. 1, 1997.

BB87   UK Department for Education and Employment, *Guidelines for Environmental Design in Schools*. Building Bulletin 87 (Revision of Design Note 17) ISBN 011 271013, London: DfEE Architects and Building Branch, 1997.

BBC02   BBC, Killer Lakes – BBC Two Thursday 4th April 2002 (accessed at http://www.bbc.co.uk/science/horizon/2001/killerlakes.shtml, August 2006).

BBC04a BBC, http://news.bbc.co.uk/onthisday/hi/witness/july/6/newsid_3036000/3036510. stm (accessed July 2006).

BE00   British Energy, *Eggborough Power Station*, British Energy Publications, 2000.

BE05   British Energy, http://www.industcards.com/nuclear-uk.htm (accessed June 2006).

BE05a   British Energy, *the Power Game*, CD, British Energy, 2005.

BEI05   Statoil, http://www.statoil.com/STATOILCOM/SVG00990.nsf?opendatabase&lang=en&artid=F36765A6833828C0C1256FEF003BAB1B (accessed August 2006).

BEL02   Kruse, B., Grinna, S. and Buch, C., *Hydrogen*, Report 6:2002, the Bellona Foundation, Oslo, Norway, pp1–52, 2002.

BEP00a   British Energy, *Heysham 2 Power Station*, British Energy Publications, 2000.

BEP00b   British Energy, *Sizewell B Power Station*, British Energy Publications, 2000.

BET26   Betz, A., *Wind Energy*, 1926.

BOE05   Boeing Corporation, http://www.boeing.com/news/feature/apollo11/gallery/apollo06.html (accessed June 2006).

BOE99   Boeker, E. and Grondelle, R., *Environmental Physics*, Wiley, 1999.

BOY03   Boyle, G., Everett, B., and Ramage, J., *Energy Systems and Sustainability*, the Open University, Oxford University Press, 2003.

BOY96   Boyle, G., *Renewable Energy: Power for a Sustainable Future*, Oxford University Press, 1996.

BP04   BP plc, *BP Statistical Review of World Energy*, BP, 2004.

BP05   BP plc, *Putting Energy in the Spotlight*, BP, 2005.

BRI02   *Encyclopaedia Britannica*, Encyclopaedia Britannica (UK) Ltd, 2002.

BRO78   Brosche, P. and Sündermann, J.(eds.), *Tidal Friction and the Earth's Rotation*, Springer Verlag, 1978.

BRO99   Brown K.A., *et al.*, *Methane Emissions from UK Landfills*, final report for the DfRA, report AEAT-5217, AEA, 1999.

BRU87   Brundtland Commission, 1987. From http://commpres.env.state.ma.us/content/glossary.asp (accessed June 2006).

BUC37   Buck, J.L., *Land Utilisation in China*, University of Nanking, China, 1937.

BUD74   Budyko, M.I., *Climate Changes*, American Geophysical Union, Washington, DC, USA, 1974.

BUR01   Burton, T., *Wind Energy Handbook*, Wiley, 2001.

CAD01   CADDET, Technical Brochure 140, ETSU, Harwell, Oxfordshire, UK, 2001.

CAD02   CADDET, Technical Brochure 162, ETSU, Harwell, Oxfordshire, UK, 2002.

CAD03   CADDET, Technical Brochure 177, ETSU, Harwell, Oxfordshire, UK, 2003.

CAD109   CADDET Technical Brochure 109, ETSU, Harwell, Oxfordshire, UK, 2000.

CAD430   CADDET Result 430, ETSU, Harwell, Oxfordshire, UK, 2002.

CAD431   CADDET Result 431, ETSU, Harwell, Oxfordshire, UK, 2002.

CAD437   CADDET Result 437, ETSU, Harwell, Oxfordshire, UK, 2002.

CAD60   CADDET Technical Brochure 60, ETSU, Harwell, Oxfordshire, UK, 1997.

CAD95   CADDET Technical Brochure 19, ETSU, Harwell, Oxfordshire, UK, 1995.

CAD97   CADDET Technical Brochure 97, http://www.caddet.org/public/uploads/pdfs/Brochure/no97.pdf (accessed July 2006).

CAD99   CADDET, Technical Brochure 101, ETSU, Harwell, Oxfordshire, UK, 1999.

CAR04   Bice, D., http://www.acad.carleton.edu/curricular/GEOL/DaveSTELLA/Carbon/long_term_carbon.htm (accessed June 2006).

CCP05   UK Department of Environment, Food and Rural Affairs, *Review Of the UK Climate Change Program*, Consultation Paper, DEFRA, London, 2004.

CGH04   Centre for gas hydrate research, http://www.pet.hw.ac.uk/research/hydrate/hydrates_why.htm (accessed June 2006).

CH205   California Hydrogen Business Council, http://www.ch2bc.org/index5.htm (accessed June 2006).

CIA04   US Central Intelligence Agency, the World Fact Book, http://www.cia.gov/cia/publications/factbook/ (accessed June 2006).

CIA04a US Central Intelligence Agency, the World Fact Book, http://www.cia.gov/cia/publications/factbook/geos/uk.html (accessed June 2006).

CIA04b US Central Intelligence Agency, the World Fact Book, http://www.cia.gov/cia/publications/factbook/geos/ch.html (accessed June 2006).

CIA05 US Central Intelligence Agency, the World Fact Book, https://www.cia.gov/cia/publications/factbook/index.html (accessed August 2006).

CIB86 CIBSE, *Guide A: Environmental Design*, Chartered Institution of Building Services Engineers, London, 1986.

CIBSE99 CIBSE, *Guide A: Environmental Design*, Chartered Institution of Building Services Engineers, London, 1999.

CNN05 CNN, http://www.cnn.com (accessed June 2006).

COO96 Copper, D.J., *et al.*, Large Decrease in Ocean Surface Carbon Dioxide Fugacity in Response to In-Situ Iron Fertilisation, *Nature* vol. 383, pp511–13, 1996.

DAN05 Danish Wind Industry Association, http://www.windpower.org (accessed June 2006).

DEB93 de Beaune, S.A., and White, R., Ice Age Lamps, *Scientific American* vol. 266 no. 3, pp108–13, 1993.

DEF05 UK Department of Environment, Food and Rural Affairs/Atmospheric Research and Information Centre, http://www.ace.mmu.ac.uk/Resources/gcc/2-8.html (accessed June 2006).

DEFRA01 UK Department of Environment, Food and Rural Affairs, Planting and growing Miscanthus – best practice guidelines, DEFRA, London, 2001.

DES06 Dessler, A.E. and Parson, E.A., *The science and politics of global climate change: a guide to the debate*, Cambridge University Press, 2006.

DEZ78 DeZeeuw, J.W., Peat and the Dutch Golden Age, *AAG Bijdragen* no. 21, pp3–31, 1978.

DFRA03 UK Department of Environment, Food and Rural Affairs, *Methane Emissions from Landfill Sites in the UK*, DEFRA, London, 2003.

DFRA06 UK Department of Environment, Food and Rural Affairs, http://www.defra.gov.uk/environment/statistics/airqual/aqmainap.htm (accessed August 2006).

DFT05 UK Department for Transport, *Transport statistics Great Britain 2004*, HMSO, London, 2005.

DID69 Diderot, D. and D'Alembert, J.L., *L'Encyclopédie ou Dictionnaire Raisonné des Sciences des Arts et des Métiers*, André Le Breton, pp1769–72, 1751.

DRA00 Drax Power Ltd, www.draxpower.com (accessed June 2006).

DRA05 Drax Power Ltd, http://www.draxpower.com/press.php?page=photos&category= Landscape (accessed June 2006).

DTI00 Royal Commission on Environmental Pollution, *Energy – the Changing Climate*, report to parliament, CM4749, the Stationary Office, London, 2000.

DTI01b UK Department of Trade and Industry, Energy Flowchart, http://www.dti.gov.uk/files/file11248.pdf (accessed August 2006).

DTI03 UK Department of Trade and Industry, *Energy in Brief*, the Stationary Office, London, 2003.

DTI03a UK Department of Trade and Industry, *Our Energy Future – Creating a Low Carbon Economy*, Energy White Paper, the Stationary Office, London, 2003.

DTI04 UK Department of Trade and Industry, *UK Energy in Brief*, the Stationary Office, London, 2004. Also at http://www.dti.gov.uk/energy/statistics/publications/in-brief/page17222.html (accessed August 2006).

DTI04a   UK Department of Trade and Industry, *Digest of UK Energy Statistics*, http://www.
          dti.gov.uk/energy/statistics/publications/dukes/page29812.html (accessed August
          2006).

DTI05   UK Department of Trade and Industry, http://www.dti.gov.uk/files/file21352.pdf
          (accessed August 2006).

DTI05a   UK Department of Trade and Industry, http://www.dti.gov.uk/energy/ (accessed
          August 2006).

DTI05b   UK Department of Trade and Industry, *Energy Indicators (E12)*, http://www.dti.
          gov.uk/files/file29698.pdf?pubpdfdload=06%2F193 (accessed August 2006).

DTI05c   UK Department of Trade and Industry, http://www.dti.gov.uk/energy/sources/
          renewables/renewables-explained/wave-and-tidal/wave-energy/case-studies/page
          17051.html#Deep-water%20wave%20energy%20%E2%80%93%20Pelamis%
          20sea%20snake (accessed August 2006).

DTI05d   UK Department of Trade and Industry, *UK Energy in Brief*, the Stationary Office,
          London, 2005.

DUKES04   UK Department of Trade and Industry, *Digest of UK Energy Statistics*, DTI, London,
          2004.

EAS90   Eastop, T.D. and Croft, D.R., *Energy Efficiency for Engineers and Technologists*, Long-
          man, 1990.

ECN04   Cleveland, C. (ed.),*Encyclopedia of Energy*, Academic Press/Elsevier Science, 2004.

EESD02   Centre for Renewable Energy Sources, *Wave Energy Utilisation in Europe*, EESD,
          Greece, 2002.

EIA02   Energy Information Administration, *International Energy Outlook 2002*, EIA, 2002.

EIA04   Energy Information Administration, *International Energy Outlook 2004*, EIA, 2004.

EIS81   Eisenberg, J.F., *the Mammalian Radiations*, University of Chicago Press, 1981.

ELE04   Electricity Association, http://www.energynetworks.org/spring/indexpages/enaindex_
          default.asp (accessed August 2006).

ELS05   Elsam A/S, http://www.hornsrev.dk (accessed June 2006).

ENE02   EnergieNed, *Energy in the Netherlands: Facts and Figures*, Federation of Energy
          Companies in the Netherlands, Arnhem, 2002.

ENE04   BP plc, Energy in Focus, *BP Statistical Review of World Energy*, BP, 2004.

EPA05   US Environmental Protection Agency, http://www.epa.gov/airnow/elozono/images/
          smog.jpg, (accessed June 2006).

ETSU96   ETSU, *Birds and Wind Turbines, Can They Co-Exist*, ETSU-R-97, Harwell, 1996.

ETSU97   ETSU, *Investigation of the Interactions Between Wind Turbines and Radio Systems Aimed
          at Establishing Co-Siting Guidelines*, Report W/13/00477/REP, Harwell, 1997.

EU05   INOGATE, http://www.inogate.org/russian/maps/map_n_go.html (accessed June
          2006).

EUE01   European Commission ExternE Programme, *Externalities of Energy*, report DG12,
          L-2920, Luxembourg, 2001.

EXX04   Exxon Valdez Oil Spill Trustee Council, http://www.evostc.state.ak.us/Gallery/gallery-
          spill.htm (accessed August 2006).

FEY63   Feynman, R., *Lectures in Physics*, Addison Wesley (paperback 1998), 1963.

FOL92   Foley, R.A. and Lee, C., Ecology and Energetics of Encephalization in Hominid Evolu-
          tion. In Whiten, A. and Widdowson, E.D. (eds.), *Foraging Strategies and Natural Diet
          of Monkeys Apes and Humans*, Clarendon Press, Oxford, pp63–71, 1992.

FOR65   Forbes, R.J, *Studies in Ancient Technology*, vol. 2, E.J Brill, 1967.

FOX88   Fox, R.E., *Energy and the Evolution of Life*, W.H. Freeman, p166, 1988.

GEL87   Geller, H., *et al.*, *Annual Review of Energy*, vol.12, p357, 1987.

GIP01   Gipe, P., Wind Related Deaths Database, *Windstats Newsletter*, vol.14, no.4, 2001.

GIP95   Gipe, P., *Wind Energy Comes of Age*, Wiley, 1995.

GLO04   University of Michigan, http://www.globalchange.umich.edu/globalchange1/current/lectures/samson/global_warming_potential/ (accessed June 2006).

GOU92   Goudsblom, J., *Fire and civilization*, Allen Lane, 1992.

GUA03   Guardian, London, 12th December 2003.

HAL88   Halliday, D. and Resnick, R., *Fundamentals of Physics*, John Wiley pA1, 1988.

HAN83   Hanna, J.M. and Brown, D.E., Human Heat Tolerance: an Anthropological Perspective, *Annual Review of Anthropology*, vol. 12, pp259–84, 1983.

HAR00   Harvey, L.D.D, *Climate and Global Environmental Change*, Prentice Hall, 2000.

HAR74   Harris, J.R, the Rise of Coal Technology, *Scientific American*, vol. 233, no. 2, pp92–7, 1974.

HEA05   Health and Safety Executive, http://www.hse.gov.uk/press/2003/c03065.htm (accessed June 2006).

HEAT05   Heatpumps Ltd, www.heatpumps.co.uk (accessed June 2006).

HER04   Herzol, H. and Golomb, D., Carbon Capture and Storage from Fossil Fuel Use. In Cleveland, C. (ed.),*Encyclopedia of Energy*, Academic Press/Elsevier Science, 2004.

HEW00   Hewitt, G.F. and Collier, J.G., *Introduction to Nuclear Power* (2nd edition), Taylor and Francis, 2000.

HIL04   Hillman, M., *How We Can Save the Planet*, Penguin Books, 2004.

HIL84   Hill, D., *a History of Engineering in Classical and Medieval Times*, La Salle Ill., open court publications, 1984.

HOL00   Holden, J.P., and Smith, K.R., Energy, the Environment and Health. In Goldemberg, J. (ed.) *World Energy Assessment: Energy and the Challenge of Sustainability*, New York, UN development programme, UN Department of Economics and Social Affairs and World Energy Council, pp61–100, 2000.

HOP69   Hopfen, H.J., *Farm Implements for Arid and Tropical Regions*, FAO, Rome, 1969.

HOU03   Houghton, J., Guardian, London, July 28th, 2003.

HOW05   HowStuffWorks, http://www.howstuffworks.com (accessed June 2006).

HOW79   Howell, Y. and Bereny, J.A., *Engineer's Guide to Solar Energy*, Solar Energy Information Services, San Mateo, CA, USA, 1979.

HYD04   Hydro Generation, http://www.hydrogeneration.co.uk (accessed June 2006).

HYW05   Wurster, R., *Hydrogen Application in Urban Vehicles*, presentation in front of the Norsk Hydrogenforum, Oslo, June 2, 1997 (accessed at http://www.hyweb.de/Wissen/NHF97.htm, August 2006).

IAEA04   International Atomic Energy Agency, http://www.iaea.org/NewsCenter/Focus/NuclearPower/table_of-reactors.pdf (accessed June 2006).

IAEA05   International Atomic Energy Agency, http://www.iaea.org/OurWork/ST/NE/Pess/assets/RDS1-24%20printfriendly.pdf (accessed June 2006).

ICEEE   Raninger, B., Li, R. and Feng, L., *Activities to Apply the Europen* (sic) *Experience on Anaerobic Digestion of Bioorganic Municipal Waste from Source Separation in China*, APCAEM, http://unapcaem.org/Activities%20Files/A01/Activities%20to%20apply%20the%20europen%20experience%20on%20Anaerobic%20digestion%20of%20Bio organic%20MUNICIPAL%20Waste%20from%20source%20separation%20in%20-China.pdf (accessed August 2006).

IEA00　　International Energy Agency, *Energy Policies of IEA Countries: the Netherlands 2000 Review*, OECD and the IEA, France, pp1–116, 2000.

IEA05　　International Energy Agency, *Renewables Information 2004*, IEA, France, 2005.

IPC01　　Watson, R. (ed.), *Climate Change 2001 – Synthesis Report*, third assessment report of the IPCC, Cambridge University Press, 2001.

IPC05　　Intergovernmental Panel on Climate Change (IPCC), http://www.grida.no/climate/ ipcc_tar/wg1/222.htm (accessed June 2006).

IPC95　　Houghton, J.T. (ed.), *Climate Change 1995: the Science of Climate Change*, contribution of Working Group I to the second assessment report of the IPCC, Cambridge University Press, 1996.

IPCC00　　IPCC, *Emissions Scenarios: Summary for Policy Makers*, IPCC, 2000.

IPCC00a　IPCC, *Land Use, Land Use Change and Forestry*, Cambridge University Press, 2000.

IPCC01　　IPCC, Climate Change 2001 – Working Group I – the Scientific Basis, IPCC, 2001.

IPCC01a　Watson, R. (ed.), *Climate Change 2001 – Synthesis Report*, third assessment report of the IPCC, Cambridge University Press, 2001.

IPCC99　　Penner, J.E., *et al.* (eds.), *Aviation and the Global Atmosphere*, a special report of the IPCC, Cambridge University Press, 1999.

ITA04　　Usina Hidrelétrica de Itaipu, http://www.itaipu.gov.br/ (accessed June 2006).

ITA05　　Usina Hidrelétrica de Itaipu, http://www.itaipu.gov.br/ (accessed June 2006).

JAC89　　Jackson, T. and Roberts, S., *Getting Out of the Greenhouse: an Agenda For UK Action on Energy Policy*, Friends of the Earth, London, 1989.

JAM89　　James, S. R., Hominid Use of Fire in the Lower and Middle Pleistocene. A Review of the Evidence. *Current Anthropology*, vol. 30, pp1–26, 1989.

JET04　　Joint European Torus, http://www.jet.efda.org/pages/content/tokamak-description. html (and other www.jet.efda.org pages, accessed June 2006).

JOC76　　Jochim, M.A., *Hunter-Gatherer Subsistence and Settlement: a Predictive Model*, Academic Press, 1976.

KEE89　　Keeping, B. and Kats, G., the Efficiency–Renewable Synergism, *Energy Policy*, vol. 17, pp614–16, 1989.

KIL54　　Kilma, B., Paleolithic Huts at Dolni Vestonice, Czechoslovakia, *Antiquity* vol. 28, pp4–14, 1954.

KIN71　　King Hubbert, M., The Energy Resources of the Earth. *Scientific American*, no. 225, pp60–70, 1971.

KOL05　　Kolbert, E., Telegraph Magazine, p24–37, June 2005.

KRA88　　Krause, F., *Analysis of Michigan's Demand-Side Electricity Resources in the Residential Sector*, Lawerence Berkeley Laboratory, Berkeley, CA, April 1988.

LEA02　　Leathwick, J.R., Wilson, G. and Stephens, R.T.T., Landcare Research Contract Report: LC9798/126, 2002, http://www.landcareresearch.co.nz/databases/lenz/downloads/ Climate_Surfaces_for_ New_Zealand_revised.pdf (accessed June 2006).

LEA92　　Leakey, R. and Lewin, R., *Origins Reconsidered*, Doubleday, 1992.

LEG90　　Leggett, J., *Global Warming: the Greenpeace Report*, Oxford University Press, 1990.

LON04　　Accuweather, http://www.onlineweather.com/v4/uk/climate/London.html (accessed June 2006)

MAL04　　Malvern Boilers, http://www.malvernboilers.co.uk/brochures_manuals.html (accessed June 2006).

MAN05　　Managing Automation, http://www.managingautomation.com/maonline/magazine/ read.jspx?id=738 (accessed June 2006)

MAT96   Matthews, B., *Climate Engineering: a Critical Review of Proposals, their Scientific and Political Context and Possible Impacts*, University of East Anglia, UK, 1996.

MCT03   Marine Current Turbines Ltd, http://www.marineturbines.com/technical.htm (accessed June 2006).

McV84   McVeigh, J.C., *Energy around the World*, Pergamon Press, Chapter 4, 1984.

MEM05   Peace, S., Another Approach to Wind, *Mechanical Engineering*, http://www.memagazine.org/backissues/jun04/features/apptowind/apptowind.html (accessed June 2006).

MET97   UK Department of the Environment, Transport and the Regions, *Climate Change and its Impacts – a Global Perspective*, the Met. Office, 1997.

MEY00   Meyer, A., *Contraction and Convergence*, Green Books, 2000.

MIT05   Mitchell, T.A., *Exeter Local Air Quality Review 2004*, Centre for Energy and the Environment Report, University of Exeter, UK, 2005.

MON05   Historic Monroe, http://www.historicmonroe.org/Images/h001.JPG (accessed June 2006).

MON66   Montet, P., *Egipto Eterno*, Ediciones Guadarrama, 1966.

NAS05   NASA/ESA, http://nssdc.gsfc.nasa.gov/planetary/factsheet/sunfact.html (accessed June 2006).

NAS05a NASA, http://nssdc.gsfc.nasa.gov/planetary/factsheet/marsfact.html (accessed June 2006).

NAT02   National Statistics, *Annual Abstract of Statistics*, ONS, p139, 2002.

NAT99   National Atmospheric Deposition Program, http://nadp.sws.uiuc.edu/isopleths/maps1999/phfield.gif (accessed June 2006).

NEW02   Giant Antarctic Ice Sheet Breaks Off, *New Scientist*, 19th March 2002.

NEW02a *New Scientist*, vol. 17, no. 2329, p36, 2002.

NEW03   *New Scientist*, p40, 22nd February 2003.

NEW04   *New Scientist*, p8, 2nd October 2004.

NEW05   *New Scientist*, p10, 7th May 2005.

NEW05a *New Scientist*, p33, 3rd September 2005.

NEW06   Sceptics forced into climate climb-down, *New Scientist*, no. 2513, 20th August 2005.

NIC04   Ask.com, http://web.ask.co.uk/web?q=climate+in+Nice%2c+France&qsrc=1&o=0&rb=0 (accessed June 2006).

NOAA05 National oceanic and atmospheric administration, http://tidesandcurrents.noaa.gov/restles3.html (accessed June 2006).

NOS05   UK Office for National Statistics, http://www.statistics.gov.uk/census2001/pyramids/pages/18uc.asp (accessed June 2006).

NRE04a NREL, http://www.nrel.gov/buildings/highperformance/serf_gallery.html (accessed June 2006).

NRE04b NREL, http://www.nrel.gov/buildings/highperformance/zion_gallery.html (accessed June 2006).

NRE04c NREL, http://www.nrel.gov/buildings/highperformance/visitors_center.html (accessed June 2006).

NRE04d NREL, http://www.nrel.gov/data/pix/Jpegs/00036.jpg (accessed June 2006).

NRE04e NREL, http://www.nrel.gov/data/pix/Jpegs/09113.jpg (accessed June 2006).

OCE05   Ocean Power Delivery Ltd, http://www.oceanpd.com/docs/OPD%20Brochure%202005.pdf (accessed June 2006) and http://www.oceanpd.com/LatestNews/default.html (accessed June 2006).

ODPM05   Office of the Deputy Prime Minister, http://www.planningportal.gov.uk/england/ professionals/en/1115314110382.html (accessed August 2006).

ODU72   Odum, E., *Ecology*, Holt-Reinhardt and Winston, 1972.

OPE94   OPET-India, http://static.teriin.org/opet/articles/art3.htm (accessed August 2006).

ORN04   ORNL Reporter, www.ornl.gov/info/reporter/no16/methane.htm (accessed June 2006).

OSB85   Osborn, P.D., *Handbook of Energy Data and Calculations*, Butterworths, 1985.

OU91   Smith, P.M. and Warr, K., *Global Environmental Issues*, the Open University, 1991.

PAC04   Pacala, S. and Socolow, R., Stabilization Wedges: Solving the Climate Problem for the Next 50 Years with Current Technologies, *Science*, no. 305, p968–72, 2004.

PAD99   Padro, C.E.G. and Putsche, V., *Survey of the Economics of Hydrogen Technologies*, National Renewable Energy Laboratory, US Department of Energy, Colorado, USA, pp1–52, 1999.

PAT95   Patel, T., Burnt Stones and Rhino Bones Hint at Earliest Fire, *New Scientist*, June 17th, p5, 1995.

PEA03   Peake, S. and Smith, J., *Climate Change: from Science to Sustainability*, the Open University, 2003.

PED03   Pedersen, E. and Halmsted, H., *Noise Annoyance from Wind Turbines – a Review*, Swedish Environmental Protection Agency, Report 5308, August 2003.

PHY05   Elert, G. (ed.), *Physics Fact Book*, http://hypertextbook.com/facts/2001/MarinaStasenko.shtml (accessed June 2006).

PRO86   Protzen, J., Inca Stonemasonry, *Scientific American*, vol. 254, no. 2, pp94–105, 1986.

PUM05   Puma race engines, http://www.pumaracing.co.uk/chip.htm (accessed June 2006).

RAM97   Ramage, J.R., *Energy a Guidebook*, Oxford University Press, 1997.

RC94   Royal Commission on Environmental Pollution, HMSO, UK, 1994.

RHC04   Department of Geology, Royal Holloway College, University of London, http://www.gl.rhbnc.ac.uk/palaeo/2coal_use.html (accessed June 2006).

RIS99   Ristinen, R.A. and Kraushaar, J.J., *Energy and the Environment*, Wiley, 1999.

ROP05   Ropatec, Srl., http://www.ropatec.com (accessed June 2006).

ROS75   Rosenbery, N., America's Rise to Woodworking Leadership. In Hindle, B. (ed.), *America's Wooden Age*, Sleepy Hollow Restorations, p37–62, 1975.

ROT99   Rothrock, D.A., Yu, Y. and Maykut, G.A., Thinning of the Arctic Sea-Ice Cover, *Geophysical Research Letters*, vol.26, no. 23, p1–5, 1st December 1999.

ROY00   The Royal Commission On Environmental Pollution, *Energy – The Changing Climate*, 22nd Report, the Stationary Office, London, 2000.

ROY05   Hore-Lacy, I., *The Future of Nuclear Energy*, Paper presented at the Royal College of Physicians Conference, Adelaide 4th May 2000, http://www.uic.com.au/ opinion6.html (accessed June 2006).

SAU98   Sausen, R., *et al.*, A Diagnostic Study of the Global Distribution of Contrails, Part I: Present Day Climate, *Theoretical and Applied Climatology*, no.61, pp 127–41, 1998.

SCI71   *Energy and Power*, Scientific American, Inc., p101, 1971.

SHE97   Shepherd, W. and Shepherd, D.W., *Energy Studies*, World Scientific, 1997.

SHI05   Shipley Mill, http://www.shipleywindmill.org.uk (accessed June 2006).

SIN05   Sintef Ltd, http://www.math.sintef.no/ns/research/img/tapchan.jpg (accessed June 2006).

SMA04   Smalley, R., the Hydrogen Backlash, *Science*, No. 305, p961, 2004.

SMI92   Smil, V., Agricultural Energy Cost. In Fluck, R.C. (ed.) *Energy In Farm Production,* Elsevier, p85–100, 1992.

SMI94   Smil, V., *Energy in World History,* Westview Press, 1994.

SMI99   Smil, V., *Energies: an Illustrated Guide to the Biosphere and Civilization,* MIT Press, 1999.

SOH05   NASA/ESA, http://sohowww.nascom.nasa.gov (accessed June 2006).

SOR79   Sorensen, B., *Renewable Energy,* Academic Press, 1979.

SRES00   IPCC, *Technical Summary: Mitigation,* IPCC, 2000.

SUL95   Sullivan, L., The Three Gorges Project, *Current History,* pp266–9, September 1995.

SYN05   Syncrude Canada Ltd, http://www.syncrude.ca/users/folder.asp?FolderID=5703 (accessed August 2006).

TAY02   Taylor, S.J., *the Severn Barrage – Definition Study for a New Appraisal of the Project: Appendices,* ETSU T/09/00212/REP/A, DTI/Pub URN 02/644A, UK DTI, 2002.

THE05   Thermal Environments, http://www.tak2000.com/data/planets/mars.htm (accessed June 2006).

TIA37   Yingxing, S., *Exploitation of the Works of Nature (Tian Gong Kai Wu),* 1637.

TIP99   Tipler, P., *Physics for Scientists and Engineers,* 4th ed., Freman Worth, 1999.

TOR98   Torp, T.A., *Capture and Re-Injection of Carbon Dioxide in a Saline Aquifer at Sleipner Field and the Future Potential of this Technology,* Statoil Publication, Statoil, Norway, 1998.

TOT87   Toth, N., the First Technology, *Scientific American,* vol. 256, no. 4, pp112–21.

TOY05   Toyota Corporation, http://www.toyota.com/prius/index.html (accessed June 2006).

TUR96   Turner, S.M., *et al.,* Increased Dimethyl Sulphide Concentrations in Seawater from In-Situ Iron Enrichment, *Nature* vol. 383, pp513–17.

TWI86   Twidell, J. and Weir, T., *Renewable Energy Resources,* E. and F. Spon Ltd, 1986.

UKR05   Ukranian Web, http://www.ukrainianweb.com/chernobyl_ukraine.htm, (accessed June 2006).

UMN04   University of Minnesota, http://www.me.umn.edu/education/courses/me8390/essay5.html (accessed June 2006).

UNE05   University of Newfoundland, http://www.library.mun.ca/qeii/cns/photos/geog800.php (accessed June 2006).

UNU05   United Nations University, http://www.unu.edu/unupress/unupbooks/uu24ee/uu24ee0i.htm (accessed June 2006).

USA05   US Army Space and Missile Defence Command, http://www.smdc.army.mil/kwaj/Media/Photo/EastReef1.jpg (accessed June 2006).

USB05   US Board of Reclamation, http://www.usbr.gov/lc/hooverdam/gallery/damviews.html (accessed June 2006).

USE05   US Department of Energy, http://www.newton.dep.anl.gov/askasci/gen99/gen99256.htm, (accessed June 2006).

USG05   US Geological Society, http://woodshole.er.usgs.gov/project-pages/hydrates/ (accessed August 2006).

USGS05   US Geological Society, http://pubs.usgs.gov/gip/dynamic/inside.html (accessed June 2006).

UST05   University of Strathclyde, UK, http://www.esru.strath.ac.uk/EandE/Web_sites/01-02/RE_info/hec.htm (accessed June 2006).

VIN99   Vinnikov, K. Y., *et al.,* Global Warming and Northern Hemisphere Sea Ice Extent, *Science,* vol. 286, no. 5446, pp1934–7, 3rd Dec 1999.

WAS89   Washington, W.M., and Meehl, G.A., Climate Sensitivity Due to Increased CO_2 – Experiments with a Coupled Atmospheric and Ocean General Circulation Model, *Climate Dynamics*, no. 4, p1–38, 1989.

WAT94   Watson, A.J. *et al.*, Minimal Effect of Iron Fertilisation on Sea-Surface Carbon Dioxide Concentrations, *Nature*, no. 379, pp143–5, 1994.

WAV05   Wavegen Ltd, http://www.wavegen.co.uk/what_we_offer_limpet_islay.htm (accessed June 2006).

WHI04   Solar Systems Ltd, http://www.solarsystems.com.au/whitecliffs.htm (access July 2006).

WHI43   White, L.A., Energy and the Evolution of Culture, *American Anthropologist*, vol. 45, pp335–56, 1943.

WHI92   Whiten, A. and Widdowson, E.M. (eds.), *Foraging Strategies and Diet of Monkeys, Apes and Humans*, Clarendon Press, 1992.

WHO04   Ocean and Climate Change Institute, http://www.whoi.edu/institutes/occi/ current-topics/abruptclimate_15misconceptions.html (accessed July 2006).

WIN77   Winterhalder, B., *et al.*, Dung as an Essential Resource in a Highland Peruvian Community, *Human Ecology*, no. 2, pp89–104, 1977.

WOR05   World Coal Institute, http://www.worldcoal.org/ (accessed July 2006).

YOU92   Young, H.D., *University Physics*, 8th ed., Addison Wesley, 1992.

Appendix 1
National energy data

In the following tables national energy data is presented. This is taken from the 2005 edition of BP's Statistical Review of World Energy [BP05] and covers the year 2004. Later versions of this document can be found via the book's website. Some chapters use data from the 2004 version of this table, so you might find small discrepancies between this appendix and the main text. Country groupings are defined in Table A1.6.

A1.1 Oil (2004)

	Proved reserves			Production (a)		Consumption (b)	
	Thousand Mt	Share of total (%)	R/P ratio	Mt	Share of total (%)	Mt	Share of total (%)
USA	3.6	2.5	11.1	329.8	8.5	937.6	24.9
Canada	2.4	1.4	14.9	147.6	3.8	99.6	2.6
Mexico	2.0	1.2	10.6	190.7	4.9	85.2	2.3
Total North America	**8.0**	**5.1**	**11.8**	**668.0**	**17.3**	**1122.4**	**29.8**
Argentina	0.4	0.2	9.7	37.9	1.0	18.7	0.5
Brazil	1.5	0.9	19.9	76.5	2.0	84.2	2.2
Chile						10.7	0.3
Colombia	0.2	0.1	7.6	27.3	0.7	10.1	0.3
Ecuador	0.7	0.4	25.8	27.3	0.7	6.3	0.2

(*Continued*)

Energy and Climate Change   David A. Coley
© 2008 John Wiley & Sons, Ltd

A1.1 (*Continued*)

	Proved reserves			Production (a)		Consumption (b)	
	Thousand Mt	Share of total (%)	R/P ratio	Mt	Share of total (%)	Mt	Share of total (%)
Peru	0.1	0.1	27.3	4.4	0.1	7.2	0.2
Trinidad & Tobago	0.1	0.1	17.5	7.4	0.2		
Venezuela	11.1	6.5	70.8	153.5	4.0	26.3	0.7
Other S. & Cent. America	0.2	0.1	26.9	7.7	0.2	58.3	1.5
Total S. & Cent. America	**14.4**	**8.5**	**40.9**	**342.0**	**8.8**	**221.7**	**5.9**
Austria						13.7	0.4
Azerbaijan	1.0	0.6	60.2	15.7	0.4	4.5	0.1
Belarus						7.5	0.2
Belgium & Luxembourg						38.1	1.0
Bulgaria						4.6	0.1
Czech Republic						9.5	0.3
Denmark	0.2	0.1	9.2	19.3	0.5	9.1	0.2
Finland						10.6	0.3
France						94.0	2.5
Germany						123.6	3.3
Greece						20.0	0.5
Hungary						6.3	0.2
Iceland						0.9	w
Republic of Ireland						8.7	0.2
Italy	0.1	0.1	19.3	5.4	0.1	89.5	2.4
Kazakhstan	5.4	3.3	83.6	60.5	1.6	9.6	0.3
Lithuania						2.5	0.1
Netherlands						46.2	1.2
Norway	1.3	0.8	8.3	149.9	3.9	9.6	0.3
Poland						21.3	0.6
Portugal						15.7	0.4
Romania	0.1	w	10.8	5.7	0.1	10.1	0.3
Russian Federation	9.9	6.1	21.3	458.7	11.9	128.5	3.4
Slovakia						3.5	0.1
Spain						77.6	2.1
Sweden						15.3	0.4
Switzerland						12.0	0.3
Turkey						32.0	0.8
Turkmenistan	0.1	w	7.4	10.1	0.3	4.9	0.1
Ukraine						17.4	0.5
United Kingdom	0.6	0.4	6.0	95.4	2.5	80.8	2.1
Uzbekistan	0.1	w	10.6	6.6	0.2	6.0	0.2
Other Europe & Eurasia	0.3	0.2	13.8	23.4	0.6	24.0	0.6
Total Europe & Eurasia	**19.0**	**11.7**	**21.6**	**850.7**	**22.0**	**957.3**	**25.4**
Iran	18.2	11.1	88.7	202.6	5.2	73.3	1.9
Iraq	15.5	9.7	*	99.7	2.6		
Kuwait	13.6	8.3	*	119.8	3.1	13.7	0.4

A1.1 (*Continued*)

	Proved reserves			Production (a)		Consumption (b)	
	Thousand Mt	Share of total (%)	R/P ratio	Mt	Share of total (%)	Mt	Share of total (%)
Oman	0.8	0.5	19.4	38.9	1.0		
Qatar	2.0	1.3	42.0	44.9	1.2	3.3	0.1
Saudi Arabia	36.1	22.1	67.8	505.9	13.1	79.6	2.1
Syria	0.4	0.3	16.1	26.7	0.7		
United Arab Emirates	13.0	8.2	*	125.8	3.3	15.6	0.4
Yemen	0.4	0.2	18.2	20.3	0.5		
Other Middle East	w	w	4.6	2.2	0.1	65.4	1.7
Total Middle East	**100.0**	**61.7**	**81.6**	**1186.6**	**30.7**	**250.9**	**6.7**
Algeria	1.5	1.0	16.7	83.0	2.1	10.7	0.3
Angola	1.2	0.7	24.3	49.0	1.3		
Cameroon				3.2	0.1		
Chad	0.1	0.1	14.6	8.8	0.2		
Rep. of Congo (Brazzaville)	0.3	0.2	20.3	12.4	0.3		
Egypt	0.5	0.3	13.8	35.0	0.9	26.7	0.7
Equatorial Guinea	0.2	0.1	10.0	17.4	0.4		
Gabon	0.3	0.2	26.6	11.8	0.3		
Libya	5.1	3.3	66.5	75.8	2.0		
Nigeria	4.8	3.0	38.4	122.2	3.2		
South Africa						24.9	0.7
Sudan	0.9	0.5	57.3	14.9	0.4		
Tunisia	0.1	0.1	25.2	3.3	0.1		
Other Africa	0.1	w	8.6	4.4	0.1	62.1	1.6
Total Africa	**14.9**	**9.4**	**33.1**	**441.1**	**11.4**	**124.3**	**3.3**
Australia	0.5	0.3	20.4	22.9	0.6	38.8	1.0
Bangladesh						4.2	0.1
Brunei	0.1	0.1	13.6	10.3	0.3		
China	2.3	1.4	13.4	174.5	4.5	308.6	8.2
China Hong Kong SAR						15.3	0.4
India	0.7	0.5	18.6	38.0	1.0	119.3	3.2
Indonesia	0.7	0.4	11.5	55.1	1.4	54.7	1.5
Japan						241.5	6.4
Malaysia	0.6	0.4	12.9	40.3	1.0	23.3	0.6
New Zealand						7.0	0.2
Pakistan						14.4	0.4
Philippines						15.8	0.4
Singapore						38.1	1.0
South Korea						104.8	2.8
Taiwan						41.5	1.1
Thailand	0.1	w	6.3	9.0	0.2	43.6	1.2
Vietnam	0.4	0.2	19.0	20.8	0.5		
Other Asia Pacific	0.1	0.1	13.2	8.6	0.2	19.5	0.5
Total Asia Pacific	**5.5**	**3.5**	**14.2**	**379.5**	**9.8**	**1090.5**	**28.9**

(*Continued*)

A1.1 (*Continued*)

	Proved reserves			Production (a)		Consumption (b)	
	Thousand Mt	Share of total (%)	R/P ratio	Mt	Share of total (%)	Mt	Share of total (%)
TOTAL WORLD	**161.9**	**100.0**	**40.5**	**3867.9**	**100.0**	**3767.1**	**100.0**
Of which: OECD	10.9	7.0	10.9	976.7	25.3	694.5	18.4
OPEC	121.5	74.9	73.9	1588.2	41.1	2252.3	59.8
Non-OPEC^	23.8	14.9	13.5	1720.8	44.5	186.0	4.9
Former Soviet Union	16.5	10.2	28.9	558.9	14.4	1328.8	35.3

Notes:

Proved reserves of oil – generally taken to be those quantities that geological and engineering information indicates with reasonable certainty can be recovered in the future from known reservoirs under existing economic and operating conditions.

Reserves/production (R/P) ratio – if the reserves remaining at the end of any year are divided by the production in that year, the result is the length of time that those remaining reserves would last if production were to continue at that level.

(a) Includes crude oil, shale oil, oil sands and NGLs (natural gas liquids – the liquid content of natural gas where this is recovered separately. Excludes liquid fuels from other sources such as coal derivatives)
(b) Inland demand plus international aviation and marine bunkers and refinery fuel and loss

* Over 100 years
w Less than 0.05
^ Excludes Former Soviet Union

A1.2 Gas (2004)

	Proved reserves			Production (a)		Consumption (b)	
	Trillion m^3	Share of total (%)	R/P ratio	Mt_{oe}	Share of total (%)	Mt_{oe}	Share of total (%)
USA	5.29	2.9	9.8	488.6	20.2	582.0	24.0
Canada	1.60	0.9	8.8	164.5	6.8	80.5	3.3
Mexico	0.42	0.2	11.3	33.4	1.4	43.3	1.8
Total North America	**7.32**	**4.1**	**9.6**	**686.5**	**28.3**	**705.9**	**29.2**
Argentina	0.61	0.3	13.5	40.4	1.7	34.1	1.4
Bolivia	0.89	0.5	*	7.7	0.3		
Brazil	0.33	0.2	29.5	10.0	0.4	17.0	0.7
Chile						7.4	0.3
Colombia	0.11	0.1	17.3	5.7	0.2	5.7	0.2
Ecuador						w	w
Peru	0.25	0.1	*			0.8	w
Trinidad & Tobago	0.53	0.3	19.2	24.9	1.0		
Venezuela	4.22	2.4	*	25.3	1.0	25.3	1.0
Other S. & Cent. America	0.17	0.1	*	2.3	0.1	15.9	0.7
Total S. & Cent. America	**7.10**	**4.0**	**55.0**	**116.2**	**4.8**	**106.2**	**4.4**
Austria						8.5	0.4
Azerbaijan	1.37	0.8	*	4.2	0.2	7.7	0.3
Belarus						16.6	0.7
Belgium & Luxembourg						14.7	0.6
Bulgaria						2.8	0.1
Czech Republic						8.0	0.3
Denmark	0.09	w	9.3	8.5	0.4	4.9	0.2
Finland						3.9	0.2
France						40.2	1.7
Germany	0.20	0.1	12.1	14.7	0.6	77.3	3.2
Greece						2.2	0.1
Hungary						11.7	0.5
Iceland						–	–
Republic of Ireland						3.6	0.2
Italy	0.17	0.1	12.8	11.7	0.5	66.0	2.7
Kazakhstan	3.00	1.7	*	16.6	0.7	13.7	0.6
Lithuania						2.8	0.1
Netherlands	1.49	0.8	21.7	61.9	2.6	39.1	1.6
Norway	2.39	1.3	30.4	70.6	2.9	4.1	0.2
Poland	0.12	0.1	26.4	3.9	0.2	11.9	0.5
Portugal						2.8	0.1
Romania	0.30	0.2	22.3	11.9	0.5	16.9	0.7
Russian Federation	48.00	26.7	81.5	530.2	21.9	361.8	15.0
Slovakia						6.1	0.3

(*Continued*)

A1.2 (*Continued*)

	Proved reserves			Production (a)		Consumption (b)	
	Trillion m³	Share of total (%)	R/P ratio	Mt$_{oe}$	Share of total (%)	Mt$_{oe}$	Share of total (%)
Spain						24.6	1.0
Sweden						0.7	w
Switzerland						2.7	0.1
Turkey						19.9	0.8
Turkmenistan	2.90	1.6	53.1	49.2	2.0	13.9	0.6
Ukraine	1.11	0.6	60.6	16.5	0.7	63.6	2.6
United Kingdom	0.59	0.3	6.1	86.3	3.6	88.2	3.6
Uzbekistan	1.86	1.0	33.3	50.3	2.1	44.4	1.8
Other Europe & Eurasia	0.45	0.2	40.9	9.8	0.4	12.3	0.5
Total Europe & Eurasia	**64.02**	**35.7**	**60.9**	**946.4**	**39.1**	**997.7**	**41.2**
Bahrain	0.09	0.1	9.2	8.8	0.4		
Iran	27.50	15.3	*	77.0	3.2	78.4	3.2
Iraq	3.17	1.8	*				
Kuwait	1.57	0.9	*	8.7	0.4	8.7	0.4
Oman	1.00	0.6	56.5	15.8	0.7		
Qatar	25.78	14.4	*	35.3	1.5	13.6	0.6
Saudi Arabia	6.75	3.8	*	57.6	2.4	57.6	2.4
Syria	0.37	0.2	72.0	4.6	0.2		
United Arab Emirates	6.06	3.4	*	41.2	1.7	35.6	1.5
Yemen	0.48	0.3	*				
Other Middle East	0.05	w	31.7	2.9	0.1	24.0	1.0
Total Middle East	**72.83**	**40.6**	*	**251.9**	**10.4**	**218.0**	**9.0**
Algeria	4.55	2.5	55.4	73.8	3.0	19.1	0.8
Egypt	1.85	1.0	69.1	24.2	1.0	23.1	1.0
Libya	1.49	0.8	*	6.3	0.3		
Nigeria	5.00	2.8	*	18.5	0.8		
South Africa						–	–
Other Africa	1.18	0.7	*	7.8	0.3	19.6	0.8
Total Africa	**14.06**	**7.8**	**96.9**	**130.6**	**5.4**	**61.8**	**2.6**
Australia	2.46	1.4	69.9	31.7	1.3	22.1	0.9
Bangladesh	0.44	0.2	33.0	11.9	0.5	11.9	0.5
Brunei	0.34	0.2	28.3	10.9	0.4		
China	2.23	1.2	54.7	36.7	1.5	35.1	1.5
China Hong Kong SAR						2.0	0.1
India	0.92	0.5	31.3	26.5	1.1	28.9	1.2
Indonesia	2.56	1.4	34.9	66.0	2.7	30.3	1.3
Japan						64.9	2.7
Malaysia	2.46	1.4	45.7	48.5	2.0	29.9	1.2
Myanmar	0.53	0.3	71.0	6.6	0.3		
New Zealand				3.2	0.1	3.2	0.1
Pakistan	0.80	0.4	34.4	20.9	0.9	23.1	1.0

A1.2 (*Continued*)

	Proved reserves			Production (a)		Consumption (b)	
	Trillion m³	Share of total (%)	R/P ratio	Mt_oe	Share of total (%)	Mt_oe	Share of total (%)
Papua New Guinea	0.43	0.2	*				
Philippines						2.2	0.1
Singapore						7.0	0.3
South Korea						28.4	1.2
Taiwan						9.1	0.4
Thailand	0.43	0.2	21.1	18.2	0.8	25.9	1.1
Vietnam	0.24	0.1	56.5	3.7	0.2		
Other Asia Pacific	0.38	0.2	38.4	6.0	0.2	7.1	0.3
Total Asia Pacific	**14.21**	**7.9**	**43.9**	**290.8**	**12.0**	**330.9**	**13.7**
TOTAL WORLD	**179.53**	**100.0**	**66.7**	**2422.4**	**100.0**	**2420.4**	**100.0**
Of which: European Union 25	2.75	1.5	12.8	193.7	8.0	420.2	17.4
OECD	15.02	8.4	13.7	988.7	40.8	1265.5	52.3
Former Soviet Union	58.51	32.6	78.9	667.2	27.5	531.0	21.9
Other EMEs				766.5	31.6	623.8	25.8

Notes:

Because of rounding some totals may not agree exactly with the sum of their component parts.

The difference between these world consumption figures and the world production statistics is due to variations in stocks at storage facilities and liquefaction plants, together with unavoidable disparities in the definition, measurement or conversion of gas supply and demand data.

Proved reserves of natural gas – generally taken to be those quantities that geological and engineering information indicates with reasonable certainty can be recovered in the future from known reservoirs under existing economic and operating conditions.

Reserves/production (R/P) ratio – if the reserves remaining at the end of any year are divided by the production in that year, the result is the length of time that those remaining reserves would last if production were to continue at that level.

(a) Excluding gas flared or recycled

* Over 100 years
w Less than 0.05

A1.3 Coal (2004)

	Proved reserves (Mt)				R/P ratio	Production (a)		Consumption (a)	
	Anthracite and bituminous	Sub-bituminous and lignite	Total	Share of total (%)		Mt$_{oe}$	Share of total (%)	Mt$_{oe}$	Share of total (%)
USA	111338	135305	246643	27.1	245	567.2	20.8	564.3	20.3
Canada	3471	3107	6578	0.7	100	34.9	1.3	30.5	1.1
Mexico	860	351	1211	0.1	135	4.3	0.2	9.0	0.3
Total North America	**115669**	**138763**	**254432**	**28.0**	**235**	**606.3**	**22.2**	**603.8**	**21.7**
Argentina								0.7	w
Brazil	–	10113	10113	1.1	*	1.6	0.1	11.4	0.4
Chile								2.5	0.1
Colombia	6230	381	6611	0.7	120	35.8	1.3	2.7	0.1
Ecuador								–	–
Peru								0.4	w
Venezuela	479	–	479	0.1	53	6.6	0.2	0.1	w
Other S. & Cent. America	992	1698	2690	0.3	*	0.2	w	1.0	w
Total S. & Cent. America	**7701**	**12192**	**19893**	**2.2**	**290**	**44.1**	**1.6**	**18.7**	**0.7**
Austria								3.5	0.1
Azerbaijan								w	w
Belarus								0.1	w
Belgium & Luxembourg								6.1	0.2
Bulgaria	4	2183	2187	0.2	84	4.4	0.2	7.2	0.3
Czech Republic	2094	3458	5552	0.6	90	23.5	0.9	20.4	0.7
Denmark								4.4	0.2
Finland								5.2	0.2
France	15	–	15	w	17	0.5	w	12.5	0.4

Germany	183	6556	6739	0.7	32	54.7	2.0	85.7	3.1
Greece	–	3900	3900	0.4	55	9.5	0.3	9.3	0.3
Hungary	198	3159	3357	0.4	240	2.9	0.1	3.0	0.1
Iceland								0.1	w
Republic of Ireland								1.8	0.1
Italy								17.1	0.6
Kazakhstan	28151	3128	31279	3.4	360	44.4	1.6	27.5	1.0
Lithuania								0.1	w
Netherlands								9.1	0.3
Norway								0.6	w
Poland	14000	–	14000	1.5	87	69.8	2.6	57.7	2.1
Portugal								3.9	0.1
Romania	22	472	494	0.1	16	6.9	0.3	7.2	0.3
Russian Federation	49088	107922	157010	17.3	*	127.6	4.7	105.9	3.8
Slovakia								4.2	0.2
Spain	200	330	530	0.1	26	6.7	0.2	21.1	0.8
Sweden								2.4	0.1
Switzerland								0.1	w
Turkey	278	3908	4186	0.5	87	10.2	0.4	23.0	0.8
Turkmenistan								–	–
Ukraine	16274	17879	34153	3.8	424	41.9	1.5	39.4	1.4
United Kingdom	220	–	220	w	9	15.3	0.6	38.1	1.4
Uzbekistan								1.2	w
Other Europe & Eurasia	1529	21944	23473	2.6	341	16.4	0.6	19.5	0.7
Total Europe & Eurasia	**112256**	**174839**	**287095**	**31.6**	**242**	**434.4**	**15.9**	**537.2**	**19.3**
Iran								1.1	w
Kuwait								–	–

(Continued)

A1.3 (Continued)

	Proved reserves (Mt)					Production (a)		Consumption (a)	
	Anthracite and bituminous	Sub-bituminous and lignite	Total	Share of total (%)	R/P ratio	Mt_{oe}	Share of total (%)	Mt_{oe}	Share of total (%)
Qatar								–	–
Saudi Arabia								–	–
United Arab Emirates								–	–
Other Middle East								8.0	0.3
Total Middle East	**419**	–	**419**	**w**	**399**	**0.6**	**w**	**9.1**	**0.3**
Algeria								0.8	w
Egypt								0.7	w
South Africa	48750	–	48750	5.4	201	136.9	5.0	94.5	3.4
Zimbabwe	502	–	502	0.1	154	2.1	0.1		
Other Africa	910	174	1084	0.1	490	1.4	w	6.8	0.2
Total Africa & Middle East	**50162**	**174**	**50336**	**5.6**	**202**	**140.3**	**5.1**	**102.8**	**3.7**
Australia	38600	39900	78500	8.6	215	199.4	7.3	54.4	2.0
Bangladesh								0.4	w
China	62200	52300	114500	12.6	59	989.8	36.2	956.9	34.4
China Hong Kong SAR								6.6	0.2
India	90085	2360	92445	10.2	229	188.8	6.9	204.8	7.4
Indonesia	740	4228	4968	0.5	38	81.4	3.0	22.2	0.8
Japan	359	–	359	w	268	0.7	w	120.8	4.3
Malaysia								5.7	0.2
New Zealand	33	538	571	0.1	115	3.0	0.1	1.8	0.1
North Korea	300	300	600	0.1	21				
Pakistan	–	3050	3050	0.3	*	1.3	w	3.2	0.1
Philippines								5.0	0.2
Singapore								–	–

South Korea	–	80	w	25	1.4	0.1	53.1	1.9
Taiwan							36.8	1.3
Thailand	150	1354	0.1	67	5.8	0.2	10.2	0.4
Vietnam	–	150	w	6	14.8	0.5		
Other Asia Pacific	97	312	w	34	19.7	0.7	24.8	0.9
Total Asia Pacific	**192564**	**296889**	**32.7**	**101**	**1506.3**	**55.1**	**1506.6**	**54.2**
TOTAL WORLD	**478771**	**909064**	**100.0**	**164**	**2732.1**	**100.0**	**2778.2**	**100.0**
Of which: European Union 25	430293						307.0	11.0
OECD	172363	373220	41.1	180	1006.9	36.9	1163.2	41.9
Former Soviet Union	94513	227254	25.0	*	214.9	7.9	175.0	6.3
Other EMEs	211895	308590	33.9	102	1510.3	55.3	1440.1	51.8

Notes:

Because of rounding some totals may not agree exactly with the sum of their component parts.

Proved reserves of coal – generally taken to be those quantities that geological and engineering information indicates with reasonable certainty can be recovered in the future from known deposits under existing economic and operating conditions.

Reserves/production (R/P) ratio – if the reserves remaining at the end of the year are divided by the production in that year, the result is the length of time that those remaining reserves would last if production were to continue at that level.

(a) Commercial solid fuels only, i.e. bituminous coal and anthracite (hard coal), and lignite and brown (sub-bituminous) coal

* More than 500 years

w Less than 0.05

A1.4 Electricity (2004)

	Nuclear: consumption			Hydroelectricity: consumption		Hydroelectricity: consumption		Electricity generation (gross output)	
	TWh	Mt$_{oe}$	Share of total (%)	TWh	Share of total (%)	Mt$_{oe}$	Share of total (%)	TWh	Share of total (%)
USA	830.1	187.9	30.1	264.1	9.4	59.8	9.4	4150	23.8
Canada	90.4	20.5	3.3	337.7	12.0	76.4	12.0	568	3.3
Mexico	9.2	2.1	0.3	25.3	0.9	5.7	0.9	210	1.2
Total North America	**929.7**	**210.4**	**33.7**	**627.2**	**22.4**	**141.9**	**22.4**	**4928**	**28.2**
Argentina	7.9	1.8	0.3	30.1	1.1	6.8	1.1	99	0.6
Brazil	11.6	2.6	0.4	320.1	11.4	72.4	11.4	386	2.2
Chile	–	–	–	21.5	0.8	4.9	0.8	50	0.3
Colombia	–	–	–	38.1	1.4	8.6	1.4	47	0.3
Ecuador	–	–	–	7.3	0.3	1.7	0.3		
Peru	–	–	–	17.5	0.6	4.0	0.6		
Venezuela	–	–	–	70.6	2.5	16.0	2.5	106	0.6
Other S. & Cent. America	–	–	–	78.5	2.8	17.8	2.8	220	1.3
Total S. & Cent. America	**19.5**	**4.4**	**0.7**	**583.7**	**20.8**	**132.1**	**20.8**	**908**	**5.2**
Austria	–	–	–	32.2	1.1	7.3	1.1	60	0.3
Azerbaijan	–	–	–	2.8	0.1	0.6	0.1	21	0.1
Belarus	–	–	–	w	w	w	w	27	0.2
Belgium & Luxembourg	48.3	10.9	1.8	2.2	0.1	0.5	0.1	90	0.5
Bulgaria	16.8	3.8	0.6	2.2	0.1	0.5	0.1	45	0.3
Czech Republic	26.3	6.0	1.0	2.6	0.1	0.6	0.1	84	0.5
Denmark	–	–	–	w	w	w	w	40	0.2
Finland	24.2	5.5	0.9	14.9	0.5	3.4	0.5	86	0.5
France	448.2	101.4	16.2	65.4	2.3	14.8	2.3	572	3.3
Germany	167.1	37.8	6.1	27.0	1.0	6.1	1.0	607	3.5
Greece	–	–	–	4.9	0.2	1.1	0.2	60	0.3
Hungary	11.9	2.7	0.4	0.2	w	w	w	33	0.2
Iceland	–	–	–	7.1	0.3	1.6	0.3	9	w
Republic of Ireland	–	–	–	0.9	w	0.2	w	26	0.1
Italy	–	–	–	48.7	1.7	11.0	1.7	300	1.7
Kazakhstan	–	–	–	8.8	0.3	2.0	0.3	67	0.4
Lithuania	15.1	3.4	0.5	1.0	w	0.2	w	19	0.1
Netherlands	3.8	0.9	0.1	0.1	w	w	w	98	0.6
Norway	–	–	–	109.3	3.9	24.7	3.9	110	0.6
Poland	–	–	–	3.7	0.1	0.8	0.1	154	0.9
Portugal	–	–	–	12.2	0.4	2.8	0.4	46	0.3

A1.4 (*Continued*)

	Nuclear: consumption			Hydroelectricity: consumption		Hydroelectricity: consumption		Electricity generation (gross output)	
	TWh	Mt$_{oe}$	Share of total (%)	TWh	Share of total (%)	Mt$_{oe}$	Share of total (%)	TWh	Share of total (%)
Romania	5.5	1.3	0.2	17.0	0.6	3.8	0.6	57	0.3
Russian Federation	143.0	32.4	5.2	176.9	6.3	40.0	6.3	931	5.3
Slovakia	17.0	3.9	0.6	4.2	0.1	0.9	0.1	30	0.2
Spain	63.2	14.3	2.3	35.0	1.3	7.9	1.3	278	1.6
Sweden	76.7	17.3	2.8	56.1	2.0	12.7	2.0	148	0.8
Switzerland	27.0	6.1	1.0	35.5	1.3	8.0	1.3	66	0.4
Turkey	–	–	–	46.1	1.6	10.4	1.6	152	0.9
Turkmenistan	–	–	–	–	–	–	–	12	0.1
Ukraine	87.1	19.7	3.2	11.8	0.4	2.7	0.4	181	1.0
United Kingdom	79.8	18.1	2.9	7.5	0.3	1.7	0.3	400	2.3
Uzbekistan	–	–	–	7.5	0.3	1.7	0.3	50	0.3
Other Europe & Eurasia	7.9	1.8	0.3	72.2	2.6	16.3	2.6	157	0.9
Total Europe & Eurasia	**1268.8**	**287.2**	**46.0**	**815.9**	**29.1**	**184.7**	**29.1**	**5017**	**28.7**
Iran	–	–	–	11.9	0.4	2.7	0.4	156	0.9
Kuwait	–	–	–	–	–	–	–	42	0.2
Qatar	–	–	–	–	–	–	–	11	0.1
Saudi Arabia	–	–	–	–	–	–	–	164	0.9
United Arab Emirates	–	–	–	–	–	–	–	46	0.3
Other Middle East	–	–	–	5.5	0.2	1.2	0.2	155	0.9
Total Middle East	**–**	**–**	**–**	**17.5**	**0.6**	**4.0**	**0.6**	**573**	**3.3**
Algeria	–	–	–	0.3	w	0.1	w	31	0.2
Egypt	–	–	–	14.6	0.5	3.3	0.5	98	0.6
South Africa	15.0	3.4	0.5	3.7	0.1	0.8	0.1	245	1.4
Other Africa	–	–	–	68.9	2.5	15.6	2.5	164	0.9
Total Africa	**15.0**	**3.4**	**0.5**	**87.4**	**3.1**	**19.8**	**3.1**	**537**	**3.1**
Australia	–	–	–	16.7	0.6	3.8	0.6	236	1.4
Bangladesh	–	–	–	1.2	w	0.3	w	22	0.1
China	50.1	11.3	1.8	328.0	11.7	74.2	11.7	2187	12.5
China Hong Kong SAR	–	–	–	–	–	–	–	37	0.2
India	16.7	3.8	0.6	83.8	3.0	19.0	3.0	651	3.7
Indonesia	–	–	–	10.8	0.4	2.5	0.4	120	0.7
Japan	286.3	64.8	10.4	100.1	3.6	22.6	3.6	1110	6.4
Malaysia	–	–	–	6.2	0.2	1.4	0.2	91	0.5
New Zealand	–	–	–	27.7	1.0	6.3	1.0	43	0.2
Pakistan	2.0	0.5	0.1	26.9	1.0	6.1	1.0	88	0.5
Philippines	–	–	–	8.5	0.3	1.9	0.3	56	0.3

A1.4 (*Continued*)

	Nuclear: consumption			Hydroelectricity: consumption		Hydroelectricity: consumption		Electricity generation (gross output)	
	TWh	Mt$_{oe}$	Share of total (%)	TWh	Share of total (%)	Mt$_{oe}$	Share of total (%)	TWh	Share of total (%)
Singapore	–	–	–	–	–	–	–	37	0.2
South Korea	130.7	29.6	4.7	5.9	0.2	1.3	0.2	374	2.1
Taiwan	39.5	8.9	1.4	6.6	0.2	1.5	0.2	218	1.3
Thailand	–	–	–	8.1	0.3	1.8	0.3	123	0.7
Other Asia Pacific	–	–	–	41.0	1.5	9.3	1.5	97	0.6
Total Asia Pacific	**525.4**	**118.9**	**19.0**	**671.5**	**24.0**	**152.0**	**24.0**	**5489**	**31.5**
TOTAL WORLD	**2758.4**	**624.3**	**100.0**	**2803.2**	**100.0**	**634.4**	**100.0**	**17452**	**100.0**
Of which: European Union 25	987.0	223.4	35.8	325.7	11.6	73.7	11.6	3171	18.2
OECD	2340.1	529.6	84.8	1293.2	46.1	292.7	46.1	10141	58.1
Former Soviet Union	247.6	56.0	9.0	248.6	8.9	56.3	8.9	1372	7.9
Other EMEs	170.7	38.6	6.2	1261.3	45.0	285.5	45.0	5940	34.0

Notes:
Values given in t$_{oe}$ are converted on the basis of thermal equivalence assuming 38 per cent conversion efficiency in a modern thermal power station.

w Less than 0.05

A1.5 Primary energy (2005)

				Mtoe			
	Oil	Natural gas	Coal	Nuclear energy	Hydro electric	Total	Share of total (%)
USA	937.6	582.0	564.3	187.9	59.8	**2331.6**	22.8
Canada	99.6	80.5	30.5	20.5	76.4	**307.5**	3.0
Mexico	85.2	43.3	9.0	2.1	5.7	**145.3**	1.4
Total North America	**1122.4**	**705.9**	**603.8**	**210.4**	**141.9**	**2784.4**	**27.2**
Argentina	18.7	34.1	0.7	1.8	6.8	**62.0**	0.6
Brazil	84.2	17.0	11.4	2.6	72.4	**187.7**	1.8
Chile	10.7	7.4	2.5	–	4.9	**25.5**	0.2
Colombia	10.1	5.7	2.7	–	8.6	**27.1**	0.3
Ecuador	6.3	w	–	–	1.7	**8.0**	0.1
Peru	7.2	0.8	0.4	–	4.0	**12.4**	0.1
Venezuela	26.3	25.3	0.1	–	16.0	**67.6**	0.7
Other S. & Cent. America	58.3	15.9	1.0	–	17.8	**92.9**	0.9
Total S. & Cent. America	**221.7**	**106.2**	**18.7**	**4.4**	**132.1**	**483.1**	**4.7**
Austria	13.7	8.5	3.5	–	7.3	**33.0**	0.3
Azerbaijan	4.5	7.7	w	–	0.6	**12.8**	0.1
Belarus	7.5	16.6	0.1	–	w	**24.2**	0.2
Belgium & Luxembourg	38.1	14.7	6.1	10.9	0.5	**70.3**	0.7
Bulgaria	4.6	2.8	7.2	3.8	0.5	**18.9**	0.2
Czech Republic	9.5	8.0	20.4	6.0	0.6	**44.5**	0.4
Denmark	9.1	4.9	4.4	–	w	**18.4**	0.2
Finland	10.6	3.9	5.2	5.5	3.4	**28.6**	0.3
France	94.0	40.2	12.5	101.4	14.8	**262.9**	2.6
Germany	123.6	77.3	85.7	37.8	6.1	**330.4**	3.2
Greece	20.0	2.2	9.3	–	1.1	**32.7**	0.3
Hungary	6.3	11.7	3.0	2.7	w	**23.7**	0.2
Iceland	0.9	–	0.1	–	1.6	**2.6**	w
Republic of Ireland	8.7	3.6	1.8	–	0.2	**14.3**	0.1
Italy	89.5	66.0	17.1	–	11.0	**183.6**	1.8
Kazakhstan	9.6	13.7	27.5	–	2.0	**52.8**	0.5
Lithuania	2.5	2.8	0.1	3.4	0.2	**9.0**	0.1
Netherlands	46.2	39.1	9.1	0.9	w	**95.3**	0.9
Norway	9.6	4.1	0.6	–	24.7	**39.0**	0.4
Poland	21.3	11.9	57.7	–	0.8	**91.8**	0.9
Portugal	15.7	2.8	3.9	–	2.8	**25.0**	0.2
Romania	10.1	16.9	7.2	1.3	3.8	**39.3**	0.4
Russian Federation	128.5	361.8	105.9	32.4	40.0	**668.6**	6.5
Slovakia	3.5	6.1	4.2	3.9	0.9	**18.6**	0.2
Spain	77.6	24.6	21.1	14.3	7.9	**145.5**	1.4
Sweden	15.3	0.7	2.4	17.3	12.7	**48.4**	0.5
Switzerland	12.0	2.7	0.1	6.1	8.0	**29.0**	0.3
Turkey	32.0	19.9	23.0	–	10.4	**85.3**	0.8
Turkmenistan	4.9	13.9	–	–	–	**18.8**	0.2

(*Continued*)

A1.5 *(Continued)*

	Mt_{oe}						
	Oil	Natural gas	Coal	Nuclear energy	Hydro electric	Total	Share of total (%)
Ukraine	17.4	63.6	39.4	19.7	2.7	**142.8**	1.4
United Kingdom	80.8	88.2	38.1	18.1	1.7	**226.9**	2.2
Uzbekistan	6.0	44.4	1.2	–	1.7	**53.2**	0.5
Other Europe & Eurasia	24.0	12.3	19.5	1.8	16.3	**73.9**	0.7
Total Europe & Eurasia	**957.3**	**997.7**	**537.2**	**287.2**	**184.7**	**2964.0**	**29.0**
Iran	73.3	78.4	1.1	–	2.7	**155.5**	1.5
Kuwait	13.7	8.7	–	–	–	**22.5**	0.2
Qatar	3.3	13.6	–	–	–	**16.9**	0.2
Saudi Arabia	79.6	57.6	–	–	–	**137.2**	1.3
United Arab Emirates	15.6	35.6	–	–	–	**51.3**	0.5
Other Middle East	65.4	24.0	8.0	–	1.2	**98.5**	1.0
Total Middle East	**250.9**	**218.0**	**9.1**	**–**	**4.0**	**481.9**	**4.7**
Algeria	10.7	19.1	0.8	–	0.1	**30.6**	0.3
Egypt	26.7	23.1	0.7	–	3.3	**53.8**	0.5
South Africa	24.9	–	94.5	3.4	0.8	**123.7**	1.2
Other Africa	62.1	19.6	6.8	–	15.6	**104.0**	1.0
Total Africa	**124.3**	**61.8**	**102.8**	**3.4**	**19.8**	**312.1**	**3.1**
Australia	38.8	22.1	54.4	–	3.8	**119.0**	1.2
Bangladesh	4.2	11.9	0.4	–	0.3	**16.6**	0.2
China	308.6	35.1	956.9	11.3	74.2	**1386.2**	13.6
China Hong Kong SAR	15.3	2.0	6.6	–	–	**23.8**	0.2
India	119.3	28.9	204.8	3.8	19.0	**375.8**	3.7
Indonesia	54.7	30.3	22.2	–	2.5	**109.6**	1.1
Japan	241.5	64.9	120.8	64.8	22.6	**514.6**	5.0
Malaysia	23.3	29.9	5.7	–	1.4	**60.3**	0.6
New Zealand	7.0	3.2	1.8	–	6.3	**18.3**	0.2
Pakistan	14.4	23.1	3.2	0.5	6.1	**47.2**	0.5
Philippines	15.8	2.2	5.0	–	1.9	**25.0**	0.2
Singapore	38.1	7.0	–	–	–	**45.1**	0.4
South Korea	104.8	28.4	53.1	29.6	1.3	**217.2**	2.1
Taiwan	41.5	9.1	36.8	8.9	1.5	**97.8**	1.0
Thailand	43.6	25.9	10.2	–	1.8	**81.5**	0.8
Other Asia Pacific	19.5	7.1	24.8	–	9.3	**60.7**	0.6
Total Asia Pacific	**1090.5**	**330.9**	**1506.6**	**118.9**	**152.0**	**3198.8**	**31.3**
TOTAL WORLD	**3767.1**	**2420.4**	**2778.2**	**624.3**	**634.4**	**10224.4**	**100.0**
Of which: European Union 25	694.5	420.2	307.0	223.4	73.7	**1718.8**	16.8
OECD	2252.3	1265.5	1163.2	529.6	292.7	**5503.3**	53.8
Former Soviet Union	186.0	531.0	175.0	56.0	56.3	**1004.3**	9.8
Other EMEs	1328.8	623.8	1440.1	38.6	285.5	**3716.8**	36.4

Notes:
In this Appendix, primary energy comprises commercially traded fuels only. Excluded, therefore, are fuels such as wood, peat and animal waste, which though important in many countries, are unreliably documented in terms of consumption statistics.

w Less than 0.05

A1.6 Definitions

North America
USA (excluding Puerto Rico), Canada and Mexico.

South and Central America
Caribbean (including Puerto Rico), Central and South America.

Europe
European members of the OECD plus Albania, Bosnia-Herzegovina, Bulgaria, Croatia, Cyprus, Former Yugoslav Republic of Macedonia, Gibraltar, Malta, Romania, Slovenia, Yugoslavia.

Former Soviet Union
Armenia, Azerbaijan, Belarus, Estonia, Georgia, Kazakhstan, Krgyzstan, Latvia, Lithuania, Moldova, Russian Federation, Tajikistan, Turkmenistan, Ukraine, Uzbekistan.

Europe and Eurasia
This includes all countries listed above under the headings *Europe* and *Former Soviet Union.*

Middle East
Arabian Peninsula, Iran, Iraq, Israel, Jordan, Lebanon, Syria.

North Africa
Territories on the north coast of Africa from Egypt to Western Sahara.

West Africa
Territories on the west coast of Africa from Mauritania to Angola, including Cape Verde, Chad.

East and Southern Africa
Territories on the east coast of Africa from Sudan to Republic of South Africa. Also Botswana, Madagascar, Malawi, Namibia, Uganda, Zambia, Zimbabwe.

Asia Pacific
Brunei, Cambodia, China, China Hong Kong SAR*, Indonesia, Japan, Laos, Malaysia, Mongolia, North Korea, Philippines, Singapore, South Asia (Afghanistan, Bangladesh,

India, Myanmar, Nepal, Pakistan and Sri Lanka), South Korea, Taiwan, Thailand, Vietnam, Australia, New Zealand, Papua New Guinea and Oceania.

Australasia
Australia, New Zealand.

Country groupings are made purely for statistical purposes and are not intended to imply any judgement about political or economic standings.

OECD members
Europe: Austria, Belgium, Czech Republic, Denmark, Finland, France, Germany, Greece, Hungary, Iceland, Republic of Ireland, Italy, Luxembourg, Netherlands, Norway, Poland, Portugal, Slovakia, Spain, Sweden, Switzerland, Turkey, UK.

Other member countries: Australia, Canada, Japan, Mexico, New Zealand, South Korea, USA.

OPEC members
Middle East: Iran, Iraq, Kuwait, Qatar, Saudi Arabia, United Arab Emirates (Abu Dhabi, Dubai, Ras-al-Khaimah and Sharjah). **North Africa:** Algeria, Libya. **West Africa:** Nigeria. **Asia Pacific:** Indonesia. **South America:** Venezuela.

(Since Ecuador and Gabon have withdrawn from OPEC, they are excluded from all OPEC totals.)

European Union members (25)
Austria, Belgium, Cyprus, Czech Republic, Denmark, Estonia, Finland, France, Germany, Greece, Hungary, Republic of Ireland, Italy, Latvia, Lithuania, Luxembourg, Malta, Netherlands, Poland, Portugal, Slovakia, Slovenia, Spain, Sweden, UK.

Other EMEs (Emerging Market Economies)
South and Central America, Africa, Middle East, Non-OECD Asia and Non-OECD Europe.

Other terms
Tonnes: Metric tons.

---

* Special Administrative Region.

Percentages: Calculated before rounding of actuals. All annual changes and shares of totals are on a weight basis except on pages 4, 12, 16, 18 and 20.

Rounding differences: Because of rounding, some totals – including the 2004 share of total – may not agree exactly with the sum of their component parts.

www.bp.com/statisticalreview

Appendix 2
Answers to in-text problems

Please see the book's website for any corrections/clarifications to these answers. Please send any comments or corrections to d.a.coley@ex.ac.uk. Your thoughts are most welcome and due acknowledgement will be given in subsequent editions.

As was stated in the Introduction, students are encouraged to make approximate estimations in these problems, therefore the following are only possible solutions and you may discover better ways of answering some of them.

Chapter 2

Problem 2.1

Energy required $= 1000 \times (200 - 20) \times 0.5 = 90\,000$ kJ $= 90\,000\,000$ J.

From Equation (2.2), $m = \dfrac{90\,000\,000}{(3 \times 10^8)^2} = 1 \times 10^{-9}$ kg.

Problem 2.2

From Table A4.2, 2.72 kJ $= \dfrac{2730}{1.356} = 2006$ foot-pounds.

So the weight would be raised 2000 feet (or 600 metres)!

Energy and Climate Change   David A. Coley
© 2008 John Wiley & Sons, Ltd

Problem 2.5

$$E_{kinetic}(car) = \frac{1}{2} \times 1000 \left(\frac{100 \times 1000}{60 \times 60}\right)^2 = 385\ 802\ J = 386\ kJ.$$

$$E_{kinetic}(pen) = \frac{1}{2} \times 0.01 \left(\frac{40064 \times 1000}{60 \times 60}\right)^2 = 2\ 477\ 000 = 2477\ kJ.$$

So the pen has greater kinetic energy than the car.

Problem 2.6

E_{pot} (before fall) $= mgh = 1000 \times 9.81 \times 10 = 9800.$

Therefore $E_{kinetic}$ (at bottom of fall) $= 9800 = \frac{1}{2}\ mv^2 = \frac{1}{2} \times 1000 \times v^2.$

So, $v = \sqrt{\dfrac{98000 \times 2}{1000}} = 14$ m/s.

Problem 2.7a

From Equation (2.5b),

$\delta E_{th}(steel) = 10 \times 0.5 \times (50 - 20) = 150$ kJ.

$\delta E_{th}(wood) = 10 \times 1.7 \times (50 - 20) = 510$ kJ.

So wood is the better thermal store per unit mass.

Problem 2.7b

We have $E_{kinetic} = \frac{3}{2}\ (1/N_A)\ RT$ so $\frac{1}{2}\ mv^2 = \frac{3}{2}\ (1/N_A)\ RT.$

Therefore $\frac{1}{2}\ mv^2 = \frac{3}{2}(1/6.02 \times 10^{23}) \times 8.31 \times (50 + 273) = 6.69 \times 10^{-21}$ J.
And the average speed is $[2 \times (6.69 \times 10^{-21})/(3.35 \times 10^{-27})]^{1/2} = 1999$ m/s.

Problem 2.8

From Equation (2.7), $E_{th} = 16 \times 0.00017 = 0.00272$ MJ $= 2.72$ kJ.

Problem 2.9

We need 1 kJ of energy per second to be provided by the falling water, i.e.

$E_{pot} = mgh = 1000$ J.

Therefore, $m = \dfrac{1000}{gh} = \dfrac{1000}{9.81 \times 10} = 10.2$ kg/s.

Problem 2.10

From Table 2.1, 1 kWh $= 3.6 \times 10^6$ J,

therefore we need $\dfrac{3.6 \times 10^6}{3.2 \times 10^{-11}} = 1.125 \times 10^{17}$ nuclei.

At 2.5×10^{21} nuclei per gram, we therefore need $\dfrac{1.125 \times 10^{17}}{2.5 \times 10^{21}} = 0.000045$ g of U^{235}, or $0.000045/0.007 = 0.0064$ g of natural uranium.

Problem 2.11

An adult's dietary intake is about 2000 kcal or 8370 kJ per day. There are $60 \times 60 \times 24$, or 86 400 seconds in the day, therefore an adult has an output of $8370 \times 10^3/(86\,400)$, or 97 W. Assuming an adult weighs 65 kg, this is 97/65, or **1.5 W/kg**. For the sun we have $4 \times 10^{26}/(2 \times 10^{30})$, or only **0.0002 W/kg**, 7500 times smaller, and not very impressive at all!

Chapter 3

Problem 3.1

$E = mc^2$, therefore $m = E/c^2$. So,

mass lost per second $= m = \dfrac{3.846 \times 10^{26}}{(3 \times 10^8)^2} = 4.27 \times 10^9$ kg, i.e. a huge mass.

Problem 3.2

Surface area of sun $= 4\pi r^2 = 4 \times 3.14(696\,000\,000)^2 = 6.08 \times 10^{18}$ m^2.

Power through 1 m$^2 = \dfrac{3.846 \times 10^{26}}{6.08 \times 10^{18}} = 63$ MW.

For a light bulb, surface area is approximately 0.02 m^2. So power per m^2 equals around $100/0.02 = 5$ kW.

Problem 3.3

Ignoring the difference in surface area of the planet and the atmosphere, annual energy received by the planet $= F_s \times$ number of seconds in one year $\times$ surface area of planet $= 342 \times 32 \times 10^6 \times 4 \times 3.14 \times (6370000)^2 = 5.58 \times 10^{24}$ J.

Problem 3.4

Total latent heat flux $= 80$ W/m$^2 \times$ surface area of the planet $= 80 \times 510 \times 10^{12} = 41 \times 10^{15}$ W.

Problem 3.5

1 km$^2 = 1$ million m^2; density of water $= 1$ kg/l $= 1000$ kg/m^3.

$E_{pot} = mgh = 50\,000 \times 10^9 \times 1000 \times 9.81 \times 850$ J $= 417$ EJ.

Dividing by number of seconds in a year gives 1.3×10^{13} W.

Problem 3.6

50 B over 2 973 000 km$^2 = 50 \times 10^{-6} \times 2\,973\,000 \times 10^6 = 1.49 \times 10^8$ m^3.

Average density $= 2.5$ g/cm$^3 = 2500$ kg/m^3; therefore total mass loss $= 1.49 \times 10^8 \times 2500 = 3.716 \times 10^{11}$ kg.

So, $E_{pot} = mgh = 3.716 \times 10^{11} \times 9.81 \times 850 = 3 \times 10^{15}$ J.

Problem 3.7

Volume of oceans $=$ depth $\times$ surface area (2/3 of which is water) $= 3800 \times 510 \times 10^{12} = 1.292 \times 10^{18}$ m^3.

Density of water $= 1000$ kg/m^3.

Therefore total thermal energy stored as heat $= 1.292 \times 10^{18} \times 1000 \times 4.2 = 5.43 \times 10^{21}$ kJ per K.

 If average temp. of oceans $= 10\,^\circ$C, then this is $5.43 \times 10^{21} \times (273 + 10) = 1.537 \times 10^{24}$ kJ.

In addition, there is heat stored from the ice/water phase change = latent heat of fusion × mass of water = 334 (Table A4.3) × $1.292 \times 10^{18} \times 1000 = 4.32 \times 10^{23}$ kJ.

So total stored energy = 1.537×10^{24} kJ + 4.32×10^{23} kJ = 1.97×10^{24} J.

Note: this resource could not be fully exploited unless a thermal sink near absolute zero could be accessed.

Problem 3.8

$\log_{10} E_{kinetic} = a + bR$, therefore $E_{kinetic} = 10^{a+bR}$.

$$\text{Ratio} = \frac{E_{kinetic}(R = 8)}{E_{kinetic}(R = 6)} = \frac{10^{a+8b}}{10^{a+6b}} = \frac{10^{8b}}{10^{6b}}.$$

If $b = 1.2$, Ratio = 251.

If $b = 2$, Ratio = 10000.

Either way, the difference is several orders of magnitude.

Problem 3.9

(Tambora release)/(world primary energy use) = 80 EJ/410 EJ = 1/5.

Problem 3.10

	TW	EJ	Percentage of world primary energy use
Solar			
Incoming solar radiation	174000	5490000	1339000
Directly reflected	52000	1640000	400000
Direct conversion to heat	82000	2590000	631000
Evaporation	40000	1260000	300000
Wind, waves and convection	370	11700	2853
Photosynthesis	40	1260	307
Tides	3	94.6	23
Internal			
Conduction	32	1010	246
Convection (volcanoes and hot springs)	0.3	9.46	2.3

Chapter 4

Problem 4.1

100 W for 24 hours $= 100 \times 60 \times 60 \times 24 = 8.64 \times 10^6$ J.

Burning 0.5 kg of wood releases $0.5 \times 16 \times 10^6 = 8 \times 10^6$ J.

That is, about the same amount of energy.

Problem 4.2

If 400 t was required for a 0.2 ha field, then 1 km^2 would require $\dfrac{100 \times 400}{0.2} = 200\,000$ t of water.

If two workers could lift 400 t in 80 hours, then $\dfrac{200\,000}{400} = 500$ pairs of workers could lift 200 000 t in 80 hours. We have four weeks, or 336 hours (at 12 hours of work per day). So $\dfrac{500 \times 80}{336} = 119$ pairs were needed, or 238 individual workers: a large communal effort.

Problem 4.3

Total output $= \dfrac{200 \times 127 \times 10^6 \times 2 \times 10^6}{350} = 145 \times 10^6$ MW.

If the turbines only produce for one quarter of the year, this equals 3.63×10^{13} J, whereas 231 TWh equals an average demand of $231 \times 10^{12}/(60 \times 60) = 6.41 \times 10^{10}$ W. So, such an investment in wind power (if possible) would yield $3.63 \times 10^{13}/6.41 \times 10^{10} = 566$ times more than the national requirement for electricity.

Problem 4.4

From Table 4.1, output of horse $= 700$ W, therefore to provide 1.5 MW (1 500 000)/700 $= 2143$ horses would be needed, although these could not be worked 24 hours per day and therefore the true number would be greater. It would have also been difficult to harness these around a single machine.

Problem 4.5

From Appendix 1, Italy's daily oil use $= 89.5/365 = 0.245$ Mt $= 10$ PJ. From the text, the pyramids required between 0.8 and 1.6 PJ. As the pyramids where built over a 20-year period, the energy expended seems to be minor compared to that which a nation now uses on even a daily basis.

Problem 4.6

From the text, 12.5 kt of TNT $\equiv$ 52.5 TJ, therefore 1 t of TNT $\equiv 52.5 \times 10^{12}/(12500) = 4.2 \times 10^9$ J. So a single large nuclear weapon equates to $100 \times 10^6 \times 4.2 \times 10^9$ J $= 4.2 \times 10^{17}$ J, whereas world primary energy consumption is 4.1×10^{20} J (Table A4.1).

Total energy capacity of nuclear arsenal $= 10\,000 \times 10^6 \times 4.2 \times 10^9$ J $= 4.2 \times 10^{19}$ J, i.e. one tenth of primary energy demand. This analysis once more demonstrates the difference between energy and power, with the destructive effect of such weapons being due to the very short time taken to effect the transformation from mass-energy to heat/kinetic-energy and the relatively small area over which they act.

Problem 4.7

From the text, output of a computer chip $= 1$ MW/m^2. Therefore output from 1 km^2 of chip $= (1\,000\,000) \times 1$ MW $= 1 \times 10^{12}$ W. World electricity consumption (Table A4.1) $= 60 \times 10^{18}$ J and number of seconds in a year $= 3.2 \times 10^7$. So 60×10^{18} J equates to $60 \times 10^{18}/3.2 \times 10^7 = 1.88 \times 10^{12}$ W, i.e. nearly twice world electricity demand.

Chapter 5

Problem 5.1

Area of a sphere centred on the sun of radius equal to the distance of Mars $= 4 \times \pi \times (228 \times 10^9)^2 = 6.533 \times 10^{23}$ m. Area of a disk of radius of Mars $= 3.6 \times 10^{13}$ m. Therefore Mars intercepts $\dfrac{3.6 \times 10^{13} \times 3.91 \times 10^{26}}{6.533 \times 10^{23}} = 2.155 \times 10^{16}$ W, or

$\dfrac{2.155 \times 10^{16}}{4\pi \, (3385 \times 10^3)^2} = 149.7$ W/m^2. So, from Equation (5.2),

$$T = \left(\frac{149.7}{5.67 \times 10^{-8}}\right)^{1/4} = 226.7 \text{ K, or } -46\,^\circ\text{C}.$$

Problem 5.3

Approximate mass of the atmosphere $= 40 \times 4\pi(6370 \times 10^3) \times 0.2 = 40.76 \times 10^{14}$ kg. Mass of water with the same thermal capacity $= 40.76 \times 10^{14}/4.2 = 9.705 \times 10^{14}$ kg. Therefore required depth $= \dfrac{9.705 \times 10^{14}}{4\pi \, (6370 \times 10^3)^2} = 1.9$ m.

Problem 5.4

From Figure 5.7 we have the following approximate values:

$C_0 = 280$, $C = 360$; $M_0 = 700$, $M = 1750$; $N_0 = 270$, $N = 310$.

This calculation is best carried out with the aid of a spreadsheet to give an answer of $2.4 \, W/m^2$.

Problem 5.5

Current regime

Total cost of ice cream = (Factory A) + (Tax) + (Factory B)

$= (2 \times 10^6 \times \pounds1) + (2 \times 10^6 \times \pounds2) + (0.5 \times 10^6 \times \pounds2) = \pounds7 \times 10^6$.

The polluter pays

Factory A now needs to charge more for its ice cream to cover its increased costs ($\pounds1$ per kg + $\pounds2$ per kg to clean the river = $\pounds3$ per kg). Therefore Factory B will now sell more ice cream than Factory A. And if we assume they sell twice as much the total cost would be:

$(0.83 \times 10^6 \times \pounds3) + (1.67 \times 10^6 \times \pounds2) = \pounds5.8 \times 10^6$.

That is, the country has reduced its expenditure on ice cream by £1.2m and created a more efficient use of wealth and resources. The river is also being less highly contaminated.

Chapter 6

Problem 6.1

We have: $-0.2 = $ percentage change in passengers/15%.

So the reduction in passengers $= -3\%$.

Three per cent of 10 000 = 300 fewer passengers.

Chapter 7

Problem 7.1

$C_3H_8 + 5O_2 \rightarrow 3CO_2 + 4H_2O$.

Therefore burning 1 mol of propane releases 3 mols of CO_2. One gram of propane is $1/[(3 \times 12) + (8 \times 1)] = 1/44$ mols of propane, and 1 kg of propane is 1000/44 mols. So, mass of CO_2 released $= 3 \times 1000 \times [12 + (2 \times 16)]/44 = 3000$ g $= 3$ kg.

Problem 7.2

From Problem 7.1, 1 kg of propane $= 1000/44$ mols.

As $C_3H_8 + 5O_2 \rightarrow 3CO_2 + 4H_2O$, combusting 1 mol of propane needs 5 mols of oxygen and 1 kg needs $5 \times 1000/44$ moles of oxygen, or $32 \times 5 \times 1000/44 = 3636$ g. As air is only 23.3 per cent oxygen, $3636/0.233 = 15607$ g $= 15.6$ kg of air is needed.

Problem 7.3

From Equation (2.5a), and because thermal energy before mixing $=$ thermal energy after mixing:

$$m_1 C_p T_1 + m_2 C_p T_2 = (m_1 + m_2) C_p T_3 \text{ (where } T \text{ is in Kelvin)}$$

or

$$m_1 T_1 + m_2 T_2 = (m_1 + m_2) T_3 \text{ (where } T \text{ is in Kelvin),}$$

$$1 \times (80 + 273) + 0.5 \times (50 + 273) = 1.5 \ T_3.$$

Therefore $T_3 = 343$ K or 70 °C.

Problem 7.4

From Equation (7.7),

$$\eta_{carnot} = 1 - \frac{20 + 273}{300 + 273} = 0.49 \text{ or } 49 \text{ per cent.}$$

From Table 7.2, C_v(gross) of coal $= 29.65$ MJ/kg, therefore Carnot efficiency reduces this to $0.49 \times 29.65 = 14.52$ MJ.

Or $14.52 \times 10^6/(1000 \times 60 \times 60) = 4$ kWh.

Problem 7.5

From Equation (7.10), at $-20°C$, $\eta_{carnot} = \dfrac{20 + 273}{(20 + 273) - (-20 + 273)} = 7.3.$

At $+20°C$, $\eta_{carnot} = \dfrac{20 + 273}{(20 + 273) - (10 + 273)} = 29.3.$

So, the efficiency is far higher when the source and sink temperatures are similar.

Problem 7.6

$1 \text{ kWh} = 1000 \times 60 \times 60 = 3.6 \text{ MJ}$.

Assuming an efficiency of 35 per cent for electricity generation means that we need to burn $3.6/(4 \times 0.35) = 2.57$ MJ of coal. If we burn gas in the home, we will need to use $3.6/0.85 = 4.24$ MJ of gas. Again, from Table A4.1, the CO_2 emission factor of coal is 98 kg/GJ and for gas 56 kg/GJ. So, using gas will emit $4.24 \times 0.056 = 0.237$ kg of CO_2, whereas using electricity and a heat pump, $2.57 \times 0.098 = 0.252$ kg: about the same.

Problem 7.7

From Equation (7.11),

$$V = \frac{i}{ne\,A} = \frac{1}{8.5 \times 10^{28} \times 1.6 \times 10^{-19}\pi \times 0.005^2} = 9.36 \times 10^{-7} \text{ m/s}.$$

Problem 7.8

From Problem 7.7, we have 8.5×10^{28} free electrons per cubic metre, so required volume $= 6.4 \times 10^{9}/8.5 \times 10^{28} = 7.53 \times 10^{-20}$ m^3 $= 7.53 \times 10^{-11}$ mm^3.

Problem 7.9

$P = iV = 4 \times 230 = 920$ W.

The energy consumption at 920 watts for a year is $920 \times 60 \times 60 \times 24 \times 365 = 2.9 \times 10^{10}$ J $= 29$ GJ.

Chapter 8

Problem 8.1

2.6 Mt of coal plus 2.6 Mt of water $= 5.2$ Mt of slurry. Assuming this has approximately the same density as water implies 5.2×10^6 m^3 is being moved per annum, or $5.2 \times 10^6/32 \times 10^6 = 0.1625$ m^3 per second. The pipe has a cross-sectional area of $\pi(0.46/2)^2 = 0.166$ m^2. So, the flow must be $0.1625/0.166 = 0.98$ m/s. If we assume walking speed is 4 km/hour, this is 1.1 m/s and the slurry is flowing at around walking speed.

Problem 8.2

2600 MT_{oe} is $2600 \times 10^6 \times 42 = 1092 \times 10^{11}$ GJ (Table A4.2). At 98 kg of CO_2 per GJ (from the text, or Table A4.1), this implies that the combustion of coal releases $98 \times 1092 \times 10^{11} = 1.07 \times 10^{16}$ kg of CO_2.

Chapter 9

Problem 9.1

3.4 Mbl $= 3.4 \times 10^6 \times 5.694$ GJ $= 1.94 \times 10^7$ GJ per day, or 7.07×10^9 GJ per annum. Assuming a 30 per cent efficiency implies 2.1×10^9 GJ of electricity. Dividing by the number of seconds in a year gives watts $= 65$ GW. (Note this is far greater than the installed generating capacity of the country.)

Problem 9.2

Number of people required $= 65 \times 10^9/(6000/60) = 650$ million. Assuming people could only work for eight hours a day, gives 1.95 billion people, i.e. over 400 times the population of the country: hence the usefulness of fossil fuels.

Problem 9.4

From the text, or Table A4.1, the emission factor of oil is 74 kg of CO_2 per GJ; so 152 EJ releases $74 \times 152 \times 10^{18}/1 \times 10^9 = 11.2 \times 10^{12}$ kg, or 11.2 Gt.

Problem 9.5

Equation (9.1) shows that each mol of octane needs 12 $\frac{1}{2}$ mols of air. A mol of octane weighs $(8 \times 12) + (18 \times 1) = 114$ g: 12 $\frac{1}{2}$ mols of air weigh $12.5 \times (2 \times 16) = 400$ g. So the stoichiometric ratio is $114/400 = 0.285$, whereas the common ratio is $1/15 = 0.067$, so the mix is air rich by a factor of over four.

Problem 9.6

We have $4 \times 3000/2 = 6000$ cylinder firings per minute. Over 70 km this is $60 \times 6000 = 360\,000$ firings, using $70/14 = 5$ litres of fuel. So, each firing uses $5/(360\,000) = 1.39 \times 10^{-5}$ litres or (Table 9.2) $1.39 \times 10^{-5} \times 0.8 \times 10^{-3} = 0.011$ g.

Problem 9.7

If the average person drives around 13 000 miles per annum at 50 km per hour, cars are in use for only $13000/(50 \times 365) = 0.71$ hours, or 43 minutes per day. Very approximately for the UK we might expect at least one third of people to own a car, so total number of cars $\approx$20 million. If the average value of a new car is, say, £10 000, the total cost of the fleet is £200 billion.

The engines could be used to provide heat or electricity (or both if the engine were replaced by a fuel cell).

Chapter 10

Problem 10.1

From Table 2.2, density of liquid natural gas $= 422$ kg/m^3.

So 120 000 m^3 is 5.06×10^7 kg of gas, or (Table 2.1) 2.48×10^9 MJ.

Assume 100 000 people equates to three people per household, with five 100 W bulbs on for eight hours a day, $(100\ 000)/3 \times 5 \times 100 \times 8 \times 60 \times 60 = 4.8 \times 10^{11}$ J, or 4.8×10^5 MJ.

So the tanker contains over 5000 times as much energy.

Problem 10.2

1 $t_{oe} = 42$ GJ (Table A4.2), therefore 2.3 Gt$_{oe} = 97 \times 10^9$ GJ.

At 56 kg CO$_2$ per GJ, $97 \times 10^9/56 = 1.73 \times 10^9$ kg of CO$_2$.

From Section 7.1, burning 1 gram of methane produces 2.25 g of water and 53 kJ of energy. So burning 97×10^9 GJ will release $97 \times 10^9 \times 10^6 \times 2.25/53 = 4.12 \times 10^{15}$ g $= 4.12$ Gt.

Problem 10.3

We have:

$$CH_4 + 2O_2 \rightarrow CO_2 + 2H_2O.$$

Molecular weight of methane is 16 and of carbon dioxide is 44, so mass of the methane used $= 0.201 \times 16/44 = 0.731$ kg.

CO_2 saving from using methane $= 0.266 - 0.201 = 0.065$ kg.

A mass of 0.065 kg of CO_2 is equivalent (in global warming terms) to $0.065/20 = 0.00325$ kg of methane.

0.00325 is $(0.00325/0.103) \times 100$ per cent of mass used, i.e. 3.2 per cent.

Chapter 11

Problem 11.1

We have $270 \times 10^9 \times 100 = 270 \times 10^{12}$ litres. Density of oil (Table A4.1) ≈ 0.8 kg/l, so 270×10^{12} litres $= 2.16 \times 10^{14}$ kg $= 2.16 \times 10^5$ Mt.

Therefore the Green River deposits could, in theory, supply the world with oil for 59 years.

Problem 11.2

From the text: hydrate resource $= 1.43 \times 10^6$ EJ; tar sand resource >6580 EJ ($Q\infty$, Chapter 9); oil shale resource >9000 EJ (Problem 11.1). Therefore total resource $>1.45 \times 10^6$ EJ, with hydrates dominating the resource.

Current world primary energy use $= 410$ EJ per annum. Therefore non-traditional hydrocarbons could (in theory) meet world energy demand for over 3500 years.

Chapter 12

Problem 12.1

Making use of Avogadro's constant (Table A4.1), 1 kg of ^{235}U contains $(1000/235) \times 6.02 \times 10^{23} = 2.56 \times 10^{24}$ atoms.

As the fission of one atom of ^{235}U produces 3.2×10^{-11} J (from text), the fission of 1 kg produces $2.56 \times 10^{24} \times 3.2 \times 10^{-11} = 8.2 \times 10^{13}$ J. This is equivalent to the heat released by burning 2343 t of oil (Table A4.1).

Problem 12.2

As $k = 1.005$, each reaction produces 1.005 additional neutrons and destroys the initialising neutron. If the time between reactions is of the order of 0.001 s (maximum life-time of prompt neutrons given in the text), then we will have 1 neutron when $t = 0$ s, 1.005 when $t = 0.001$ s, 1.005^2 when $t = 0.002$ s and 1.005^3 when $t = 0.003$ s.

So when $t = 1$ s we will have $1.005^{1000} = 147$, but when $t = 10$ s we will have $1.005^{10000} = 4.6 \times 10^{21}$.

Problem 12.3

From Table A4.2, 1 eV $= 1.6 \times 10^{-19}$ J, therefore 1 MeV $= 1.6 \times 10^{-13}$ J.

From Chapter 2, $E_{kinetic} = \frac{1}{2} mv^2$, or $v = (2 E_{kinetic}/m)^{1/2}$.

So at 1 eV, $v = [2 \times 1.6 \times 10^{-19}/(1.6749 \times 10^{-27})]^{1/2} = 13\ 822$ m/s, and at 1 MeV, $v = [2 \times 1.6 \times 10^{-13}/(1.6749 \times 10^{-27})]^{1/2} = 13\ 822\ 000$ m/s.

Problem 12.4

Assume mean sea water temperature is around 10 °C,

$$\eta_{carnot} = 1 - \frac{10 + 273}{635 + 273} = 68.8 \text{ per cent.}$$

So efficiency is reduced from 68.8 per cent to 40 per cent.

Problem 12.5

From Table 12.4, core volume $= 8.31\pi (9.46/2)^2 = 584$ m^3.

Thermal output (Table 12.4) $= 1.596$ GW per reactor, therefore power density $= 1.596 \times 10^9/584 = 2.7$ MW/m^3.

Problem 12.6

From Problem 12.4, η_{Carnot} (Heysham) $= 68.8$ per cent.

For Sizewell B, $\eta_{Carnot} = 1 - \dfrac{10 + 273}{300 + 273} = 50.6$ per cent.

Problem 12.7

Rate $= 5000/(30 \times 365) = 0.4566$ per day, or approximately one every two days.

Chapter 13

Problem 13.1

$P(20 \text{ m head}) = 45ah^{3/2} = 45a20^{3/2}$,

$P(200 \text{ m head}) = 45a200^{3/2}$.

Therefore a 200 m head gives $(45a200^{3/2})/(45a20^{3/2}) = 200^{3/2}/20^{3/2} = 32$ times more power.

Problem 13.2

From the text, area of catchment $= 820\ 000 \text{ km}^2$ and rainfall $= 1400$ mm per annum. So, volume of water $= (820\ 000) \times 10^6 \times 1.4 = 1.148 \times 10^{12} \text{ m}^3$, or 1.148×10^{15} kg.

From the text, average head is very approximately $(128 - 84)/2 = 22$ m.

So using Equation (13.1), $E_{pot} = 22 \times 9.81 \times 1.148 \times 10^{15} = 2.48 \times 10^{17}$ J.

Assuming an overall efficiency of 77 per cent (see text) then gives 1.9×10^{17} J.

This compares with a world annual electricity production of 60 EJ (see Table A4.1).

Problem 13.3

Itaipu produces 93 000 GWh of electricity per annum (from text), or 3.348×10^{17} J. Assuming a calorific value of 29 MJ/kg and an efficiency of 33 per cent for coal powered stations implies an answer of 3.5×10^{10} kg.

The carbon dioxide emission factor of coal is 98 kg/GJ (Table A4.1) so this implies Itaipu saves the emission of 3.416 Mt of carbon dioxide.

Problem 13.4

First estimate CO_2 emission from coal-fired generation. The CO_2 emission factor for coal is 98 kg/GJ of heat (Table A4.1). If the power station is 33 per cent efficient, this is $98/0.33 = 297$ kg/GJ$_e$, or 0.891 kg/kWh. As methane has a GWP of 23, we would need only the emission of $0.891/23 = 0.0387$ kg/kWh of methane for hydropower to have a similar climate change impact to coal-fired generation.

Chapter 14

Problem 14.1

From the text 10 people die per day in the UK from traffic accidents, or 3650 per annum. If we assume average life expectancy to be 75, then $(60\ 000\ 000)/75 = 800\ 000$ people die per annum.

So, $(800\ 000)/3650$ or 1 in 219 die in road accidents.

Problem 14.2

Assume average life expectancy is 75, therefore mortality is $(60\ 000\ 000)/75 = 800\ 000$ per annum.

$(10\ 000)/(800\ 000) = 0.0125$ or 1.25 per cent.

Chapter 15

Problem 15.1

From Figure 15.16, cost (1980) $\approx$ 75 p/l: cost (2003) $\approx$ 73 p/l.

Adjusting for fuel efficiency, cost (1980) $\approx$ 7.5 p/km: cost (2003) $\approx$ 4.29 p/km.

(a) Therefore reduction in price $= 7.5 - 4.29 = 3.21$ p/km, or 1.9 per cent per annum (no compounding).

(b) For the cost to be equivalent the price has to be 7.5 p/km; which at 17 km/l equates to $17 \times 7.5 = 127.5$ p/l, whereas the current price is only 90 p/l (2005).

Chapter 19

Problem 19.1

From Table A4.1 we have 504.8 $Mt_{oe} = 21.2 \times 10^{19}$ J, 5 per cent of which is 1.06×10^{19} J.

500 kW continuously for half the year is $500 \times 10^3 \times 60 \times 60 \times 24 \times 365/2 = 7.884 \times 10^{12}$ J.

Therefore required number of turbines is $1.06 \times 10^{19}/7.884 \times 10^{12} = 1.3$ million.

Problem 19.2

Using Equation (19.2), and assuming a brick measures 20 by 10 by 10 cm.

$P = 0.93 \times 0.1 \times 5.67 \times 10^{-8} \times (T_1^4 - T_2^4)$.

When $T_2 = 15 + 273$ this is $5 \times 10^{-9} \times (4.9 \times 10^8) = 2.45$ W, or 2.1×10^5 J per day.

When $T_2 = -10C + 273$ this is 12.9 W, or 11.1×10^5 J per day.

(This answer ignores the inefficiencies in Japan's power stations, i.e. the true number of turbines would be less.)

Problem 19.3

U-value of walls $= 0.37$ (Table 19.7).

U-value of roof $= 0.35$.

Area of roof $\times$ U-value $= 12 \times 0.35 = 4.2$.

Area of walls $\times$ U-value $= 42 \times 0.37 = 15.54$.

Therefore heat loss from the building $= 19.74$ W/K.

As the internal-external temperature difference is 18 °C, the total fabric loss is 355 W and a heating system of this size is required.

Problem 19.4

Volume of building $= 3 \times 3 \times 4 = 36$ m^2.

Heat loss $=$ mass $\times C_p =$ air changes $\times$ density $\times$ volume $\times C_p \times \delta T$.

2 air changes an hour $= 2/(60 \times 60)$ air changes a second.

Therefore heat loss $= 2/(60 \times 60) \times 1 \times 36 \times 1000 \times 18 = 360$ W.

Whereas at 0.1 air changes per hour, heat loss $= 18$ W.

So, when occupied ventilation losses are likely to be of the same order as fabric ones. However, when the building is unoccupied and tightly sealed ventilation losses are negligible.

Problem 19.5

Assume three domestic light bulbs per person, that these are 100 W tungsten and that they are on for five hours per day.

Total electrical energy used by these is then:

$3 \times 4.4 \times 10^6 \times 100 \times 5 \times 60 \times 60 \times 365 = 8.67 \times 10^{15}$ J.

As the same amount of light is still required, assume these are replaced by 28 W (1200 lm) compact fluorescents (Table 19.9), so total annual energy use will be:

$3 \times 4.4 \times 10^6 \times 28 \times 5 \times 60 \times 60 \times 365 = 2.43 \times 10^{15}$ J.

Therefore difference in energy use is $8.67 \times 10^{15} - 2.43 \times 10^{15} = 6.14 \times 10^{15}$ J.

Assuming the electricity is produced from oil burning power stations with an efficiency of 33 per cent, then (Table A4.1) this implies a CO_2 saving of:

$6.14 \times 10^{15} \times [74/(1 \times 10^9)] \times (1/0.33) = 1.38 \times 10^9$ kg per annum.

Chapter 20

Problem 20.1

Assume house is $10 \times 10 \times 6$ m with four 1×1 m windows on each side.

Window area $= 16$ m^2.

Wall area $= 4 \times 10 \times 6 - 16 = 224$ m^2.

Roof area $= 100$ m^2.

Heat loss per °C for the poorly insulated design (see Table 19.7 for U-values) $= (16 \times 5.9) + (224 \times 1.4) + (100 \times 2.6) = 668$ W/°C.

Heat loss per °C for the well insulated design

$= (16 \times 1.4) + (224 \times 0.37) + (100 \times 0.35) = 140$ W/°C.

Problem 20.2

2000 kcal is (Table A4.2) $2000 \times 1000 \times 4.186 = 8.372 \times 10^6$ J.

Therefore for two people over 14 hours per day for six months we have $2 \times 8.372 \times 10^6 \times (14/24) \times (365/2) = 1.783 \times 10^9$ J, (or Table A4.2) 495 kWh.

Which at €0.11 per kWh equals €54 (£39.60, US$69.30).

Problem 20.3

Doubling the windows to four 2×2 m windows on the south side gives a total of 16 m^2 of glazing on this facade. In the South of France or London this gives $1.5 \times 16 = 24$ kWh per day (Figure 20.2).

The house has an approximate heat loss of 140 W/°C, or 1.4 kW for a 10 °C temperature difference. Over 24 hours this is $1.4 \times 24 = 33.6$ kWh. So the free heat from the windows will not fully heat the house, but it will greatly help.

Problem 20.4

Assume efficiency of electricity generation = 33 per cent.

Assume efficiency of alternative heating plant = 85 per cent (natural gas).

Therefore the requirement is that $COP \times$ efficiency of electrical generation > 0.85, i.e. $COP \times 0.33 > 0.85$, so $COP > 2.6$.

Problem 20.5

Assume each house houses three people, number of houses = 4/3 million.

Assume average area of roof = 100 m².

In summer, total free energy = $(4/3) \times 10^6 \times 100 \times 20 = 2.67 \times 10^9$ MJ.

In winter, total free energy = $(4/3) \times 10^6 \times 100 \times 7 = 9.33 \times 10^8$ MJ.

For comparison, daily primary energy demand (Appendix 1) of New Zealand = $18.3/365 = 0.0502$ $Mt_{oe} = 2.12 \times 10^9$ MJ, i.e. comparable to the summertime figure.

Problem 20.6

Assume water temperature = 15 °C and that this needs to be raised to 40 °C, therefore temperature difference = $40 - 15 = 25$ °C. As C_p of water = 1 kJ/kg (Table A4.1) and density of water is 1000 kg/m³, the required energy is

$25 \times 1 \times 0.24 \times 1000 = 6000$ kJ, or (Table A4.2) 1.67 kWh.

Therefore bath water would take $1.67/4.8 = 0.35$ hours in summer and $1.67/0.6 = 2.8$ hours in winter to warm up.

Problem 20.7

$$\eta_c = 0.8 - 8\frac{(T_c - T_a)}{R}, \text{ here } \eta_c = 0.8 - 8\frac{(T_c - 20)}{1000}.$$

Using a spreadsheet to plot η_c against T_c we have:

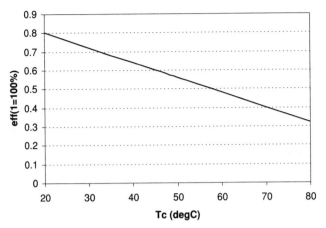

Data from, Putting Energy in the Spotlight, BP Plc, 2005

Problem 20.8

Current world coal demand = 2778 Mt_{oe} (Appendix 1).

Therefore current demand is $2778/300 = 9.26$ times that of 1913.

Therefore area of collector required = 52 000 (from the text) $\times 9.26/4 = 120\,380$ km^2.

Problem 20.9

Equation (20.5) states: $P_{th} = \dfrac{I\,A\eta_{Carnot}}{16}$, therefore $A = \dfrac{16\,P_{th}}{R\eta_{Carnot}}$.

From Chapter 7, $\eta_{Carnot} = 1 - T_{sink}/T_{source} = 1 - (100 + 273)/(600 + 273) = 0.57$.

So, $A = 250 \times 10^6 \times 16/(0.57 \times 1000) = 7.0 \times 10^6$ m^2.

Problem 20.10

$\eta_{Carnot} = 1 - T_{sink}/T_{source} = 1 - (5 + 273)/(25 + 273) = 0.067 = 6.7$ per cent.

The true efficiency will be even lower than this because of pumping cost and engineering constraints.

Chapter 21

Problem 21.1

Allowing for the fact that the array will only generate for 23 hours out of 24, we will require:

$$\frac{24}{23} \times \frac{1 \times 10^9}{1370} \times \frac{1}{0.14} = 5.44 \times 10^6 \text{ m}^2 = 5.44 \text{ km}^2 \text{ of PV.}$$

Problem 21.2

Assume roof area equals 10×10 m $= 100$ m^2, half of which faces south.

Assume insolation $= 240$ W/m^2 (Chapter 3).

Therefore one roof gives $50 \times 240 = 12\,000$ W, or 1680 W if we assume an efficiency of 14 per cent. Therefore total number of houses required $= 1 \times 10^9/1680 \approx 600\,000$.

Problem 21.3

$3880/365 = 10.63$ kWh.

80 amp-hours $\times$ 12 volts $= 960$ Wh.

Therefore number of batteries $= 10630/960 = 12$.

Chapter 22

Problem 22.1

$P_{kinetic} = \frac{1}{2} A\rho v^3 = \frac{1}{2}\pi (30/2)^2 \times 1.2 \times v^3 = 424v^3$.

Therefore $70\,000 = 424v^3$, or v $= 5.49$ m/s.

This is a common, relatively gentle wind speed.

Problem 22.3

a. From Figure 22.17, mean specific power $\approx 600/1500 = 0.4$ kW/m^2.

b. At 0.4 kW/m^2 we would need $1.5/0.4 = 3.75$ m^2 of swept area.

c. Equation (22.4) implies $E_{annual} = 2.5 \times 3.75 \times 8^3 = 4800$ kWh. Dividing by the number of hours in a year gives a mean power of only 0.548 kW. So the fickle nature of the wind has reduced the power to only 0.36 of its rated value (i.e. by 2/3). However Chapter 21 gives the average domestic UK electricity demand (for non-heating purposes) as 440 kW. So such a machine would still be useful.

d. $A = \pi r^2$, so $r = (3.75/\pi)^{0.5} = 1.09$, or a diameter of 2.18 m. This is likely to be considered unsightly and therefore unlikely to find favour.

Problem 22.4

For 60 m diameter blades, the circumference of the swept area will be $2\pi r = 2\pi (60/2) = 188.5$ m. So at 65 m/s we will have $65/188.5 = 0.345$ revolutions per second or 20 per minute.

Problem 22.5

Assume a 72 m diameter turbine of 2 MW (i.e. that in Figure 22.8), and allow five rotor diameters between machines.

$200/2 = 100$ machines; $(100)^{1/2} = 10$. So assume the site consists of 10 rows of 10 machines, $5 \times 72 = 360$ m apart.

Therefore the wind farm occupies a square of, $(10 - 1) \times 360 = 3.24$ km, or 10.5 km^2 in total.

Problem 22.6

From Problem 22.4, or simple arithmetic, 65 m/s for a 60 m turbine implies 20 rotations per minute, or 0.345 per second.

As we have three blades, the flicker frequency will be a maximum of $3 \times 0.345 = 1.035$ Hz.

Problem 22.7

From the text, India's electricity production from fossil fuels $= 457 \times 10^9$ kWh, or 1.645×10^{18} J, i.e. 51 000 MW continuously.

Assume 0.74 birds killed per MW of installed capacity.

Assume the turbines produce 50 per cent of their rated capacity continuously.

To meet 20 per cent of demand, India therefore needs $(51\,000) \times 0.2 \times 2 = 20\,400$ MW of installed capacity.

Therefore the number of birds killed per annum might be expected to be of the order of $(20\,400) \times 0.74$ (maximum value from Table 22.4) $= 15\,000$.

Problem 22.8

From Equation (22.11) for one turbine, at 300 m:

$$L_p = 100 - 10\log_{10}(4\pi \times 300^2) = 39.5 \text{ dB(A)},$$

and at 600 m:

$$L_p = 100 - 10\log_{10}(4\pi \times 600^2) = 29 \text{ dB(A)}.$$

Whereas 600 km from 10 turbines Equation (22.10) gives:

$$L_p = 10\log_{10}(10 \times 10^{29/10}) = 39 \text{ dB(A), i.e. less than a single turbine at 300 m.}$$

Chapter 23

Problem 23.1

$$P_{elec} = 0.18\,h^2 T = 0.18 \times 7^2 \times 10 = 88.2 \text{ kW/m.}$$

$100\,000$ homes require $100\,000 \times 440 = 4.4 \times 10^7$ W, or 4.4×10^4 kW.

Therefore total required length of machines $= 4.4 \times 10^4/88.2 = 499$ m.

Chapter 24

Problem 24.1

From Equation (22.3): $P_{kinetic} = 0.5 \times A\rho v^3$.

So we require $0.5 \times A\rho_{water}v_{water}^3 = 0.5 \times A\rho_{air}v_{air}^3$,

$$\text{or } v_{air} = \left(\frac{\rho_{water}}{\rho_{air}}v_{water}^3\right)^{1/3} = \left(\frac{1000}{1}1^3\right)^{1/3} = 10 \text{ m/s.}$$

Problem 24.2

Capacity is $24 \times 10 = 240$ MW$_e$. This is $240 \times 365 \times 24 = 2.1 \times 10^6$ MWh.

Therefore 'efficiency' is $480/2100 = 0.23$, or 23 per cent.

Chapter 25

Problem 25.1

One mol of carbohydrate weighs $12 + (2 \times 1) + 16 = 30$ g.

Therefore 1 kg equates to $1000/30 = 33.3$ moles or

$33.3 \times 6.02 \times 10^{23} = 2.0 \times 10^{25}$ molecules.

At 8×10^{-19} J per molecule (or unit of carbohydrate) (from text), this is

$8 \times 10^{-19} \times 2.0 \times 10^{25} = 16 \times 10^6$ J $= 16$ MJ.

The calorific value of oil $= 45.6$ MJ/kg (Table A4.1), so $16/45.6$ or 0.35 kg of oil contains the same amount of chemical energy as 1 kg of carbohydrate.

Problem 25.2

Density of methane (at atmospheric pressure, Table A4.1) $= 0.73$ kg/m^3, and calorific value of methane $= 48$ MJ/kg.

Therefore 3500 m^3/h $= 122\,640$ MJ/h or 34 MJ/s (MW).

So this output equates to 17 MW$_e$ turbines or twice the number if they are only likely to produce this output half the time.

Problem 25.3

From the text, we have a production rate of 320g/m^2 per annum. So for Indonesia we have $0.32 \times 1.8 \times 10^{12} = 5.76 \times 10^{11}$ kg of biomass per annum, which at 4 GJ/t (from text) $= 2.3 \times 10^{18}$ J $= 5.5 \times 10^7$ t$_{oe}$.

For the UK: $0.32 \times 242 \times 10^9 \times 0.004 = 3.1 \times 10^{17}$ J $= 7.4 \times 10^6$ t$_{oe}$.

So Indonesia could (in theory) replace $55/105 = 52$ per cent of its fossil fuel use with natural biomass, but the UK could only replace $7.4/202 = 3.7$ per cent of its fossil fuel use.

Problem 25.4

From the text, we could get 200 EJ per annum from 400×10^{10} m^2 of land. This is $(200 \times 10^{18})/(400 \times 10^{10}) = 5 \times 10^7$ J/m^2.

Therefore for Indonesia we have $1.8 \times 10^{12} \times 5 \times 10^7 = 9 \times 10^{19}$ J or 2110 Mt$_{oe}$, and for the UK we have $242 \times 10^9 \times 5 \times 10^7 = 1.21 \times 10^{19}$ J or 290 Mt$_{oe}$.

So, both Indonesia and the UK could replace their use of fossil fuels with biomass. In the case of Indonesia this would require only $105/2110 = 5$ per cent of its land area; however, for the UK $202/290 = 70$ per cent of its land area would be required.

Chapter 26

Problem 26.1

Power density $= (32 \times 10^{12}$ W$)/(510 \times 10^{12}$ m$^2) = 0.063$ W/m^2.

Chapter 27

Problem 27.1

Mass of oceans $\approx 4\pi r^2 \times (2/3) \times$ ocean depth $\times$ density of water

$= 510 \times 10^{12} \times (2/3) \times 3180 \times 1000 = 1.08 \times 10^{21}$ kg.

Water is H_2O, so the mass ratio of hydrogen to oxygen is 2:16 (Chapter 7), or 1:8. So mass of hydrogen $= (1.08 \times 10^{21})/8 = 1.35 \times 10^{20}$ kg.

Therefore total number of atoms of hydrogen

$=$ Avogadro's number $(6.02 \times 10^{23}) \times 1.35 \times 10^{20} = 8.136 \times 10^{43}$.

Therefore total number of deuterium atoms $= (8.136 \times 10^{43})/7000 = 1.16 \times 10^{40}$.

From Equation (27.1), this could provide $1.16 \times 10^{40} \times 2.8 \times 10^{-12} = 3.25 \times 10^{28}$ J.

One per cent of which is 3.25×10^{26} J, or enough to meet world primary energy demand for $(3.25 \times 10^{26})/(410 \times 10^{18}) = 790\,000$ years.

Chapter 28

Problem 28.1

Current world CO_2 emissions are 29 Gt per annum (Table A4.1). Figure 16.4 suggests only about one quarter of which is from the developing world. If one third of industrialized-world CO_2 emissions are from transport, then allowing equal access across the world to fossil-fuel based transport world increase world CO_2 emissions by:

$29 + (1 - 1/4) \times 29 \times (1/3) \times (3/2) = 10.9$ Gt or 38 per cent, if we assume the developing world includes 2/3 of the world's population (an underestimate) and has few CO_2 emissions accountable to transport.

Problem 28.2

Assume average width of road = 10 metres.

Total area of road = $0.01 \times 392\,000 = 3920$ km^2.

Or $3920/(242000) = 0.016$ or 1.6 per cent (note, for reference, [DFT05] gives a total road area of 3304 km^2).

Area needed for housing = population/number of occupants per house $\times$ size of house $\approx (60\,000\,000)/3 \times 10 \times 10 = 2 \times 10^9$ m^2 or 2000 km^2.

So road and housing area are of similar magnitude.

Problem 28.3

As the vehicle is neither on a hill nor accelerating the only forces are drag and roll. Assuming the frontal area is 2 m^2 and the mass 1000 kg,

$F_{drag} = 1.2 \times 0.35 \times 2 \times [(120\,000)/(60 \times 60)]^2 \times 0.5 = 467$ N.

$F_{roll} = 1000 \times 9.81 \times 0.015 = 147$ N.

$P = Fv = (467 + 147) \times (120\,000)/(60 \times 60) = 20467$ W.

Calorific value of oil (Table A4.1) = 43 MJ/kg or 36 MJ/l.

So 20 467 W requires $(20467)/(36 \times 10^6) = 5.69 \times 10^{-4}$ l/s to be burnt, or 1 l in 1759 s, in which time the car will have travelled 58.6 km. So, the fuel efficiency is 58.6 km/l.

Problem 28.4

Mass of petrol = $50/1.2 = 41.7$ kg.

Therefore energy stored = $41.7 \times 43 = 1792$ MJ.

Therefore weight of batteries required = $1792/(90 \times 10^{-3}) = 19\,911$ kg.

Which is twice the weight of the car.

Problem 28.5

Assume refuelling a 50-litre tank takes 1 minute, therefore energy transfer rate = $(50/1.2) \times (42.8/60) = 29.7$ MW.

$P = IV$, so $I = P/V = (29.7 \times 10^6)/240 = 124\,000$ A.

Which is clearly impossible in a domestic setting.

Problem 28.6

For the flywheel: $v = \dfrac{d}{t} = \dfrac{n2\pi r}{1}$; if $n = 500$, $v = 1570$ m/s then

$$v = \frac{1}{2}mv^2 = \frac{1}{2} \times 100 \times 1570^2 = 123 \text{ MJ}.$$

Which is equivalent to 3.4 litres of petrol, enough to power a car approximately 48 km (30 miles). However, as we saw, an internal combustion engine is only around 20 per cent efficient, whereas an electric motor might be 80 per cent efficient. Drive chain, air resistance and rolling losses will be the same. Therefore the range of the flywheel vehicle might extend to $48 \times (80/20) = 192$ km.

Chapter 29

Problem 29.1

World carbon emission $= 8$ GtC (Table A4.1).

Therefore capacity of the various storage options is:

Option	Approximate worldwide capacity (GtC)	Approximate worldwide capacity (years)
Ocean	1000–10 000+	125–1250+
Deep saline aquifers	100–10 000	12.5–1250
Depleted oil and gas reservoirs	100–1000	12.5–125
Coal seams	10–1000	1.25–125
Terrestrial (reforestation)	10–100	1.25–12.5
Utilization	Currently <0.1 GtC per annum	<0.0125

Problem 29.2

Given a 33 per cent efficiency in coal-fired generation, the amount of CO_2 (kg) released per GJ of generation is:

emission factor for coal (Table A4.1)/$0.33 = 98/0.33 = 297$ kg/GJ.

But 1 kWh $= 3.6$ MJ (Table A4.2), so the generation of 1 kWh releases

$(297 \times 3.6 \times 10^6)/(1 \times 10^9) = 1.07$ kg of carbon.

A sequestration cost of €77 per tC would therefore add $77 \times 1.07/1000 = €0.082$ per kWh, i.e. an increase in cost of 82 per cent.

Problem 29.3

The CO_2 emission factor for natural gas is 56 kg/GJ (Table A4.1) and 1 t_{oe} equates to 42 GJ (Table A4.2).

Therefore the consumption of 976 Mt_{oe} of natural gas releases:

$976 \times 10^6 \times 42 \times 56 = 2.3 \times 10^{12}$ kg $= 2.3$ Gt.

Storage capacity of Europe (from the example application) $= 800$ Gt.

Therefore $800/2.3 = 348$ years worth of CO_2 could be stored.

Appendix 3
Bibliography and suggested reading

The following texts provide excellent sources of additional information. The book's website contains links to other, web-based, sources.

Boyle, G., Everett, B., and Ramage, J., *Energy Systems and Sustainability*, the Open University, Oxford University Press, 2003. Concentrates on energy systems. Lots of good background information and excellent graphics.

Boyle, G., *Renewable Energy: Power for a Sustainable Future* (2nd edition), the Open University, Oxford University Press, 2004. An excellent review of the various technologies.

British Petroleum plc, *BP Statistical Review of World Energy*. The best source of annual energy data for most countries in the world. Available in printed form and as Excel spreadsheets from the BP website.

Burton, T., *Wind Energy Handbook*, Wiley, 2001. Very detailed and highly technical in places.

Eastop, T.D. and Croft, D.R., *Energy Efficiency for Engineers and Technologists*, Longman, 1990. An in-depth introduction aimed at engineers.

Hewitt, G.F. and Collier, J.G., *Introduction to Nuclear Power* (2nd edition), Taylor and Francis, 1997. An in-depth introduction.

IPCC, *Technical Summary: Mitigation*, IPCC, 2000. This, along with connected documents from the IPCC, gives a full account of climate change science and policies.

Meyer, A., *Contraction and Convergence*, Green Books, 2000. A personal account of the subject and of climate negotiations.

Ristinen, R.A. and Kraushaar, J.J., *Energy and the Environment*, Wiley, 1999. A highly accessible concise account; USA centred.

Shepherd, W. and Shepherd, D.W., *Energy Studies*, World Scientific, 1997. A slightly more technical presentation on current and future energy systems.

Energy and Climate Change   David A. Coley
© 2008 John Wiley & Sons, Ltd

Smil, V., *Energy in World History*, Westview Press, 1994. The best account available on the subject.

Smil, V., *Energies: an Illustrated Guide to the Biosphere and Civilization*, MIT Press, 1999. A fascinating and slightly alternative account of how many of the processes found in nature can be connected through the concept of energy.

Sorensen, B., *Renewable Energy* (3rd edition), Elsevier Press, 2004. An in-depth and mathematical review of energy flows and possible technologies.

Appendix 4
Useful data

A4.1 Useful data and constants

Object	Value
Physical constants	
Stefan-Boltzmann's constant, σ	5.67×10^{-8} W/m^{2}K^4
Avogadro's constant, N_A	6.02×10^{23} mol^{-1}
Universal gas constant, R	8.31 J/molK
Planck's constant, h	6.63×10^{-34} Jsec
Speed of light, c	3.00×10^8 m/s
Universal gravitational constant, G	6.67×10^{-11} Nm^2/kg^2
Local gravitational constant, g	9.81 ms^{-2}
Mass and thermal	
Molecular weight of air	28.97 g/mol
Latent heat of vaporization of water (at 100 °C)	2260 kJ/kg
Latent heat of vaporization of water (at 20 °C)	2453 kJ/kg
Latent heat of fusion of water	334 kJ/kg
Density of air	1.2 kg/m^3
Density of water	1000 kg/m^3
Density of natural gas	Liquid: 422 kg/m^3, gaseous: 0.76 kg/m^3
Density of gas oil (at 15 °C)	0.84 kg/l
Specific heat capacity of water	4.184 kJ/kgK
Specific heat capacity of steel	0.45 kJ/kgK
Specific heat capacity of wood	1.7 kJ/kgK
Specific heat capacity of air	1.006 kJ/kgK
Calorific value of wood (dry)	16 MJ/kg
Calorific value of charcoal	33 MJ/kg

(Continued)

Energy and Climate Change   David A. Coley
© 2008 John Wiley & Sons, Ltd

A4.1 (*Continued*)

Object	Value
Mass and thermal	
Calorific value of natural gas	48.16 MJ/kg (net), 53.42 MJ/kg (gross)
Calorific value of oil (gas oil)	42.8 MJ/kg (net), 45.6 MJ/kg (gross)
Calorific value of coal (anthracite)	28.95 MJ/kg (net), 29.65 MJ/kg (gross)
Sun, earth and moon	
Surface area of earth	$4\pi r^2 = 510$ million km^2 = 510×10^{12} m^2
Average depth of oceans	3.8 km
Solar luminosity	3.846×10^{26} W
Distance from earth to the sun	1.50×10^{11} m
Distance from earth to the moon	3.82×10^{8} m
Radius of the earth	6370 km
Radius of the sun	696 000 km
Mass of the earth	5.98×10^{24} kg
Mass of the sun	1.99×10^{30} kg
Mass of the moon	7.36×10^{22} kg
Energy	
World annual electricity production (2003)	60.0 EJ (= 16 663 TWh)
World annual primary energy use (2003)	409.1 EJ (equates to 12.97 TW continuously), 9741 Mt$_{oe}$
World annual oil use (2003)	153 EJ, 3637 Mt$_{oe}$
World annual coal use (2003)	108 EJ, 2578 Mt$_{oe}$
World annual gas use (2003)	97.9 EJ, 2332 Mt$_{oe}$
World hydropower consumption (2003)	595 Mt$_{oe}$ (primary energy, converted on the basis of a 38 per cent thermal efficiency); actual electricity, 0.2255 EJ, 2630.7 TWh
World nuclear power consumption (2003)	599 Mt$_{oe}$ (converted on the basis of a 38 per cent thermal efficiency)
CO_2 emission factors	
Natural gas	56 kg/GJ, 0.201 kg/kWh
Coal	98 kg/GJ (anthracite), 0.35 kg/kWh
Oil	74 kg/GJ (gas oil), 0.266 kg/kWh
Electricity (UK value, but typical for a fossil fuel-based nation with some nuclear and hydro)	0.43 kg/kWh (0.117 kgC/kWh), 120 kg/GJ
Miscellaneous	
Volume of a sphere	$(4/3)\pi r^3$
Surface area of a sphere	$4\pi r^2$
Number of seconds in a year	31 536 000
World population	6.4 billion
Annual increase in world carbon dioxide emissions from fossil fuel burning (1990–2000)	Mean 0.06 Gt or 1%

A4.1 (*Continued*)

Object	Value
	Miscellaneous
Annual increase in carbon dioxide concentration (1975–2004)	Mean 1.7 ppm$_v$ or 0.5 %
Annual world carbon emission (2005)	8 GtC (approx.) or 1.2 tC per person
Annual world carbon dioxide emission (2005)	29 GtC (approx.) or 4.4 tC per person
Annual UK carbon emission	2.5 tC per person

A4.2 Conversions

Length, volume and mass	
1 km = 0.6214 miles	1 mile = 1.609 km
1 m = 39.37 inches	1 ft = 0.3048 m
1 km/h = 0.6214 miles/h	1 mile/h = 1.609 km/h
1 kg = 2.205 lb	1 lb = 0.4536 kg
1 hectare = 10 000 m^2 (or 2.47 acres)	1 m^2 = 0.0001 hectares
1 litre = 0.22 imperial gallons	1 imperial gallon = 4.546 litres
1 litre = 0.264 US gallons	1 US gallon = 3.785 litres
1 litre = 0.00629 barrels	1 barrel = 159 litres
1 kg CO_2 = 0.273 kg of carbon	1 kg of carbon = 3.67 kg of CO_2
1 mile per gallon = 0.35 km per litre	1 km per litre = 2.86 miles per gallon

Heat, energy, temperature, power and pressure	
1 atmosphere = 1.013×10^5 Pa (pascal)	
1 bar = 100 000 Pa	
1 J = 0.2389 cal	1 cal = 4.186 J
1 eV = 1.60×10^{-19} J	
1 kWh = 3.6×10^6 J	1 MJ = 0.278 kWh
1 GWh = 3.6×10^{12} J	
1 t$_{oe}$ = 42 GJ	1 GJ = 0.024 t$_{oe}$
1 foot-pound = 1.356 J	
1 BTU = 1055 J	
1 therm = 105.5 MJ	
1 horse power = 0.7457 kW	
1 cubic foot of natural gas ≡ 1.09 MJ	
°C to K add 273	K to °C subtract 273

Currency (as of 29th March 2005)	
€1 (euro) = £0.69 = US$1.29	

Global warming potentials	
CH_4 (100 year horizon)	23 per kg
N_2O (100 year horizon)	296 per kg

A4.3 List of symbols

Symbol	Meaning	Units
c	Speed of light	3.00×10^8 m/s
C_m	Thermal capacity	J
COP	Coefficient of performance	Dimensionless
COP_{Carnot}	Carnot (maximum) coefficient of performance of a heat engine	Dimensionless
C_p	Specific heat capacity	J/kgK
C_v	Calorific value	J/K
$E_{kinetic}$, E_{th}, E_{pot}, E_{rad}, E_{mass},	Kinetic, thermal, potential, radiative and mass (nuclear) energy, respectively	J
F	Solar collector heat removal efficiency factor (0.7–0.95)	Dimensionless
G	Universal gravitational constant	6.67×10^{-11} Nm^2/kg^2
g	Local gravitational constant	9.81 ms^{-2}
h	Planck's constant	6.63×10^{-34} Jsec
h_p	Effective head of pressure in rock	Metres of water
H_{Carnot}	Carnot maximum theoretical efficiency	Dimensionless
I	Current	Amps
K_h	Hydraulic conductivity of rock	m/day
K_q	Thermal conductivity of rock	Wm^{-1} °C^{-1}
L	Solar collector heat loss temperature coefficient	W/m^2 °C
m	Mass	J
MtC etc.	Million tonnes of carbon, i.e. reported as mass of carbon, not carbon dioxide	
MtC_{eq} etc.	Million tonnes of carbon equivalent, i.e. emissions adjusted for global warming potential and reported as mass of carbon, not carbon dioxide	
N_A	Avogadro's constant	6.02×10^{23} mol^{-1}
P	Power	Watts
Q (also E_{th})	Heat energy	J
Q_∞	Total reserve plus all extractions to date	J, or appropriate energy unit for the fuel being considered
R	Incident radiation of a solar collector	W/m^2
R	Universal gas constant	8.31 J/molK
r	Electrical resistance	Ohms (Ω)
R_{hp}	Heat/power ratio in a cogeneration scheme	Dimensionless
S	Entropy	J/K.kg
T	Temperature	°C or K
t	Time	Seconds
t_∞	Date at which reserve is exhausted at current extraction rate	
U	Internal energy	J

A4.3 (*Continued*)

Symbol	Meaning	Units
V	Voltage	Volts
W	Work	J
W_e	Watts of electrical power	
W_{th}	Watts of thermal power	
X	Exergy	J
ε	Emissivity	Dimensionless
ε	Elasticity (economics)	Dimensionless
η	Efficiency	Dimensionless
σ	Stefan-Boltzmann's constant	5.67×10^{-8} W/m^{2}K^4
Ω_∞	Total resource plus all extractions to date	J, or appropriate energy unit for the fuel being considered

A4.4 Common prefixes (see Chapter 2 for others)

Prefix	Symbol	Multiply by
kilo	(k)	10^3
mega	(M)	10^6
giga	(G)	10^9
tera	(T)	10^{12}
peta	(P)	10^{15}
exa	(E)	10^{18}

Index